COMPUTER-
CONTROLLED
SYSTEMS

PRENTICE HALL INFORMATION AND SYSTEM SCIENCES SERIES

Thomas Kailath, *Editor*

COMPUTER-CONTROLLED SYSTEMS
Theory and Design

SECOND EDITION

Karl J. Åström

Björn Wittenmark

 PRENTICE HALL, Englewood Cliffs, New Jersey 07632

Library of Congress Cataloging-in-Publication Data

Åström, Karl J. (Karl Johan)
    Computer controlled systems : theory and design / Karl Åström,
Björn Wittenmark. -- 2nd ed.
        p.   cm.
    Includes bibliographical references.
    ISBN 0-13-168600-3
    1. Automatic control--Data processing.   I. Wittenmark, Björn.
    II. Title.
    TJ213.A78 1990
    629.8'9--dc20   89-22809
                                                                  CIP

To **Bia** and **Karin**

Editorial/production supervision: Christina Burghard
Cover design: Wanda Lubelska Design
Manufacturing buyer: Donna Douglass

© 1990, 1984 by Prentice-Hall, Inc.
A Division of Simon & Schuster
Englewood Cliffs, New Jersey 07632

Printed in the United States of America
10 9 8 7 6 5 4 3 2 1

ISBN   0-13-168600-3

PRENTICE-HALL INTERNATIONAL (UK) LIMITED, *London*
PRENTICE-HALL OF AUSTRALIA PTY, LIMITED, *Sydney*
PRENTICE-HALL CANADA INC., *Toronto*
PRENTICE-HALL HISPANOAMERICANA, S.A., *Mexico*
PRENTICE-HALL OF INDIA PRIVATE LIMITED, *New Delhi*
PRENTICE-HALL OF JAPAN, INC., *Tokyo*
SIMON & SCHUSTER ASIA PTE. LTD., *Singapore*
EDITORA PRENTICE-HALL DO BRASIL, LTDA., *Rio de Janeiro*

CONTENTS

3 COMPUTER-ORIENTED MATHEMATICAL MODELS: DISCRETE-TIME SYSTEMS   41

4 PROCESS-ORIENTED MODELS   86

5 ANALYSIS OF DISCRETE-TIME SYSTEMS   114

Contents

Contents

PREFACE

In the Sixties a control engineer had to master analog computing technology because of its use in analog computers in the simulation of control systems and as the major tool for computations. Analog technology in mechanics, pneumatics, and electronics was also used in control systems.

The scene is now rapidly changing due to the dramatic developments in digital computers and microelectronics. Digital computers were originally used as components in complicated process-control systems. However, because of their small size and low price, digital computers are now also being used in regulators for individual control loops. In several areas digital computers are now outperforming their analog counterparts and are cheaper as well.

Digital computers are also being used increasingly as tools for analysis and design of control systems. The control engineer thus has much more powerful tools available now than in the past. Digital computers are still in a state of rapid development because of the progress in Very Large-Scale Integration (VLSI) technology. Thus substantial technological improvements can be expected in the future.

Because of these developments, the approach to analysis, design, and implementation of control systems is changing drastically. Originally it was only a matter of "translating" the earlier analog designs into the new technology. However, it has been realized that there is much to be gained by exploiting the full potential of the new technology. Fortunately, control theory has also developed substantially over the past 35 years. For a while it was quite unrealistic to implement the type of regulators that the new theory produced except in a few exotic mostly in aerospace or advanced process control. However, due to the revolutionary development of microelectronics, advanced regulators can be implemented even for basic applications. It is also possible to do analysis and design at a reasonable cost with the interactive design tools that are becoming increasingly available.

The purpose of this book is to present control theory that is relevant to the analysis and design of computer-controlled systems, with an emphasis on basic concepts and ideas. It is assumed that a digital computer with a reasonable software is available for computations and simulations so that many tedious details can be left to the computer. The control-system design is also carried out up to the stage of implementation in the form of computer programs in a high-level language.

The book is organized as follows: An overview of the development of computer control is given in Chapter 1. A survey of the development of the theory is also given in order to provide some perspective. (Those who do not know history are bound to repeat it.)

Sampling, which is a fundamental property of computer-controlled systems, is discussed in Chapter 2. The basic mathematical models needed are given in Chapters 3, 4, and 6. Chapter 3 gives the models as seen from the computer, while Chapter 4 treats the models as seen from the process. Without disturbances there are no control problems; it is therefore important to find suitable ways to characterize disturbances, which is done in Chapter 6.

In Chapter 5 the major tools for analysis and simulation are given. Simulation plays an important role because there are many detailed questions that are very hard to answer through analysis alone. Simnon, an interactive simulation language that is used throughout the book, is presented in an appendix. It is not very difficult to translate the programs into other simulation languages. The fact that a powerful simulation tool is available makes a drastic change in attitudes and techniques. It is very important that the simulations be accompanied by analysis that can give order-of-magnitude estimates to ensure that the simulation results are reasonable. At the same time it is not necessary to provide tools for very accurate calculations because these can easily be done by the computer. Chapters 7 through 12 are devoted to the design problem. An overview is given in Chapter 7. Translation of analog design methods is discussed in Chapter 8. State-space design techniques for deterministic systems based on pole assignment are discussed in Chapter 9. The same problem is discussed in Chapter 10 using input-output models. Optimal design methods based on Kalman filters, linear quadratic, and linear quadratic Gaussian control are treated in Chapter 11 based on state-space models and in Chapter 12 using input-output models.

A characteristic feature of many of the new design methods is that a model of the process and its disturbances is needed. Chapter 13 discusses how such models can be obtained. A brief treatment of parameter-adaptive control systems is given in Chapter 14. This may be viewed as a combination of the design methods in Chapters 9 to 12 with the recursive identification methods in Chapter 13. Chapter 15 discusses different aspects of implementation of computer-control algorithms.

The theory is organized in such a way that all models and specifications are given in continuous time. This makes applications easier because of the close connections with physics. Multivariable systems are covered whenever state-space techniques are used; however, the treatment of input-output models using the polynomial approach is limited to the single-input–single-output case. Both deterministic and stochastic aspects of the analysis and the design problem are given.

When designing a system it is often advantageous to see a problem from several viewpoints. Since the goal of the book is to give a good foundation for design of computer-controlled systems, it is necessary to cover a wide range of topics. A reasonable balance between detail and overview has been achieved; however, Chapters 6, 11, 12, 13, and 14 require complete books to cover each topic fully.

In sampled-data theory it has been the custom to let the same symbol z denote both a complex variable and a forward-shift operator. We have found this practice confusing for students and have therefore introduced the symbol q to denote the forward-shift operator. This is analogous to the use of s as a complex variable and $p = d/dt$ as a differential operator for continuous-time systems. The notation q^{-1} is used to denote the backward-shift operator.

The delta operator is defined as $\delta = (q - 1)/h$ is an alternative to the shift operator. It has the advantage that is an approximation of the differential operator.

From the very beginning of the development of sampled-data theory there has been a discussion of the relative merits of shift and delta operators. This discussion is renewed every now and then. It is our opinion that it is useful to keep the shift operator and the z-transform and to view the delta operators as a useful numerical device. One reason for this is that z-transforms naturally appear in many branches of applied mathematics such as analysis of difference equations and generalized functions in probability theory.

This book can be used in many different ways. Chapters 2, 3, 5, 7, 8, 9, 10, and 15 and Sections 6.1–6.3 are suited for an undergraduate course in sampled-data systems. A detailed treatment of Chapters 4, 6, 7, and 9 through 15 can form the core of a graduate course in design of computer-controlled systems. We have given courses to industrial audiences based on Chapters 3, 4, 5, 8, 9, 10, 13, 14, and 15. In all cases we have found it very advantageous to have access to computer simulation and to supplement lectures and exercises with laboratory experiments. Some suggestions for this are given in the solutions manual.

Many students and colleagues have given very good suggestions for improving the book. We are grateful for this feedback which has resulted in significant changes in this second edition. Some modifications were also motivated by the fact that we have learned more ourselves about computer-controlled systems. The major changes have been made in Chapters 8–12, which deal with control-system design. Large sections have been rewritten, larger design examples are introduced. This reflects our belief that more emphasis should be given to control-system design. To compare different design techniques, they have been applied to the same examples. Many issues of practical relevance have been added. A typical case is antialiasing filters and their effect on control design. Chapter 15 has also been modified significantly by including more material on realization of digital controllers. Numerics and quantization are also treated in more detail. These are more relevant now when digital control systems are implemented using digital signal processors and custom VLSI.

The general availability of tools for Computer-Aided Control Engineering has drastically changed the way we can teach automatic control. For instance, it allows us to solve more realistic examples. Since much of the routine work can be done using the computer, we can devote more time to the fundamentals. It is, however, absolutely essential to instill an attitude of sound criticism in the minds of students so that they can judge the results of the computations. The specific tools needed for computer-controlled systems are software for calculating with matrices and polynomials and a nonlinear simulator for verifying the results. We are currently using Matlab for the calculations and the interactive simulation program Simnon. All the graphs in the book showing time responses were generated using Simnon. The graphs were generated by macros which are available to our students. To recreate the simulation of Figure 10.9, the student simply types fig109. In this way it is possible to have students explore many things that were previously not possible. There are also convenient interfaces between Matlab and Simnon so that a design can be carried out in Matlab and transferred to Simnon.

Simnon is available for MS-DOS computers, Vax, and Sun work-stations. More information about the program can be obtained using the tear-out card at the end of the book or by writing directly to the authors. For instructors, the macros are available for a handling charge.

Acknowledgments

While writing this book we have had the pleasure and privilege of interaction with many persons. For the second edition we are particularly grateful to: Per Hagander, Bo Bernhardsson, Kjell Gustafsson, Tore Hägglund, and Bengt Lennartson. We also want to thank Eva Dagnegård and Agneta Tuszyński for the excellent typing and editing of the new material. Finally, we thank Tim Bozik and the staff of Prentice Hall for their support and encouragement.

Karl J. Åström
Björn Wittenmark
Department of Automatic Control
Lund Institute of Technology
Box 118, S-221 00
Lund, Sweden

COMPUTER CONTROL

GOAL

To Introduce the Subject and to Give Some Historical Background on the Development of Computer-Control Technology and Theory.

1.1 Introduction

Digital computers are increasingly being used to implement control systems. It is therefore important to understand computer controlled systems well. One can view computer-controlled systems as approximations of analog-control systems, but this is a poor approach because the full potential of computer control is not used. At best the results are only as good as those obtained with analog control. Alternatively, one can learn about computer-controlled systems, so that the full potential of computer control is used. The main goal of this book is to provide the required background.

A computer-controlled system can be schematically described as in Fig. 1.1. The output from the process $y(t)$ is a continuous-time signal. The output is converted into digital form by the analog-to-digital (A-D) converter. The A-D converter can be included in the computer or regarded as a separate unit, according to one's preference. The conversion is done at the sampling times, t_k. The computer interprets the converted signal, $\{y(t_k)\}$, as a sequence of numbers, processes the measurements using an algorithm, and gives a new sequence of numbers, $\{u(t_k)\}$. This sequence is converted to an analog signal by a digital-to-analog (D-A) converter. Notice that the system runs open loop in the interval between the A-D and the D-A conversion. The events are synchronized by the real-time clock in the computer. The digital computer operates sequentially in time and each

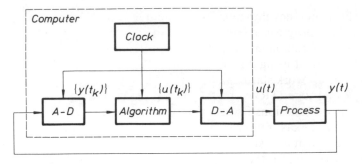

Figure 1.1 Schematic diagram of a computer-controlled system.

operation takes some time. The D-A converter must, however, produce a continuous-time signal. This is normally done by keeping the control signal constant between the conversions. The computer-controlled system contains both continuous-time signals and *sampled*, or *discrete-time*, signals. Such systems have traditionally been called *sampled-data systems*, and this term will be used here as a synonym for *computer-controlled systems*.

The mixture of different types of signals sometimes causes difficulties. In most cases it is, however, sufficient to describe the behavior of the system at the sampling instants. The signals are then of interest only at discrete times. Such systems will be called *discrete-time systems*. Discrete-time systems deal with sequences of numbers, so a natural way to represent these systems is to use difference equations.

The purpose of the book is to present the control theory that is relevant to the analysis and design of computer-controlled systems. This chapter provides some background. A brief overview of the development of computer-control technology is given in Sec. 1.2. The need for a suitable theory is discussed in Sec. 1.3. Examples are used to demonstrate that computer-controlled systems cannot be *fully* understood by the theory of linear, time-invariant, continuous-time systems. An example shows not only that computer-controlled systems can be designed using continuous-time theory and approximations, but also that substantial improvements can be obtained by other techniques that use the full potential of computer control. Sec. 1.4 gives some examples of inherently sampled systems. The development of the theory of sampled-data systems is outlined in Sec. 1.5.

1.2 Computer Technology

The idea of using digital computers as components in control systems emerged around 1950. Applications in missile and aircraft control were investigated first. Studies showed that there was no potential for using the general-purpose digital computers that were available at that time. The computers were too big, they consumed too much power, and they were not sufficiently reliable. For this reason special-purpose computers—digital differential analyzers (DDA)—were developed for the early aerospace applications.

The major developments in computer control occurred in the process industries. The progress of these developments is illustrated in Fig. 1.2, which shows the growth of computers used for process control over a period of 25 years.

The idea of using digital computers for process control emerged in the midfifties. Serious work started in March 1956 when the aerospace company Thomson Ramo Woodridge (TRW) contacted Texaco to set up a feasibility study. After preliminary discussions it was decided to investigate a polymerization unit at the Port Arthur, Texas, refinery. A group of engineers from TRW and Texaco made a thorough feasibility study, which required about 30 people-years. A computer-controlled system for the polymerization unit was designed based on the RW-300 computer. The control system went on-line March 12, 1959. The system controlled 26 flows, 72 temperatures, 3 pressures, and 3 compositions. The essential functions were to minimize the reactor pressure, to determine an optimal distribution among the feeds of 5 reactors, to control the hot-water inflow based on measurement of catalyst activity, and to determine the optimal recirculation.

The pioneering work done by TRW was noticed by many computer manufacturers, who saw a large potential market for their products. Many different feasibility studies were initiated and vigorous development was started. The results of these efforts are reflected in the growth shown in Fig. 1.2.

To discuss the dramatic developments, it is useful to introduce five periods.

Pioneering period	≈ 1955
Direct-digital-control period	≈ 1962
Minicomputer period	≈ 1967
Microcomputer period	≈ 1972
General use of digital control	≈ 1980

It is difficult to give precise dates, because the development was highly diversified. There was a wide difference between different application areas and different industries; there was also considerable overlap. The dates given refer to the first appearance of new ideas.

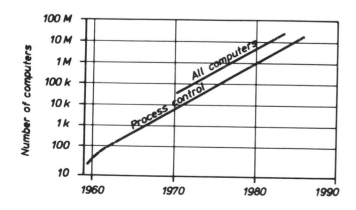

Figure 1.2 Growth of computers used for industrial process control. For comparison the total number of computers is also given. The picture is compiled from several sources: Control Engineering, A. D. Little, Frost and Sullivan, and Diebold. (Redrawn from data published in Control Engineering, © 1980, Technical Publishing Co., with permission.)

The Pioneering Period

The work done by TRW and Texaco evoked substantial interest at process industries, among computer manufacturers, and in research organizations. The industries saw a potential tool for increased automation, the computer industries saw new markets, and universities saw a new research field. Many feasibility studies were initiated by the computer manufacturers because they were eager to learn the new technology and were very interested in knowing what a proper process-control computer should look like. Feasibility studies continued throughout the sixties.

The computer systems that were used were slow, expensive, and unreliable. The earlier systems used vacuum tubes. Typical data for a computer around 1958 were an addition time of 1 ms, a multiplication time of 20 ms, and a Mean Time Between Failures (MTBF) for a central processing unit of 50–100 h. To make full use of the expensive computers, it was necessary to have them perform many tasks. Because the computers were so unreliable, they controlled the process by printing instructions to the process operator or by changing the set points of analog regulators. These supervisory modes of operation were referred to as *operator guide* and *set-point control.*

The major tasks of the computer were to find the optimal operating conditions, to perform scheduling and production planning, and to give reports about production and raw-material consumption. The problem of finding the best operating conditions was viewed as a static optimization problem. Mathematical models of the processes were necessary in order to perform the optimization. The models used—which were quite complicated—were derived from physical models and from regression analysis of process data. Attempts were also made to carry out on-line optimization.

Progress was often hampered by lack of process knowledge. It also become clear that it was not sufficient to view the problems simply as static optimization problems; dynamic models were needed. A significant proportion of the effort in many of the feasibility studies was devoted to modeling, which was quite time consuming because there was a lack of good modeling methodology. This stimulated research into system-identification methods.

A lot of experience was gained during the feasibility studies. It became clear that process control puts special demands on computers. The need to respond quickly to demands from the process led to development of the *interrupt feature,* which is a special hardware device that allows an external event to interrupt the computer in its current work so that it can respond to more urgent process tasks. Many sensors that were needed were not available. There were also several difficulties in trying to introduce a new technology into old industries.

The progress made was closely monitored at conferences and meetings and in journals. A series of articles describing the use of computers in process control was published in the journal *Control Engineering.* By March 1961 thirty-seven systems had been installed. A year later the number of systems had grown to 159. The applications involved control of steel mills and chemical industries and generation of electric power. The development progressed at different rates in different industries. Feasibility studies continued through the sixties and the seventies.

Direct Digital Control

The early installations of control computers operated in supervisory mode, either as operator guide or as set-point control. The ordinary analog-control equipment was needed in both cases. A drastic departure from this approach was made by Imperial Chemical Industries (ICI) in England in 1962. A complete analog instrumentation for process control was replaced by one computer, a Ferranti Argus. The computer measured 224 variables and controlled 129 valves directly. This was the beginning of a new era in process control: Analog technology was simply replaced by digital technology; the function of the system was the same. The name *Direct Digital Control* (DDC) was coined to emphasize that the computer controlled the process directly. In 1962 a typical process-control computer could add two numbers in 100 μs and multiply them in 1 ms. The MTBF was around 1000 h.

Cost was the major argument for changing the technology. The cost of an analog system increased linearly with the number of control loops; the initial cost of a digital system was large, but the cost of adding an additional loop was small. The digital system was thus cheaper for large installations. Another advantage was that the operator communication could be changed drastically; an operator communication panel could replace a large wall of analog instruments. The panel used in the ICI system was very simple—a digital display and a few buttons.

Flexibility was another advantage of the DDC systems. Analog systems were changed by rewiring; computer-controlled systems were changed by reprogramming. Digital technology also offered other advantages. It was easy to have interaction among several control loops. The parameters of a control loop could be made functions of operating conditions. The programming was simplified by introducing special DDC languages. A user of such a language did not need to know anything about programming, but simply introduced inputs, outputs, regulator types, scale factors, and regulator parameters into tables. To the user the systems thus looked like a connection of ordinary regulators. A drawback of the systems was that it was difficult to do unconventional control strategies. This certainly hampered development of control for many years.

DDC was a major change of direction in the development of computer-controlled systems. Interest was focused on the basic control functions instead of the supervisory functions of the earlier systems. Considerable progress was made in the years 1963–65. Specifications for DDC systems were worked out jointly between users and vendors. Problems related to choice of sampling period and control algorithms, as well as the key problem of reliability, were discussed extensively. The concept DDC was quickly accepted in spite of the fact that DDC systems often turned out to be more expensive than the corresponding analog systems.

The Minicomputer Period

There was substantial development of digital computer technology in the sixties. The requirements on a process-control computer were neatly matched with progress in integrated circuit technology. The computers became smaller, faster, more reliable, and cheaper. The term *minicomputer* was coined for the new computers that emerged. It was possible to design efficient process-control systems by using minicomputers.

The development of minicomputer technology combined with the increasing knowledge gained about process control with computers during the pioneering and DDC periods caused a rapid increase in applications of computer control. Special process-control computers were announced by several manufacturers. A typical process computer of the period had a word length of 16 bits. The primary memory was 8–124k words. A disc drive was commonly used as a secondary memory. The CDC 1700 was a typical computer of this period, with an addition time of 2 μs and a multiplication time of 7 μs. The MTBF for a central processing unit was about 20,000 h.

An important factor in the rapid increase of computer control in this period was that digital computer control now came in a smaller "unit." It was thus possible to use computer control for smaller projects and for smaller problems. Because of minicomputers, the number of process computers grew from about 5000 in 1970 to about 50,000 in 1975.

Microcomputers

The minicomputer was still a fairly large system. Even as performance continued to increase and prices to decrease, the price of a minicomputer mainframe in 1975 was still about $10,000. This meant that a small system rarely cost less than $100,000. Computer control was still out of reach for a large number of control problems. But with the development of the microcomputer in 1972, the price of a card computer with the performance of a 1975 minicomputer dropped to $500 in 1980. Another consequence was that digital computing power in 1980 came in quanta as small as $50. This meant, of course, that computer control could now be considered as an alternative, no matter how small the application.

Since there are even more drastic developments in microelectronics to come with the very large scale integration (VLSI) technology in the eighties, it is a safe guess that there will be a large increase in computer-control applications then. Microcomputers have already made an impact on control equipment: Microcomputers are replacing analog hardware even as single-loop controllers; smaller DDC systems have been made using microcomputers; operator communication has been vastly improved in these systems with the introduction of color video-graphics displays; hierarchical control systems with a large number of microprocessors have been constructed; and special-purpose regulators based on microcomputers have been designed.

General Use of Digital Control

From about 1980 digital control has been the standard technique for implementing new control systems. This applies to dedicated, simple, single-loop controllers as well as large distributed control systems. Fig. 1.3 shows an example of a single-loop controller for process control. The particular controller is a PID regulator where the PID function has been implemented digitally. With digital control it is also possible to obtain added functionality. In this particular case, the regulator is provided with automatic tuning and gain scheduling in the form of a table where different regulator parameters are stored.

Digital control has also had a significant impact on standard equipment for process control of large plants. New system architectures for process control were introduced in the late 1970s and early 1980s. The systems may be viewed as a natural development of DDC and minicomputer-based systems. Figure 1.4 shows the structure of these systems, which are called distributed control systems. The systems are composed of many digital computers that perform different control functions such as feedback control, logic, and sequencing, e.g., for start-up and shutdown and human-machine interaction. The different computers communicate over a computer network. There is also a data base which may be either central or distributed where all systems variables are stored.

The distributed system has several advantages over the minicomputer solutions used in the 1970s. The system no longer relies on a single computer to work. The system can be expanded in a modular fashion by adding hardware as the need arises. The advantages of digital control can be exploited fully. The cost

Figure 1.3 A standard single loop controller for process control. (By courtesy of Satt Control Instruments, Stockholm, Sweden).

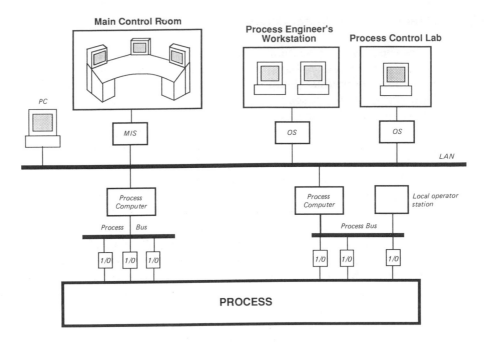

Figure 1.4 Architecture of a distributed control system.

of wiring can be reduced significantly. Honeywell was the first vendor that developed a distributed control system, the TDC2000, which was announced in 1975. After a few years most of the vendors of process control equipment then announced their version. A drawback with the distributed control systems is that their software tend to be quite complex, which has led to many delays in deliveries. Another drawback is that the processing and communication may give rise to delays in the system. Some systems have also suffered from the fact that many practical aspects of digital control like integrator antiwindup are not properly handled. An interesting aspect is that the recently announced systems integrate DDC with logic and sequencing found in programmable logic controllers (PLCs).

The Future

Based on the dramatic developments in the past it is tempting to speculate about the future. There are four areas that are important for the development of computer process control.

Process knowledge.
Measurement technology.
Computer technology.
Control theory.

Knowledge about process control and process dynamics is increasing slowly but steadily. The possibilities of learning about process characteristics are increasing substantially with the installation of process-control systems because it is then easy to collect data, perform experiments, and analyze the results. Progress in system identification and data analysis has also provided valuable information.

Progress in measurement technology is hard to predict. Many things can be done using existing techniques. The possibility of combining outputs of several different sensors with mathematical models is interesting. It is also possible to obtain automatic calibration with a computer. The advent of new sensors will, however, always offer new possibilities.

Spectacular developments are expected in computer technology with the introduction of the VLSI. The ratio of price to performance will continue to drop substantially. The microcomputers of the nineties are expected to have computing power greater than the large mainframes of the late eighties. Substantial improvements are also expected in display techniques and in communications.

Programming has so far been one of the bottlenecks. There were only marginal improvements in productivity in programming from 1950 to 1970. At the end of the seventies many computer-controlled systems were still programmed in assembler code. In the computer-control field, it has been customary to overcome some of the programming problems by providing table-driven software. A user of a DDC system is thus provided with a so-called DDC package that allows the user to generate a DDC system simply by filling in a table, so very little effort is needed to generate a system. The widespread use of packages hampers development, however, because it is very easy to use DDC but it is a major effort to do something else. So only the well-proven methods are tried.

Control theory made substantial progress in the period 1955–80. Very little of this theory had, however, made its way into existing computer-control systems, even though feasibility studies have indicated that significant improvements can be made. One reason for this is the cost of programming. As already mentioned, it requires little effort to use a package provided by a vendor. It is, however, a major effort to try to do something else. Several signs show that this situation can be expected to change. Personal computers with interactive high-level languages are starting to be used for process control. With an interactive language it is very easy to try new things. It is, however, unfortunately very difficult to write *safe* real-time control systems. This will change as better interactive systems become available.

Thus there are many signs that point to interesting developments in the field of computer-controlled systems. A good way to be prepared is to learn the theory presented in this book!

1.3 Computer-Control Theory

A schematic diagram of a computer-controlled system is shown in Fig. 1.1. The system contains essentially five parts: the process, the A-D and D-A converters, the control algorithm, and the clock. Its operation is controlled by the clock. The times when the measured signals are converted to digital form are called the *sampling instants*; the time between successive samplings is called the *sampling period* and is denoted by *h*. Periodic sampling is normally used but there are, of course, many other possibilities. For example, it is possible to sample when the output signals have changed by a certain amount. It is also possible to use different sampling periods for different loops in a system. This is called *multirate sampling*.

The only difference between a computer-controlled system and an ordinary analog-feedback system is that the control law is implemented using a digital computer, so the class of control laws that can be used conveniently is greatly increased. For example, it is easy to use nonlinear calculations, to incorporate logic, and to perform substantial calculations in the controller. Tables can be used to store data in order to accumulate knowledge about the properties of the system.

Is There a Need for a Theory for Computer-Controlled Systems?

A good theory should make it possible to understand how a system like the one in Fig. 1.1 works and how it should be designed. It seems clear that a sampled system would behave as a continuous-time system if the sampling period were sufficiently small. This is certainly true under very reasonable assumptions. Is there then any need for a special theory for computer-controlled systems?

Some examples will be used to show that the system in Fig. 1.1 cannot be *fully* understood in terms of the theory of time-invariant, linear systems even if the process to be controlled is a linear, time-invariant, continuous-time system.

Example 1.1—Time dependence

Suppose that we want to implement a compensator that is simply a first-order lag. Such a compensator can be implemented using A-D conversion, a digital computer, and D-A conversion. The first-order differential equation is approximated by a first-order difference equation. The step response of such a system is shown in Fig. 1.5. The figure clearly shows that the sampled system is not time invariant because the response depends on the time when the step occurs. If the input is delayed, then the output is delayed by the same amount only if the delay is a multiple of the sampling period. □

The phenomenon illustrated in Fig. 1.5 depends on the fact that the system is controlled by a clock (compare with Fig. 1.1). The response of the system to an external stimulus will then depend on how the external event is synchronized with the internal clock of the computer system.

A computer-controlled system with periodic sampling is a *periodic system*. The effects of the periodicity can be made arbitrarily small by choosing a sufficiently high sampling rate; however, it is necessary to consider the periodic nature of sampled systems to understand fully how they work. This is further illuminated by another example.

Example 1.2—Higher-order harmonics

It is well known that a sinusoidal input signal applied to a stable, linear, time-invariant, continuous-time system will—after a transient—give signals in the system

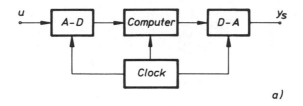

a)

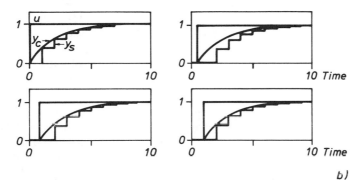

b)

Figure 1.5 a. Block diagram of a digital filter.
b. Step responses, y_s, of a digital computer implementation of a first-order lag for different delays in the input step compared with the first sampling instant. For comparison the response of the corresponding continuous-time system, y_c, is also shown.

that are sinusoidal with the same frequency as that of the input. Figure 1.6 shows what can happen when a computer-controlled system is subjected to a periodic excitation. A sinusoidal signal of frequency 4.9 Hz is applied to the system in Example 1.1. It is clear that the phenomena shown in Fig. 1.6 cannot be explained in terms of linear, time-invariant systems. The beats in the output of the computer-controlled system are due to interference between the input frequency and a higher frequency generated through the sampling process. □

There are many aspects of sampled systems that can indeed be understood by linear, time-invariant theory. The examples given indicate, however, that the sampled systems cannot be fully understood within that framework. It is thus useful to have other tools for analysis.

In solving the problem of controller design, it is certainly possible to use ordinary continuous-time control theory to design a control law and then to use a reasonable discrete-time approximation. An example follows.

Example 1.3—Discrete-time approximation

A double-integrator plant can easily be controlled by state feedback. A straightforward way to do this is simply to calculate the feedback gains using continuous-time

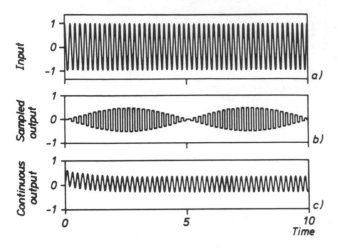

Figure 1.6 Sinusoidal excitation of the sampled system in Example 1.1.
a. Input sinusoidal with frequency 4.9 Hz.
b. Sampled-system output. The sampling period is 0.1 s.
c. Output of the corresponding continuous-time system.

theory and then to implement the state feedback using computer control. If the sampling period is sufficiently small, digital control can be expected to have the same properties, for all practical purposes, as continuous-time control. Figure 1.7 illustrates that this is indeed the case. The agreement with the continuous-time regulator can be made even better by reducing the sampling period. ☐

Based on the results of Example 1.3 it would be tempting to conclude that we do *not* need a theory for sampled systems. This is, however, not correct, because computer-controlled systems can indeed perform better than their continuous-time equivalents. This is the main reason why a theory for sampled systems is useful! An example illustrates this further.

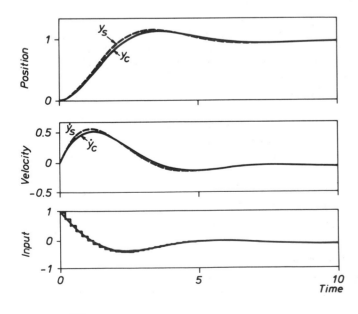

Figure 1.7 Step response of a double-integrator plant with state feedback, continuous-time solution y_c, and sampled approximation y_s, when the sampling period is 0.2 s.

Example 1.4—Deadbeat control

Consider the double-integrator plant of Example 1.3. The result of computer-controlled state feedback with a special linear-feedback strategy is shown in Fig. 1.8. The strategy is of the same form as the control strategy used in Example 1.3, i.e., a linear feedback from the sampled values of the states. The feedback coefficients and the sampling period are, however, different. The particular control strategy is called *deadbeat* control.

A comparison of Figs. 1.7 and 1.8 shows that the system in Fig. 1.8 settles faster than the continuous-time system, even if the largest magnitude of the control signals is the same in both cases. The magnitude of the velocity is, however, larger in Fig. 1.8. The sampled system does not have any overshoot. Observe also that the signals settle on constant values after a finite time. This cannot happen with continuous-time systems because the solutions to such systems are sums of functions that are products of polynomials and exponential functions. Notice also that the sampling period used in Fig. 1.8 is five times longer than the sampling period used in the approximation of the continuous-time control system in Fig. 1.7.   □

The example demonstrates clearly that *even in the linear case* something better than just an approximation of a continuous-time regulator is possible.

1.4 Inherently Sampled Systems

Sampled models are natural descriptions for many phenomena. The theory of sampled-data systems, therefore, has many applications outside the field of computer control. A few examples follow.

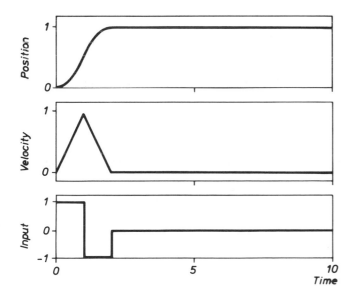

Figure 1.8 Computer control of the double-integrator plant with a deadbeat strategy. The sampling period is 1 s.

Discrete-Time Systems as Models for Computer Algorithms

Algorithms in computers can be described as discrete-time systems. This will be illustrated with an iterative algorithm and a real-time application.

Example 1.5—Iterative solution

Iterative algorithms are examples of inherently sampled systems. Assume that the solution to an equation of the form

$$x - f(x) = 0$$

is desired. One way to find the solution is to guess an initial value and then use Picard's algorithm, i.e., to use the iterative scheme

$$x(k + 1) = f[x(k)]$$

where $x(k)$ is the kth iteration. The numerical algorithm can thus be interpreted as a discrete-time system in which the time represents the number of iterations. Specifically, assume that $f(x) = 3 - \sqrt{x}$. In this case it is easy to show that the solution is $x = (7 - \sqrt{3})/2 \approx 1.697$. The sequence of numbers shown in Fig. 1.9 is obtained if one starts with the initial guess $x(0) = 0$. □

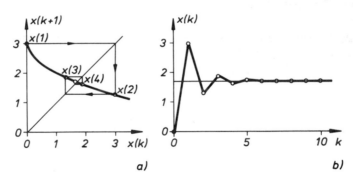

Figure 1.9 Two graphic illustrations of the iterative scheme in Example 1.5.

Example 1.6—Control algorithm

A simple computer algorithm for a proportional and integral (PI) controller follows:

```
uc:=adin(in1)        {read reference value}
y:=adin(in2)         {read process value}
e:=uc − y
u:=k*(e + i)
dout(u)              {output control signal}
i:=i + h*e/ti
```

The program is executed every sampling period by a scheduling program, as illustrated in Fig. 1.10. The computer code is equivalent to the following difference equations:

$$e(k) = uc(k) - y(k)$$
$$u(k) = k[e(k) + i(k - 1)]$$
$$i(k) = i(k - 1) + \frac{he(k)}{ti}$$

□

Sampling Due to the Measurement System

In many cases, sampling will occur naturally in connection with the measurement procedure. A few examples follow.

Example 1.7—Radar

When a radar antenna rotates, information about range and direction is naturally obtained once per revolution of the antenna. A sampled model is thus the natural way to describe a radar system. Attempts to describe radar systems were, in fact, one of the starting points of the theory of sampled systems. □

Example 1.8—Analytical Instruments

In process-control systems, there are many variables that cannot be measured on-line, so a sample of the product is analyzed off-line in an analytical instrument such as a mass spectrograph or a chromatograph. □

Example 1.9—Economic systems

Accounting procedures in economic systems are often tied to the calendar. Although transactions may occur at any time, information about important variables is accumulated only at certain times—e g , daily, weekly, monthly, quarterly, or yearly. □

Sampling Due to Pulsed Operation

Many systems are inherently sampled because information is transmitted using pulsed information. Electronic circuits are a prototype example. They were also

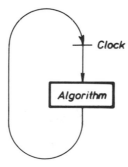

Figure 1.10 Scheduling of a computer program.

one source of inspiration for the development of sampled-data theory. Other examples follow.

Example 1.10—Thyristor control

Power electronics using thyristors are sampled systems. Consider the circuit in Fig. 1.11. The current can be switched on only when the voltage is positive. When the current is switched on, it remains on until the current has a zero crossing. The current is thus synchronized to the periodicity of the main's supply. □

Example 1.11—Biological systems

Biological systems are fundamentally sampled because the signal transmission in the nervous system is in the form of pulses. □

Example 1.12—Internal combustion engines

An internal combustion engine is a sampled system. The ignition can be viewed as a clock that synchronizes the operation of the engine. A torque pulse is generated at each ignition. □

The systems in these examples are periodic because of their pulsed operation. Periodic systems are quite difficult to handle, but they can be considerably simplified by studying the systems at instants synchronized with the pulses—that is, by using sampled-data models. The processes can then be described as time-invariant, discrete-time systems at the sampling instants. Examples 1.10 and 1.12 are of this type.

1.5 How Theory Developed

Although the major applications of the theory of sampled systems are currently in computer control, many of the problems were encountered earlier. In this section some of the main ideas in the development of the theory are discussed.

The Sampling Theorem

Since all computer-controlled systems operate on values of the process variables at discrete times only, it is very important to know the conditions under which a signal can be recovered from its values in discrete points only. The key issue was explored by Nyquist, who showed that to recover a sinusoidal signal from its samples, it is necessary to sample at least twice per period. A complete solution was given in an important work by Shannon in 1949.

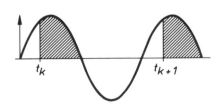

Figure 1.11 Thyristor control circuit.

Difference Equations

The first germs of a theory for sampled systems appeared in connection with analyses of specific control systems. The behavior of the chopper-bar galvanometer, investigated in Oldenburg and Sartorius (1948), was one of the earliest contributions to the theory. It was shown that many properties could be understood by analyzing a linear time-invariant difference equation. The difference equation replaced the differential equations in continuous-time theory. For example, stability could be investigated by the Schur-Cohn method, which is equivalent to the Routh-Hurwitz criterion.

Transform Methods

During and after World War II, a lot of activity was devoted to analysis of radar systems. These systems are naturally sampled because a position measurement is obtained once per antenna revolution. Since transform theory had been so useful for continuous-time systems, it was natural to try to develop a similar theory for sampled systems. The first steps in this direction were taken by Hurewicz (1947). He introduced the transform of a sequence $\{f(kh)\}$, defined by

$$\mathscr{L}\{f(kh)\} = \sum_{k=0}^{\infty} z^{-k} f(kh)$$

This transform is similar to the *generating function,* which had been used so successfully in many branches of applied mathematics. The transform was later defined as the *z-transform* by Ragazzini and Zadeh (1952). Transform theory was developed independently in the Soviet Union, in the United States, and in Great Britain. Tsypkin (1950) called the transform the *discrete Laplace transform* and developed a systematic theory for pulse-controlled systems based on the transform. The transform method was also independently developed by Barker (1952) in England.

In the United States the transform was further developed in a Ph.D. dissertation by Jury at Columbia University. Jury developed tools both for analysis and design. He also showed that sampled systems could be better than their continuous-time equivalents. (See Example 1.4 in Section 1.3.) Jury also emphasized that it was possible to obtain a closed-loop system that exactly achieved steady state in finite time. In later works he also showed that sampling can cause cancellation of poles and zeros. A closer investigation of this property later gave rise to the notions of observability and reachability.

The z-transform theory leads to comparatively simple results. A limitation of the theory, however, is that it tells what happens to the system only at the sampling instants. The behavior between the sampling instants is not just an academic question, because it was found that systems could exhibit *hidden oscillations*. These oscillations are zero at the sampling instants, but very noticeable in between.

Another approach to the theory of sampled system was taken by Linvill

(1951). Following ideas due to MacColl (1945), he viewed the sampling as an amplitude modulation. Using a describing-function approach Linvill effectively described intersample behavior. Yet another approach to the analysis of the problem was the *delayed z-transform,* which was developed by Tsypkin in 1950, Barker in 1951, and Jury in 1956. It is also known as the *modified z-transform.*

Much of the development of the theory was done by a group at Columbia University in the United States, led by Ragazzini. Jury, Kalman, Bertram, Zadeh, Franklin, Friedland, Kranc, Freeman, Sarachik, and Sklansky all did their Ph.D. work for Ragazzini.

Toward the end of the fifties, the *z*-transform approach to sampled systems had matured, and several textbooks appeared almost simultaneously: Ragazzini and Franklin (1958), Jury (1958), Tsypkin (1958), and Tou (1959). This theory, which was patterned after the theory of linear, time-invariant, continuous-time systems, gave good tools for analysis and synthesis of sampled systems. A few modifications had to be made because of the time-varying nature of sampled systems. For example, all operations in a block-diagram representation do not commute!

State-Space Theory

A very important event in the late fifties was the development of state-space theory. The major inspiration came from mathematics and the theory of ordinary differential equations and from mathematicians such as Lefschetz, Pontryagin, and Bellman. Kalman deserves major credit for the state-space approach to control theory. He formulated many of the basic concepts and solved many of the important problems.

Several of the fundamental concepts grew out of an analysis of the problem of whether it would be possible to get systems in which the variables achieved steady state in finite time. The analysis of this problem led to the notions of reachability and observability. Kalman's work also led to a much simpler formulation of the analysis of sampled systems: The basic equations could be derived simply by starting with the differential equations and integrating them under the assumption that the control signal is constant over the sampling period.

Optimal and Stochastic Control

There were also several other important developments in the late fifties. Bellman (1957) and Pontryagin and others (1962) showed that many design problems could be formulated as optimization problems. For nonlinear systems this led to non-classical calculus of variations. An explicit solution was given for linear systems with quadratic loss functions by Bellman and others (1958). Kalman (1960) showed in a celebrated paper that the linear quadratic problem could be reduced to a solution of a Riccati equation. Kalman also showed that the classical Wiener filtering problem could be reformulated in the state-space framework. This permitted a "solution" in terms of recursive equations, which were very well suited to computer calculation.

In the beginning of the sixties, a stochastic variational problem was formulated by assuming that disturbances were random processes. The optimal control problem for linear systems could be formulated and solved for the case of quadratic loss functions. This led to the development of *stochastic control theory*. The work resulted in the so-called Linear Quadratic Gaussian (LQG) theory. This is now a major design tool for multivariable linear systems.

Algebraic System Theory

The fundamental problems of linear system theory were reconsidered at the end of the sixties and the beginning of the seventies. The algebraic character of the problems was reestablished, which resulted in a better understanding of the foundations of linear system theory. Techniques to solve specific problems using polynomial methods were another result [see Kalman and others (1969), Rosenbrock (1970), Wonham (1974), Blomberg and Ylinen (1983), and Kučera (1979)].

System Identification

All techniques for analysis and design of control system are based on the availability of appropriate models for the process dynamics. The success of classical control theory that almost exclusively builds on Laplace transforms was largely due to the fact that the transfer function of a process can be determined experimentally using frequency response. The development of digital control was accompanied by a similar development of system identification methods. These allow experimental determination of the pulse transfer function or the difference equations that are the starting point of analysis and design of digital control systems. Good sources of information on these techniques are Åström and Eykhoff (1971), Ljung (1987), Söderström and Stoica (1989), and Norton (1986).

Adaptive Control

When digital computers are used to implement a controller it is possible to implement more complicated control algorithms. A natural step is to include both parameter estimation methods and control design algorithms. In this way it is possible to obtain adaptive control algorithms that determine the mathematical models and perform control system design on-line. Research on adaptive control began in the mid-fifties. Significant progress was made in the seventies when feasibility was demonstrated in industrial applications. The advent of the microprocessor made the algorithms cost-effective, and commercial adaptive regulators appeared in the early eighties. This has stimulated vigorous research on theoretical issues and significant product development. See, for instance, Åström and Wittenmark (1973), (1980), Åström (1983), (1987), Goodwin and Sin (1984), and Åström and Wittenmark (1989).

Automatic Tuning

Controller parameters are often tuned manually. Experience has shown that it is difficult to adjust more than two parameters manually. From the user point of view it is therefore helpful to have tuning tools built into the controllers. Such systems are similar to adaptive controllers. They are, however, easier to design and use. With computer based controllers it is easy to incorporate tuning tools. Such systems also started to appear industrially in the mid-eighties. See Åström and Hägglund (1988).

1.6 References

To acquire mature knowledge about a field it is useful to know its history and to read some of the original papers. The papers

JURY, E. I. (1980): "Sampled-data Systems, Revisited: Reflections, Recollections, and Reassessments," *Trans. of the ASME, Journal of Dynamic Systems, Measurement, and Control,* 102 (December 1980) 208–16.

JURY, E. I. and Y. Z. TSYPKIN (1971): "On the Theory of Discrete Systems," *Automatica* 7, 89–107.

written by two of the originators of sampled-data theory give a useful perspective. Early work on sampled systems is found in

MacCOLL, L. A. (1945): *Fundamental Theory of Servomechanisms.* New York: D. Van Nostrand.

OLDENBURG, R. C. and H. SARTORIUS (1948): *The Dynamics of Automatic Control.* New York: ASME.

HUREWICZ, W. (1947): "Filters and Servo Systems with Pulsed Data," in *Theory of Servomechanisms,* ed. H. M. James, N. B. Nichols, and R. S. Philips. New York: McGraw-Hill.

The sampling theorem was given in

SHANNON, C. E. (1949): "Communication in Presence of Noise," *Proc. IRE,* 37, 10–21.

and

KOTELNIKOV, V. A. (1933): "On the Transmission Capacity of 'Ether' and Wire in Electrocommunications," *Proc. First All-union Conference on Questions of Communication.* Moscow.

Major contributions to the early theory of sampled data systems were obtained in England by

BARKER, R. H. (1952): "The Pulse Transfer Function and Its Applications to Sampling Servosystems," *Proc. IEE,* 99, pt. IV, 302–17.

LAWDEN, D. F. (1951): "A General Theory of Sampling Servomechanisms," *Proc. IEE,* 98, pt. IV (October) 31–36.

in the United States by

LINVILL, W. K. (1951): "Sampled-Data Control Systems Studied Through Comparison Sampling with Amplitude Modulation," *AIEE Trans.,* 70, pt. II, 1778–88.

RAGAZZINI, J. R. and L. A. ZADEH (1952): "The Analysis of Sampled-Data Systems," *AIEE Trans.*, 71, pt. II (November) 225–34.

JURY, E. I. (1956): "Synthesis and Critical Study of Sampled-Data Control Systems," *AIEE Trans.*, 75, pt. II, 141–51.

and in the Soviet Union by

TSYPKIN, Y. Z. (1949) and (1950): "Theory of Discontinuous Control," *Avtomat i Telemekh.*, 3 (1949), 5 (1949), 5 (1950).

The first textbooks on sampled data theory appeared toward the end of the fifties. They were

RAGAZZINI, J. R. and G. F. FRANKLIN (1958): *Sampled-Data Control Systems*. New York: McGraw-Hill.

TSYPKIN, Y. Z. (1958): *Theory of Impulse Systems*. Moscow: State Publisher for Physical Mathematical Literature.

JURY, E. I. (1958): *Sampled-Data Control Systems*. New York: John Wiley.

TOU, J. T. (1959): *Digital and Sampled-Data Control Systems*. New York: McGraw-Hill.

A large number of textbooks have appeared since then. Among the more common ones we can mention

ACKERMANN, J. (1972): *Abtastregelung*. Berlin: Springer-Verlag.

ISERMANN, R. (1977): *Digitale Regelsysteme*. Eng. trans. (1981): *Digital Control Systems*. Berlin: Springer-Verlag.

KUO, B. C. (1980): *Digital Control Systems*. Tokyo: Holt-Saunders.

FRANKLIN, G. F. and J. D. POWELL (1989): *Digital Control of Dynamic Systems*. Reading, Mass.: Addison-Wesley.

The idea of formulating control problems in the state space also resulted in a reformulation of sampled-data theory. The paper

KALMAN, R. E. (1969): "On the General Theory of Control Systems," Proc. First IFAC Congress, Moscow, *Butterworths*, 1, 481–92.

is seminal.

Some basic references on optimal and stochastic control are

ÅSTRÖM, K. J. (1970): *Introduction to Stochastic Control Theory*. New York: Academic Press.

BELLMAN, R. (1957): *Dynamic Programming*. Princeton, N.J.: Princeton University Press.

BELLMAN, R., I. GLICKSBERG, and O. A. GROSS (1958): "Some Aspects of the Mathematical Theory of Control Processes," Report R-313. Santa Monica, Calif.: The RAND Corporation.

PONTRYAGIN, L. S., V. G. BOLTYANSKII, R. V. GAMKRELIDZE, and E. F. MISCHENKO (1962): *The Mathematical Theory of Optimal Processes*. New York: John Wiley.

KALMAN, R. E. (1960): "Contributions to the Theory of Optimal Control," *Boletin de la Sociedad Matematica Mexicana*, 5, 102–19.

The algebraic system approach is discussed in

KALMAN, R. E., P. L. FALB, and M. A. ARBIB (1969): *Topics in Mathematical System Theory*. New York: McGraw-Hill.

ROSENBROCK, H. H. (1979): *State-Space and Multivariable Theory*. London: Nelson.

WONHAM, W. M. (1974): *Linear Multivariable Control: A Geometric Approach*. New York: Springer-Verlag.

BLOMBERG, H. and R. YLINEN (1983): *Algebraic Theory for Multivariable Linear Systems*. New York: Academic Press.

KUČERA, V. (1979): *Discrete Linear Control*. Prague: Academia.

System identification is surveyed in

ÅSTRÖM, K. J. and P. E. EYKHOFF (1971): "System Identifcation: A Survey," *Automatica*, 7, 123–62.

LJUNG, L. and T. SÖDERSTRÖM (1983): *Theory and Practice of Recursive Identification*. Cambridge, Mass.: MIT Press.

LJUNG, L. (1987): *System Identification: Theory for the User,* Englewood Cliffs, N.J.: Prentice Hall, Inc.

NORTON, J. P. (1986): *An Introduction to Identification*. London: Academic Press.

SÖDERSTRÖM, T. and P. STOICA (1989): *System Identification,* Hemel Hempstead, U.K.: Prentice Hall International.

Adaptive control is discussed in

BELLMAN, R. (1961): *Adaptive Control: A Guided Tour*. Princeton, N.J.: Princeton University Press.

ÅSTRÖM, K. J. and B. WITTENMARK (1973): "On Self-tuning Regulators," *Automatica,* 9, 185–99.

ÅSTRÖM, K. J. and B. WITTENMARK (1980): "Self-tuning Controllers Based on Pole-Zero Placement" *Proc., IEE* pt. D, 127, 120–30.

ÅSTRÖM, K. J. (1983): "Theory and Applications of Adaptive Control," *Automatica,* 19, 471–86.

ÅSTRÖM, K. J. (1987): "Adaptive Feedback Control," *Proc. IEEE,* 75, 185–217.

ÅSTRÖM, K. J. and T. HÄGGLUND (1988): *Automatic Tuning of PID Regulators*. Research Triangle Park, N.C.: ISA.

ÅSTRÖM, K. J. and B. WITTENMARK (1989): *Adaptive Control*. Reading, Mass.: Addison-Wesley.

GOODWIN, G. C. and K. S. SIN (1984): *Adaptive Filtering, Prediction and Control*. Englewood Cliffs, N.J.: Prentice Hall, Inc.

GUPTA, M. M., ed., (1986): *Adaptive Methods for Control System Design*. New York: IEEE Press.

A survey of distributed computer systems is found in

LUCAS, M. P. (1986): *Distributed Control Systems—Their Evaluation and Design*. New York: Van Nostrand Reinhold.

Many additional references references are given in the following sections. We also recommend the proceedings of the IFAC Symposia on Digital Computer Applications to Process Control and on Identification and System Parameter Estimation, which are published by Pergamon Press.

2
SAMPLING OF CONTINUOUS-TIME SIGNALS

GOAL

To Understand the Sampling Mechanism and to Introduce Some of the Fundamental Concepts and Notations for Sampled-Data Systems. To Illustrate the Problem of Aliasing.

2.1 Introduction

According to some dictionaries, *sampling* means "the act or process of taking a small part or quantity of something as a sample for testing or analysis." In the context of control and communication, sampling means that *a continuous-time signal is replaced by a sequence of numbers,* which represent the values of the signal at certain times.

Sampling is a fundamental property of computer-controlled systems because of the discrete-time nature of the digital computer. Consider, for example, the system shown in Fig. 1.1. The process variables are sampled in connection with the analog conversion and then converted to digital representation for processing. The continuous-time signal that represents the time variation of the process variables is thus converted to a sequence of numbers, which is processed by the digital computer. The processing gives a new sequence of numbers, which is converted to a continuous-time signal and applied to the process. In the system shown in Fig. 1.1, this is handled by the D-A converter. The process of converting a sequence of numbers into a continuous-time signal is called *signal reconstruction*.

Because sampling is a basic property of computer-controlled systems, it is necessary to have a good understanding of the sampling process. Sampling of continuous-time signals is discussed in this chapter; descriptions of systems that involve sampled signals are discussed in Chapter 3. The sampling mechanism is discussed in Sec. 2.2. While the theory includes many different types of sampling, most of it is devoted to the special case of periodic sampling. Sec. 2.3 deals with the problem that occurs of representing a continuous-time signal exactly by its sampled values. Shannon's sampling theorem gives the result for periodic sampling. Different ways to reconstruct a continuous-time signal from its sampled values are discussed in Sec. 2.4. Shannon's reconstruction and first- and higher-order holds are discussed.

A peculiar effect called *aliasing* appears if the signal sampled contains frequencies that are higher than half the sampling frequency. Sec. 2.5 treats aliasing or frequency folding and the use of prefilters to avoid it. Practical aspects of the choice of the sampling period are discussed in Sec. 2.6.

2.2 Description of the Sampling Mechanism

For the purpose of analysis, it is useful to have a mathematical description of sampling. Sampling a continuous-time signal simply means to replace the signal by its values in a discrete set of points. Let Z be the positive and negative integers $Z = \{\ldots, -1, 0, 1, \ldots\}$ and let $\{t_k: k \in Z\}$ be a subset of the real numbers called the sampling instants. The sampled version of the signal f is then the sequence $\{f(t_k): k \in Z\}$. Sampling is a linear operation. The sampling instants are often equally spaced in time, i.e.,

$$t_k = k \cdot h$$

This case is called *periodic sampling* and h is called the sampling period, or the sampling time. The corresponding frequency $f_s = 1/h$ (Hz) is called the *sampling frequency*.

More complicated sampling schemes can also be used. For instance, different sampling periods can be used for different control loops. This is called *multirate sampling* and can be considered to be the superposition of several periodic sampling schemes.

The case of periodic sampling is well understood. Most theory is devoted to this case, but systems with multirate sampling are becoming more important because of the increased use of multiprocessor systems. With modern software for concurrent processes, it is also possible to design a system as if it were composed of many different processes running asynchronously. There are also technical advantages in using different sampling rates for different variables.

2.3 The Sampling Theorem

Very little is lost by sampling a continuous-time signal if the sampling instants are sufficiently close, but much of the information about a signal can be lost if the sampling points are too far apart. This is illustrated in Fig. 2.1, in which a sine wave is sampled. The figure shows that the sampled sine curve cannot be

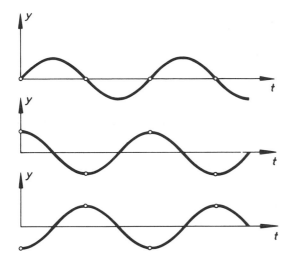

Figure 2.1 Loss of information due to slow sampling. The sine curve is sampled at the rate of two samples per period.

distinguished from the zero signal if the frequency of the sine curve is half the sampling frequency.

It is, of course, essential to know precisely when a continuous-time signal is uniquely given by its sampled version. The following theorem gives the conditions for the case of periodic sampling.

Theorem 2.1—Shannon's sampling theorem. A continuous-time signal with a Fourier transform that is zero outside the interval $(-\omega_0, \omega_0)$ is given uniquely by its values in equidistant points if the sampling frequency is higher than $2\omega_0$. The continuous-time signal can be computed from the sampled signal by the interpolation formula

$$f(t) = \sum_{k=-\infty}^{\infty} f(kh) \frac{\sin \omega_s(t - kh)/2}{\omega_s(t - kh)/2} = \sum_{k=-\infty}^{\infty} f(kh) \sin c \frac{\omega_s(t - kh)}{2} \quad (2.1)$$

where ω_s is the sampling angular frequency in radians per second (rad/s).

Proof. Let the signal be f and let F be its Fourier transform.

$$F(\omega) = \int_{-\infty}^{\infty} e^{-i\omega t} f(t) \, dt \quad (2.2)$$

$$f(t) = \frac{1}{2\pi} \int_{-\infty}^{\infty} e^{i\omega t} F(\omega) \, d\omega \quad (2.3)$$

Introduce

$$F_s(\omega) = \frac{1}{h} \sum_{k=-\infty}^{\infty} F(\omega + k\omega_s) \quad (2.4)$$

The proof is based on the observation that the samples $f(kh)$ can be regarded as the coefficients of the Fourier series of the periodic function $F_s(\omega)$. This is shown by a direct calculation. The Fourier expansion of F_s is

$$F_s(\omega) = \sum_{k=-\infty}^{\infty} C_k e^{-ikh\omega} \tag{2.5}$$

where the coefficients are given by

$$C_k = \frac{1}{\omega_s} \int_0^{\omega_s} e^{ikh\omega} F_s(\omega)\, d\omega$$

Using the definition of the Fourier coefficients, given in (2.3) and (2.4), it is straightforward to show that

$$C_k = f(kh) \tag{2.6}$$

It thus follows that the sampled signal $\{f(kh),\ k = \ldots, -1, 0, 1, \ldots\}$, uniquely determines the function $F_s(\omega)$. Under the assumptions of the theorem the function F is zero outside the interval $(-\omega_0, \omega_0)$. If $\omega_s > 2\omega_0$ it follows from (2.4) that

$$F(\omega) = \begin{cases} hF_s(\omega) & |\omega| \le \dfrac{\omega_s}{2} \\[2ex] 0 & |\omega| > \dfrac{\omega_s}{2} \end{cases} \tag{2.7}$$

The Fourier transform of the continuous-time signal is thus uniquely given by F_s, which in turn is given by the sampled function $\{f(kh),\ k = \ldots, -1, 0, 1, \ldots\}$. The first part of the theorem is thus proved. To show formula (2.1), notice that it follows from (2.3) and (2.7) that

$$f(t) = \frac{1}{2\pi} \int_{-\infty}^{\infty} e^{i\omega t} F(\omega)\, d\omega$$

$$= \frac{h}{2\pi} \int_{-\omega_s/2}^{\omega_s/2} e^{i\omega t} F_s(\omega)\, d\omega$$

$$= \frac{h}{2\pi} \int_{-\omega_s/2}^{\omega_s/2} e^{i\omega t} \sum_{k=-\infty}^{\infty} e^{-ikh\omega} f(kh)\, d\omega$$

where the last equality follows from (2.5) and (2.6). Interchanging the order of integration and summation,

$$f(t) = \sum_{k=-\infty}^{\infty} f(kh) \frac{h}{2\pi} \int_{-\omega_s/2}^{\omega_s/2} e^{i\omega t - i\omega kh}\, d\omega$$

$$= \sum_{k=-\infty}^{\infty} f(kh) \frac{h}{2\pi i(t - kh)} e^{i\omega t - i\omega kh} \Big|_{-\omega_s/2}^{\omega_s/2}$$

$$= \sum_{k=-\infty}^{\infty} f(kh) \frac{\sin \omega_s(t - kh)/2}{\pi(t - kh)/h}$$

Because $\omega_s h = 2\pi$, Equation (2.1) now follows. $\qquad\qquad\square$

Remark 1. The frequency $\omega_N = \omega_s/2$ plays an important role. This frequency is called the *Nyquist frequency.*

Remark 2. Notice that Equation (2.1) defines the reconstruction of signals whose Fourier transforms vanish for frequencies larger than the Nyquist frequency $\omega_N = \omega_s/2$.

Remark 3. Because of the factor $1/h$ in Equation (2.4), the sampling operation has a gain of $1/h$.

2.4 Reconstruction

The inversion of the sampling operation, i.e., the conversion of a sequence of numbers $\{f(t_k): k \in Z\}$ to a continuous-time function $f(t)$ is called *reconstruction.* In computer-controlled systems, it is necessary to convert the control actions calculated by the computer as a sequence of numbers to a continuous-time signal that can be applied to the process. In digital filtering, it is similarly necessary to convert the representation of the filtered signal as a sequence of numbers into a continuous-time function. Some different reconstructions are discussed in this section.

Shannon Reconstruction

For the case of periodic sampling of band-limited signals, it follows from the sampling theorem that a reconstruction is given by (2.1). This reconstruction is called the *Shannon reconstruction.* Equation (2.1) defines an inverse of the sampling operation, which can be considered as a linear operator. It is, however, not a causal operator because the value of f at time t is expressed in terms of past values $\{f(kh): k \le t/h\}$ as well as future values $\{f(kh): k > t/h\}$. The characteristics of the Shannon reconstruction are given by the function

$$h(t) = \frac{\sin \omega_s t/2}{\omega_s t/2} \tag{2.8}$$

See Fig. 2.2. This reconstruction will introduce a delay. The weight is 10% after about three samples and less than 5% after six samples. The delay implies that the Shannon reconstruction is not useful in control applications. It is, however, sometimes used in communication and signal-processing applications, where the delay can be acceptable. One example where Shannon reconstruction is used is in compact disk players. Other drawbacks of the Shannon reconstruction are that it is complicated and that it can be applied only to periodic sampling. It is therefore useful to have other reconstructions.

Zero-Order Hold (ZOH)

A simple causal reconstruction is given by

$$f(t) = f(t_k), \qquad t_k \le t < t_{k+1} \tag{2.9}$$

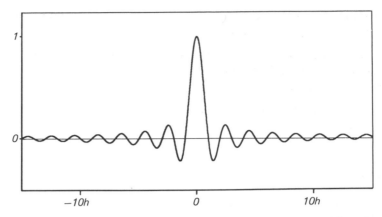

1

0

−10h   0   10h

Figure 2.2 The impulse response of the Shannon reconstruction given by (2.8).

This means that the reconstructed signal is piecewise constant, continuous from the right, and equal to the sampled signal at the sampling instants. The reconstructed value is thus held constant until the next sampling instant.

Because of its simplicity, the zero-order hold is very common in computer-controlled systems. The standard D-A converters are often designed in such a way that the old value is held constant until a new conversion is ordered.

The zero-order hold also has the advantage that it can be used for nonperiodic sampling. Notice, however, that the reconstruction in (2.9) gives an exact inverse of the sampling operation only for signals that are right continuous and piecewise constant over the sampling intervals. For all other signals, the reconstruction of (2.9) gives an error (see Fig. 2.3). For periodic sampling of a signal with a smooth first derivative, the following estimate of the error is obtained. The largest value of the error is given by

$$e_{ZOH} = \max_k | f(t_{k+1}) - f(t_k) | \leq h \max_t | f'(t) | \tag{2.10}$$

where f' is the derivative of f.

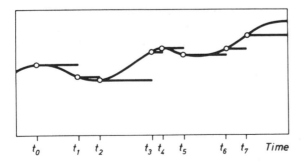

t_0 t_1 t_2 t_3 t_4 t_5 t_6 t_7 *Time*

Figure 2.3 Sampling and zero-order-hold reconstruction of a continuous-time signal.

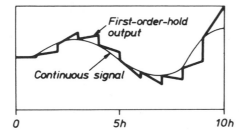

Figure 2.4 Sampling and first-order-hold reconstruction of a continuous time signal.

Higher-Order Holds

The zero-order hold can be regarded as an extrapolation using a polynomial of degree zero. For smooth functions it is possible to obtain smaller reconstruction errors by extrapolation with higher-order polynomials. A first-order causal polynomial extrapolation gives

$$f(t) = f(t_k) + \frac{t - t_k}{t_k - t_{k-1}} [f(t_k) - f(t_{k-1})], \qquad t_k \le t < t_{k+1}$$

The reconstruction is thus obtained by drawing a line between the two most recent samples. The first-order hold is illustrated in Fig. 2.4.

The largest error when using a first-order hold is given by

$$e_{FOH} = \max_k \max_t \left| f(t) - f(t_k) - \frac{t - t_k}{t_k - t_{k-1}} [f(t_k) - f(t_{k-1})] \right| \qquad (2.11)$$

For periodic sampling of signals with a smooth second derivative, the error can be estimated by

$$e_{FOH} \le h^2 \max_t | f''(t) | \qquad (2.12)$$

For signals with smooth higher-order derivatives, it is possible to construct extrapolation polynomials of higher order. Reconstructions of this type are not very common in practice, however, because they are complicated to implement.

2.5 Aliasing or Frequency Folding

If a continuous-time signal that has the Fourier transform F is sampled periodically, it follows from (2.5) and (2.6) that the sampled signal $f(kh)$, $k = \ldots, -1, 0, 1, \ldots$, can be interpreted as the Fourier coefficients of the function F, defined by (2.4).

The function F_s can thus be interpreted as the Fourier transform of the sampled signal. The function of (2.4) is periodic with a period equal to the sampling frequency ω_s. If the continuous-time signal has no frequency components higher than the Nyquist frequency, the Fourier transform is simply a periodic repetition of the Fourier transform of the continuous-time signal (see Fig. 2.5).

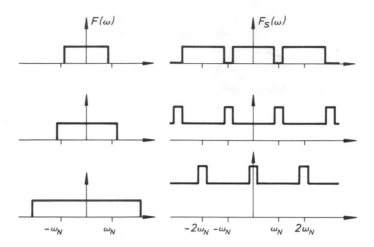

Figure 2.5 The relationship between the Fourier transform for continuous and sampled signals for different sampling frequencies. For simplicity it has been assumed that the Fourier transform is real.

It follows from (2.4) that the value of the Fourier transform of the sampled signal at ω is the sum of the values of the Fourier transform of the continuous-time signal at the frequencies $\omega + n\omega_s$. After sampling, it is thus no longer possible to separate the contributions from these frequencies. The frequency ω can thus be considered to be the *alias* of $\omega + n\omega_s$. It is customary to consider only positive frequencies. The frequency ω is then the alias of $\omega_s - \omega$, $\omega_s + \omega$, $2\omega_s - \omega$, $2\omega_s + \omega$, ..., where $0 \le \omega < \omega_N$. After sampling, a frequency thus cannot be distinguished from its aliases. The fundamental alias for a frequency $\omega_1 > \omega_N$ is given by

$$\omega = |(\omega_1 + \omega_N) \bmod(\omega_s) - \omega_N| \qquad (2.13)$$

Notice that although sampling is a linear operation, it is not time invariant. This explains why new frequencies will be created by the sampling. This is discussed further in Chapter 4.

An illustration of the aliasing effect is shown in Fig. 2.6. Two signals with the frequencies 0.1 Hz and 0.9 Hz are sampled with a sampling frequency of 1

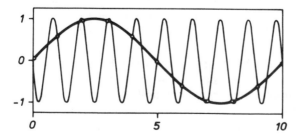

Figure 2.6 Two signals with different frequencies, 0.1 Hz and 0.9 Hz, may have the same value at all sampling instants.

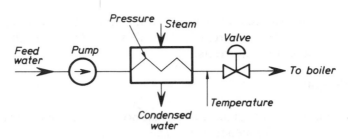

Figure 2.7 Process diagram for a feed-water heating system of a boiler.

Hz ($h = 1$ s). The figure shows that the signals have the same values at the sampling instants.

The aliasing problem is illustrated with an example.

Example 2.1—Aliasing

Figure 2.7 is a process diagram of feed-water heating in a ship's boiler. A valve controls the flow of water. There is a backlash in the valve positioner due to wear. This causes the temperature and the pressure to oscillate. Figure 2.8 shows a sampled recording of the temperature and a continuous recording of the pressure. From the temperature recording one might believe that there is an oscillation with a period of about 38 min. The pressure recording reveals, however, that the oscillation in pressure has a period of 2.11 min. Physically the two variables are coupled and should oscillate with the same frequency. The temperature is sampled every other minute. The sampling frequency is $\omega_s = 2\pi/2 = 3.142$ rad/min and the frequency of the pressure oscillation is $\omega_0 = 2\pi/2.11 = 2.978$ rad/min. The lowest aliasing frequency is $\omega_s - \omega_0 = 0.1638$ rad/min. This corresponds to a period of 38 min, which is the period of the recorded oscillation in the temperature. □

Frequency Folding

Equation (2.4) can also be given another interpretation. The graph of the spectrum of the continuous-time signal is first drawn on a paper. The paper is then folded

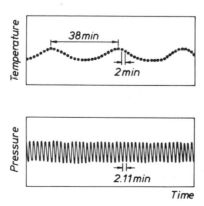

Figure 2.8 Recordings of temperature and pressure.

at abscissas that are odd multiples of the Nyquist frequency, as indicated in Fig. 2.9. The sampled spectrum is then obtained by adding the contributions with proper phase from all sheets.

Prefiltering

A practical difficulty is that real signals do not have Fourier transforms that vanish outside a given frequency band. The high-frequency components may appear to be low-frequency components due to aliasing. The problem is particularly serious if there are periodic high-frequency components. To avoid the alias problem, it is necessary to filter the analog signals before sampling. This may be done in many different ways.

Practically all analog sensors have some kind of filter, but the filter is seldom chosen for a particular control problem. It is therefore often possible to modify the filter so that the signals obtained do not have frequencies above the Nyquist frequency.

Sometimes the simplest solution is to introduce an analog filter in front of the sampler. A standard analog circuit for a second-order filter with a transfer function

$$G_f(s) = \frac{\omega^2}{s^2 + 2\zeta\omega s + \omega^2} \tag{2.14}$$

is shown in Fig. 2.10.

Higher-order filters are obtained by cascading first- and second-order systems. Examples of filters are given in Table 2.1. The table gives filters with band-

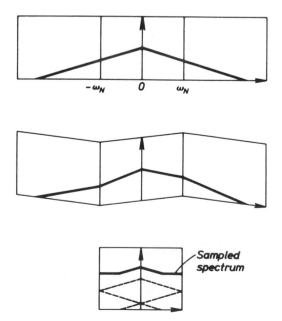

Figure 2.9 Frequency folding.

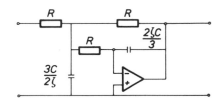

Figure 2.10 An operational amplifier realization of a second-order filter. The transfer function of the filter is given by (2.13) with $\omega = 1/RC$.

width $\omega_B = 1$. The filters get the bandwidth ω_B by changing the factors (2.14) to

$$\frac{\omega^2}{(s/\omega_B)^2 + 2\zeta\omega(s/\omega_B) + \omega^2} \qquad (2.15)$$

where ω and ζ are given by Table 2.1.

The Bessel filter has a linear phase curve, which means that the shape of the signal is not distorted much. The Bessel filters are therefore common in high-performance systems.

The filter must be taken into account in the design of the regulator if the desired crossover frequency is larger than about $\omega_B/10$, where ω_B is the bandwidth of the filter. The Bessel filter can, however, be approximated with a time delay, since the filter has linear phase for low frequencies. Table 2.2 shows the delay for different orders of the filter. Figure 2.11 shows the Bode plot of a sixth-order Bessel filter and a time delay of $2.7/\omega_B$. This property implies that the sampled data model including the antialiasing filter can be assumed to contain an additional time delay compared to the process. Assume that the bandwidth of the filter is chosen as

$$|G_{aa}(i\omega_N)| = \beta$$

where ω_N is the Nyquist frequency and $G_{aa}(s)$ is the transfer function of the antialiasing filter. Table 2.3 gives some values of T_d as a function of β. First the attenuation β is chosen. The table then gives the bandwidth of the filter in relation to the Nyquist frequency. The delay measured in the units of the sampling period is also obtained, i.e., if a small value of β is desired, then bandwidth of the filter must be low and the corresponding delay is long.

TABLE 2.1 Damping ζ and natural frequency ω for Butterworth, ITAE (Integral Time Absolute Error), and Bessel filters. The higher-order filters with arbitrary bandwidth ω_B are obtained by cascading filters of the form (2.15).

Order	Butterworth ω	ζ	ITAE ω	ζ	Bessel ω	ζ
2	1	0.71	0.99	0.71	1.27	0.87
4	1	0.38	1.49	0.32	1.60	0.62
	1	0.92	0.84	0.83	1.43	0.96
6	1	0.26	1.51	0.24	1.90	0.49
	1	0.71	1.13	0.60	1.69	0.82
	1	0.97	0.92	0.93	1.61	0.98

TABLE 2.2 Approximate time delay T_d of Bessel filters of different orders.

Order	$\omega_B T_d$
2	1.3
4	2.1
6	2.7

Example 2.2—Prefiltering

The usefulness of a prefilter is illustrated in Fig. 2.12. An analog signal composed of a square wave with a superimposed sinusoidal perturbation (0.9 Hz) is shown in (a). The result of sampling the analog signal with a period of 1 Hz is shown in (c). The Nyquist frequency is 0.5 Hz. The disturbance with the frequency 0.9 Hz has the alias 0.1 Hz [see (2.13)]. This signal is clearly noticeable in the sampled signal. The output of a prefilter, a sixth-order Bessel filter with a bandwidth of 0.25 Hz, is shown in (b), and the result obtained by sampling with the prefilter is shown in (d). Thus the amplitude of the disturbance is reduced significantly by the prefilter. □

Example 2.3—Product stream sampling

In process control there is one situation in which prefiltering cannot be used: namely when a process stream is sampled and sent to an instrument for analysis. Examples are samples taken for mass spectrographs, gas chromatographs, and laboratory anal-

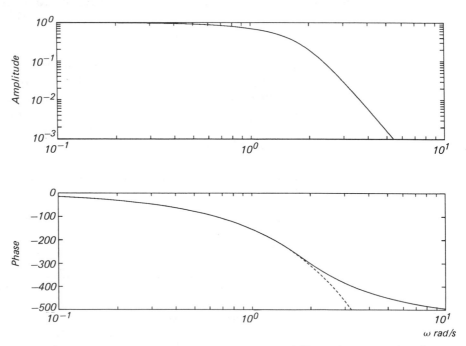

Figure 2.11 Bode plot of a sixth-order Bessel filter when $\omega_B = 1$ and of a time delay $T_d = 2.7$.

TABLE 2.3 The time delay T_d as a function of the desired attenuation β at the Nyquist frequency for fourth- and sixth-order Bessel filters. The sampling period is denoted h.

β	4th order		6th order	
	ω_B/ω_N	T_d/h	ω_B/ω_N	T_d/h
0.001	0.1	5.6	0.2	4.8
0.01	0.2	3.2	0.3	3.1
0.05	0.3	2.1	0.4	2.3
0.1	0.4	1.7	0.4	2.0
0.2	0.5	1.4	0.5	1.7
0.5	0.7	0.9	0.7	1.2
0.7	1.0	0.7	1.0	0.9

ysis. In such a case it is advisable to take many samples and to mix them thoroughly before sending them to the analyzer. This is equivalent to taking several samples and taking the mean value. □

Postsampling Filters

The signal from the D-A converter is piecewise constant. This may cause difficulties for systems with weakly damped oscillatory modes because they may be

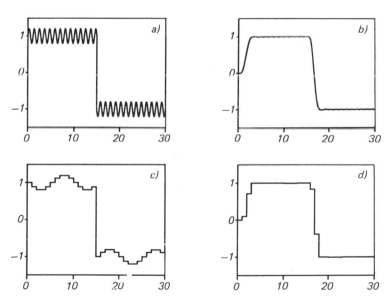

Figure 2.12 Usefulness of a prefilter.
a. Signal plus sinusoidal disturbance.
b. The signal is filtered through a sixth-order Bessel filter.
c. Sample and hold of the signal in (a).
d. Sample and hold of the signal in (b).

excited by the small steps in the signal. In such a case it is useful to introduce a special postsampling filter that smoothes the signal before applying it to the actuator. In some cases this can be achieved by suitable modification of the actuator dynamics. In extreme cases it may be advisable to design special D-A converters that will give piecewise linear control signals.

2.6 Practical Aspects of the Choice of the Sampling Period

Selection of the sampling period in sampled systems is a fundamental problem that will be discussed several times in this book. The proper choice depends on the properties of the signal, the reconstruction method, and the purpose of the system. In a pure signal-processing problem, the purpose is simply to record a signal digitally and to recover it from its samples. A reasonable criterion for selection may then be the size of the error between the original signal and the reconstructed signal.

Shannon's sampling theorem gives a very simple rule in the ideal case when a long time delay in the reconstruction can be accepted and when the frequency content of the signal is limited to a given frequency band. In practice it is often necessary to have restrictions on the time delay of the reconstructed signal. It is also important to recognize that the signals may be contaminated by high-frequency disturbances.

Choice of Sampling Period for Signal Processing

Consider a signal-processing problem whose purpose can be expressed in terms of the error in reconstruction of the signal. Assume that the Fourier transform of the signal is zero for $|\omega| \geq \omega_0$. If a delay in the reconstruction is acceptable, Shannon's theorem says that a sampling rate of $2\omega_0$ is sufficient. Considerably higher sampling rates are required, however, if the time delay in the reconstruction is restricted. It is then necessary to use a causal reconstruction like a first- or zero-order hold. The reconstruction error can then be estimated by (2.10) or (2.12). Example 2.4 illustrates typical orders of magnitude.

Example 2.4—Sampling sinusoidal signals

Assume that the signal is a pure sine wave with frequency ω. The maximum errors of the peak-to-peak amplitude for zero- and first-order hold reconstruction are then given by

$$e_0 = \frac{\omega h}{2} = \frac{\pi}{N}$$

$$e_1 = \frac{(\omega h)^2}{2} = \frac{2\pi^2}{N^2}$$

where N is the number of samples per period. Some typical numerical values are given in Table 2.4.

To get a relative error of 1% with a zero-order hold reconstruction it is necessary

TABLE 2.4 Relative errors obtained when sampling and reconstructing a sine curve using different sampling rates.

Samples per period N	Maximum relative error	
	Zero-order hold	First-order hold
2	1.5	2.5
5	0.6	0.8
10	0.3	0.19
20	0.15	0.05
50	0.06	0.008
100	0.03	0.002
200	0.015	$5 \cdot 10^{-4}$
500	0.006	$8 \cdot 10^{-5}$

to have about 300 samples per period. The table also shows that the effect of using a first-order hold is significantly better only if N is larger than 20.

The results are similar for sampling and reconstruction of other signals. $\square$

It follows from the example that sampling rates of several hundred per period could well be justified in signal-processing applications.

Closed-Loop Control

A rational choice of the sampling rate in a closed-loop control system should be based on an understanding of its influence on the performance of the control system. It seems reasonable that the highest frequency of interest should be closely related to the bandwidth of the closed loop system. The selection of sampling rates can then be based on the bandwidth or, equivalently, on the rise time of the closed-loop system. Reasonable sampling rates are ten to thirty times the bandwidth, or four to ten per rise time, which may seem slow in relation to the typical signal-processing problem. Comparatively low sampling rates can be used in control problems because the dynamics of many controlled systems are of low-pass character and their time constants are typically larger than the closed-loop response times. The contribution to the output from one sampling period then depends on the pulse area; it is comparatively insensitive to the pulse shape. It is important to use an antialiasing filter with a bandwidth that is related to the Nyquist frequency.

The choice of sampling period will be discussed several times later in the book.

2.7 Summary

Sampling of continuous time signals has been discussed in this chapter. To sample a signal means to replace it by a sequence of its values for a discrete set of times.

For periodic sampling, the signal can be recovered from its samples if the following conditions are met:

The continuous-time signal has a Fourier transform that vanishes for $|\omega| > \omega_1$.

The sampling frequency ω_s is larger than $2\omega_1$.

When dealing with sampled signals, it is thus important to choose a sampling frequency that is sufficiently high compared with the highest frequency of interest.

It is also important to filter a continuous-time signal before it is sampled, so that frequency components above the Nyquist frequency $\omega_N = \omega_s/2$ are not transformed to low-frequency components via aliasing or frequency folding.

2.8 Problems

2.1. Sketch the impulse response of the Shannon reconstructor given by (2.1).

2.2. The signal

$$f(t) = a_1 \sin 2\pi t + a_2 \sin 20t$$

is the input to a zero-order, sample-and-hold circuit. Which frequencies are there at the output if the sampling period is $h = 0.2$?

2.3. A signal that is going to be sampled has the spectrum shown in Fig. 2.13. Of interest are the frequencies in the range from 0 to f_1 Hz. A disturbance has a fixed known frequency with $f_2 \approx 5f_1$. Discuss choice of sampling interval and presampling filter.

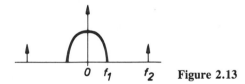

$0 \quad f_1 \qquad f_2$ **Figure 2.13**

2.4. Show that the system in Fig. 2.14 is an implementation of a first-order hold and determine its response to a pulse of unit magnitude and a duration of one sampling interval.

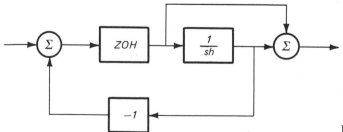

Figure 2.14

     Sampling of Continuous-Time Signals   Chap. 2

2.5. The magnitude of the spectrum of a signal is shown in Fig. 2.15. Sketch the magnitude of the spectrum when the signal has been sampled with (a) $h = 2\pi/10$ s, (b) $h = 2\pi/20$ s, and (c) $h = 2\pi/50$ s.

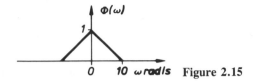

0   10 ω rad/s   Figure 2.15

2.6. Consider the signal in Problem 2.5, but let the spectrum be centered around $\omega = 100$ rad/s and with (a) $\omega_s = 120$ rad/s and (b) $\omega_s = 240$ rad/s.

2.7. A camera is used to get a picture of a rotating wheel with a mark on it. The wheel rotates with r revolutions per second. The camera takes one frame each h seconds. Discuss how the picture will appear when shown on a screen. (Compare with what you see in western movies.)

2.8. The signal $y(t) = \sin 3\omega t$ is sampled with the sampling period h. Determine h such that the sampled signal is periodic.

2.9. An amplitude modulated signal

$$u(t) = \sin(4\omega_0 t) \cos(2\omega_0 t)$$

is sampled with $h = \pi/(3\omega_0)$. Determine the frequencies f, $0 \le f \le 3\omega_0/(2\pi)$ that are represented in the sampled signal.

2.9 References

The fact that a sinusoid can be retrieved from its sampled values if it is sampled at least twice per period was stated in

NYQUIST, H. (1928): "Certain Topics in Telegraph Transmission Theory," *AIEE Trans.*, 47, 617–44.

The sampling theorem in the form presented in this chapter was introduced in

SHANNON, C. E. (1949): "Communications in the Presence of Noise," *Proc. IRE*, 37, 10–21.

where the implications for communication were emphasized. The results had, however, been known earlier as a theorem in mathematics. In the Soviet communication literature, the theorem was introduced by

KOTELNIKOV, V. A. (1933): "On the Transmission Capacity of 'Ether' and Wire in Electrocommunications," *Proc. First All-union Conference on Questions of Communications*. Moscow.

A review of the sampling theorem with many references is given in

JERRI, A. J. (1977): "The Shannon Sampling Theorem—Its Various Extensions and Applications: A Tutorial Review," *Proc. IEEE*, 65, 1565–95.

There are many ways of sampling. A review of different schemes is given in

JURY, E. I. (1961): "Sampling Schemes in Sampled-Data Control Systems," *IRE Trans.,* AC-6, 88–90.

Different types of hold circuits are discussed in more detail in

RAGAZZINI, J. R. and G. F. FRANKLIN (1958): *Sampled-Data Control Systems.* New York: McGraw-Hill.

Selection of the sampling period for signal processing is discussed in

GARDENHIRE, L. W. (1964): "Selection of Sample Rates," *ISA Journal* (April) 59–64.

Choice of the sampling interval for control purposes will be discussed later and further references will be given then. The different trade-offs in the areas of control and signal processing may lead to very different rules for choosing the sampling rate.

COMPUTER-ORIENTED MATHEMATICAL MODELS: DISCRETE-TIME SYSTEMS

GOAL

To Give Mathematical Descriptions of Computer-Controlled Systems from the View of the Computer. To Develop Concepts and Tools for Working with Discrete-Time Systems.

3.1 Introduction

In this chapter mathematical models for computer-controlled systems are introduced. The system is studied as seen from the computer. The computer receives measurements from the process at discrete times and transmits new control signals at discrete times. The goal then is to describe the change in the signals from sample to sample and disregard the behavior between the samples. It should be emphasized that computer-oriented mathematical models give the behavior only at the sampling points—the physical process is still a continuous-time system. Looking at the problem this way, however, will greatly simplify the treatment. A description of process-oriented models, which takes the continous-time behavior into account, is given in Chapter 4.

One point that must be treated with some caution is that the sampled-data system is time dependent (see Example 1.1). This problem is discussed in Chapter 4 also. In this chapter the problem of time dependence is avoided by studying the signals at time instances that are synchronized with the clock in the computer. This is sometimes called the *stroboscopic model,* since only intermittent observations of the signals are obtained. This gives models described by difference equations in state-space and input-output forms. Section 3.2 treats the problem of finding the discrete-time representation of a continuous-time, state-space model by using zero-order-hold devices. The general solution of forced difference equations and the inverse problem of finding the continuous-time system that corresponds to a given discrete-time system are also treated in Sec. 3.2. Sections 3.3 and 3.4 deal with transformation of state-space models and the connection between state-space and input-output models. Shift operators are used to describe input-output models. Shift-operator calculus is equivalent to the use of differential operators for continuous-time systems. The discrete-time equivalent to the Laplace transform is the z-transform, which is covered in Sec. 3.5.

The treatment of state-space models in Sec. 3.2 covers the multivariable case. The discussion of input-output models is, however, restricted to single-input-single-output systems. Extensions to the multivariable case are possible, but are not used in this book because they require the mathematics of polynomial matrices.

In order to design computer-controlled systems, it is important to understand how poles and zeros of continuous-time and discrete-time models are related. This is treated in Sec. 3.6. The selection of sampling period is discussed in Sec. 3.7. Rules of thumb based on the appearances of transient responses are given in terms of samples per rise time.

3.2 Sampling of a Continuous-Time, State-Space System

A fundamental problem is how to describe a continuous-time system connected to a computer via A-D and D-A converters. Consider the system shown in Fig. 3.1. The signals in the computer are the sequences $\{u(t_k)\}$ and $\{y(t_k)\}$. The key problem is to find the relationship between these sequences. To find the discrete-time equivalent of a continuous-time system is called *sampling of a continuous-time system*. The model obtained is also called a stroboscopic model because it gives a relationship between the system variables at the sampling instants only. To obtain the desired descriptions, it is necessary to describe the converters and the system. Assume that the continuous-time system is given in the following state-space form:

$$\frac{dx}{dt} = Ax(t) + Bu(t)$$

$$y(t) = Cx(t) + Du(t)$$

$$(3.1)$$

The system has r inputs, p outputs, and order n.

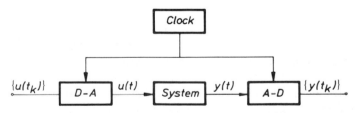

Figure 3.1 Block diagram of a continuous-time system connected to A-D and D-A converters.

Zero-Order-Hold Sampling of a System

A common situation in computer control is that the D-A converter is so constructed that it holds the analog signal constant until a new conversion is commanded. It is then natural to choose the sampling instants, t_k, as the times when the control changes. Since the control signal is discontinuous, it is necessary to specify its behavior at the discontinuities. The convention that the signal is continuous from the right is adopted. The control signal is thus represented by the sampled signal $\{u(t_k): k = \ldots, -1, 0, 1, \ldots\}$.

The relationship between the system variables at the sampling instants will now be determined. Given the state at the sampling time t_k the state at some future time t is obtained by solving (3.1).

$$x(t) = e^{A(t-t_k)}x(t_k) + \int_{t_k}^{t} e^{A(t-s')}Bu(s')\, ds' \tag{3.2}$$

The state at the next sampling time t_{k+1} is thus given by

$$x(t_{k+1}) = e^{A(t_{k+1}-t_k)}x(t_k) + \int_{t_k}^{t_{k+1}} e^{A(t_{k+1}-s')}Bu(s')\, ds'$$

$$= e^{A(t_{k+1}-t_k)}x(t_k) + \int_{t_k}^{t_{k+1}} e^{A(t_{k+1}-s')}\, ds'\, Bu(t_k)$$

$$= \Phi(t_{k+1}, t_k)x(t_k) + \Gamma(t_{k+1}, t_k)u(t_k)$$

The second equality follows from the fact that u is constant between the sampling instants.

The state vector at time t_{k+1} is thus a linear function of $x(t_k)$ and $u(t_k)$. If the A-D and D-A converters in Fig. 3.1 are perfectly synchronized and if the conversion times are negligible, the input u and the output y can be regarded as being sampled at the same instants. The system equation of the sampled system is then

$$\begin{cases} x(t_{k+1}) = \Phi(t_{k+1}, t_k)x(t_k) + \Gamma(t_{k+1}, t_k)u(t_k) \\ y(t_k) = Cx(t_k) + Du(t_k) \end{cases} \tag{3.3}$$

where

$$\Phi(t_{k+1}, t_k) = e^{A(t_{k+1}-t_k)} \tag{3.4}$$

$$\Gamma(t_{k+1}, t_k) = \int_{0}^{t_{k+1}-t_k} e^{As}\, ds\, B \tag{3.5}$$

The relationship between the sampled signals can thus be expressed by the linear difference equation, (3.3). Notice that Equation (3.3) does not involve any approximations. It gives the *exact* values of the state variables and the output at the sampling instants because the control signal is constant between the sampling instants. The model in (3.3) is therefore called a *zero-order-hold sampling of the system* in (3.1). The system in (3.3) can also be called the zero-order-hold equivalent of (3.1).

In most cases $D = 0$. One reason for this is because in computer-controlled systems, the output y is first measured and the control signal $u(t_k)$ is then generated as a function of $y(t_k)$. In practice it often happens that there is a significant delay between the A-D and D-A conversions. However, it is easy to make the necessary modifications. The state vector at times between sampling points can also be computed using (3.2). This makes it possible to investigate the intersample behavior of the system. Notice that the responses between the sampling points are parts of step responses, with initial conditions, for the system. This implies, for instance, that the system is running in open loop between the sampling points.

For periodic sampling with period h,

$$t_k = k \cdot h$$

the model of (3.3) simplifies to the time-invariant system

$$x(kh + h) = \Phi x(kh) + \Gamma u(kh) \tag{3.6}$$
$$y(kh) = Cx(kh) + Du(kh)$$

where

$$\begin{cases} \Phi = e^{Ah} \\ \Gamma = \int_0^h e^{As} \, ds \, B \end{cases} \tag{3.7}$$

Notice that it follows from (3.7) that

$$\frac{d\Phi(t)}{dt} = A\Phi(t) = \Phi(t)A$$

$$\frac{d\Gamma(t)}{dt} = \Phi(t)B$$

The matrices Φ and Γ therefore satisfy the equation

$$\frac{d}{dt} \begin{pmatrix} \Phi(t) & \Gamma(t) \\ 0 & I \end{pmatrix} = \begin{pmatrix} \Phi(t) & \Gamma(t) \\ 0 & I \end{pmatrix} \begin{pmatrix} A & B \\ 0 & 0 \end{pmatrix}$$

where I is a unit matrix of the same dimension as the number of inputs. The matrices $\Phi(h)$ and $\Gamma(h)$ for the sampling period h can therefore be obtained from the block matrix

$$\begin{pmatrix} \Phi(h) & \Gamma(h) \\ 0 & I \end{pmatrix} = \exp \left\{ \begin{pmatrix} A & B \\ 0 & 0 \end{pmatrix} h \right\} \tag{3.8}$$

How to Compute Φ and Γ

The calculations required to sample a continuous-time system are evaluation of a matrix exponential and integration of a matrix exponential. This can be done in many different ways, for instance, by using the following:

Series expansion of the matrix exponential.

The Laplace transform—the Laplace transform of $\exp(At)$ is $(sI - A)^{-1}$.

Cayley-Hamilton's theorem (see Appendix B).

Transformation to Jordan form.

Hand calculations are feasible for low-order systems, $n \le 2$, and for high-order systems with special structures. One way to simplify the computations is to compute

$$\Psi = \int_0^h e^{As} \, ds = Ih + \frac{Ah^2}{2!} + \frac{A^2h^3}{3!} + \cdots + \frac{A^ih^{i+1}}{(i+1)!} + \cdots \qquad (3.9)$$

The matrices Φ and Γ are given by

$$\begin{cases} \Phi = I + A\Psi \\ \Gamma = \Psi B \end{cases} \qquad (3.10)$$

Computer evaluation can be done using several different numerical algorithms.

Example 3.1—First-order system

Consider the system

$$\frac{dx}{dt} = \alpha x + \beta u$$

Applying the formulas (3.7) we get

$$\Phi = e^{\alpha h}$$

$$\Gamma = \int_0^h e^{\alpha h} \, ds \, \beta = \frac{\beta}{\alpha}(e^{\alpha h} - 1)$$

The sampled system thus becomes

$$x(kh + h) - e^{\alpha h}x(kh) + \frac{\beta}{\alpha}(e^{\alpha h} - 1)u(kh) \qquad \qquad \square$$

Example 3.2—Double integrator

The double integrator (see Appendix A) is described by

$$\dot{x} = \begin{bmatrix} 0 & 1 \\ 0 & 0 \end{bmatrix} x + \begin{bmatrix} 0 \\ 1 \end{bmatrix} u$$

$$y = [1 \quad 0]x \qquad (3.11)$$

Hence

$$\Phi = e^{Ah} = I + Ah + A^2h^2/2 + \cdots$$

$$= \begin{bmatrix} 1 & 0 \\ 0 & 1 \end{bmatrix} + \begin{bmatrix} 0 & h \\ 0 & 0 \end{bmatrix} = \begin{bmatrix} 1 & h \\ 0 & 1 \end{bmatrix}$$

$$\Gamma = \int_0^h \begin{bmatrix} s \\ 1 \end{bmatrix} ds = \begin{bmatrix} \dfrac{h^2}{2} \\ h \end{bmatrix}$$

The discrete-time model of (3.11) is

$$x(kh + h) = \begin{bmatrix} 1 & h \\ 0 & 1 \end{bmatrix} x(kh) + \begin{bmatrix} \dfrac{h^2}{2} \\ h \end{bmatrix} u(kh)$$

$$y(kh) = \begin{bmatrix} 1 & 0 \end{bmatrix} x(kh)$$

(3.12)

Example 3.3—Motor

A simple, normalized model of an electrical DC motor (see Example A.2 in Appendix A) is given by

$$\dot{x} = \begin{bmatrix} -1 & 0 \\ 1 & 0 \end{bmatrix} x + \begin{bmatrix} 1 \\ 0 \end{bmatrix} u$$

$$y = \begin{bmatrix} 0 & 1 \end{bmatrix} x.$$

The Laplace transform method gives

$$(sI - A)^{-1} = \begin{bmatrix} s + 1 & 0 \\ -1 & s \end{bmatrix}^{-1} = \frac{1}{s(s + 1)} \begin{bmatrix} s & 0 \\ 1 & s + 1 \end{bmatrix}$$

$$= \begin{bmatrix} \dfrac{1}{s + 1} & 0 \\ \dfrac{1}{s(s + 1)} & \dfrac{1}{s} \end{bmatrix}$$

Hence

$$\Phi = e^{Ah} = \mathcal{L}^{-1}(sI - A)^{-1} = \begin{bmatrix} e^{-h} & 0 \\ 1 - e^{-h} & 1 \end{bmatrix}$$

and

$$\Gamma = \int_0^h \begin{bmatrix} e^{-v} \\ 1 - e^{-v} \end{bmatrix} dv = \begin{bmatrix} 1 - e^{-h} \\ h - 1 + e^{-h} \end{bmatrix}$$

where $\mathcal{L}^{-1}$ is the inverse of the Laplace transform. ☐

Solution of the System Equation

Time-variant, discrete-time systems can be described by the difference equation

$$x(k + 1) = \Phi x(k) + \Gamma u(k)$$

$$y(k) = Cx(k)$$

(3.13)

For simplicity the sampling time is used as the time unit, $h = 1$. Assume that $x(k_0)$ and the input signals $u(k_0)$, $u(k_0 + 1)$, . . . are given. How is the state then evolving? It is possible to solve (3.13) by simple iterations.

$$x(k_0 + 1) = \Phi x(k_0) + \Gamma u(k_0)$$

$$x(k_0 + 2) = \Phi x(k_0 + 1) + \Gamma u(k_0 + 1)$$

$$= \Phi^2 x(k_0) + \Phi \Gamma u(k_0) + \Gamma u(k_0 + 1) \tag{3.14}$$

$$\vdots$$

$$x(k) = \Phi^{k-k_0} x(k_0) + \Phi^{k-k_0-1}\Gamma u(k_0) + \cdots + \Gamma u(k - 1)$$

$$= \Phi^{k-k_0} x(k_0) + \sum_{j=k_0}^{k-1} \Phi^{k-j-1}\Gamma u(j)$$

The solution consists of two parts: One depends on the initial condition, and the other is a weighted sum of the input signals.

The Inverse of Sampling

Sampling of a system defines a map from continuous-time systems, as in (3.1), to discrete-time systems, as in (3.6). A natural question is if and when it is possible to get the corresponding continuous-time system from a discrete-time description.

Example 3.4—Inverse sampling

Consider the first-order difference equation

$$x(kh + h) = ax(kh) + bu(kh)$$

From Example 3.1 we find that the corresponding continuous-time system is obtained from

$$e^{\alpha h} = a$$

$$\frac{\beta}{\alpha}(e^{\alpha h} - 1) = b$$

This gives

$$\alpha = \frac{1}{h}\ln a$$

$$\beta = \frac{1}{h}\ln a \cdot \frac{b}{a - 1}$$

The example indicates that there may be difficulties when a is negative. This implies that the model (3.6) is more general than (3.1). $\square$

Inverse sampling in the general case can now be discussed. It follows from (3.8) that

$$\begin{pmatrix} A & B \\ 0 & 0 \end{pmatrix} = \frac{1}{h}\ln\begin{pmatrix} \Phi & \Gamma \\ 0 & I \end{pmatrix}$$

where $\ln(\cdot)$ is the matrix logarithmic function. The continuous-time system is thus obtained by taking the matrix logarithm function of a block matrix. Computation of matrix logarithm is discussed in Appendix B. From the Cayley-Hamilton theorem it follows that it must be assumed that Φ does not have any eigenvalues on the negative real axis to be able to compute the matrix logarithm. There is also a nonuniqueness in the matrix logarithmic function for complex arguments. This is illustrated by the following example.

Example 3.5—Harmonic oscillator

The discrete-time system

$$x(kh + h) = \begin{bmatrix} \cos \alpha h & \sin \alpha h \\ -\sin \alpha h & \cos \alpha h \end{bmatrix} x(kh) + \begin{bmatrix} 1 - \cos \alpha h \\ \sin \alpha h \end{bmatrix} u(kh)$$

can be obtained by sampling a continuous-time system with

$$A = \begin{bmatrix} 0 & \omega \\ -\omega & 0 \end{bmatrix} \quad \text{and} \quad B = \begin{bmatrix} 0 \\ \omega \end{bmatrix}$$

where

$$\omega = \alpha + \frac{2\pi}{h} \cdot n \qquad n = 0, 1, \ldots$$

In this case the inverse problem has many solutions (Compare Examples A.3 and B.1). This is generally the case if the matrix Φ has complex eigenvalues. Notice that there always exists a unique ω in the interval $-\omega_N \leq \omega \leq \omega_N$, where $\omega_N = \pi/h$ is the Nyquist frequency associated with the sampling period h. The example gives another view of the aliasing problem discussed in Sec. 2.5. ☐

Sampling a System with Time Delay

Time delays are common in mathematical models of industrial processes. The theory of continuous-time systems with time delays is complicated because the systems are infinite dimensional.

A system composed of a time delay and a continuous-time, state-space system will now be sampled. Let the system be described by

$$\dot{x} = Ax(t) + Bu(t - \tau) \tag{3.15}$$

It is assumed initially that the time delay τ is less than the sampling period. The zero-order-hold sampling of the system (3.15) will now be calculated.

Integration of (3.15) over one sampling period gives

$$x(kh + h) = e^{Ah}x(kh) + \int_{kh}^{kh+h} e^{A(kh+h-s')}Bu(s' - \tau) \, ds' \tag{3.16}$$

Because the signal $u(t)$ is piecewise constant over the sampling interval, the delayed signal $u(t - \tau)$ is also piecewise constant. The delayed signal will, however, change between the sampling instants (see Fig. 3.2). To evaluate the integral of (3.16), it is then convenient to split the integration interval into two parts so that

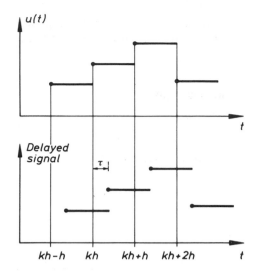

u(t)

Delayed signal

τ

kh−h kh kh+h kh+2h t

Figure 3.2 The relationship among $u(t)$, the delayed signal $u(t - \tau)$, and the sampling instants.

$u(t - \tau)$ is constant in each part. Hence

$$\int_{kh}^{kh+h} \cdots - \int_{kh}^{kh+\tau} e^{A(kh+h-s')}B \ ds' \ u(kh - h)$$

$$+ \int_{kh+\tau}^{kh+h} e^{A(kh+h-s')}B \ ds' \ u(kh)$$

$$= \Gamma_1 u(kh - h) + \Gamma_0 u(kh)$$

Sampling of the continuous-time system of (3.15) thus gives

$$x(kh + h) = \Phi x(kh) + \Gamma_0 u(kh) + \Gamma_1 u(kh - h) \qquad (3.17)$$

where

$$\Phi = e^{Ah}$$

$$\Gamma_0 = \int_0^{h-\tau} e^{As} \ ds \ B \qquad (3.18)$$

$$\Gamma_1 = e^{A(h-\tau)} \int_0^{\tau} e^{As} \ ds \ B \qquad (3.19)$$

A state-space model of (3.17) is given by

$$\begin{bmatrix} x(kh + h) \\ u(kh) \end{bmatrix} = \begin{bmatrix} \Phi & \Gamma_1 \\ 0 & 0 \end{bmatrix} \begin{bmatrix} x(kh) \\ u(kh - h) \end{bmatrix} + \begin{bmatrix} \Gamma_0 \\ I \end{bmatrix} u(kh)$$

Notice that r extra state variables $u(kh - h)$, which represent the past values of the control signal, are introduced. The continuous-time system of (3.15) is infinite dimensional; the corresponding sampled system however, is a finite-dimensional system. Thus time delays are considerably simpler to handle if the system is sampled, for the following reason: To specify the state of the system, it is necessary to store the input over a time interval equal to the time delay. With zero-order-hold reconstruction, the input signal can always be represented by a finite number of values.

Example 3.6—First-order system with time delay

Consider the system

$$\dot{x}(t) = \alpha x(t) + \beta u(t - \tau)$$

Assume that the system is sampled with period h, where $0 \le \tau < h$. Equations (3.7), (3.18), and (3.19) give

$$\Phi = a = e^{\alpha h}$$

$$\Gamma_0 = b_0 = \int_0^{h-\tau} e^{\alpha s} \beta \, ds = \frac{\beta}{\alpha} (e^{\alpha(h-\tau)} - 1)$$

$$\Gamma_1 = b_1 = e^{-\alpha(h-\tau)} \int_0^{\tau} e^{\alpha s} \beta \, ds = \frac{\beta}{\alpha} (e^{\alpha h} - e^{\alpha(h-\tau)})$$

The sampled system is thus

$$x(kh + h) = ax(kh) + b_0 u(kh) + b_1 u(kh - h) \qquad \square$$

Example 3.7—Double integrator with delay

Consider the double integrator in Example 3.2 and introduce a time delay $0 \le \tau \le h$. Then

$$\Phi = e^{Ah} = \begin{bmatrix} 1 & h \\ 0 & 1 \end{bmatrix}$$

$$\Gamma_1 = e^{A(h-\tau)} \int_0^{\tau} e^{As} \, ds \, B = \begin{bmatrix} 1 & h-\tau \\ 0 & 1 \end{bmatrix} \begin{bmatrix} \dfrac{\tau^2}{2} \\ \tau \end{bmatrix} = \begin{bmatrix} \tau\left(h - \dfrac{\tau}{2}\right) \\ \tau \end{bmatrix}$$

$$\Gamma_0 = \int_0^{h-\tau} e^{As} \, ds \, B = \begin{bmatrix} \dfrac{(h-\tau)^2}{2} \\ h - \tau \end{bmatrix} \qquad \square$$

Longer Time Delays

If the time delay is longer than h, then the previous analysis has to be modified a little. If

$$\tau = (d - 1)h + \tau' \qquad 0 < \tau' \le h$$

where d is an integer, the following equation is obtained:

$$x(kh + h) = \Phi x(kh) + \Gamma_0 u(kh - dh + h) + \Gamma_1 u(kh - dh) \qquad (3.20)$$

where Γ_0 and Γ_1 are given by (3.18) and (3.19) with τ replaced by τ'. The corresponding state-space description is

$$\begin{bmatrix} x(kh + h) \\ u(kh - dh + h) \\ \vdots \\ u(kh - h) \\ u(kh) \end{bmatrix}$$

$$= \begin{bmatrix} \Phi & \Gamma_1 & \Gamma_0 & \cdots & 0 \\ 0 & 0 & I & \cdots & 0 \\ \vdots & \vdots & \vdots & & \vdots \\ 0 & 0 & 0 & \cdots & I \\ 0 & 0 & 0 & \cdots & 0 \end{bmatrix} \begin{bmatrix} x(kh) \\ u(kh - dh) \\ \vdots \\ u(kh - 2h) \\ u(kh - h) \end{bmatrix} + \begin{bmatrix} 0 \\ 0 \\ \vdots \\ 0 \\ I \end{bmatrix} u(kh) \quad (3.21)$$

Notice that if $\tau > 0$ then $d \cdot r$ extra state variables are used to describe the delay, where r is the number of inputs. The characteristics polynomial of the state-space description is $\lambda^{dr} A(\lambda)$, where $A(\lambda)$ is the characteristic polynomial of Φ.

An example illustrates use of the general formula.

Example 3.8—Simple paper-machine model

Determine the zero-order-hold sampling of the system

$$\dot{x}(t) = -x(t) + u(t - 2.5)$$

with sampling interval $h = 1$. In this case $d = 3$ and $\tau' = 0.5$, and (3.17) will be modified as follows:

$$x(k + 1) = \Phi x(k) + \Gamma_0 u(k - 2) + \Gamma_1 u(k - 3)$$

where

$$\Phi = e^{-1} \approx 0.37$$

$$\Gamma_0 = \int_0^{0.5} e^{-s} \, ds = 1 - e^{-0.5} \approx 0.39$$

$$\Gamma_1 = e^{-0.5} \int_0^{0.5} e^{-s} \, ds = e^{-0.5} - e^{-1} \approx 0.24 \qquad \square$$

System with Internal Time Delay

In the derivation above it is assumed that the time delay of the system is at the input (or the output) of the system. Many physical systems have the structure shown in Figure 3.3, i.e., the time delay is internal. Let the system be described by the equations

$$S_1: \dot{x}_1 = A_1 x_1 + B_1 u$$

$$y_1 = C_1 x_1 + D_1 u \qquad (3.22)$$

$$S_2: \dot{x}_2 = A_2 x_2 + B_2 u_2$$

$$y = C_2 x_2 + D_2 u_2 \qquad (3.23)$$

$$u_2(t) = y_1(t - \tau) \qquad (3.24)$$

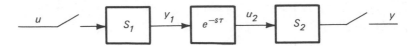

Figure 3.3 System with inner time delay.

It is assumed that $u(t)$ is piecewise constant over the sampling interval h. We now want to find the recursive equations for $x_1(kh)$ and $x_2(kh)$.

Sampling (3.22) to (3.24) when $\tau = 0$ gives

$$\begin{pmatrix} x_1(kh + h) \\ x_2(kh + h) \end{pmatrix} = \begin{pmatrix} \Phi_1(h) & 0 \\ \Phi_{21}(h) & \Phi_2(h) \end{pmatrix} \begin{pmatrix} x_1(kh) \\ x_2(kh) \end{pmatrix} + \begin{pmatrix} \Gamma_1(h) \\ \Gamma_2'(h) \end{pmatrix} u(kh) \quad (3.25)$$

where

$$\Phi_i(t) = e^{A_i t}$$

$$\Gamma_i(t) = \int_0^t e^{A_i s} B_i \, ds$$

We now have the following theorem:

Theorem 3.1—Inner time delay. Periodic sampling of the system (3.22) to (3.24) with the sampling interval h and with $0 < \tau \le h$ gives the sampled data representation

$$x_1(kh + h) = \Phi_1(h)x_1(kh) + \Gamma_1(h)u(kh)$$

$$x_2(kh + h) = \Phi_{\bar{21}}x_1(kh - h) + \Phi_2(h)x_2(kh) \quad (3.26)$$

$$+ \Gamma_2^- u(kh - h) + \Gamma_2'(h - \tau)u(kh)$$

where

$$\Phi_i(t) = e^{A_i t}$$

$$\Gamma_i(t) = \int_0^t e^{A_i s} B_i \, ds$$

$$\Phi_{\bar{21}} = \Phi_{21}(h)\Phi_1(h - \tau)$$

$$\Gamma_2^- = \Phi_{21}(h)\Gamma_1(h - \tau) + \Phi_{21}(h - \tau)\Gamma_1(\tau) + \Phi_2(h - \tau)\Gamma_2'(\tau)$$

Reference to proof of the theorem is given in the end of the chapter.

Remark. The sampled data system (3.26) for the time delay τ is obtained by sampling (3.22) to (3.24) without any time delay for the sampling intervals h, $h - \tau$, and τ. This gives Φ_1, Φ_2, Φ_{21}, Γ_1, and Γ_2' for the needed sampling intervals. This implies that standard software for sampling systems can be used to obtain (3.26).

Other Types of Hold Devices for the Control Signal

In order to calculate the sampled system from the continuous-time system, the shape of the control signal over the sampling interval must be known. These calculations are particularly easy to do when using zero-order-hold because the control signal is piecewise constant. The zero-order-hold sampling is also most common in practice because it is easily implemented using ordinary D-A converters. Examples using other types of hold devices are given in the exercises. A typical case is a system with hydraulic actuators, where smooth control signals are desired in order to avoid pressure transients.

The systems to be controlled often have low pass characteristics. The shape of the control signal over a sampling period is then not crucial. If the sampling period is short compared with the time constants of the system, the response is essentially determined by the integral of the control signal over the sampling period.

It follows from (3.5) that a change in the shape of the control signal will change Γ, but not the transition matrix Φ. This is illustrated by an example.

Example 3.9

Consider a double integrator that is sampled with a period h. Assume that the hold circuit is such that the control signal is $\alpha u(k)$ during the first half of the sampling period and $\beta u(k)$ during the second half. It then follows that

$$\Phi = e^{Ah} = \begin{bmatrix} 1 & h \\ 0 & 1 \end{bmatrix}$$

$$\Gamma = \alpha \int_0^{h/2} e^{A(h-s)} B \, ds + \beta \int_{h/2}^{h} e^{A(h-s)} B \, ds$$

$$-\frac{1}{8} \begin{bmatrix} 3\alpha h^2 + \beta h^2 \\ 4\alpha h + 4\beta h \end{bmatrix}$$

□

Intersample Behavior

The discrete-time models (3.3) and (3.6) give the values of the state variables and the outputs at the sampling instants $\{t_k\}$. The values of the variables between the sampling points are also of interest. These values are given by (3.2), which can be written as

$$x(t) = \Phi(t, t_k) x(t_k) + \Gamma(t, t_k) u(t_k) \qquad (3.27)$$

where

$$\Phi(t, t_k) = e^{A(t-t_k)} \qquad (3.28)$$

$$\Gamma(t, t_k) = \int_0^{t-t_k} e^{As} \, ds \, B \qquad (3.29)$$

3.3 Transformation of State-Space Models

In Sec. 3.2 a discrete-time system is described by the model of (3.13). As for continuous-time systems, new coordinates can be introduced in the state space.

Assume that T is a nonsingular matrix and define a new state vector $z(k) = Tx(k)$. Then

$$z(k + 1) = Tx(k + 1) = T\Phi x(k) + T\Gamma u(k) = T\Phi T^{-1}z(k) + T\Gamma u(k)$$
$$= \tilde{\Phi}z(k) + \tilde{\Gamma}u(k)$$

and

$$y(k) = Cx(k) + Du(k) = CT^{-1}z(k) + Du(k) = \tilde{C}z(k) + Du(k)$$

The matrices Φ, Γ, and C thus depend on the coordinate system chosen to represent the state. The invariants under the transformation are of interest.

Theorem 3.2. The characteristic equation

$$\det[\lambda I - \Phi] = 0$$

is invariant when new states are introduced through the nonsingular transformation matrix T.

Proof.

$$\det[\lambda I - \tilde{\Phi}] = \det[\lambda TT^{-1} - T\Phi T^{-1}] = \det T \det[\lambda I - \Phi] \det T^{-1}$$
$$= \det[\lambda I - \Phi] \qquad\qquad \square$$

Coordinates can be chosen to give simple forms of the system equations.

Diagonal Form

Assume that Φ has distinct eigenvalues. Then there exists a T such that

$$T\Phi T^{-1} = \begin{bmatrix} \lambda_1 & & 0 \\ & \ddots & \\ 0 & & \lambda_n \end{bmatrix}$$

where λ_i are the eigenvalues of Φ. The computation of T is discussed in Sec. 5.3. In this case a set of decoupled, first-order difference equations is obtained.

$$z_1(k + 1) = \lambda_1 z_1(k) + \beta_1 u(k)$$
$$\vdots$$
$$z_n(k + 1) = \lambda_n z_n(k) + \beta_n u(k)$$
$$y(k) = \gamma_1 z_1(k) + \cdots + \gamma_n z_n(k)$$

The solution to the system of equations is now simple. Each mode will have the solution

$$z_i(k) = \lambda_i^k z_i(0) + \sum_{j=0}^{k-1} \lambda_i^{k-j-1} \beta_i u(j) \tag{3.30}$$

If Φ has multiple eigenvalues, then it is generally not possible to diagonalize Φ. Any matrix can, however, be transformed to a *Jordan form*. In this form the transformed matrix, $\tilde{\Phi}$, has the eigenvalues in the diagonal and some 1s in the superdiagonal.

Controllable Form

Assume that Φ has the characteristic equation

$$\det[\lambda I - \Phi] = \lambda^n + a_1 \lambda^{n-1} + \cdots + a_n = 0 \tag{3.31}$$

and that

$$W_c - [\Gamma \quad \Phi\Gamma \quad \cdots \quad \Phi^{n-2}\Gamma \quad \Phi^{n-1}\Gamma] \tag{3.32}$$

is nonsingular. Then there exists a transformation such that the transformed system is

$$z(k+1) - \begin{bmatrix} -a_1 & -a_2 & \cdots & -a_{n-1} & -a_n \\ 1 & 0 & \cdots & 0 & 0 \\ 0 & 1 & \cdots & 0 & 0 \\ \vdots & & & & \\ 0 & 0 & \cdots & 1 & 0 \end{bmatrix} z(k) + \begin{bmatrix} 1 \\ 0 \\ 0 \\ \vdots \\ 0 \end{bmatrix} u(k) \tag{3.33}$$

$$y(k) = [b_1 \quad \cdots \quad b_n] z(k)$$

which is called the *controllable canonical form*. The advantage of this form is that it is easy to compute the input-output model and to compute a state feedback-control law. There are simple ways of finding transformations between different representations of a system. Let

$$x(k+1) = \Phi x(k) + \Gamma u(k)$$

be one representation and let

$$z(k+1) = \tilde{\Phi} z(k) + \tilde{\Gamma} u(k)$$

be another representation obtained by the coordinate transformation $z = Tx$. Consider the matrices

$$W_c = [\Gamma \quad \Phi\Gamma \ldots \Phi^{n-1}\Gamma]$$

and

$$\tilde{W}_c = [\tilde{\Gamma} \quad \tilde{\Phi}\tilde{\Gamma} \ldots \tilde{\Phi}^{n-1}\tilde{\Gamma}]$$

Since

$$\tilde{\Gamma} = T\Gamma$$

$$\tilde{\Phi} = T\Gamma T^{-1}$$

it follows that

$$\tilde{W}_c = TW_c$$

The transformation matrix is thus

$$T = \tilde{W}_c W_c^{-1} \tag{3.34}$$

The transformation exists if the matrix W_c is invertible. An explicit expression for the transformation matrix is given in Section 5.3.

Observable Form

Assume that the characteristic equation of Φ is (3.31) and that

$$W_o = \begin{bmatrix} C \\ C\Phi \\ \vdots \\ C\Phi^{n-1} \end{bmatrix} \tag{3.35}$$

is nonsingular. Then there exists a transformation matrix such that the transformed system is

$$z(k+1) = \begin{bmatrix} -a_1 & 1 & 0 & \cdots & 0 \\ -a_2 & 0 & 1 & \cdots & 0 \\ \vdots & & & & \\ -a_{n-1} & 0 & 0 & \cdots & 1 \\ -a_n & 0 & 0 & \cdots & 0 \end{bmatrix} z(k) + \begin{bmatrix} b_1 \\ b_2 \\ \vdots \\ b_{n-1} \\ b_n \end{bmatrix} u(k) \tag{3.36}$$

$$y(k) = [1 \quad 0 \quad \cdots \quad 0]z(k)$$

which is called the *observable canonical form*. This form has the advantage that it is easy to find the input-output model and to determine a suitable observer. The transformation is in this case given by

$$T = \tilde{W}_o^{-1} W_o$$

where W_0 is the matrix given by (3.35) using the representation (3.36). An explicit expression for the transformation matrix is given in Sec. 5.3.

Remark. The observable and controllable forms are also called *companion forms*.

3.4 Input-Output Models

A dynamic system can be described using either internal models or external models. Internal models—for instance, the state-space models discussed in Sec. 3.2—describe all internal couplings among the system variables. The external models give only the relationship between the input and the output of the system.

In this section, it is first shown that the input-output relationship for a general linear system can be expressed by a pulse-response function. It is then shown that shift-operator calculus can be used to derive input-output relationships directly, which leads to characterization of the input-output behavior in terms of pulse-transfer operators.

The Pulse Response

Consider a discrete-time system with one input and one output. The input and output signals over a finite interval can be represented as finite-dimensional vectors

$$U = [u(t_0) \ldots u(t_{N-1})]^T$$

$$Y = [y(t_0) \ldots y(t_{N-1})]^T$$

The general linear model that relates Y to U can then be expressed as

$$Y = \overline{H}U + Y_p$$

where $\overline{H}$ is an $N \times N$ matrix. Y_p accounts for the initial conditions. If the relation between U and Y is *causal*, the matrix $\overline{H}$ must be lower triangular. The *km*th element, $h(k, m)$, of H is thus zero if $m > k$. The input-output relationship for a general linear system can be written as

$$y(t_k) = \sum_{m=0}^{k} \overline{h}(k, m)u(t_m) + y_p(t_k) \tag{3.37}$$

where the term y_p is introduced to account for initial conditions in the system. The function $\overline{h}(k, m)$ is called the *pulse response,* or the *weighting function,* of the system. The pulse response function is a convenient representation, because it can easily be measured directly by injecting a pulse of unit magnitude and the width of the sampling interval and recording the output. For zero initial conditions the value $\overline{h}(k, m)$ of the pulse response gives the output at time t_k for a unit pulse at time t_m. For systems with many inputs and outputs, the pulse response is simply a matrix-valued function.

For time-invariant systems, the pulse response is a function of $k - m$ only; i.e.,

$$\overline{h}(k, m) = h(k - m)$$

It is easy to compute the pulse response of the system defined by the state-space model in (3.13). It follows from (3.14) that

$$y(k) = C\Phi^{k-k_0}x(k_0) + \sum_{j=k_0}^{k-1} C\Phi^{k-j-1}\Gamma u(j)$$

The pulse-response function for the discrete-time system is thus

$$h(k) = \begin{cases} 0 & k < 1 \\ C\Phi^{k-1}\Gamma & k \geq 1 \end{cases} \tag{3.38}$$

The pulse response is a sum of functions of the form

$$\text{Re}\{P(k)\lambda_i^k\}$$

where P is a polynomial in k and λ_i are the eigenvalues of the matrix Φ.

The pulse response has the following property.

Theorem 3.3. The pulse response of (3.38) is invariant with respect to coordinate transformations of the state-space model.

Proof. Introduce new coordinates $z = Tx$. The pulse response of the transformed system is then

$$\bar{h}(k) = \tilde{C}\tilde{\Phi}^{k-1}\tilde{\Gamma} = (CT^{-1})(T\Phi T^{-1})^{k-1}T\Gamma$$

$$= (CT^{-1})T\Phi^{k-1}T^{-1}T\Gamma = C\Phi^{k-1}\Gamma = h(k) \qquad \square$$

Shift-Operator Calculus

Differential-operator calculus is a convenient tool for manipulating linear differential equations with constant coefficients. An analogous operator calculus can be developed for systems described by linear difference equations with constant coefficients. In the development of operator calculus, the systems are viewed as operators that map input signals to output signals. To specify an operator it is necessary to give its range—i.e., to define the class of input signals and to describe how the operator acts on the signals. In operator calculus, all signals are considered as doubly infinite sequences $\{f(k): k = \ldots -1, 0, 1, \ldots\}$. For convenience the sampling period is chosen as the time unit.

The *forward-shift operator* is denoted by q. It has the property

$$qf(k) = f(k + 1)$$

If the norm of a signal is defined as

$$\| f \| = \sup_k | f(k) |$$

or

$$\| f \|^2 = \sum_{k=-\infty}^{\infty} f^2(k)$$

it follows that the shift operator has unit norm. This means that the calculus of shift operators is simpler than differential calculus, because the differential operator is unbounded. The inverse of the forward-shift operator is called the *backward-shift operator* or the *delay operator* and is denoted by q^{-1}. Hence

$$q^{-1}f(k) = f(k - 1)$$

Notice that it is important for the range of the operator to be doubly infinite sequences; otherwise the inverse of the forward-shift operator may not exist. In discussions of problems related to the characteristic equation of a system, such as stability and system order, it is most convenient to use the forward-shift operator. In discussion of problems related to causality, it is more convenient to use the backward-shift operator.

Operator calculus gives compact descriptions of systems and makes it easy to derive relationships among system variables, because manipulation of difference equations is reduced to a purely algebraic problem. In many textbooks z is used for the shift operator as well as for the complex variable in the z-transform. However, it is convenient to have different notations of the two notions. This is the same separation that is normally done between the complex variable s in the Laplace transform and the differential operator $p = d/dt$.

The shift operator is used to simplify the manipulation of higher-order difference equations. Consider the equation

$$y(k + na) + a_1 y(k + na - 1) + \cdots + a_{na} y(k)$$
$$= b_0 u(k + nb) + \cdots + b_{nb} u(k) \tag{3.39}$$

where $na > nb$.

Use of the shift operator gives

$$(q^{na} + a_1 q^{na-1} + \cdots + a_{na}) y(k) = (b_0 q^{nb} + \cdots + b_{nb}) u(k)$$

With the introduction of the polynomials

$$A(z) = z^{na} + a_1 z^{na-1} + \cdots + a_{na}$$

and

$$B(z) = b_0 z^{nb} + b_1 z^{nb-1} + \cdots + b_{nb}$$

the difference equation can be written as

$$A(q) y(k) = B(q) u(k) \tag{3.40}$$

When necessary, the degree of a polynomial can be indicated by a subscript, e.g., $A_{na}(q)$.

Equation (3.40) can also be expressed in terms of the backward-shift operator. Notice that (3.40) can be written as

$$y(k) + a_1 y(k - 1) + \cdots + a_{na} y(k - na) = b_0 u(k - d) + \cdots + b_{nb} u(k - d - nb)$$

where $d = na - nb$ is the *pole excess* of the system.

The polynomial

$$A^*(z) = 1 + a_1 z + \cdots + a_{na} z^{na} = z^{na} A(z^{-1})$$

is obtained from the polynomial A by reversing the order of the coefficients. This polynomial is called the *reciprocal polynomial*. Introduction of the reciprocal polynomials allows the system in (3.39) to be written as

$$A^*(q^{-1}) y(k) = B^*(q^{-1}) u(k - d) \tag{3.41}$$

Some care must be exercised when operating with reciprocal polynomials because

A^{**} is not necessarily the same as A. The polynomial

$$A(z) = z$$

has the reciprocal

$$A^*(z) = 1$$

The reciprocal of this is

$$A^{**}(z) = 1$$

which is apparently different from A. A polynomial $A(z)$ is called *self-reciprocal* if

$$A^*(z) = A(z)$$

A Difficulty

The goal of algebraic system theory is to convert manipulations of difference equations to purely algebraic problems. It follows from the definition of the shift operator that the difference equation of (3.40) can be multiplied by powers of q which simply means a forward shift of time. Equations for shifted times can also be multiplied by real numbers and added, which corresponds to multiplying Equation (3.40) by a polynomial in q. If (3.40) holds, it is thus also true that

$$C(q)A(q)y(k) = C(q)B(q)u(k)$$

To obtain a convenient algebra, it is also useful to be able to divide an equation like (3.40) with a polynomial in q. For example, if

$$A(q)y(k) = 0$$

it is then possible to conclude that

$$y(k) = 0$$

If division is possible, an equation like (3.40) can be solved with respect to $y(k)$. A simple example shows that it is not possible to divide by a polynomial in q unless special assumptions are made.

Example 3.10

Consider the difference equation

$$y(k + 1) - ay(k) = u(k)$$

where $|a| < 1$. In operator notation the equation can be written as

$$(q - a)y(k) = u(k)$$

If $y(k_0) = y_0$, it follows from (3.14) that the solution can be written as

$$y(k) = a^{k-k_0}y_0 + \sum_{j=k_0}^{k-1} a^{k-j-1}u(j)$$

$$= a^{k-k_0}y_0 + \sum_{i=1}^{k-k_0} a^{i-1}u(k - i)$$

(3.42)

A formal solution of the operator equation can be obtained as follows:

$$y(k) = \frac{1}{q - a}\, u(k) = \frac{q^{-1}}{1 - aq^{-1}}\, u(k)$$

Because q^{-1} has unit norm, the right-hand side can be expressed as a convergent series.

$$y(k) = q^{-1}(1 + aq^{-1} + a^2 q^{-2} + \cdots)u(k) \qquad (3.43)$$

$$= \sum_{i=1}^{\infty} a^{i-1} u(k - i)$$

It is clear that solutions in (3.42) and (3.43) are the same only if additional assumptions are made. $\qquad\square$

It is possible to develop an operator algebra that allows division by an arbitrary polynomial in q if it is assumed that there is some k_0 such that all sequences are zero for $k \leq k_0$. This algebra then allows the normal manipulations of multiplication and division of equations by polynomials in the shift operator as well as addition and subtraction of equations. However, the assumption does imply that all initial conditions for the difference equation are zero, which is the convention used in this book. (Compare with the example.)

If no assumptions on the input sequences are made, it is possible to develop a slightly different shift operator algebra that allows division only by polynomials with zeros inside the unit disc. This corresponds to the fact that effects of initial conditions on stable modes will eventually vanish. This algebra is slightly more complicated because it does not allow normal division.

The Pulse-Transfer Operator

Use of operator calculus allows the input-output relationship to be conveniently expressed as a rational function in either the forward or the backward shift operator. This function is called the *pulse-transfer operator* and is easily obtained from any system description by eliminating internal variables using purely algebraic manipulations.

Consider, for example, the state-space model of (3.6). To obtain the input-output relationship, the state vector must be eliminated. It follows from (3.6) that

$$x(k + 1) = qx(k) = \Phi x(k) + \Gamma u(k)$$

Hence

$$(qI - \Phi)x(k) = \Gamma u(k)$$

This gives

$$y(k) = Cx(k) + Du(k) = [C(qI - \Phi)^{-1}\Gamma + D]u(k)$$

The pulse-transfer operator for the system (3.6) is thus given by

$$H(q) = C(qI - \Phi)^{-1}\Gamma + D$$

The pulse-transfer operator can also be expressed in terms of the backward-shift operator.

$$H^*(q^{-1}) = C(I - q^{-1}\Phi)^{-1}q^{-1}\Gamma + D = H(q)$$

The pulse-transfer operator for the system of (3.6) is thus a matrix whose elements are rational functions in q. For a system with one input and one output,

$$H(q) = C(qI - \Phi)^{-1}\Gamma + D = B(q)/A(q) \tag{3.44}$$

If the state vector is of dimension n and if the polynomials $A(q)$ and $B(q)$ do not have common factors, then the polynomial A is of degree n. It follows from (3.44) that the polynomial A is also the characteristic polynomial of the matrix Φ, which means that the input-output model can be written as

$$y(k) + a_1 y(k - 1) + \cdots + a_n y(k - n) = b_0 u(k) + \cdots + b_n u(k - n)$$

The most common case in computer-control systems is that $b_0 = 0$; i.e., there is no direct term in the discrete-time model. Usually $y(k)$ is measured first, and then $u(k)$ is determined. Then $y(k)$ cannot be influenced by $u(k)$ even if the continuous-time system has a direct term.

Example 3.11

Consider the double integrator in Example 3.2 when $h = 1$. From (3.44)

$$H(q) = \begin{bmatrix} 1 & 0 \end{bmatrix} \begin{bmatrix} q - 1 & -1 \\ 0 & q - 1 \end{bmatrix}^{-1} \begin{bmatrix} 0.5 \\ 1 \end{bmatrix} = \frac{0.5(q + 1)}{(q - 1)^2}$$

$$= \frac{0.5(q^{-1} + q^{-2})}{1 - 2q^{-1} + q^{-2}} \qquad \qquad \square$$

Example 3.12

Use $h = 1$ for the double integrator and introduce a time delay of 0.5 s. Then from (3.17) and Example 3.7,

$$H(q) = C(qI - \Phi)^{-1}(\Gamma_0 + \Gamma_1 q^{-1})$$

$$= \begin{bmatrix} 1 & 0 \end{bmatrix} \frac{\begin{bmatrix} q - 1 & 1 \\ 0 & q - 1 \end{bmatrix}}{(q - 1)^2} \begin{bmatrix} 0.125 + 0.375q^{-1} \\ 0.5 + 0.5q^{-1} \end{bmatrix}$$

$$= \frac{0.125(q^2 + 6q + 1)}{q(q^2 - 2q + 1)} = \frac{0.125(q^{-1} + 6q^{-2} + q^{-3})}{1 - 2q^{-1} + q^{-2}} \qquad \square$$

Example 3.13

Consider the following system, which is written in observable canonical form:

$$x(k + 1) = \begin{bmatrix} -a_1 & 1 \\ -a_2 & 0 \end{bmatrix} x(k) + \begin{bmatrix} b_1 \\ b_2 \end{bmatrix} u(k)$$

$$y(k) = \begin{bmatrix} 1 & 0 \end{bmatrix} x(k)$$

The pulse-transfer operator is

$$H(q) = \begin{bmatrix} 1 & 0 \end{bmatrix} \begin{bmatrix} q + a_1 & -1 \\ a_2 & q \end{bmatrix}^{-1} \begin{bmatrix} b_1 \\ b_2 \end{bmatrix}$$

$$= \frac{1}{q^2 + a_1 q + a_2} \begin{bmatrix} 1 & 0 \end{bmatrix} \begin{bmatrix} q & 1 \\ -a_2 & q + a_1 \end{bmatrix} \begin{bmatrix} b_1 \\ b_2 \end{bmatrix}$$

$$= \frac{b_1 q + b_2}{q^2 + a_1 q + a_2} = \frac{b_1 q^{-1} + b_2 q^{-2}}{1 + a_1 q^{-1} + a_2 q^{-2}}$$

Thus the a_i's and b_i's in the canonical form are defining the polynomials A and B, respectively. This is true for nth-order systems also, in both observable and controllable form. □

Section 3.3 shows that different state-space representations can be used. Of course, this does not change the input-output model.

Theorem 3.4.   The pulse-transfer operator $H(q)$ for the state-space model (3.6) is independent of the state-space representation.

Proof. Given the pulse-transfer operator

$$H(q) = C(qI - \Phi)^{-1} \Gamma$$

and a transformation matrix T. In the new coordinates

$$\tilde{H}(q) = \tilde{C}(qI - \tilde{\Phi})^{-1} \tilde{\Gamma} = CT^{-1}(qTT^{-1} - T\Phi T^{-1})^{-1} T\Gamma$$

$$= CT^{-1}[T(qI - \Phi)T^{-1}]^{-1} T\Gamma = CT^{-1} T(qI - \Phi)^{-1} T^{-1} T\Gamma$$

$$= C(qI - \Phi)^{-1} \Gamma = H(q)$$ □

The input-output models of a system with a zero-order hold can be obtained by using (3.7) and (3.44). In order to simplify the computation of the pulse-transfer operator $H(q)$, it is convenient to use Table 3.1, which gives $H(q)$ for some standard systems.

Poles and Zeros

The *poles* of a system are the zeros of the denominator of $H(q)$, the characteristic polynomial. The *zeros* are obtained from $B(z) = 0$, the poles of the inverse system. For instance, the system in Example 3.11 has one zero in -1; the system has two poles in 1.

Time delay in a system gives rise to poles at the origin. The system in Example 3.12 has three poles: two in 1, and one at the origin. There are two zeros: $-3 \pm \sqrt{8}$.

TABLE 3.1 Zero-order hold sampling of a continuous-time system, $G(s)$.

The table gives the zero-order-hold equivalent of the continuous time system, $G(s)$, preceded by a zero-order hold. The sampled system is described by its pulse-transfer operator. The pulse-transfer operator is given in terms of the coefficients of

$$H(q) = \frac{b_1 q^{n-1} + b_2 q^{n-2} + \cdots + b_n}{q^n + a_1 q^{n-1} + \cdots + a_n}$$

$G(s)$	$H(q)$ or the coefficients in $H(q)$
$\dfrac{1}{s}$	$\dfrac{h}{q-1}$
$\dfrac{1}{s^2}$	$\dfrac{h^2(q+1)}{2(q-1)^2}$
$\dfrac{1}{s^m}$	$\dfrac{q-1}{q} \lim_{a \to 0} \dfrac{(-1)^m}{m!} \dfrac{\partial^m}{\partial a^m}\left(\dfrac{q}{q-e^{-ah}}\right)$
e^{-sh}	q^{-1}
$\dfrac{a}{s+a}$	$\dfrac{1-\exp(-ah)}{q-\exp(-ah)}$

For $\dfrac{a}{s(s+a)}$:

$$b_1 = \frac{1}{a}(ah - 1 + e^{-ah}) \qquad b_2 = \frac{1}{a}(1 - e^{-ah} - ahe^{-ah})$$
$$a_1 = -(1 + e^{-ah}) \qquad a_2 = e^{-ah}$$

For $\dfrac{a^2}{(s+a)^2}$:

$$b_1 = 1 - e^{-ah}(1 + ah) \qquad b_2 = e^{-ah}(e^{-ah} + ah - 1)$$
$$a_1 = -2e^{-ah} \qquad a_2 = e^{-2ah}$$

For $\dfrac{s}{(s+a)^2}$:

$$\frac{(q-1)he^{-ah}}{(q-e^{-ah})^2}$$

For $\dfrac{ab}{(s+a)(s+b)}$, $a \neq b$:

$$b_1 = \frac{b(1 - e^{-ah}) - a(1 - e^{-bh})}{b-a}$$
$$b_2 = \frac{a(1 - e^{-bh})e^{-ah} - b(1 - e^{-ah})e^{-bh}}{b-a}$$
$$a_1 = -(e^{-ah} + e^{-bh})$$
$$a_2 = e^{-(a+b)h}$$

For $\dfrac{(s+c)}{(s+a)(s+b)}$, $a \neq b$:

$$b_1 = \frac{e^{-bh} - e^{-ah} + (1 - e^{-bh})c/b - (1 - e^{-ah})c/a}{a-b}$$
$$b_2 = \frac{c}{ab}e^{-(a+b)h} + \frac{b-c}{b(a-b)}e^{-ah} + \frac{c-a}{a(a-b)}e^{-bh}$$
$$a_1 = -e^{-ah} - e^{-bh} \qquad a_2 = e^{-(a+b)h}$$

For $\dfrac{\omega_0^2}{s^2 + 2\zeta\omega_0 s + \omega_0^2}$:

$$b_1 = 1 - \alpha\left(\beta + \frac{\zeta\omega_0}{\omega}\gamma\right) \qquad \omega = \omega_0\sqrt{1-\zeta^2} \qquad \zeta < 1$$
$$b_2 = \alpha^2 + \alpha\left(\frac{\zeta\omega_0}{\omega}\gamma - \beta\right) \qquad \alpha = e^{-\zeta\omega_0 h}$$

TABLE 3.1 (*CONT.*)

$$a_1 = -2\alpha\beta \qquad\qquad \beta = \cos(\omega h)$$

$$a_2 = \alpha^2 \qquad\qquad \gamma = \sin(\omega h)$$

$$\frac{s}{s^2 + 2\zeta\omega_0 s + \omega_0^2} \qquad b_1 = \frac{1}{\omega} e^{-\zeta\omega_0 h} \sin(\omega h) \qquad b_2 = -b_1 \qquad \omega = \omega_0\sqrt{1 - \zeta^2}$$

$$a_1 = -2e^{-\zeta\omega_0 h}\cos(\omega h) \qquad a_2 = e^{-2\zeta\omega_0 h}$$

$$\frac{a^2}{s^2 + a^2} \qquad\qquad b_1 = 1 - \cos ah \qquad\qquad b_2 = 1 - \cos ah$$

$$a_1 = -2\cos ah \qquad\qquad a_2 = 1$$

$$\frac{s}{s^2 + a^2} \qquad\qquad b_1 = \frac{1}{a}\sin ah \qquad\qquad b_2 = -\frac{1}{a}\sin ah$$

$$a_1 = -2\cos ah \qquad\qquad a_2 = 1$$

$$\frac{a}{s^2(s + a)} \qquad\qquad b_1 = \frac{1 - a}{a^2} + h\left(\frac{h}{2} - \frac{1}{a}\right) \qquad \alpha = e^{-ah}$$

$$b_2 = (1 - \alpha)\left(\frac{h^2}{2} - \frac{2}{a^2}\right) + \frac{h}{a}(1 + \alpha)$$

$$b_3 - -\left(\frac{1}{a^2}(\alpha - 1) + \alpha h\left(\frac{h}{2} + \frac{1}{a}\right)\right)$$

$$a_1 = -(\alpha + 2)$$

$$a_2 = 2\alpha + 1$$

$$a_3 = -\alpha$$

The Order of a System

The *order* of a system is the same as the dimension of a state-space representation or, equivalently, the number of poles of the system. Notice that it is important to use the forward-shift form to determine the order because of the time delays. The determination of the poles, zeros, and order of a system are occasions when it is important to use the forward-shift form.

3.5 The *Z*-Transform

The discrete-time analogy of the Laplace transform is the *z*-transform—a convenient tool to study linear difference equations with or without initial conditions. The *z*-transform maps a *sequence of semi-infinite time sequence* into a function of a complex variable. Notice the difference in range for the *z*-transform and the operator calculus. The variable *z* is a complex variable and should be distinguished from the operator *q*.

Definition 3.1—z-transform.   Consider the discrete-time signal $\{f(kh): k = 0, 1, \ldots\}$. The z-transform of $f(kh)$ is defined as

$$F(z) = \sum_{k=0}^{\infty} f(kh)z^{-k} \tag{3.45}$$

where z is a complex variable. The z-transform of f is denoted by Zf or F. The inverse transform is given by

$$f(kh) = \frac{1}{2\pi i} \oint F(z)z^{k-1}\, dz \tag{3.46}$$

where the contour of integration encloses all singularities of $F(z)$.

Example 3.14

Let $y(kh) = kh$ for $k \geq 0$. Then

$$Y(z) = 0 + hz^{-1} + 2hz^{-2} + \cdots = h(z^{-1} + 2z^{-2} + \cdots) = \frac{hz}{(z-1)^2} \qquad \square$$

Some properties of the z-transform are collected in Table 3.2. Notice that the formulas for forward and backward time shifts are not the same. This is a consequence of the assumption that the time sequences are semi-infinite.

The z-transform can be used to solve difference equations, for instance

$$x(k+1) = \Phi x(k) + \Gamma u(k)$$

$$y(k) = Cx(k)$$

If the z-transform of both sides is taken,

$$\sum_{k=0}^{\infty} z^{-k}x(k+1) = z\left[\sum_{k=0}^{\infty} z^{-k}x(k) - x(0)\right]$$

$$= \sum_{0}^{\infty} \Phi z^{-k}x(k) + \sum_{0}^{\infty} \Gamma z^{-k}u(k)$$

Hence

$$z[X(z) - x(0)] = \Phi X(z) + \Gamma U(z)$$

$$X(z) = (zI - \Phi)^{-1}[zx(0) + \Gamma U(z)]$$

and

$$Y(z) = C(zI - \Phi)^{-1}zx(0) + C(zI - \Phi)^{-1}\Gamma U(z)$$

The *pulse-transfer function* can now be introduced.

$$H(z) = C(zI - \Phi)^{-1}\Gamma \tag{3.47}$$

which is the same as (3.44) with q replaced by z. The time sequence $y(k)$ can now be obtained using the inverse transform. The following theorem is analogous to that of continuous-time systems.

TABLE 3.2 Some properties of the z-transform.

1. Definition.

$$F(z) = \sum_{k=0}^{\infty} f(kh)z^{-k}$$

2. Inversion.

$$f(kh) = \frac{1}{2\pi i} \oint F(z)z^{k-1}\, dz$$

3. Linearity.

$$\mathscr{Z}(\alpha f + \beta g) = \alpha \mathscr{Z} f + \beta \mathscr{Z} g$$

4. Time shift.

$$\mathscr{Z} q^{-n} f = z^{-n} F$$
$$\mathscr{Z}\{q^n f\} = z^n (F - F_1)$$

where $\Gamma_1(z) = \sum_{j=0}^{n-1} f(jh)z^{-i}$

5. Initial-value theorem.

$$f(0) = \lim_{z\to\infty} F(z)$$

6. Final-value theorem.

If $(1 - z^{-1})F(z)$ does not have any poles on or outside the unit circle, then
$$\lim_{k\to\infty} f(kh) = \lim_{z\to 1} (1 - z^{-1})F(z)$$

7. Convolution

$$\mathscr{Z}(f * g) = \mathscr{Z} \sum_{n=0}^{k} f(n)g(k - n) = (\mathscr{Z}f)(\mathscr{Z}g)$$

   Theorem 3.5.   The pulse response of (3.38) and the pulse-transfer function (3.47) are a z-transform pair.

Computation of the Pulse-Transfer Function

It is possible to determine the pulse-transfer function directly from the continuous-time transfer function. Let the system be described by the transfer function $G(s)$ preceded by a zero-order hold (See Fig. 3.4). The pulse-transfer function is uniquely determined by the response to a given signal. Consider, for instance, a unit-step input. The sequence $\{u(kh)\}$ is then a sequence of ones, and the signal $u(t)$ is then also a unit step. Let $Y(s)$ denote the Laplace transform of $y(t)$, i.e.,

$$Y(s) = \frac{G(s)}{s}$$

Let the sampled output $\{y(kh)\}$ have the z-transform $\tilde{Y} = \mathscr{Z}\mathscr{L}^{-1}Y$. Division of

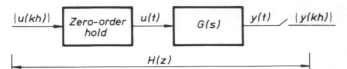

Figure 3.4 Sampling of a continuous-time system.

$\tilde{Y}$ by the pulse-transfer function of the input, which is $z/(z - 1)$, gives

$$H(z) = (1 - z^{-1})\tilde{Y}(z) \tag{3.48}$$

The pulse-transfer function is now obtained as follows:

1. Determine the step response of the system with the transfer function $G(s)$.
2. Determine the corresponding z-transform of the step response.
3. Divide by the z-transform of the step function.

Using this procedure the following formula can be derived:

$$H(z) = \frac{z - 1}{z} \frac{1}{2\pi i} \int_{\gamma - i\infty}^{\gamma + i\infty} \frac{e^{sh}}{z - e^{sh}} \frac{G(s)}{s} ds \tag{3.49}$$

If the transfer function $G(s)$ goes to zero at least as fast as $|s|^{-1}$ for a large s and has distinct poles, none of which are at the origin, we get

$$H(z) = \sum_{s = s_i} \frac{1}{z - e^{sh}} \operatorname{Res} \left\{ \frac{e^{sh} - 1}{s} G(s) \right\} \tag{3.50}$$

where s_i are the poles of $G(s)$. A proof of this formula is given in Section 4.5.

Table 3.3 shows some time functions and the corresponding Laplace and z-

TABLE 3.3 Some time functions and corresponding Laplace and z-transforms. *Warning*: Use the table only as prescribed!

f		$\mathscr{L}f$	$\mathscr{Z}f$
1	$k \geq 0$ (step)	$\dfrac{1}{s}$	$\dfrac{z}{z - 1}$
kh		$\dfrac{1}{s^2}$	$\dfrac{hz}{(z - 1)^2}$
$\dfrac{1}{2}(kh)^2$		$\dfrac{1}{s^3}$	$\dfrac{h^2 z(z + 1)}{2(z - 1)^3}$
$e^{-kh/T}$		$\dfrac{T}{1 + sT}$	$\dfrac{z}{z - e^{-h/T}}$
$1 - e^{-kh/T}$		$\dfrac{1}{s(1 + sT)}$	$\dfrac{z(1 - e^{-h/T})}{(z - 1)(z - e^{-h/T})}$
$\sin \omega kh$		$\dfrac{\omega}{s^2 + \omega^2}$	$\dfrac{z \sin \omega h}{z^2 - 2z \cos \omega h + 1}$

transforms. The table can thus be used to combine steps 1 and 2. Tables in textbooks are usually found in this form.

Warning. Notice that $\mathscr{Z}f$ in Table 3.3 does not give the zero-order-hold sampling of a system with the transfer function $\mathscr{L}f$. Examine Table 3.1. It is a very common mistake to believe that it does. The desired pulse-transfer function is obtained through the procedure given.

Shift Operator Calculus and *z*-transforms

There are strong formal relations between shift operator calculus and calculation with *z*-transforms. When manipulating difference equations we can use either. The expressions obtained look formally very similar. In many textbooks the same notation is in fact used for both. The situation is very similar to the difference between the differential operator $p = d/dt$ and the Laplace transform s for continuous-time systems. First we may notice that q is an operator that acts on sequences and z is a complex variable. From a purely mathematical point of view it clearly makes sense to make a distinction between such different objects. There is, however, also a good system theoretic reason for making a distinction. We illustrate this by an example.

Example 3.15

Consider the difference equation

$$y(k + 1) + ay(k) = u(k + 1) + au(k) \qquad (3.51)$$

If (3.51) is considered as a dynamical system its pulse transfer function is obtained as

$$H(z) = \frac{z + a}{z + a} = 1$$

The last equality is obtained because z is a complex variable. We may thus be misled to believe that the system (3.51) is identical to

$$y(k) = u(k)$$

This is clearly not true because the difference equation (3.51) has the solution

$$y(k) = a^k y(0) + u(k)$$

which is identical to (3.51) *only if the initial condition y(0) is zero.* It may be reasonable to neglect the initial conditions if $|a| < 1$, but not reasonable if $|a| \geq 1$. We thus have the situation that from a system theoretic point of view the expression

$$\frac{z + a}{z + a}$$

can be considered equal to one if $|a| < 1$ but not otherwise. If equation (3.51) is solved using shift-operator calculus we obtain formally

$$(q + a)y(k) = (q + a)u(k)$$

Notice that we cannot divide by $q + a$ since q is an operator. $\qquad \square$

The conclusion that we can draw from the simple example is that the algebra of z-transforms and shift operators are different. In z-transforms calculus we can divide with an arbitrary expression, but this is not allowed in shift-operator calculus. The system theoretic interpretation is that we may throw away some modes in the system with z-transform calculus by cancellation factors. This may make sense if the cancelled factors correspond to stable modes, but it may be strongly misleading if the cancelled factors are unstable. Another manifestation of this effect will be given in the discussion of the notions of observability and controllability in Chapter 4.

3.6 Poles and Zeros

A rational transfer function can be characterized by its poles (the zeros of the denominator) and zeros (the zeros of the numerator). The poles and zeros have good system theoretic interpretation for a continuous-time system. The poles correspond to the free motion of the system. They are equal to the eigenvalues of the matrix A in Equation (3.1). The zeros depend on how the inputs and outputs are coupled. If a system has a zero at $s = a$, it means that the signal transmission of the signal e^{at} is blocked. For continuous-time systems the zeros can be defined as

$$\det \begin{pmatrix} sI - A & -B \\ C & D \end{pmatrix} = 0$$

Poles and zeros have an analogous role in discrete-time systems. For single-input single-output finite dimensional systems, poles and zeros can be conveniently defined as the poles and zeros of the pulse-transfer function. Poles and zeros have the same interpretation as for continuous-time systems. A pole $z = a$ thus corresponds to a free mode of the system associated with the time function $z(kh) = a^k$. A zero $z = a$ means that the transmission of the input signal $u(kh) = a^k$ is blocked by the system.

Poles and zeros can also be defined directly in terms of the state-space description equation (3.6). The poles are simply the eigenvalues of the matrix Φ. Using the signal-blocking property we find that if z is a zero then there exist x and u such that

$$\begin{pmatrix} zI - \Phi & -\Gamma \\ C & D \end{pmatrix} \begin{pmatrix} x \\ u \end{pmatrix} = 0$$

Poles

Consider a continuous-time system described by the nth order state-space model

$$\dot{x} = Ax + Bu$$
$$y = Cx \tag{3.52}$$

The poles of the system are the eigenvalues of A, which we denote by $\lambda_i(A)$, $i = 1, \ldots, n$. The zero-order-hold sampling of (3.52) gives the discrete-time system

$$x(kh + h) = \Phi x(kh) + \Gamma u(kh) \qquad (3.53)$$
$$y(kh) = Cx(kh)$$

Its poles are the eigenvalues of Φ, $\lambda_i(\Phi)$, $i = 1, \ldots, n$. Because $\Phi = \exp(Ah)$ it follows from the properties of matrix functions (see Appendix B) that

$$\lambda_i(\Phi) = e^{\lambda_i(A)h} \qquad (3.54)$$

Equation (3.54) gives the mapping from the continuous-time poles to the discrete-time poles. Figure 3.5 illustrates a mapping of the complex s-plane into the z-plane, when $z = \exp(sh)$. For instance, the left half of the s-plane is mapped into the unit disc of the z-plane. The map is not bijective—several points in the s-plane are mapped into the same point in the z-plane (see Fig. 3.6). This is another illustration of the aliasing effect discussed in Sec. 2.5. For poles inside the fundamental strip S_0 in Fig. 3.6, there is a simple relationship between continuous- and discrete-time poles. (Also compare with Example 3.5.)

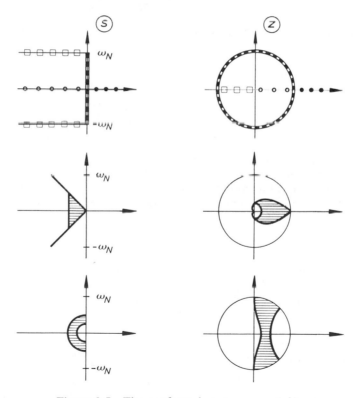

Figure 3.5   The conformal map $z = \exp(sh)$.

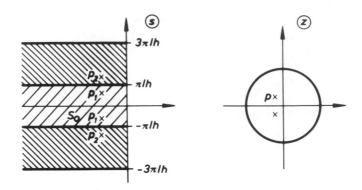

Figure 3.6 Each strip in the left half of the s-plane is mapped into the unit disc. This means that the pair of poles p_1 and p_2 are both mapped into the pair p.

Example 3.16

Consider the continuous-time system

$$\frac{\omega_0^2}{s^2 + 2\zeta\omega_0 s + \omega_0^2} \tag{3.55}$$

The poles of the corresponding discrete-time system are given by the characteristic equation

$$z^2 + a_1 z + a_2 = 0$$

where

$$a_1 = -2e^{-\zeta\omega_0 h} \cos(\sqrt{1 - \zeta^2}\,\omega_0 h)$$

$$a_2 = e^{-2\zeta\omega_0 h}$$

(Compare with Table 3.1.) Figure 3.7 shows the step responses of the discrete-time system for different values of the sampling interval when $\omega_0 = 1.83$ and $\zeta = 0.5$. Figure 3.8 gives a more detailed picture of how the continuous-time poles of (3.55) are mapped into the unit circle for different values of ζ and $\omega_0 h$ when the system is sampled. ☐

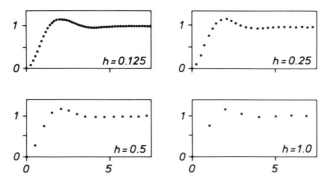

Figure 3.7 Step responses of the discrete-time system in Example 3.16 for different values of h when $\zeta = 0.5$ and $\omega_0 = 1.83$, which gives the rise time $T_r = 1$.

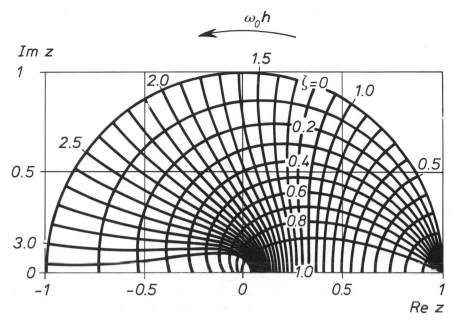

Figure 3.8 Loci of constant ζ and $\omega_0 h$ when (3.55) is sampled.

Zeros

It is not possible to give a simple formula for the mapping of zeros. If a continuous-time transfer function is viewed as a rational function, it has zeros at the zeros of the numerator polynomial and $d = r - 1$ zeros at infinity, where d is the pole excess for the continuous-time transfer function—i.e., the difference between the number of poles and the number of zeros.

For short sampling periods, a discrete-time system will have zeros in

$$z_i \approx e^{s_i h}$$

where the s_i's are the zeros of the continuous-time system. The r 1 zeros that correspond to the infinite zeros of the continuous-time system will go to the zeros of the polynomials Z_r in Table 3.4 as the sampling interval goes to zero because

TABLE 3.4 Numerator polynomials, Z_r, when sampling s^{-r}.

r	Z_r
1	1
2	$z + 1$
3	$z^2 + 4z + 1$
4	$z^3 + 11z^2 + 11z + 1$
5	$z^4 + 26z^3 + 66z^2 + 26z + 1$

for large s, the transfer function of the continuous-time system is approximately given by $G(s) \approx s^{-r}$.

Example 3.17

Consider the continuous-time transfer function

$$\frac{2}{(s + 1)(s + 2)}$$

Using Table 3.1 gives the zero of the pulse-transfer function

$$z = -\frac{(1 - e^{-2h})e^{-h} - 2(1 - e^{-h})e^{-2h}}{2(1 - e^{-h}) - (1 - e^{-2h})}$$

When h is small

$$z \approx -1 + 3h$$

and when h approaches zero the zero moves to -1. The zero moves towards the origin when h is increased.

The zero for small values of h can also be obtained from Table 3.4. The pole excess of the continuous-time system is $r = 2$. The discrete-time system will have a zero at $Z_r(z) = z + 1 = 0$ when h goes to zero. □

The zeros of a discrete-time system obtained by sampling a continuous-time system depend on the hold circuit that is used. This is illustrated by Example 3.18.

Example 3.18

Consider the double integrator in Example 3.9, in which the hold device is such that the control signal is $\alpha u(k)$ in the first half of the sampling interval and $\beta u(k)$ in the second half. The pulse-transfer function of the system is given by

$$H(z) = \begin{bmatrix} 1 & 0 \end{bmatrix}(zI - \Phi)^{-1}\Gamma$$

$$= \frac{h^2[z(3\alpha + \beta) + (\alpha + 3\beta)]}{8(z - 1)^2}$$

The low-frequency gain is determined by $\alpha + \beta$. Choose

$$\alpha + \beta = 2$$

The pulse-transfer function has zero

$$z = -\frac{\alpha + 3\beta}{3\alpha + \beta} = -\frac{3 - \alpha}{1 + \alpha}$$

The zero of the pulse-transfer function can thus be positioned at an arbitrary value by selecting α appropriately. The value $\alpha = 1$, which corresponds to a zero-order hold, gives $z = -1$, and $\alpha = 3$ gives $z = 0$. □

The results of the example can be extended to nth-order systems. It can be shown that the zeros can be positioned arbitrarily if the control signal has a constant value over each nth part of the sample interval. Such a hold circuit can be implemented by sampling the control signal at a higher rate; however, the idea has not been used much because the control signal may become highly irregular.

Systems with Unstable Inverses

A continuous-time system with a rational transfer function is nonminimum phase if it has right half-plane zeros or time delays. Analogously, a discrete-time system is often defined to be nonminimum phase if it has zeros outside the unit disc. That definition implies that a time delay does not make the system nonminimum phase. On the other hand, time delays do not pose the same severe problems as they do for continuous-time systems. For discrete-systems it is therefore more relevant to talk about systems with or without stable inverses, which are defined as follows.

Definition 3.2—Unstable inverse. A discrete-time system has an unstable inverse if it has zeros outside the unit disc.

A continuous-time system with a stable inverse may become a discrete-time system with an unstable inverse when it is sampled. It follows from Table 3.4 that the inverse system is always unstable if the pole excess of the continuous-time system is larger than two, and if the sampling period is sufficiently short. Further, a continuous-time, nonminimum-phase system will not always become a discrete-time system with an unstable inverse, as shown in the following example.

Example 3.19

The transfer function

$$G(s) = \frac{6(1 - s)}{(s + 2)(s + 3)}$$

has an unstable zero $s = 1$. Sampling the system gives a discrete-time, pulse-transfer function with a zero

$$z_1 = - \frac{8e^{-2h} - 9e^{-3h} + e^{-5h}}{1 - 9e^{-2h} + 8e^{-3h}}$$

For $h \approx 1.25$, $z_1 = -1$; for larger h the zero is always inside the unit circle and the sampled system has a stable inverse. □

3.7 Selection of Sampling Rate

The importance of a proper selection of the sampling rate is discussed in Sec. 2.6. Too long a sampling period will make it impossible to reconstruct the continuous-time signal. Too short a sampling period will increase the load on the computer. The problem of sample-rate selection is treated again later in the book. This section discusses the relationship of sampling-rate selection to the poles of the continuous-time system.

It is useful to characterize the sampling period with a variable that is dimension-free and that has a good physical interpretation. For oscillatory systems
it is natural to normalize with respect to the period of oscillation; for nonoscillatory
systems, the rise time is a natural normalization factor.

Introduce N_r as the number of sampling periods per rise time,

$$N_r = \frac{T_r}{h}$$

where T_r is the rise time. Shannon's sampling theorem gives the lowest limit. For
pure sinusoidal signals, this gives $N_r \approx 0.32$. Shannon's reconstruction is, however, quite complicated and associated with a delay.

For first-order systems, the rise time is equal to the time constant. It is then
reasonable to choose N_r between 4 and 10. For a second-order system with damping ζ and natural frequency ω_0, rise time is given by

$$T_r = \omega_0^{-1} e^{\varphi/\tan \varphi}$$

where $\zeta = \cos \varphi$. For a damping around $\zeta = 0.7$, this gives

$$\omega_0 h \approx 0.2\text{--}0.6$$

where ω_0 is in radians per second.

Figures 3.7 and 3.9 illustrate the choice of the sampling interval for different
signals. It is thus reasonable to choose the sampling period so that

$$N_r = \frac{T_r}{h} \approx 4\text{--}10$$

Example 3.20—Pole-zero variation with sampling interval

Consider the system

$$G(s) = \frac{1}{(s + 1)(s^2 + s + 1)} \tag{3.56}$$

Figure 3.10 shows the step response of the system. Assume that the system is sampled with period h. Figure 3.11 shows how the poles and zeros of the sampled data
system vary with the sampling period. Sampling intervals close to zero give three
poles close to one. Further, the continuous time system has a pole excess of three.
This implies that the zeros for short sampling intervals are close to the roots of

$$z^2 + 4z + 1 = 0$$

See Table 3.4. The poles and zeros approach the origin when the sampling interval
is increased. The sampled data system has a stable inverse if $h > 2.24$. The rules of
thumb for the choice of the sampling interval give that a reasonable choice is $h \approx$
0.5. Compare Figure 3.10. □

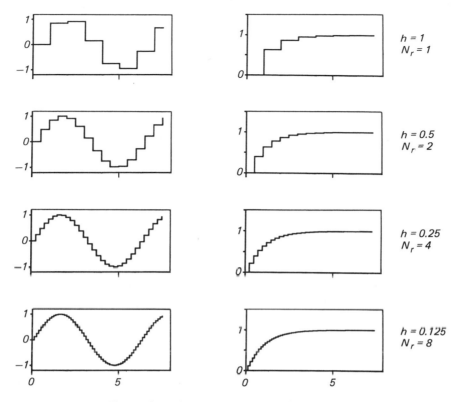

Figure 3.9 Illustration of sample and hold of a sinusoidal and an exponential signal. The rise times of the signals are $T_r = 1$.

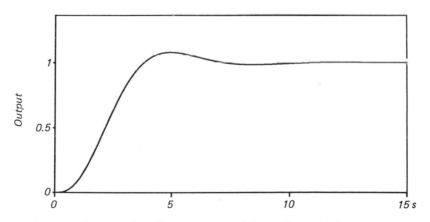

Figure 3.10 Step response of the system (3.56).

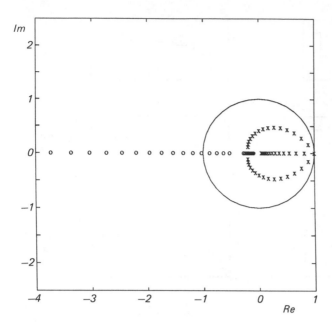

Figure 3.11 Poles (x) and zeros (0) when the system (3.56) is sampled with $h = 0, 0.2, 0.4, \ldots 3$.

3.8 Problems

3.1. Consider the system

$$\dot{x} = -ax + bu$$
$$y = cx$$

Let the input be constant over periods of length h. Sample the system and discuss how the poles of the discrete-time system vary with the sampling interval h.

3.2. Derive the discrete-time system corresponding to the following continuous-time systems when a zero-order-hold circuit is used:

(a) $\dfrac{dx}{dt} = \begin{bmatrix} 0 & 1 \\ -1 & 0 \end{bmatrix} x + \begin{bmatrix} 0 \\ 1 \end{bmatrix} u$

$y = [1 \quad 0]x$

(b) $\dfrac{d^2y}{dt^2} + 3\dfrac{dy}{dt} + 2y = \dfrac{du}{dt} + 3u$

(c) $\dfrac{d^3y}{dt^3} = u$

3.3. The following difference equations are assumed to describe continuous-time systems sampled using a zero-order-hold circuit and the sampling period h. Determine, if possible, the corresponding continuous-time systems.

(a) $y(kh) - 0.5y(kh - h) = 6u(kh - h)$

(b) $x(kh + h) = \begin{bmatrix} -0.5 & 1 \\ 0 & -0.3 \end{bmatrix} x(kh) + \begin{bmatrix} 0.5 \\ 0.7 \end{bmatrix} u(kh)$

$y(kh) = [1 \quad 1]x(kh)$

(c) $y(kh) + 0.5y(kh - h) = 6u(kh - h)$

3.4. Consider the harmonic oscillator [see Example A.3 or Problem 3.2(a).] Compute the step response at 0, h, $2h$, . . . when the sampling period is (a) $h = \pi/2$, (b) $h = \pi/4$. Compare with the continuous-time step response.

3.5. Sample the system $G(s) = 1/s$ using a first-order-hold circuit.

3.6. Consider the system in (3.1). Assume that the input is a sum of impulses at the sampling instants, i.e.,

$$u(t) = \sum \delta(t - kh)u(kh)$$

Determine the discrete-time representation.

3.7. Find the transformation matrix, T, that transforms the state-space representation of the double integrator (3.12) into controllable canonical form.

3.8. Determine the pulse-transfer function of the system

$$x(kh + h) = \begin{bmatrix} 0.5 & -0.2 \\ 0 & 0 \end{bmatrix} x(kh) + \begin{bmatrix} 2 \\ 1 \end{bmatrix} u(kh)$$

$$y(kh) = [1 \quad 0]x(kh)$$

3.9. Many physical systems can be described by the form

$$\frac{dx}{dt} = \begin{bmatrix} -a & b \\ c & -d \end{bmatrix} x + \begin{bmatrix} f \\ g \end{bmatrix} u$$

where a, b, c, and d are nonnegative. Derive a formula for the sampled-data system when using a zero-order hold.

(*Hint:* Show first that the poles of the system are real.)

3.10. Figure 3.12 shows a system of two tanks, where the input signal is the flow to the first tank and the output is the level in the second tank. Use of the levels as state variables gives the system

$$\frac{dx}{dt} = \begin{bmatrix} -0.0197 & 0 \\ 0.0178 & -0.0129 \end{bmatrix} x + \begin{bmatrix} 0.0263 \\ 0 \end{bmatrix} u$$

$$y = [0 \quad 1]x$$

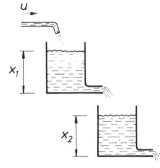

Figure 3.12 The two tank process.

(a) Sample the system with the sampling period $h = 12$.

(b) Verify that the pulse-transfer operator for the system is

$$H_0(q) = \frac{0.030q + 0.026}{q^2 - 1.65q + 0.68}$$

3.11. The normalized motor is described in Example A.2. Show that the sampled system is described by (A.6). Determine the following:

(a) The pulse-transfer function.

(b) The pulse response.

(c) A difference equation relating the input and the output.

(d) The variation of the poles and zero of the pulse-transfer with the sampling period.

3.12. A continuous-time system with the transfer function

$$G(s) = \frac{1}{s} e^{-s\tau}$$

is sampled with $h = 1$ when $\tau = 0.5$.

(a) Determine a state-space representation of the sampled system. What is the order of the sampled system?

(b) Determine the pulse-transfer function and the pulse response of the sampled sytem.

(c) Determine the poles and zeros of the sampled system.

3.13. Solve Problem 3.12 with

$$G(s) = \frac{1}{s + 1} e^{-s\tau}$$

and $\tau = 1.5$ and $h = 1$.

3.14. Consider the sampled system

$$y(k + 1) = ay(k) + b_3 u(k - 3) + b_4 u(k - 4)$$

where the sampling interval is 1 s. Show that the system may be obtained by sampling the system

$$\frac{dy(t)}{dt} = -\alpha y(t) + bu(t - \tau)$$

where

$$\tau = 4 - \frac{\ln[(ab_3 + b_4)/(b_3 + b_4)]}{\ln a}$$

3.15. Consider the system

$$y(k) - 0.5y(k - 1) = u(k - 9) + 0.2u(k - 10)$$

Determine the polynomials $A(q)$, $B(q)$, $A^*(q^{-1})$, and $B^*(q^{-1})$ in the representations

$$A(q)y(k) = B(q)u(k)$$

and

$$A^*(q^{-1})y(k) = B^*(q^{-1})u(k - d)$$

What are d and the order of the system?

3.16. A filter with the pulse-transfer operator

$$H^*(q^{-1}) = b_0 + b_1 q^{-1} + \cdots + b_n q^{-n}$$

is called a finite impulse response (FIR) filter.
(a) Determine the order of the system.
(b) Give a state-space representation on observable canonical form.

3.17. Use the z-transform to determine the output sequence of the difference equation

$$y(k + 2) - 1.5y(k + 1) + 0.5y(k) = u(k + 1)$$

when $u(k)$ is a step at $k = 0$ and when $y(0) = 0.5$ and $y(-1) = 1$.

3.18. Verify that

$$\mathcal{L}\left\{ \frac{1}{2}(kh)^2 \right\} = \frac{h^2 z(z + 1)}{2(z - 1)^3}$$

Compare with Table 3.3 and use that to determine the pulse-transfer function of the double integrator (see Example A.1).

3.19. Use (3.49) to determine the pulse transfer function of (a) the system in Problem 3.1 and (b) the normalized motor (see Example A.2).

3.20. Show that a curve of constant damping ζ in the s-plane is a logarithmic spiral in the z-plane when using the mapping $z = \exp(sh)$.

3.21. If $\beta < \alpha$, then

$$\frac{s + \beta}{s + \alpha}$$

is called a *lead network* (i.e., it gives a phase advance). Consider the discrete-time system

$$\frac{z + b}{z + a}$$

(a) Determine when it is a lead network.
(b) Simulate the step response for different pole and zero locations.

3.22. Consider the system

$$\frac{z + b}{(1 + b)(z^2 - 1.1z + 0.4)}$$

The pole location corresponds to a continuous-time system with damping $\zeta = 0.7$. Simulate the system and determine the overshoot for different values of b in the interval $(-1, 1)$.

3.23. Consider the stable continuous-time system

$$G(s) = \frac{s + b}{s + a}$$

where $a \neq b$. Sample the system with the sampling period h. Derive conditions for when the sampled system will have a stable inverse.

3.24. Consider the discrete-time system

$$H(z) = \frac{b_1 z + b_2}{z^{n+1}(z - a)}$$

This system is obtained by sampling a continous-time system with the transfer function

$$G(s) = \frac{Ke^{-sL}}{1 + sT}$$

using the sampling interval h. Show that

$$T = -h/\ln a$$

$$K = \frac{b_1 + b_2}{1 - a}$$

$$L = nh - h \ln \left(\frac{ab_1 + b_2}{a(b_1 + b_2)}\right) / \ln a$$

3.25. Use (3.49) to determine the pulse-transfer function associated with

$$G(s) = \frac{1}{s^3}$$

3.26. Use the formula (3.49) to show that the pulse-transfer function obtained with zero-order hold samplings of the transfer function

$$G(s) = \frac{1}{s^n}$$

is given by

$$H(z) = \frac{h^n}{n!} \frac{B_n(z)}{(z - 1)^n}$$

where

$$B_n(z) = b_1^n z^{n-1} + b_2^n z^{n-2} + \cdots + b_n^n$$

and

$$b_k^n = \sum_{i=1}^{k} (-1)^{k-i} i^n \binom{n + 1}{k - i} \qquad k = 1, 2, \ldots, n$$

Furthermore show that

$$B_1(z) = 1$$

$$B_2(z) = z + 1$$

$$B_3(z) = z^2 + 4z + 1$$

$$B_4(z) = z^3 + 11z^2 + 11z + 1$$

$$B_5(z) = z^4 + 26z^3 + 66z^2 + 26z + 1$$

$$B_6(z) = z^5 + 57z^4 + 302z^3 + 302z^2 + 57z + 1$$

3.27. Prove the formula (3.49).

3.28. Derive the formula (3.50) from (3.49).

3.29. Solve the difference equation

$$y(k) = y(k - 1) + y(k - 2) \qquad k = 2, 3, \ldots$$

when $y(0) = y(1) = 1$. (The numbers $y(k)$ are called the Fibonacci numbers.)

3.30. Determine the poles and zeros (with multiplicity) of the system

$$y(k) - 0.5y(k - 1) + y(k - 2) = 2u(k - 10) + u(k - 11)$$

3.31. Which of the following discrete-time systems can be obtained by sampling a causal continuous-time system using a zero-order hold?

$$H_1(q) = \frac{1}{q - 0.8} \qquad H_2(q) = \frac{1}{q + 0.8}$$

$$H_3(q) = \frac{q - 1}{(q + 0.8)^2}$$

$$H_4(q) = \frac{2q^2 - 0.7q - 0.8}{q(q - 0.8)}$$

3.32. Determine the pulse-transfer operator obtained by sampling

$$G(s) = \frac{2(s + 2)}{(s + 1)(s + 3)}$$

with $h = 0.02$.

3.33. Sample the continuous-time system

$$\dot{x}(t) = \begin{pmatrix} 1 & 0 \\ 1 & 1 \end{pmatrix} x(t) + \begin{pmatrix} 1 \\ 0 \end{pmatrix} u(t - 0.2)$$

Using the sampling interval $h = 0.3$. Determine the pulse-transfer operator.

3.34. Show that

$$\exp \begin{pmatrix} Ah & Bh \\ 0 & 0 \end{pmatrix} = \begin{pmatrix} \Phi & \Gamma \\ 0 & I \end{pmatrix}$$

where Φ and Γ are given by (3.7).

3.35. Consider a linear system with the transfer function

$$G_a(s) = \frac{a}{s + a}$$

Sampling the system gives the pulse-transfer function

$$H_a(z) = \frac{1 - e^{-ah}}{z - e^{-ah}}$$

Letting $a \to \infty$ we get

$$G_\infty(s) = \lim_{a \to \infty} G_a(s) = 1$$

and

$$H_\infty(z) = \lim_{a \to \infty} H_a(z) = \frac{1}{z}$$

Notice that $H_\infty(z)$ is *not* the pulse-transfer function obtained by sampling the system with the pulse-transfer function $G_\infty(s) = 1$. Determine conditions on the transfer function $G_a(s)$ such sampling commutes with limit operations.

3.9 References

The early texts on sampled-data systems

JURY, E. I. (1958): *Sampled-Data Control Systems*. New York: John Wiley.

RAGAZZINI, J. R. and G. F. FRANKLIN (1958): *Sampled-Data Control Systems*. New York: McGraw-Hill.

TSYPKIN, Y. Z. (1958): *Theory of Impulse Systems*. Moscow: State Publisher for Physical Mathematical Literature.

dealt exclusively with input-output models and transform theory. The state-space approach used in this chapter offers significant simplifications. With a zero-order hold, the control signal is constant over the sampling period and the discrete-time model is obtained simply by integrating the state equations over one sampling period. This problem formulation was introduced in

KALMAN, R.E. and J. E. BERTRAM (1958): "General Synthesis Procedure for Computer Control of Single and Multiloop Linear Systems," *AIEE Trans*. 77, 602–09.

It took some time before this approach found its way into textbooks. Because of its simplicity it is now the predominant approach.

Transformation of state variables and canonical forms is standard material in state-space theory. These results are very similar to the corresponding results for continuous-time systems. A more detailed treatment is given in

KAILATH, T. (1980): *Linear Systems*. Englewood Cliffs, N.J.: Prentice Hall, Inc.

Historically, the input-output approach preceded the state-space approach. A direct treatment from this point of view is given in the classic texts just mentioned. The multivariable case is discussed in

ROSENBROCK, H. H. (1970): *State-Space and Multivariable Theory*. London: Nelson.

KUČERA, V. (1979): *Discrete Linear Control*. Prague: Academia.

The z-transform is extensively discussed in

JURY, E. I. (1958): *Sampled-Data Control Systems*. New York: John Wiley. Second printing: Krieger, 1977.

JURY, E. I. (1964): *Theory and Application of the Z-Transform Method*. New York: John Wiley. Second printing: Krieger, 1973.

DOETSCH, G. (1971): *Guide to the Applications of the Laplace and Z-Transforms*. New York: Van Nostrand Reinhold.

In these references larger tables of z-transform pairs are available. A table of zero-order-hold equivalent transfer functions (compare with Table 3.1) is given in

Neuman, C. P. and C. S. Baradello (1979): "Digital Transfer Functions for Microcomputer Control," *IEEE Trans. Sys., Man. and Cybernetics,* SMC-9, 856–60.

The relationship between the zeros of continuous and sampled systems is discussed in

Åström, K. J., P. Hagander, and J. Sternby (1980): "Zeros of Sampled Systems," *Proc. 19th IEEE Conf. on Decision and Control,* Albuquerque, 1077–81.

The theorems for the limiting zeros for large and small sampling periods are given in this paper.

The observation (Example 3.18) that the zeros of a sampled system can be positioned arbitrarily by a proper choice of the hold circuit is believed to be new.

PROCESS-ORIENTED MODELS

*To Develop Mathematical Models That Give the Relationships
Between the Continuous-Time Signals in a Sampled-Data System.*

4.1 Introduction

Mathematical models for a sampled-data system *from the point of view of the
computer* are developed in Chapter 3. These models are quite simple. The variables that represent the measured signal and the control signal are considered at
the sampling instants only. These variables change in a given time sequence in
synchronization with the clock. The signals are naturally represented in the computer as sequences of numbers. Thus the time-varying nature of sampled-data
systems can be ignored, because the signals are considered only at times that are
synchronized with the clock in the system. The sampled-data system can then be
described as a *time-invariant, discrete-time system*. The model obtained is called
the stroboscopic model.

   The stroboscopic model has the great advantage of being simple. Most of the
problems in analysis and design of sampled-data systems can fortunately be handled by this model. The model will also give a complete description of the system
as long as it is observed from the computer, but sometimes this is not enough.
Therefore it is useful to have other models that give a more detailed description.
Such models are needed when the computer-controlled system is observed from
the process, for example, if a frequency response is performed by cutting the loop
on the analog side. The models required are necessarily more complicated than

those discussed in Chapter 3 because the periodic nature of the system must be dealt with explicitly.

A detailed description of the major events in a computer-controlled system is given in Sec. 4.2. The key modeling problem is the description of the sampling process. This is described using the *modulation model* in Sec. 4.3. Section 4.4 deals with frequency response of sampled-data systems—several unexpected things can happen. The results give more insight into the aliasing problem. An algebraic system theory for sampled-data systems is outlined in Sec. 4.5. Multirate systems are discussed in Sec. 4.6.

4.2 A Computer-Controlled System

A schematic diagram of a computer-controlled system is given in Fig. 4.1. In Chapter 3 the loop is cut inside the computer between the A-D and D-A converters—for example, at C in the figure. In this chapter the loop is instead cut on the analog side—for example, at A in the figure.

The discussions of this chapter require a more detailed description of the sequence of operations in a computer-controlled system. The following events take place in the computer:

1. Wait for a clock pulse.
2. Perform analog to digital conversion.
3. Compute control variable.
4. Perform digital to analog conversion.
5. Update the state of the regulator.
6. Go to step 1.

Because the operations in the computer take some time, there is a time delay between steps 2 and 4. The relationships among the different signals in the system are illustrated in Fig. 4.2. When the control law is implemented in a computer it is important to structure the code so that the calculations required in step 3 are minimized (see Chapter 15).

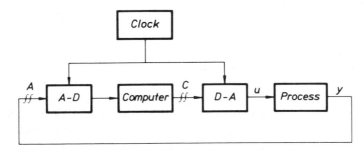

Figure 4.1 Schematic diagram of a computer-controlled system.

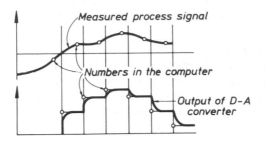

Figure 4.2 Relationships among measured signal, control signal, and their representations in the computer.

It is also important to express the synchronization of the signals precisely. For the analysis the sampling instants have been arbitrarily chosen as the time when the D-A conversion is completed. Because the control signal is discontinuous, it is important to be precise about the limit points. The convention of continuity from the right was adopted. Notice that the real input signal to the process is continuous because of the nonzero settling time of the D-A converter and the actuators.

4.3 The Modulation Model

A characteristic feature of computer-controlled systems with zero-order hold is that the control signal is constant over the sampling period. This fact is used in Chapter 3 to describe how the system changes from one sampling instant to the next by integrating the system equations over one sampling period; this section attempts to describe what happens between the sampling instants. Other mathematical models are then needed, because it is no longer sufficient to model signals as sequences (functions that map Z to R); instead they must be modeled as continuous-time functions (functions that map R to R).

The central theme is to develop the *modulation model*. This model is more complicated than the stroboscopic model discussed in Chapter 3. The main difficulty is that the periodic nature of sampled-data systems must be taken into account. The system can be described as an amplitude modulator followed by a linear system. The modulation signal is a pulse train. A further idealization is obtained by approximating the pulses by impulses. The model has its origin in early work on sampled-data systems by MacColl (1945), Linvil (1951), and others.

In the special case of computer control with a unit-gain algorithm and negligible time delay, the combined action of the A-D converter, the computer, and the D-A converter can be described as a system that samples the analog signal and produces another analog signal that is constant over the sampling periods. Such a circuit is called a *sample-and-hold circuit*. An A-D converter can also be described as a sample-and-hold. The hold circuit keeps the analog voltage constant during the conversion to a digital representation. A more detailed model for the sample-and-hold circuit will first be developed.

A Model of the Sample-and-Hold

A schematic diagram of an analog sample-and-hold circuit is shown in Fig. 4.3. It is assumed that the circuit is followed by an amplifier with very high input impedance. The circuit works as follows: When the sampling switch is closed, the capacitor is charged to the input voltage via the resistor R. When the sampling switch is opened, the capacitor holds its voltage until the next closing.

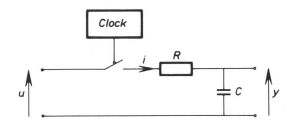

Figure 4.3 Schematic diagram of a sample-and-hold circuit.

To describe the system a function m, which describes the closing and opening of the sampling switch, is introduced. This function is defined by

$$m(t) = \begin{cases} 1 & \text{if switch is closed} \\ 0 & \text{if switch is open} \end{cases}$$

The current is then given by

$$i = \frac{u - y}{R} m$$

The current is thus *modulated* by the function m, which is called the *modulation function*. If the input impedance of the circuit that follows the sample-and-hold is high, the voltage over the capacitor is given by

$$C \frac{dy(t)}{dt} = i(t) = \frac{u(t) - y(t)}{R} m(t) \tag{4.1}$$

The differential equation of (4.1) is a linear *time-varying* system. The time variation is caused by the modulation. If the sampling period h is constant and if the switch is closed for τ seconds at each sampling, the function m has the shape shown in Fig. 4.4. Because m is a periodic function the system becomes a *periodic system*.

Once a mathematical model of the circuit is obtained the circuit's response to an input signal u can be investigated. It follows directly from Equation (4.1)

Figure 4.4 Graph of the modulation function m with period h and pulse width τ.

that the voltage across the capacitor is constant when the switch is open, i.e., when $m(t) = 0$. When the switch is closed, the voltage y approaches the input signal u as a first-order dynamic system with the time constant RC. The time constant of the RC circuit must be considerably shorter than the pulse width; otherwise there is no time to charge the capacitor to the input voltage when the switch is closed.

Results of a simulation of the sample-and-hold circuit are shown in Fig. 4.5.

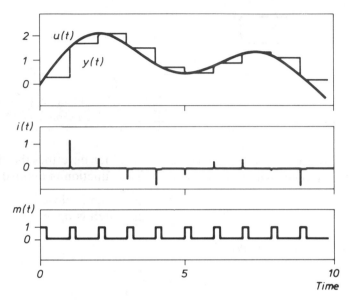

Figure 4.5 Simulation of a sample-and-hold circuit. The pulse width τ is 0.2s and the time constant is $RC = 0.01$ s.

With the chosen parameters, the pulse width τ is so long that the input signal changes significantly when the switch is closed.

Figure 4.6 shows what happens when the pulse width is shorter. The results shown in Fig. 4.6 represent a reasonable choice of parameter values. The sample-and-hold circuit quickly reaches the value of the input signal and then remains constant over the sampling period.

Practical Samplers

In practice, a sampler is not implemented as shown in Fig. 4.3. A common implementation is shown in Fig. 4.7. An operational amplifier is used to help charge the capacitor. The circuit shown in Fig. 4.7 can also be described by Equation (4.1).

To avoid difficulties with noise and ground loops, it is important to have the computer galvanically isolated from the process signals. This can be achieved with the sample-and-hold circuit shown in Fig. 4.8, which is called the *flying capacitor circuit*. The circuit combines electrical insulation with sample-and-hold

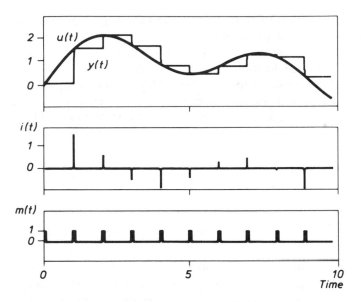

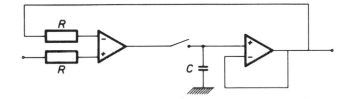

Figure 4.6 Simulation of a sample-and-hold circuit. The pulse width τ is 0.05 s and the time constant is $RC = 0.01$ s.

Figure 4.7 A practical implementation of a sample-and-hold circuit.

action in an elegant way. The capacitor is charged to the input voltage when it is connected to the input line. When the capacitor is connected to the D-A converter it holds its voltage. Electrical isolation is obtained because the capacitor is connected either to the process or to the D-A converter of the control computer. In practice it is common to charge the capacitor via an operational amplifier. The flying capacitor circuit can also be described by Equation (4.1).

A Mathematical Idealization

The pulse-modulation scheme is easy to simulate but difficult to analyze. A more easily used mathematical idealization will therefore be introduced. It seems reasonable to design the sample-and-hold circuit so that the pulse width τ is much shorter than the sampling period. It also seems reasonable to choose the time constant RC to be shorter than the pulse width. The current through the capacitor

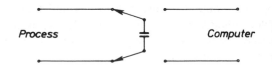

Figure 4.8 A sample-and-hold circuit based on the flying capacitor implementation.

will then consist of short pulses. Both the height and the time integral of a pulse are proportional to the difference $u - y$ between the input voltage u and the capacitor voltage y at the sampling instant.

In the idealization, the current pulses are replaced by impulses. For simplicity the integral of the impulse is chosen to be proportional to the value of the input signal u at the sampling instant. The capacitor is then replaced by an integrator. Since the pulses were chosen to be proportional to u and not to $u - y$, it is necessary to reset the integral to zero when a new pulse arrives. The current is then represented as

$$u^* = um \tag{4.2}$$

where

$$m(t) = \sum_{k=-\infty}^{\infty} \delta(t - kh) \tag{4.3}$$

and δ is a delta function [compare with (4.1)]. The signal u^* is called the *sampled representation* of the continuous signal u. It is useful to remember that u^* is related to the current through the capacitor of the sample-and-hold circuit in Fig. 4.3.

The signal u^* can be thought of as a modulation of u with a carrier signal in the form of an impulse train. The model is therefore called the *impulse-train modulation model*. The signal u^* carries the same information as the sequence $\{u(kh), k = \ldots, -1, 0, 1, \ldots\}$. Notice, however, that u^* is a (generalized) time function. The signal u^* is introduced to represent a sampled signal in a form that can be processed by linear filtering.

The Hold Circuit

The hold circuit can be represented as an integrator that is automatically reset to zero after one sampling period. Such a system has the transfer function

$$G(s) = \frac{1}{s} (1 - e^{-sh}) \tag{4.4}$$

The impulse response of the transfer function $1/s$ is a unit step and the impulse response of $(1/s) \exp(-sh)$ is a unit step that is delayed h time units. Subtraction of these impulse responses gives the impulse response as a pulse of unit height and duration h.

Notice that the steady-state gain of the hold circuit is $G(0) = h$. Section 2.3 shows that ideal sampling could be said to have a gain $1/h$. The combination of a sampler with a hold circuit thus has unit steady-state gain. For very fast sampling, the sample-and-hold circuit thus acts as a continuous-time system with unit transfer function.

The idealized model of a sample-and-hold circuit is thus obtained by combining a sampler with impulse modulation given by (4.2) and (4.3) with a hold circuit given by (4.4). A block-diagram representation of the system is shown in Fig. 4.9. Because the impulse modulator is a periodic system it follows that the sample-and-hold is also a periodic system.

Input-Output Relationships

Once a convenient representation of a sample-and-hold circuit is obtained, the response of a sampled-data system to an arbitrary input signal can be computed. Consider the system shown in Fig. 4.10(a), which is composed of a sample-and-hold circuit connected to a time-variant, linear, dynamic system with the transfer function G. This is a typical representation of a sampler and a D-A converter connected to a process. Use of the impulse-modulation model of the sample-and-hold allows the system to be represented as in Fig. 4.10(b).

Let u be the input and y the output. It follows from the impulse-modulation model (4.2) that

$$u^* = um$$

Let F be the transfer function of the combination of the zero-order-hold circuit and the process, i.e.,

$$F(s) = \frac{1}{s}(1 - e^{-sh})G(s) \tag{4.5}$$

The input-output relationship is easily determined using transform theory. The Laplace transform of u^* is given by

$$U^*(s) = \int_0^\infty e^{-st}u^*(t)\,dt = \sum_{k=0}^\infty e^{-skh}u(kh) \tag{4.6}$$

The Laplace transform of the output signal is then given by

$$Y(s) = F(s)\sum_{k=0}^\infty e^{-skh}u(kh) \tag{4.7}$$

It is thus straightforward to calculate the Laplace transform of the output signal. Notice, however, that it is not possible to factor out the Laplace transform of the signal u on the right-hand side of (4.7). This means that the input-output rela-

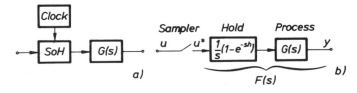

Figure 4.10 Schematic diagram of a sample-and-hold circuit connected to a linear system and its representation using the idealized model of a sample-and-hold.

tionship of the system *cannot* be characterized by an ordinary transfer function. This is because the system is not time-invariant.

4.4 Frequency Response

There are many cases in which the specifications on a control system are given in the frequency domain. Autopilots for aircraft are typical examples. To guarantee that high-frequency bending modes are not excited by the control system, it is critical to ensure that the open-loop frequency response is less than 1 in magnitude for high frequencies. One way to ensure that the specifications are fulfilled is to perform a frequency-response test on the complete system. If the regulator is implemented using a digital computer, it is important to understand phenomena that occur in the frequency response of sampled-data systems.

A Special Case

When performing the frequency-response test, it is natural to cut the loop on the analog side, e.g., at A in Fig. 4.1. To simplify the analysis consider the special case in which the output of the D-A converter is equal to the input of the A-D converter. The action of the computer on the signals can then be described as a sample-and-hold circuit. It follows from Fig. 4.9 that a sample-and-hold circuit can be represented as a sampler followed by a hold circuit. The problem is thus reduced to calculation of the response of a sampler followed by a linear, time-invariant system.

Equation (4.2) gives the sampled representation u^* of the input signal u. A formal Fourier series representation of a sequence of delta functions gives

$$m(t) = \sum_{k=-\infty}^{\infty} \delta(t - kh) = \frac{1}{h}\left[1 + 2\sum_{k=1}^{\infty}\cos k\omega_s t\right] \qquad (4.8)$$

where h is the sampling period and ω_s is the corresponding sampling frequency in radians per second.

Assume that the input to the system is

$$u(t) = \sin(\omega t + \varphi) = Im[\exp i(\omega t + \varphi)]$$

The series expansion of the output $u^* = um$ of the sampler then becomes

$$u^*(t) = \frac{1}{h}\left[\sin(\omega t + \varphi) + 2\sum_{k=1}^{\infty}\cos(k\omega_s t)\sin(\omega t + \varphi)\right]$$

$$= \frac{1}{h}\left\{\sin(\omega t + \varphi) + \sum_{k=1}^{\infty}[\sin(k\omega_s t + \omega t + \varphi) - \sin(k\omega_s t - \omega t - \varphi)]\right\}$$

The signal u^* has a component with the frequency ω of the input signal. This component is multiplied by $1/h$ because the steady-state gain of a sampler is $1/h$. The signal also has components corresponding to the *sidebands* $k\omega_s \pm \omega$. The frequency content of the output u^* of the sampler is shown in Fig. 4.11. The

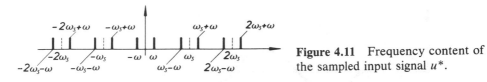

Figure 4.11 Frequency content of the sampled input signal u^*.

output signal y is obtained by linear filtering of the signal u^* with a system having the transfer function $F(s)$. The output thus has components with the fundamental frequency ω and the sidebands $k\omega_s \pm \omega$.

For $\omega \neq k\omega_N$, where ω_N is the Nyquist frequency, the fundamental component of the output is

$$y(t) = \frac{1}{h} Im[F(i\omega)e^{i(\omega t + \varphi)}]$$

For $\omega = k\omega_N$ the frequency of one of the sidebands coincides with the fundamental frequency. Two terms thus contribute to the component with frequency ω. This component is

$$y(t) = \frac{1}{h} Im\{F(i\omega)e^{i(\omega t + \varphi)} - F(i\omega)e^{i(\omega t - \varphi)}\}$$

$$= \frac{1}{h} Im\{(1 - e^{-2i\varphi})F(i\omega)\,e^{i(\omega t - \varphi)}\}$$

$$= \frac{1}{h} Im\{2e^{i(\pi/2 - \varphi)}\sin\varphi\,F(i\omega)e^{i(\omega t + \varphi)}\}$$

If the input signal is a sine wave with frequency ω, it is found that the output contains the fundamental frequency ω and the sidebands $k\omega_s \pm \omega$, $k - 1, 2, \ldots$. (Compare with the discussion of alias in Section 2.5). The transmission of the fundamental frequency is characterized by

$$\hat{F}(i\omega) = \begin{cases} \dfrac{1}{h} F(i\omega) & \omega \neq k\omega_N \\[2em] \dfrac{2}{h} F(i\omega)e^{i(\pi/2 - \varphi)}\sin\varphi & \omega = k\omega_N \end{cases} \tag{4.9}$$

For $\omega \neq k\omega_N$, the transmission is simply characterized by a combination of the transfer functions of the sample-and-hold circuit and the system G. The factor $1/h$ is due to the steady-state gain of the sampler.

The fact that the signal transmission at the Nyquist frequency ω_N critically depends on φ—i.e., how the sinusoidal input signal is synchronized with respect to the sampling instants—is illustrated in Fig. 4.12.

There may be interference between the sidebands and the fundamental frequency which can cause the output of the system to be very irregular. A typical illustration of this was given in Example 1.2. In this case the fundamental component has the frequency 4.9 Hz and the Nyquist frequency is 5 Hz. The inter-

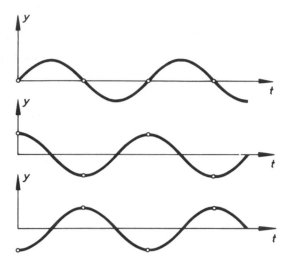

Figure 4.12 Sampling of a sinusoidal at a rate that corresponds to the Nyquist frequency. Notice that the sampled signal depends strongly on how the sine wave is synchronized to the sampling instants.

action between the fundamental component and the lowest sideband, which has the frequency 5.1 Hz, will produce beats with the frequency 0.1 Hz. This is clearly seen in Fig. 1.6.

If the sideband frequencies are filtered out, the sampled system appears as a linear, time-invariant system except at frequencies that are multiples of the Nyquist frequency, $\omega_s/2$. At this frequency the amplitude ratio and the phase lag depend on the phase shift of the input relative to the sampling instants.

If an attempt is made to determine the frequency response of a sampled system using frequency response, it is important to filter out the sidebands efficiently. Even with perfect filtering, there will be problems at the Nyquist frequency. The results depend critically on how the input is synchronized with the clock of the computer.

The General Case

It is easy to extend the analysis to the general case of the system shown in Fig. 4.1. The corresponding open-loop system is shown in Fig. 4.13. It consists of an A-D converter, the computer, a D-A converter, and the process. It is assumed

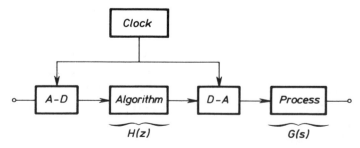

Figure 4.13 Open-loop computer-controlled system.

Process-Oriented Models   Chap. 4

that the D-A converter holds the signal constant over a sampling interval. It is also assumed that the calculations performed by the computer can be expressed by the pulse-transfer function $H(z)$ and that the process is described by the transfer function $G(s)$.

If a sinusoid

$$v(t) = \sin(\omega t + \varphi) = Im[\exp i(\omega t + \varphi)]$$

is applied to the A-D converter, then the computer will generate a sequence of numbers that in steady state can be described by

$$w(kh) = Im[H(e^{i\omega h})e^{i(\omega kh + \varphi)}], \qquad k = \ldots -1, 0, 1, \ldots$$

This sequence is applied to the D-A converter. Because the D-A converter holds the signal constant over a sampling period, the output is the same as if the signal w were applied directly to a hold circuit. The discussion of the previous section can thus be applied: The output contains the fundamental component with frequency ω and sidebands $k\omega_s \pm \omega$. The signal transmission of the fundamental component may be described by the transfer function

$$K(i\omega) = \begin{cases} \dfrac{1}{h} H(e^{i\omega h}) F(i\omega) & \omega \neq k\omega_N \\[3mm] \dfrac{2}{h} H(e^{i\omega h}) F(i\omega) e^{i(\pi/2 - \varphi)} \sin \varphi & \omega = k\omega_N \end{cases} \qquad (4.10)$$

where ω_N is the Nyquist frequency and

$$F(s) = \frac{1}{s}(1 - e^{-sh})G(s) \qquad (4.11)$$

When ω is not a multiple of the Nyquist frequency, the signal transmission of the fundamental component can be characterized by a transfer function that is a product of four terms: the gain $1/h$ of the sampler, the transfer function $[1 - \exp(-sh)]/s$ of the hold circuit, the pulse-transfer function $H[\exp(sh)]$ of the algorithm in the computer, and the transfer function $G(s)$ of the process. Notice, however, that there are other frequencies in the output of the system because of the sampling. At the Nyquist frequency the fundamental component and the lowest sideband coincide.

It follows from the discussion that the hold circuit can be interpreted as a filter. The frequency functions of zero- and first-order-hold circuits are shown in Fig. 4.14. It is clear from the figure that both the zero-order hold and the first-order hold permit significant signal transmission above the Nyquist frequency $\omega_N = \pi/h$. Notice that the phase curve is discontinuous at arguments $\omega h = 2k\pi$, $k = 1, 2, \ldots$. Because the phase is defined modulo 2π, the discontinuities may be π or $-\pi$. In the figure they are shown as π for convenience only.

The following example illustrates the calculation and interpretation of the frequency response of a sampled system.

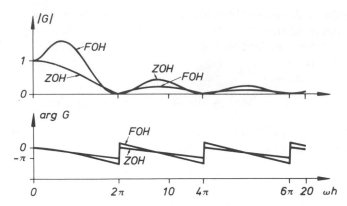

Figure 4.14 Magnitude and argument curves of the transfer function for first- and zero-order-hold circuits.

Example 4.1

Consider a system composed of a sampler and a zero-order hold, given by (4.4) followed by a linear system, with the transfer function

$$G(s) = \frac{1}{s + 1}$$

The sampling period is $h = 0.05$ s. The Nyquist frequency is thus $\pi/0.05 = 62.8$ rad/s. Figure 4.15 shows the Bode diagram of the system. For comparison, the Bode diagram of the transfer function G is also shown in the figure. The curves are very close for frequencies that are much smaller than the Nyquist frequency. The deviations occur first in the phase curve. At $\omega = 0.1\omega_N$ the phase curves differ by about 10°. There is no signal transmission at frequencies that are multiples of the sampling frequency ω_s, because the transfer function of the zero-order hold is zero for these frequencies. The phase curve is also discontinuous at these frequencies. (Compare with Fig. 4.14.) Notice also the ambiguities of the transfer function at frequencies that are multiples of the Nyquist frequency. The value of ω_N is indicated by a bar in Fig. 4.15.

The interpretation of the Bode diagram requires some care because of the modulation introduced by the sampling. If a sine wave of frequency ω is introduced, the output signal is the sum of the outputs of the sine wave and all its aliases. This is illustrated in Fig. 4.16, which shows the steady-state outputs for different frequencies. For frequencies smaller than the Nyquist frequency, the contribution from the fundamental frequency dominates. At frequencies close to the Nyquist frequency, there is a substantial interaction with the first alias, $\omega_s - \omega$. Typical beats are thus obtained. At the Nyquist frequency, the signal and its first alias have the same frequency and magnitude. The resulting signal then depends on the phase shift between the signals. For frequencies higher than the Nyquist frequency, the contribution from the alias in the frequency range $(0, \omega_N)$ dominates.

This clearly shows how important it is to filter a signal before the sampling, so that the signal transmission above the Nyquist frequency is negligible. Compare this conclusion with the discussion of aliasing in Sec. 2.5.

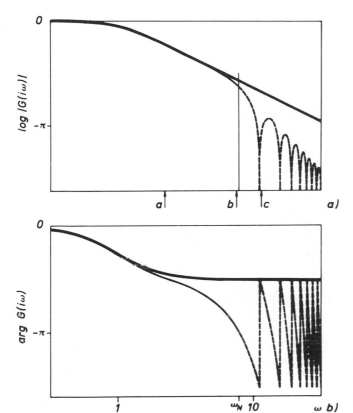

Figure 4.15 Bode diagrams for a zero-order sample-and-hold circuit followed by a first-order lag. The sampling period is 0.05 s.

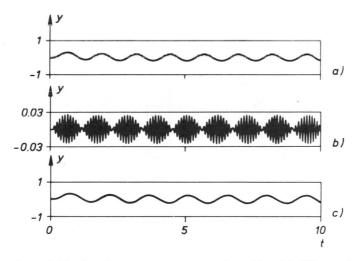

Figure 4.16 Steady-state responses to sinusoids with different frequencies for a zero-order hold followed by a first-order system with unit time constant. The sampling period is 0.05 s. The frequencies are 5 rad/s in (a), 60 rad/s in (b), and 130 rad/s in (c). They are indicated by arrows in Fig. 4.15.

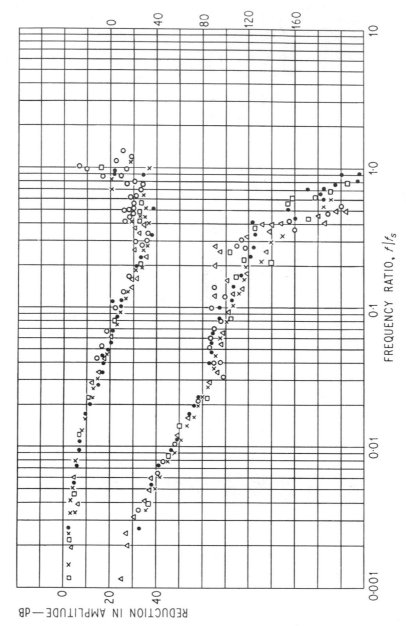

Figure 4.17 Measured frequency response for a diesel engine. The frequencies are normalized with respect to the sampling frequency. [Redrawn from D. E. Bowns, "The Dynamic Transfer Characteristics of Reciprocating Engines," *Proc. Mech. Engrs.*, 185. (1971) with permission.]

An Application

An internal-combustion engine is a typical example of a system that is inherently sampled. The sampling is caused by the ignition mechanism, and its frequency is the number of independently fired cylinders divided by the time required for a full cycle.

When an attempt was made to investigate the dynamic response of the engines, reproducible results were easily obtained for frequencies lower than the sampling frequency. For a long time, however, the results for higher frequencies were erratic: Different results were obtained at different measurements and results of experiments could not be verified when the experiments were repeated. This was due to the sampled nature of the process. For input signals with a frequency close to the Nyquist frequency, there is interference from the sidebands. At the Nyquist frequency, the results depend on how the sinusoid is synchronized to the ignition pulses.

When the source of the difficulty was finally understood, it was easy to find a solution. The sinusoid was simply synchronized to the ignition pulses; then it became possible to measure the frequency response to high frequencies. A typical result is shown in Fig. 4.17. Notice, in particular, that the measurement is done in a range of frequencies that includes the Nyquist frequency.

4.5 Pulse-Transfer Function Formalism

Linear, continuous-time systems can be conveniently described, analyzed, and synthesized using algebraic methods. When the theory of sampled-data systems was developed, it was natural to try to develop similar algebraic tools. Much of the early development of the theory of sampled-data systems went in this direction.

The approach is useful, simple, and successful if the system is viewed from the computer or if the process is observed at times that are synchronized with the computer clock, because the system is then *time-invariant*. (See Chapter 3 for the appropriate analysis.) However, when the system is analyzed from the process point of view, as is done in this chapter, the system is time-variable. The algebraic approach then loses some of its simplicity, because multiplication with time functions does not commute with differential and difference operators.

For the case of completeness, a brief description of the algebraic system theory in the more complicated case is given. The main reason is historical. Much of the theory of sampled-data systems was originally developed using this approach which is also used in many papers.

Goals

Before going into the details, it is useful to state the goals. The main purpose is to develop a formalism for manipulating the system descriptions. The formalism will have many properties in common with the transform methods for linear, time-invariant systems. Each A-D and D-A converter is associated with a sampling operation. Because sampling can be described as an amplitude modulation, the time-varying parts will be associated with these operations. The system can then be separated into different parts: Some parts are ordinary linear, time-invariant systems that can be handled by the ordinary transform methods; the other parts consist of the samplers that are intrinsically time-varying.

The Z-Transform

Section 3.5 introduces z-transforms as mappings from sequences to functions of a complex variable. A different z-transform whose domain is continuous functions can be defined as follows:

Definition 4.1—The z-transform.   The z-transform of a continuous-time function is defined as

$$\tilde{F}(z) = \sum_{k=0}^{\infty} z^{-k} f(kh) \tag{4.12}$$

The inverse transform is given by

$$f(kh) = \frac{1}{2\pi i} \oint_{\Gamma} z^{k-1} \tilde{F}(z) \, dz \tag{4.13}$$

where the contour of integration Γ encloses all singularities of the integrand.

The z-transform of a continuous-time signal is thus obtained by sampling the signal and then applying the z-transform to the sampled sequence. Because the transform depends only on the values at the sampling instants, all time functions that agree at the sampling instants have the same transform.

Notice that the transform is inherently related to the clock, which defines the sampling instants. Also notice that the inverse transform defines the function at the sampling instants only.

These properties of the z-transform of a continuous-time function are easily misunderstood and have led to much confusion and many mistakes.

Two Basic Theorems

To develop an algebra that allows formal manipulation of the systems, two theorems are needed. The first theorem tells how the z-transform of a continuous-time function is related to its Laplace transform.

Theorem 4.1.   Let the function f have the Laplace transform F and the z-transform $\tilde{F}$, and let F^* be the Laplace transform of the sampled representation f^* of f. Assume that for some $\epsilon > 0$, $|F(s)| \leq |s|^{-1-\epsilon}$ for large $|s|$ then

$$F^*(s) = \tilde{F}(e^{sh}) = \frac{1}{h} \sum_{k=-\infty}^{\infty} F(s + ik\omega_s) \qquad (4.14)$$

where $\omega_s = 2\pi/h$ is the sampling frequency.

Proof. The definition of F^* gives

$$F^*(s) = \int_0^\infty e^{-st} f^*(t)\, dt$$

$$= \int_0^\infty e^{-st} f(t) m(t)\, dt$$

$$= \int_0^\infty e^{-st} f(t) \left\{ \sum_{k=-\infty}^{\infty} \delta(t - kh) \right\} dt$$

where the last equality is obtained from (4.3). Interchange the order of integration and summation gives

$$F^*(s) = \sum_{-\infty}^{\infty} \int_0^\infty e^{-st} f(t) \delta(t - kh)\, dt$$

$$= \sum_{k=0}^{\infty} (e^{sh})^{-k} f(kh) \qquad (4.15)$$

$$= \tilde{F}(e^{sh})$$

The last equality follows from (4.12).

Because the Laplace transform of a product of two functions is a convolution of their transforms, it follows that

$$F^*(s) = F(s) * M(s) = \frac{1}{2\pi i} \int_{\gamma - i\infty}^{\gamma + i\infty} F(v) M(s - v)\, dv$$

$$= \frac{1}{2\pi i} \int_{\gamma - i\infty}^{\gamma + i\infty} F(v) \frac{1}{1 - e^{-h(s-v)}}\, dv \qquad (4.16)$$

The integration path should be to the right of all poles of F and to the left of all poles of M (see Fig. 4.18).

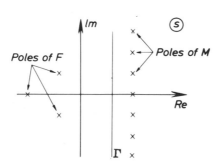

Figure 4.18 Singularities of F and M and the integration contour.

If F goes to zero faster than $|s|^{-1-\epsilon}$ as $|s| \to \infty$, the integral of FM on a large semicircle will vanish. Upon completion of the integration path by a large semicircle to the right, the integral can be evaluated with residue calculus. In the domain enclosed by the contour the integrand has simple poles at the zeros of

$$\exp h(s - v) = 1$$

i.e., at

$$v_k = s + \frac{2\pi ik}{h} = s + ik\omega_s, \qquad k = \ldots -1, 0, 1, \ldots$$

The residues at these poles are

$$\frac{1}{h} F\left(s + \frac{2\pi ik}{h}\right)$$

Summation of the residues now gives (4.14). □

Remark 1. Notice that Equation (4.14) can also be written as

$$F^*(s) = \frac{1}{h} [F(s) + F(s + i\omega_s) + F(s - i\omega_s) + \cdots] \qquad (4.17)$$

Remark 2. Notice that if F is analytic for Re $s < -\gamma_0$, the integration path in (4.16) may be closed by a large semicircle to the left. The following formula is obtained:

$$\tilde{F}(z) = \sum_{\text{Poles of } F} \text{Res}\left\{F(s) \frac{z}{z - \exp(sh)}\right\} \qquad (4.18)$$

This gives a proof of formula (3.50).

Remark 3. The theorem can be extended to the case in which the function F goes to zero as $1/|s|$ for large $|s|$. Equation (4.14) is then replaced by

$$F^*(s) = \frac{1}{h} \sum_{k=-\infty}^{\infty} F(s + ik\omega_s) + \frac{1}{2} f(0+) \qquad (4.19)$$

Remark 4. In the literature the same notation is sometimes used for the functions F^* and $\tilde{F}$. This is confusing and should be avoided.

Remark 5. Notice that Formula (4.14) is closely related to Formula (2.4) for the Fourier transform of a sampled signal.

Pulse-transfer functions. Section 4.3 shows that the input-output relationship of a sampler followed by a linear transfer function is given by Equation (4.7). This equation cannot be described by a transfer function. If a fictitious sampler is added to the system output, the configuration shown in Fig. 4.19 is

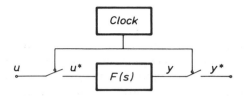

Clock

u u^* F(s) y y^*

Figure 4.19 Block diagram of a system with two samplers.

obtained. For this system it is possible to define a transfer function. The input-output relationship is given by

$$y^*(t) = [f(t)*u^*(t)]^* \qquad (4.20)$$

The following theorem is useful for obtaining the corresponding transforms.

Theorem 4.2. Let f and g be functions that have Laplace transforms and let m be the modulation function corresponding to an impulse train. Then

$$m(t)\{f(t)*[m(t)\cdot g(t)]\} = [m(t)\cdot f(t)]*[m(t)\cdot g(t)] \qquad (4.21)$$

or, equivalently,

$$[f(t)*g^*(t)]^* = f^*(t)*g^*(t) \qquad (4.22)$$

Proof. Use of the definition of a convolution allows the left-hand side of (4.21) to be written as

$$(f*g^*)^*(t) = m(t) \int_{-\infty}^{\infty} f(t - \tau)g^*(\tau)\, d\tau$$

$$= \int_{-\infty}^{\infty} m(t)f(t - \tau)m(\tau)\, g(\tau)\, d\tau$$

Similarly, the right hand side of Equation (4.22) can be written as

$$(f^**g^*)(t) = \int_{-\infty}^{\infty} m(t - \tau)f(t - \tau)m(\tau)g(\tau)\, d\tau$$

$$= \int_{-\infty}^{\infty} m(t)f(t - \tau)m(\tau)g(\tau)\, d\tau$$

The last equality holds because $m(\tau) \neq 0$ only for $\tau = nh$ and $m(t - nh) = m(t)$. ☐

Remark 1. The Laplace transformation of (4.21) gives

$$[F(s)G^*(s)]^* = F^*(s)G^*(s) \qquad (4.23)$$

Remark 2. Notice that the multiplication by m outside the brace in (4.21) can be interpreted as introduction of a fictitious sampler.

A Formalism

It is now straightforward to develop a formalism for dealing with sampled sytems. First, a system is represented by a block diagram. Each A-D converter is represented as an ideal sampler. Each D-A converter is represented as a hold circuit having the transfer function (4.4). Linear, continuous-time blocks are represented by their transfer functions, and linear calculations in the computer, by their pulse-transfer functions. The paths between the samplers can be reduced using ordinary rules for linear, time-invariant systems. The equations describing the system are then written down. Theorems 4.1 and 4.2 are then used to rewrite the equations. The procedure is illustrated by an example.

Example 4.2

Consider the standard configuration of a computer-controlled system in Fig. 4.20(a). The process is characterized by a linear transfer function G, and the calculations performed in the computer are represented by a pulse-transfer function H. The analog and digital parts of the system are, as usual, connected via D-A and A-D converters.

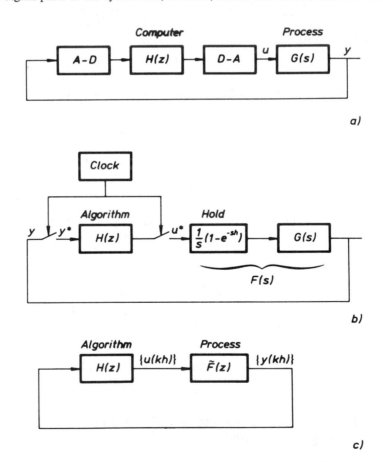

Figure 4.20 Standard configuration of a computer-controlled system.

To apply the formalism, the A-D converter is represented by an ideal sampler. The computer is represented as a system that transforms an impulse-modulated signal to another impulse-modulated signal. The D-A converter is represented by a sampler, followed by a zero-order hold. It is assumed that the samplers are perfectly synchronized. The block diagram shown in Fig. 4.20(b) is then obtained. The analog parts are thus the hold and the process. Their combined transfer function is

$$F(s) = \frac{1}{s}(1 - e^{-sh})G(s)$$

The Laplace transform Y of the output y is given by

$$Y(s) = F(s)U^*(s)$$

The sampled output has the transform

$$Y^*(s) = [F(s)U^*(s)]^* = F^*(s)U^*(s)$$

where Theorem 4.2 is used to obtain the last equality. The relationship between y^* and u^* can thus be represented by the pulse-transfer function

$$\tilde{F}(z) = F^*(s)\,\big|_{s = (\ln z)/h}$$

The calculations in the computer can furthermore be represented by the pulse-transfer function $H(z)$. If the loop is cut in the computer the pulse-transfer function is thus

$$H(z)\tilde{F}(z)$$

A block diagram of the properties of the system that can be seen from the computer is shown in Fig. 4.20(c). By considering all signals as sequences like $\{y(kh), k = \ldots -1, 0, 1, \ldots\}$ and by introducing appropriate pulse-transfer functions for the algorithm and the process with the sample-and-hold, a representation that is equivalent to the ordinary block-diagram representation of continuous-time systems was thus obtained. $\square$

A further illustration is given by a slightly more complicated example.

Example 4.3

The system illustrated in Fig. 4.21(a) has two measured analog signals, y_1 and y_2, and one analog command signal, u_c. The analog signals are scanned by a multiplexer and converted to digital form. The computer calculates the control signal, which is fed to the process via the *D-A* converter. Figure 4.21(b) is obtained by the procedure given in Example 4.2.

Introduce

$$F_1(s) = G_1(s)\frac{1}{s}(1 - e^{-sh})$$

$$F_2(s) = G_2(s)F_1(s)$$

The Laplace transforms of the output signals are then given by

$$Y_1(s) = F_1(s)U^*(s)$$

$$Y_2(s) = F_2(s)U^*(s)$$

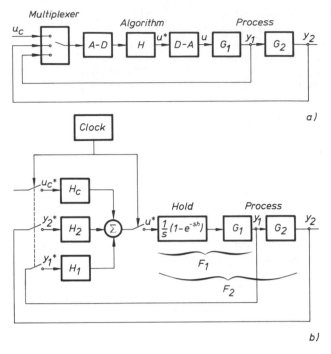

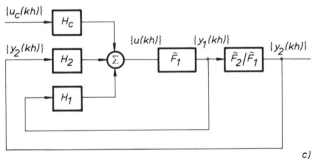

Figure 4.21 Computer-controlled system with multiplexer and two feedback loops and equivalent block diagram.

Hence

$$Y_1^*(s) = [F_1(s)U^*(s)]^* = F_1^*(s)U^*(s)$$

$$Y_2^*(s) = [F_2(s)U^*(s)]^* = F_2^*(s)U^*(s)$$

It follows from (4.14) and (4.23) that

$$Y_1(z) = \tilde{F}_1(z)U(z)$$

$$Y_2(z) = \tilde{F}_2(z)U(z)$$

Let the calculations performed by the control computer be represented by

$$U(z) = H_c(z)U_c(z) - H_1(z)Y_1(z) - H_2(z)Y_2(z)$$

The relationship between the output, Y_2, and the sampled command signal, U_c, is

Process-Oriented Models   Chap. 4

obtained by eliminating Y_1 in the preceding equations. This gives

$$Y_2(z) = \frac{H_c(z)\tilde{F}_2(z)}{1 + H_1(z)\tilde{F}_1(z) + H_2(z)\tilde{F}_2(z)} U_c(z)$$

Notice, however, that the relationship between the analog signals y_1 and u_c cannot be represented by a simple pulse-transfer function because of the periodic nature of the sampled-data system.

With the introduction of the sampled signals as sequences and pulse-transfer functions, the system can be represented as in Fig. 4.21(c). □

Modified Z-Transforms

The problem of sampling a system with a delay can be handled by the modified z-transform, which is defined as follows.

Definition 4.2—The modified z-transform. The modified z-transform of a continuous-time function is given by

$$\tilde{F}(z, m) = \sum_{k=0}^{\infty} z^{-k} f(kh - h + mh), \qquad 0 \le m \le 1$$

The inverse transform is given by

$$f(nh - h + mh) = \frac{1}{2\pi i} \int_{\Gamma} \tilde{F}(z, m) z^{n-1} \, dz$$

where the contour Γ encloses all singularities of the integrand. □

The modified z-transform is useful for many purposes—for example, the intersample behavior can easily be investigated using these transforms. There are extensive tables of modified z-transforms and many theorems about their properties (see the references).

4.6 Multirate Sampling

So far only systems in which the A-D and the D-A conversions are made at the same rates have been discussed. Example 3.18 indicates that it may be advantageous to make the D-A conversion more rapidly. There are also situations where the converse is true. It is, for example, difficult to implement antialiasing filters with long time constants using analog techniques. In such cases it is much easier to sample the signals rapidly with analog antialiasing filters and to do digital filtering afterward. In both cases the systems have two samplers that operate at different rates. This is called *multirate sampling*. Such sampling schemes may be necessary for systems with special data transmission links or special sensors and actuators and are useful for improving the responses of systems in which measurements are obtained at slow rates, e.g., when laboratory instruments are used. Multirate systems may allow better control of what happens between the sampling instants. In multivariable systems it may also be advantageous to have different sampling rates in different loops to reduce the computational load and to improve the numeric conditioning. Use of multirate sampling is also natural in multiprocessor sytems.

A detailed treatment of multirate systems is outside the scope of this book; however, a short discussion of the major ideas will be given to show how the methods presented in the book can be extended to cover multirate systems also.

State-Space Descriptions

Consider a system composed of two subsystems that are continuous, constant-coefficient, dynamic sytems. Assume that there are two periodic samplers with periods h_1 and h_2. Let the ratio of the periods be a rational number $h_1/h_2 = m_1/m_2$ where m_1 and m_2 have no common factor. Then there exists a smallest integer m and a real number h such that

$$h_1 = \frac{hm_1}{m} \tag{4.24}$$

$$h_2 = \frac{hm_2}{m} \tag{4.25}$$

and

$$m = m_1 m_2 \tag{4.26}$$

If the samplers are synchronized, it follows that the control signals will be constant over sampling periods of length h/m. Sampling with that period gives a discrete-time system that is periodic with period h. The system can then be described as a constant, discrete-time system if the values of the system variables are considered only at integer multiples of h. The ordinary discrete-time theory can then be applied. An example illustrates the idea.

Example 4.4

Consider the system shown in Fig. 4.22, which has two subsystems and two samplers with periods 0.5 and 1. It is assumed that the samplers are synchronized. It is also assumed that the hold circuits are included in the subsystems. If the subsystems are sampled with period 0.5 and 0.5 is chosen as a time unit, then

$$\begin{cases} x_1(k+1) = \Phi_1 x_1(k) + \Gamma_1 u_1(k) \\ y_1(k) = C_1 x_1(k) \end{cases}$$

$$\begin{cases} x_2(k+1) = \Phi_2 x_2(k) + \Gamma_2 u_2(k) \\ y_2(k) = C_2 x_2(k) \end{cases}$$

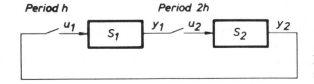

Figure 4.22   Block diagram of a simple multirate system.

The interconnection are described by

$$u_1(k) = y_2(k), \qquad k = \ldots -1, 0, 1, 2, \ldots$$

$$u_2(k) = y_1(k), \qquad k = \ldots -2, 0, 2, 4, \ldots$$

The system is periodic with a period of two sampling intervals. A time-invariant description can be obtained by considering the system variables at even sampling periods only. Straightforward calculations give

$$\begin{bmatrix} x_1(2k + 2) \\ x_2(2k + 2) \end{bmatrix} = \begin{bmatrix} \Phi_1^2 + \Gamma_1 C_2 \Gamma_2 C_1 & \Phi_1 \Gamma_1 C_2 + \Gamma_1 C_2 \Phi_2 \\ (\Phi_2 \Gamma_2 + \Gamma_2) C_1 & \Phi_2^2 \end{bmatrix} \begin{bmatrix} x_1(2k) \\ x_2(2k) \end{bmatrix} \quad (4.27)$$

This equation can be used to analyze the response of the multirate system. For example, the stability condition is that the matrix on the right-hand side of (4.27) has all its eigenvalues inside the unit disc. The values of the state variables at odd sampling periods are given by

$$\begin{bmatrix} x_1(2k + 1) \\ x_2(2k + 1) \end{bmatrix} = \begin{bmatrix} \Phi_1 & \Gamma_1 C_2 \\ \Gamma_2 C_1 & \Phi_2 \end{bmatrix} \begin{bmatrix} x_1(2k) \\ x_2(2k) \end{bmatrix} \qquad \square$$

The analysis illustrated by the example can be extended to an arbitrary number of samplers provided that the ratios of the sampling periods are rational numbers. Delayed sampling can also be handled by the methods described in Sec. 3.2.

Input-Output Methods

Multirate systems can also be investigated by input-output analysis. First, observe as before that the system is periodic with period h if the ratios of the sampling periods are rational numbers. The values of the system variables at times that are synchronized to the period can then be described as a time-invariant, dynamic system. Ordinary operator or transfer-function methods for linear systems can then be used.

The procedure for analyzing a system can be described as follows: A block diagram of the system including all subsystems and all samplers is first drawn. The period h is determined. All samplers appearing in the system then have periods h/m, where m is an integer. A trick called *switch decomposition* is then used to convert samplers with rate h/m to a combination of samplers with period h. The system can then be analyzed using the methods described in Sec. 4.5.

Switch Decomposition. To understand the concept of switch decomposition, first consider a sampler with period $h/2$. Such a sampling can be obtained by combining a sampler with period h and another sampler with period h that is delayed $h/2$. The scheme is illustrated in Fig. 4.23(a). The idea can easily be extended to sampling at rate h/m, where m is an arbitrary integer [see Fig. 4.23(b)].

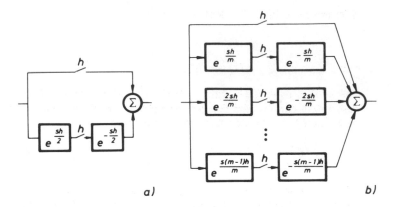

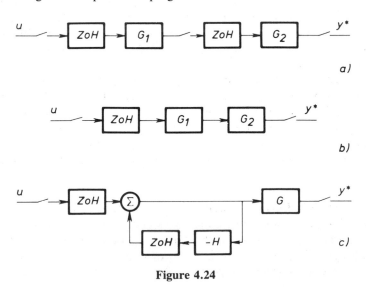

Figure 4.23 Representation of samplers with periods (a) $h/2$ and (b) h/m by switch decomposition.

Multirate Systems with Nonrational Periods

The methods described so far will work only when the ratios of the sampling periods are rational numbers. If this is not the case, it is not possible to obtain a periodic system; different techniques must then be used. The multirate techniques also lead to complicated analysis if there are many samplers with a wide range of periods.

4.7 Problems

4.1. Find Y^* for the systems in Fig. 4.24.

4.2. Write a program to compute the frequency response of a sampled-data system. Let the following be the input to the program:

Figure 4.24

(a) The polynomials in the pulse-transfer function $H(z)$.
(b) The sampling interval.
(c) The maximum and minimum frequencies.
Use the program to plot $H[\exp(i\omega h)]$ for the normalized motor sampled with a zero-order hold and compare with the continuous-time system.

4.8 References

The approach taken in this chapter corresponds to the classic treatment of sampled-data systems. The modulation model was proposed by MacColl (1945) and elaborated upon by Linvil (1951). A more detailed treatment is given in the classic texts by Ragazzini and Franklin (1958) and Jury (1958), which were mentioned in Sec. 1.6. The ideal-sampler approximation is discussed in

LI, Y. T., J. L. MEIRY, and R. E. CURRY (1972): "On the Ideal-Sampler Approximation," *IEEE Trans.*, AC-17, 167–68.

Frequency response is important from the point of view of both analysis and design. A fuller treatment of this problem is given in

LINDORFF, D. P. (1965): *Theory of Sampled-Data Control Systems.* New York: John Wiley.

Practical applications of frequency response analysis are discussed in

FLOWER, J. O., G. P. WINDETT, and S. C. FORGE (1971): "Aspects of the Frequency Response Testing of Simple Sampled Systems," *Int. Journal Control,* 14, 881–96.

More material on the z-transform is given in

JURY, E. I. (1964): *Theory and Application of the Z-Transform Method.* New York: John Wiley.

The modified z-transform is discussed in

JURY, E. I. (1958): *Sampled-Data Control Systems.* New York: John Wiley.

Tables of modified z-transforms are also given in that book. Systems with multirate sampling were first analyzed in

KRANC, G. M. (1957): "Input-Output Analysis of Multirate Feedback Systems," *IEEE Trans.,* AC-3, 21–28.

Additional results are given in

JURY, E. I. (1967): "A Note on Multirate Sampled-Data Systems," *IEEE Trans.* AC-12, 319–20.

JURY, E. I. (1967): "A General z-Transform Formula for Sampled-Data Systems," *IEEE Trans.,* AC-12, 606–08.

KONAR, A. F. and J. K. MAHESH (1978): "Analysis Methods for Multirate Digital Control Systems," *Honeywell Report No. F0636-TR1,* Honeywell Systems and Research Center. Minneapolis, Minn.

WHITBECK, R. F. (1980): "Multirate Digital Control Systems with Simulation Applications," *Report AFWAL-TR-80-3101,* vols. I, II, and III, Flight Dynamics Laboratory, Air Force Wright Aeronautical Lab., Wright-Patterson Air Force Base, Ohio.

CROCHIEVE, R. E. and L. R. RABINER (1983): *Multirate Digital Signal Processing.* Englewood Cliffs, N.J.: Prentice Hall, Inc.

ANALYSIS OF DISCRETE-TIME SYSTEMS

GOAL
────────────

To Introduce the Concepts of Stability, Reachability, and Observability. To Elaborate on Analysis of Simple Feedback Loops and Simulation.

5.1 Introduction

Previous chapters show how continuous-time systems are transformed when sampled. The concepts of stability, controllability, reachability, and observability, which are useful for understanding discrete-time systems, are discussed in Secs. 5.2 and 5.3. Simple feedback loops and their properties are treated in Sec. 5.4. Simulation is a very important tool for the analysis of sampled-data systems—for instance, in investigating intersample behavior.

Sampling of continuous-time systems is discussed in Chapter 3. The continuous-time system matrix A gives the discrete-time system matrix $\Phi = \exp Ah$. This chapter treats mainly discrete-time systems and uses the state-space model

$$x(k + 1) = \Phi x(k) + \Gamma u(k) \tag{5.1}$$
$$y(k) = Cx(k)$$

and the input-output model

$$A(q)y(k) = B(q)u(k) \tag{5.2}$$

where

$$A(q) = q^n + a_1 q^{n-1} + \cdots + a_n$$
$$B(q) = b_1 q^{n-1} + \cdots + b_n$$

5.2 Stability

Definitions

It is assumed that the notion of stability is known from basic texts in control theory. Only the basic definitions are given here. Consider the discrete-time, state-space equation (possibly nonlinear and time-varying)

$$x(k + 1) = f(x(k), k) \tag{5.3}$$

Let $x^0(k)$ and $x(k)$ be solutions of (5.3) when the initial conditions are $x^0(k_0)$ and $x(k_0)$, respectively. Further, let $\| \cdot \|$ denote a vector norm.

Definition 5.1—Stability. The solution $x^0(k)$ of (5.3) is *stable* if for given $\epsilon > 0$, there exists a $\delta(\epsilon, k_0) > 0$ such that all solutions with $\| x(k_0) - x^0(k_0) \| < \delta$ are such that $\| x(k) - x^0(k) \| < \epsilon$ for all $k \geq k_0$.

Definition 5.2—Asymptotic stability. The solution $x^0(k)$ of (5.3) is *asymptotically stable* if it is stable and if $\| x(k) - x^0(k) \| \to 0$ when $k \to \infty$ provided that $\| x(k_0) - x^0(k_0) \|$ is small enough

From the definitions, it follows that stability in general is defined for a particular *solution* and not for the system.

Stability of Linear Discrete-Time Systems

Consider the linear system

$$x^0(k + 1) = \Phi x^0(k), \qquad x^0(0) = a^0 \tag{5.4}$$

To investigate the stability of the solution of (5.4), the initial value is perturbed. Hence

$$x(k + 1) = \Phi x(k), \qquad x(0) = a$$

The difference $\tilde{x} = x - x^0$ satisfies the equation

$$\tilde{x}(k + 1) = \Phi \tilde{x}(k), \qquad \tilde{x}(0) = a - a^0 \tag{5.5}$$

This implies that if the solution x^0 is stable, then every other solution is also stable. For linear, time-invariant systems, stability is thus a property of the system and not of a special solution.

The system (5.5) has the solution

$$\tilde{x}(k) = \Phi^k \tilde{x}(0)$$

[See (3.14)]. If it is possible to diagonalize Φ, then the solution is a combination of terms λ_i^k, where λ_i, $i = 1, \ldots, n$, are the eigenvalues of Φ [See (3.30)]. In the general case, when Φ cannot be diagonalized, the solution is instead a linear combination of the terms $p_i(k)\lambda_i^k$, where $p_i(k)$ are polynomials in k of order one less than the multiplicity of the corresponding eigenvalue. To get asymptotic stability, all solutions must go to zero as k increases to infinity. The eigenvalues of Φ then have the property

$$|\lambda_i| < 1, \quad i = 1, \ldots, n$$

which is formulated as the following theorem.

Theorem 5.1—Asymptotic stability of linear systems. A discrete-time, linear, time-invariant system (5.4) is asymptotically stable if and only if all eigenvalues of Φ are strictly inside the unit disc. $\quad\square$

Stability with respect to disturbances in the initial value has already been defined. Other types of stability concepts are also of interest.

Definition 5.3—BIBO stability. A linear, time-invariant system is *bounded-input-bounded-output* (BIBO) stable if a bounded input gives a bounded output for every initial value.

From the definition it follows that asymptotic stability is the strongest concept. The following theorem is a result.

Theorem 5.2 Asymptotic stability implies stability and BIBO stability. $\quad\square$

When the word *stable* is used without further qualification in this text, it normally means asymptotic stability.

It is easy to give examples showing that stability does not imply BIBO stability, and vice versa.

Example 5.1

Consider the sampled harmonic oscillator (see Example A.3)

$$x(kh + h) = \begin{bmatrix} \cos \omega h & \sin \omega h \\ -\sin \omega h & \cos \omega h \end{bmatrix} x(kh) + \begin{bmatrix} 1 - \cos \omega h \\ \sin \omega h \end{bmatrix} u(kh)$$

$$y(kh) = [1 \quad 0]x(kh)$$

The magnitude of the eigenvalues is one. The system is stable because $\|x(kh)\| = \|x(0)\|$ if $u(kh) = 0$. Let the input be a square wave with the frequency ω rad/s. Figure 5.1 shows the input and output of the system. The output amplitude is growing, and the system is not BIBO stable.

Stability Tests

The following are some of the ways of determining the stability of a discrete-time system:

Direct computation of the eigenvalues of Φ.

Methods based on properties of characteristic polynomials.

The root-locus method.

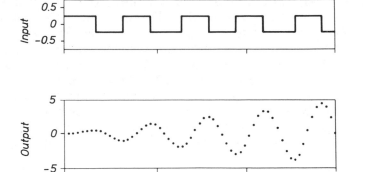

Figure 5.1 Input and output of the system in Example 5.1 when $\omega = 1$, $h = 0.5$, and the initial state is zero.

The Nyquist criterion.

Lyapunov's method.

It follows from Theorem 5.1 that a straightforward way to test the stability of a given system is to calculate the eigenvalues of the matrix Φ. There are good numerical algorithms for doing this. Well-established methods are available, for instance, in the package EISPACK, which is easily accessible in most computing centers. The routines are also included in packages like Matlab. A direct calculation of the eigenvalues cannot conveniently be done by hand for systems of order higher than two, nor can the eigenvalue method be used when the matrix has parameters in its coefficients.

In some cases it is easier to calculate the characteristic polynomial

$$A(z) = a_0 z^n + a_1 z^{n-1} + \cdots + a_n \tag{5.6}$$

and investigate the characteristic equation

$$A(z) = a_0 z^n + a_1 z^{n-1} + \cdots + a_n = 0 \tag{5.7}$$

Recall from Sec. 3.4 that the characteristic polynomial is the denominator polynomial of the pulse-transfer function. Stability tests can be obtained by investigating conditions for the zeros of a polynomial to be inside the unit disc.

The Routh-Hurwitz criterion is one method of determining if a polynomial has all its zeros in the left half-plane. The Möbius transformation

$$w = \frac{z + 1}{z - 1}$$

maps the unit disc in the z-plane to the left-half-w-plane. The transformation can be applied to (5.7) and the Routh-Hurwitz criterion can then be used.

It is, however, useful to have conditions that tell directly if a polynomial has all its zeros inside the unit disc. Such a criterion, which is the equivalent of the Routh-Hurwitz criterion, was developed by Schur, Cohn, and Jury. This test will be described in detail in the following section. The calculation of the coefficients of the characteristic polynomial from the elements of a matrix is poorly conditioned. If a matrix is given, it is therefore preferable to calculate the eigenvalues directly instead of calculating the characteristic equation.

The well-known root-locus method for continuous-time systems can be used for discrete-time systems also. The stability boundary is changed only from the imaginary axis to the unit circle. The rules of thumb for drawing the root locus are otherwise the same. The root locus method and the Nyquist criterion are used to determine the stability of the closed-loop system when the open-loop system is known.

Jury's Criterion

The following test is useful for determining if a polynomial (5.6) has all its zeros inside the unit disc.

Form the table

$$
\begin{array}{ccccc}
a_0 & a_1 & \cdots & a_{n-1} & a_n \\
a_n & a_{n-1} & \cdots & a_1 & a_0 \qquad \alpha_n = \dfrac{a_n}{a_0} \\
\hline
a_0^{n-1} & a_1^{n-1} & \cdots & a_{n-1}^{n-1} \\
a_{n-1}^{n-1} & a_{n-2}^{n-1} & \cdots & a_0^{n-1} \qquad \alpha_{n-1} = \dfrac{a_{n-1}^{n-1}}{a_0^{n-1}} \\
\hline
\vdots \\
\\
\\
\hline
a_0^0
\end{array}
$$

where

$$a_i^{k-1} = a_i^k - \alpha_k a_{k-i}^k$$

$$\alpha_k = a_k^k / a_0^k$$

The first and second rows are the coefficients in (5.6) in forward and reversed order, respectively. The third row is obtained by multiplying the second row by $\alpha_n = a_n/a_0$ and subtracting this from the first row. The last element in the third row is thus zero. The fourth row is the third row in reversed order. The scheme is then repeated until there are $2n + 1$ rows. The last row consists of only one element. The following theorem results.

Theorem 5.3—Jury's stability test. If $a_0 > 0$, then Equation (5.7) has all roots inside the unit disc if and only if all a_0^k, $k = 0, 1, \ldots , n - 1$ are positive. If no a_0^k is zero then the number of negative a_0^k is equal to the number of roots outside the unit disc. $\qquad \square$

Remark. If all a_0^k are positive for $k = 1, 2, \ldots, n - 1$, then the condition $a_0^0 > 0$ can be shown to be equivalent to the conditions

$$A(1) > 0$$

$$(-1)^n A(-1) > 0$$

These conditions constitute necessary conditions for stability and hence can be used before forming the table.

Example 5.2

Let the characteristic equation be

$$A(z) = z^2 + a_1 z + a_2 = 0 \tag{5.8}$$

Jury's scheme is

$$
\begin{array}{ccccc}
1 & & a_1 & a_2 & \\
a_2 & & a_1 & 1 & \alpha_2 = a_2 \\
\hline
1 - a_2^2 & & a_1(1 - a_2) & & \\
a_1(1 - a_2) & & 1 - a_2^2 & & \alpha_1 = \dfrac{a_1}{1 + a_2} \\
\hline
\end{array}
$$

$$1 - a_2^2 - \dfrac{a_1^2(1 - a_2)}{1 + a_2}$$

All the roots of Equation (5.8) are inside the unit circle if

$$1 - a_2^2 > 0$$

$$\frac{1 - a_2}{1 + a_2}[(1 + a_2)^2 - a_1^2] > 0$$

This gives the conditions

$$a_2 < 1$$

$$a_2 > -1 + a_1$$

$$a_2 > -1 - a_1$$

The stability area for the second-order equation is shown in Fig. 5.2.

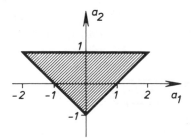

Figure 5.2 The stability area for the second-order equation (5.8) as a function of the coefficients a_1 and a_2.

The time-delay process in Example A.4 and Example 3.6 has the characteristic equation

$$z^2 - z \exp(-h) = 0$$

when the sampling period is h. The system is stable because $\exp(-h) < 1$. The inventory model in Example A.5 has the characteristic equation

$$z^2 - z = 0$$

This system is stable but not asymptotically stable because the characteristic equation has the root $z = 1$. $\qquad\square$

The Nyquist Criterion

The Nyquist criterion is a well-known stability test for continuous-time systems. It is based on the principle of arguments. The test can easily be reformulated to handle discrete-time systems. The Nyquist criterion is especially useful for determining the stability of the closed-loop system when the open-loop system is given.

Consider the discrete-time system in Fig. 5.3. The closed-loop system has the pulse-transfer function

$$H(z) = \frac{Y(z)}{U_c(z)} = \frac{H_0(z)}{1 + H_0(z)}$$

The characteristic equation of the closed-loop system is

$$1 + H_0(z) = 0 \tag{5.9}$$

The stability of the closed-loop system can be investigated from the Nyquist plot of $H_0(z)$. For discrete-time systems, the stability area in the z-plane is the unit disc instead of the left half-plane. Figure 5.4 shows the path Γ_c encircling the area outside the unit disc. The small indentation at $z = 1$ is to exclude the integrators in the open-loop system. The mapping of the infinitesimal semicircle at $z = 1$ with decreasing arguments from $\pi/2$ to $-\pi/2$ is mapped into the $H_0(z)$-plane as an infinitely large circle from $-n\pi/2$ to $n\pi/2$, where n is the number of integrators in the open-loop system. If there are poles on the unit circle other than for $z = 1$, those have to be excluded with small semicircles in the same way as for $z = 1$. The map of the unit circle is $H_0(e^{i\omega h})$ with $\omega h \in (0, 2\pi)$.

The stability of the closed-loop system can now be determined by investigating how the path Γ_c is mapped by $H_0(z)$. The map $H_0(e^{i\omega h})$ with arguments from 0 to π is called the *frequency-response curve*, or the *Nyquist curve*, of the

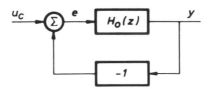

Figure 5.3 A simple unit-feedback system.

Analysis of Discrete-Time Systems   Chap. 5

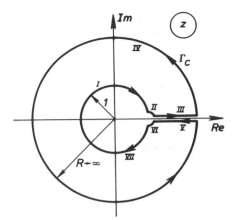

Figure 5.4 The path Γ_c encircling the area outside the unit disc.

system. The principle of arguments states that the number of encirclements N in the positive direction around $(-1, 0)$ by the map of Γ_c is equal to

$$N = Z - P$$

where Z and P are the number of zeros and poles, respectively, of $1 + H_0(z)$ outside the unit disc. Notice that if the open-loop system is stable, then $P = 0$ and thus $N = Z$. The stability of the closed-loop system is then ensured if the map of Γ_c does not encircle the point $(-1, 0)$. If $H(z) \to 0$ when $z \to \infty$ the parallel lines III and V do not influence the stability test, and it is sufficient to find the map of the unit circle and the small semicircle at $z = 1$. The Nyquist criterion can be simplified further if the open-loop system and its inverse are stable. Stability of the closed-loop system is then ensured if the point $(-1, 0)$ in the $H_0(z)$-plane is to the left of the map of $H_0(e^{i\omega h})$ for $\omega h = 0$ to π—i.e., to the left of the Nyquist curve.

Example 5.3

Consider a system with sampling period $h = 1$ and the pulse-transfer function

$$H_0(z) = \frac{0.25K}{(z - 1)(z - 0.5)}$$

Then

$$H_0(e^{i\omega}) = \frac{0.25K[1.5(1 - \cos \omega) - 2 \sin^2 \omega - i \sin \omega(2 \cos \omega - 1.5)]}{(2 - 2 \cos \omega)(1.25 + \cos \omega)}$$

The map of Γ_c is shown in Fig. 5.5. The full line is the Nyquist curve, i.e., the map of $H_0(e^{i\omega})$ for $\omega = 0$ to π. From the figure it can be found that the Nyquist curve crosses the negative real axis at -0.5. The closed-loop sytem is thus stable if $K < 2$. □

The Nyquist criterion has so far been considered only for general discrete-time systems. It will now be assumed that the pulse-transfer function is obtained by sampling a continuous-time system $G(s)$, preceded by a zero-order hold with

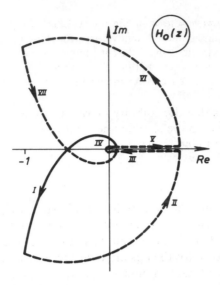

Figure 5.5 The map of Γ_c into the $H_0(z)$-plane of the system in Example 5.3, when $K = 1$. The full line is the Nyquist curve.

the sampling period h. An alternative viewpoint is obtained from (4.14) and (4.17) because

$$H_0(e^{i\omega h}) = \frac{1}{h} \sum_{k=-\infty}^{\infty} F(i\omega + ik\omega_s)$$

where

$$F(s) = \frac{1 - \exp(-hs)}{s} G(s)$$

If numerical values of $G(i\omega)$ are available, then the value of the pulse-transfer function on the unit circle can be determined using the series expansion shown. In most applications, $|G(i\omega)|$ decreases with increasing ω. The term for $k = 0$ will then dominate. This is especially true if an effective antialiasing filter is included in $G(s)$. This implies that it is sufficient to have good knowledge of $G(s)$ up to the Nyquist frequency.

The drawback of the Nyquist criterion is the difficulty of drawing the Nyquist curve, but computer programs can be used to draw the curve and make it easier to evaluate the stability of the closed-loop system.

Example 5.4

Consider the continuous-time system

$$G(s) = \frac{1}{s^2 + 1.4s + 1} \tag{5.10}$$

Zero-order hold sampling of the system gives the discrete-time system

$$H(z) = \frac{b_1 z + b_2}{z^2 + a_1 z + a_2}$$

Analysis of Discrete-Time Systems   Chap. 5

where a_i and b_i depend on the sampling interval. The frequency curve is given by $H(e^{i\omega h})$. Figure 5.6 shows the frequency curve for the continuous-time system and for the discrete-time system using different sampling intervals. The continuous-time system has an infinite amplitude margin. The amplitude margin is finite for the sampled data systems and decreases with increasing sampling period. □

The example shows that the stability margin decreases with increasing sampling intervals. This is due to the sample and hold which can be approximated as a time delay of half a sampling interval.

Robustness

It is very interesting to find the sensitivity of a system to perturbations, which may be introduced because of component tolerances. Because the designs of control systems are based on simplified models, it is also interesting to know how accurate the model has to be for the design to be successful. The Nyquist theorem can give good insight into these problems.

Consider the simple closed-loop system in Fig. 5.3. Let the true open-loop pulse-transfer function be $H^0(z)$ and let $H(z)$ be the nominal value of H^0. The closeness of H to H^0 needed to make the closed-loop system stable is of concern. The pulse-transfer function of the closed-loop system is

$$H_c(z) = \frac{H^0(z)}{1 + H^0(z)} \tag{5.11}$$

The poles of the closed-loop system are thus the zeros of the function

$$f(z) = 1 + H^0(z) = 1 + H(z) + H^0(z) - H(z)$$

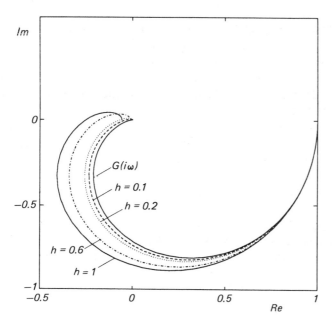

Figure 5.6 The frequency function of (5.10) and for (5.10) sampled with zero-order hold when $h = 0.1, 0.2, 0.6, 1$.

If

$$|H^0(z) - H(z)| < |1 + H(z)| \qquad (5.12)$$

on the unit circle, then it follows from the principle of variation of the argument that the differences between the number of poles and zeros outside the unit disc for the functions $1 + H$ and $1 + H^0$ are the same. The following theorem results:

Theorem 5.4. Consider the closed-loop systems S and S^0 obtained by applying unit negative feedback around systems with pulse-transfer functions H and H^0, respectively. The system S^0 is stable if the following conditions are true:

1. S is stable.
2. H and H^0 have the same number of poles outside the unit disc.
3. The inequality (5.12) is fulfilled for $|z| = 1$. □

The result shows that it is important to know the number of unstable modes in order to design a regulator for the system. The inequality also gives the frequency range in which it is important to have a good description of the process. Notice in particular that the precision requirements are very modest for the frequencies where the loop gain is large. Good precision is needed for frequencies where $H^0(z) \approx -1$.

A closely related result that gives additional insight is obtained as follows. The pulse-transfer function of the closed-loop system given in (5.11) can also be written as

$$H_c = \frac{H^0}{1 + 1/H^0}$$

The poles of the closed-loop system are thus given by the zeros of the function

$$f_c(z) = 1 + \frac{1}{H^0(z)} = 1 + \frac{1}{H(z)} + \left[\frac{1}{H^0(z)} - \frac{1}{H(z)}\right]$$

It follows from the principle of variation of the argument that the differences between the zeros and poles outside the unit disc of the functions $1 + 1/H^0$ and $1 + 1/H$ are the same if

$$\left|\frac{1}{H^0(z)} - \frac{1}{H(z)}\right| < \left|1 + \frac{1}{H(z)}\right| \qquad (5.13)$$

on the unit circle. The following result is thus obtained.

Theorem 5.5. Consider the closed-loop systems S and S^0 obtained by applying unit negative feedback around systems with the pulse-transfer functions H and H^0, respectively. The system S^0 is stable if the following conditions are true:

1. S is stable.

Analysis of Discrete-Time Systems   Chap. 5

2. H and H^0 have the same number of zeros outside the unit disc.

3. The inequality (5.13) is fulfilled for $|z| = 1$. □

The theorem indicates the importance of knowing the number of zeros outside the unit disc. The theorem shows that stbility can be maintained in spite of large differences between H and H^0 provided that the loop gain is large.

From the conclusions of Theorem 5.4 and 5.5, these rules are obtained for design of a feedback system based on approximate or uncertain models.

It is important to know the number of unstable poles and zeros.

It is not important to know the model precisely for those frequencies for which the loop gain can be made large.

It is necessary to make the loop gain small for those frequencies for which the model is uncertain.

It is necessary to have a model that describes the system precisely for those frequencies for which $H^0(z) \approx -1$.

Lyapunov's Second Method

Lyapunov's second method is a useful tool for determining the stability of nonlinear dynamic systems. Lyapunov developed the theory for differential equations, but a corresponding theory can also be derived for difference equations.

Definition 5.4—Lyapunov function. $V(x)$ is a *Lyapunov function* for the system

$$x(k + 1) = f(x(k)), \qquad f(0) = 0 \tag{5.14}$$

if:

1. $V(x)$ is continuous in x and $V(0) = 0$.
2. $V(x)$ is positive definite.
3. $\Delta V(x) = V(f(x)) - V(x)$ is negative definite.

A simple geometric illustration of the definition is given in Fig. 5.7. The level curves of a positive definite continuous function V are closed curves in the neighborhood of the origin. Let each curve be labeled by the value of the function. Condition 3 implies that the dynamics of the system are such that the solution

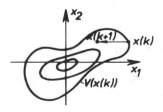

Figure 5.7 Geometric illustration of Lyapunov's theorem.

always moves toward curves with lower values. All level curves encircle the origin and do not intersect any other level curve. It thus seems reasonable that the existence of a Lyapunov function ensures asymptotic stability. The following theorem is a precise statement of this fact.

Theorem 5.6—Stability theorem of Lyapunov. The solution $x(k) = 0$ is asymptotically stable if there exists a Lyapunov function to the system (5.14). Further, if

$$0 < \varphi(\| x \|) < V(x)$$

where $\varphi(\| x \|) \to \infty$ as $\| x \| \to \infty$, then the solution is asymptotically stable for all initial conditions. ☐

The main obstacle to using the Lyapunov theory is finding a suitable Lyapunov function. This is in general a difficult problem; however, for the linear system of (5.4), it is straightforward to determine quadratic Lyapunov functions. Take $V(x) = x^T P x$ as a candidate for a Lyapunov function. The increment of V is then given by

$$\Delta V(x) = V(\Phi x) - V(x) = x^T \Phi^T P \Phi x - x^T P x$$

$$= x^T [\Phi^T P \Phi - P] x = -x^T Q x$$

For V to be a Lyapunov function, it is thus necessary and sufficient that there exists a positive definite matrix P that satisfies the equation

$$\Phi^T P \Phi - P = -Q \qquad (5.15)$$

where Q is positive definite. Equation (5.15) is called the *Lyapunov equation*. It can be shown that there is always a solution to the Lyapunov equation when the linear system is stable. The matrix P is positive definite if Q is positive definite.

5.3 Controllability, Reachability, and Observability

In this section two fundamental questions for dynamic systems are discussed. The first is whether it is possible to steer a system from a given initial state to any other state. The second is how to determine the state of a dynamic system from observations of inputs and outputs. These questions were posed and answered by Kalman (1960), who also introduced the concepts of controllability and observability. The systems are allowed to be multiple-input-multiple-output systems.

Controllability and Reachability

Consider the system of (5.1). Assume that the initial state $x(0)$ is given. The state at time n, where n is the order of the system, is given by

$$x(n) = \Phi^n x(0) + \Phi^{n-1}\Gamma u(0) + \cdots + \Gamma u(n-1) \qquad (5.16)$$
$$= \Phi^n x(0) + W_c U$$

where

$$W_c = [\Gamma \quad \Phi\Gamma \quad \ldots \quad \Phi^{n-1}\Gamma]$$
$$U = [u^T(n-1) \quad \ldots \quad u^T(0)]^T$$

[Compare with Equation (3.14).] If W_c has rank n, then it is possible to find n equations from which the control signals can be found such that the initial state is transferred to the desired final state $x(n)$. Notice that the solution is not unique if there is more than one input signal. In the literature, controllability is defined in different ways; the following definition will be used in this text:

Definition 5.5—Controllability. The system (5.1) is controllable if it is possible to find a control sequence such that the origin can be reached from any initial state in finite time.

A concept related to controllability is *reachability*, which is defined as follows.

Definition 5.6—Reachability. The system (5.1) is reachable if it is possible to find a control sequence such that an arbitrary state can be reached from any initial state in finite time.

Controllability does not imply reachability, which is easily seen from (5.16). If $\Phi^n x(0) = 0$, then the origin will be reached with zero input but the system is not necessarily reachable. The two concepts are, however, equivalent if Φ is invertible. Reachability will be discussed here, primarily.

The following theorem follows from the preceding definition and calculations.

Theorem 5.7. The system (5.1) is reachable if and only if the matrix W_c has rank n. $\qquad\square$

Remark. The matrix W_c is usually referred to as the controllability matrix because of its analogy with continuous-time systems.

Example 5.5

The system

$$x(k+1) = \begin{bmatrix} 1 & 0 \\ 0 & 1 \end{bmatrix} x(k) + \begin{bmatrix} 1 \\ 1 \end{bmatrix} u(k)$$

is not reachable because

$$W_c = \begin{bmatrix} 1 & 1 \\ 1 & 1 \end{bmatrix}$$

If instead there were two inputs with a nonsingular Γ-matrix, then this system would be reachable. □

By the Cayley-Hamilton theorem it is found from (5.16) that all states that can be reached from the origin are spanned by the columns of the controllability matrix W_c. This implies that the reachable states belong to the linear subspace spanned by the columns of W_c.

Example 5.6

Given the system

$$x(k + 1) = \begin{bmatrix} 1 & 1 \\ -0.25 & 0 \end{bmatrix} x(k) + \begin{bmatrix} 1 \\ -0.5 \end{bmatrix} u(k); \quad x(0) = \begin{bmatrix} 2 \\ 2 \end{bmatrix}$$

is it possible to find a control sequence such that $x^T(2) = [-0.5 \quad 1]$? From (5.16),

$$x(2) = \Phi^2 x(0) + \Phi\Gamma u(0) + \Gamma u(1)$$

or

$$\begin{bmatrix} -0.5 \\ 1 \end{bmatrix} = \begin{bmatrix} 3.5 \\ -1 \end{bmatrix} + \begin{bmatrix} 1 \\ -0.5 \end{bmatrix} [0.5u(0) + u(1)]$$

which gives the condition

$$0.5u(0) + u(1) = -4$$

One possible sequence of controls is $u(0) = -2$ and $u(1) = -3$. Assume instead that $x^T(2) = [0.5 \quad 1]$. This gives the system of equations

$$\begin{bmatrix} -3 \\ 2 \end{bmatrix} = \begin{bmatrix} 1 \\ -0.5 \end{bmatrix} [0.5u(0) + u(1)]$$

which does not have a solution. The reason, of course, is that the system is not reachable. The controllability matrix is

$$W_c = \begin{bmatrix} 1 & 0.5 \\ -0.5 & -0.25 \end{bmatrix}$$

Starting at the origin it is possible to reach only those points of the state space that belong to the subspace spanned by the vector $[1 \quad 0.5]^T$. In the example, it is possible to reach other points due to the effect of the initial value. □

Assume that new coordinates are introduced by a nonsingular transformation matrix T (compare with Sec. 3.3). In the new coordinates

$$\tilde{W}_c = [\tilde{\Gamma} \quad \tilde{\Phi}\tilde{\Gamma} \quad \dots \quad \tilde{\Phi}^{n-1}\tilde{\Gamma}]$$

$$= [T\Gamma \quad T\Phi T^{-1}Y\Gamma \quad \dots \quad T\Phi^{n-1}T^{-1}T\Gamma] \quad (5.17)$$

$$= TW_c$$

If W_c has rank n, then $\tilde{W}_c$ also has rank n. This means that the reachability of a system is independent of the coordinates.

Notice that the formula relating $\tilde{W}_c$ and W_c is useful for computing the transformation matrix from one form to another. Compare (3.34).

The controllable form, introduced in Sec. 3.3, can be obtained if the matrix in (3.32) has an inverse, which explains the name *controllable canonical form*.

Example 5.7

Consider the third-order system

$$x(k + 1) = \begin{bmatrix} -a_1 & -a_2 & -a_3 \\ 1 & 0 & 0 \\ 0 & 1 & 0 \end{bmatrix} x(k) + \begin{bmatrix} 1 \\ 0 \\ 0 \end{bmatrix} u(k)$$

which is in controllable form. The controllability matrix is

$$W_c = [\Gamma \quad \Phi\Gamma \quad \Phi^2\Gamma] = \begin{bmatrix} 1 & -a_1 & a_1^2 - a_2 \\ 0 & 1 & -a_1 \\ 0 & 0 & 1 \end{bmatrix}$$

The inverse is given by

$$W_c^{-1} = \begin{bmatrix} 1 & a_1 & a_2 \\ 0 & 1 & a_1 \\ 0 & 0 & 1 \end{bmatrix}$$

The example can be generalized to the nth-order case, where

$$W_c^{-1} = \begin{bmatrix} 1 & a_1 & a_2 & \cdots & a_{n-2} & a_{n-1} \\ 0 & 1 & a_1 & \cdots & a_{n-3} & a_{n-2} \\ \vdots & & \ddots & \ddots & \vdots & \vdots \\ 0 & 0 & & \cdots & 1 & a_1 \\ 0 & 0 & & \cdots & 0 & 1 \end{bmatrix} \qquad \square$$

Trajectory Following

From the preceding definitions and calculations, it is possible to determine a control sequence such that a desired state can be reached after at most n steps of time. Does reachability also imply that it is possible to follow a given trajectory in the state space? Assume that any $x(k)$ is given and that it is necessary to get to $x(k + 1)$. From (5.16) it can be seen that this is possible only if Γ has rank n; i.e., it is necessary but not sufficient to have n input signals. For a single-input-single-output system it is, in general, possible to reach desired states only at each nth sample point, provided that the desired points are known n steps ahead.

It is easier to make the output follow a given trajectory. Assume that the trajectory is given by $u_c(k)$. The control signal u should then satisfy

$$y(k) = \frac{B(q)}{A(q)} u(k) = u_c(k)$$

or

$$u(k) = \frac{A(q)}{B(q)} u_c(k) \tag{5.18}$$

Assume that there are d steps of delay in the system. The generation of $u(k)$ is then causal only if the desired trajectory is known d steps ahead. The control signal can then be generated in real time. The control signal is thus obtained by sending the desired output trajectory through the inverse system A/B. Equation (5.18) has a unique solution if the signal $u_c(k)$ is such that there exists a k_0 such that $u(k) = 0$ for all $k < k_0$ (compare with Section 3.4). The signal u is bounded if u_c is bounded and if the system has a stable inverse.

Observability

To solve the problem of finding the state of a system from observations of the output the concept of unobservable states is introduced.

Definition 5.7. The state $x^0 \neq 0$ is *unobservable* if there exists a finite $k_1 \geq n - 1$ such that $y(k) = 0$ for $0 \leq k \leq k_1$ when $x(0) = x^0$ and $u(k) = 0$ for $0 \leq k \leq k_1$.

The system in (5.1) is *observable* if there is a finite k such that knowledge of the inputs $u(0), \ldots , u(k-1)$ and the outputs $y(0), \ldots , y(k-1)$ is sufficient to determine the initial state of the system.

Consider the system in (5.1). The effect of the known input signal can always be determined, and there is no loss of generality to assume that $u(k) = 0$. Assume that $y(0), y(1), \ldots , y(n-1)$ are given. This gives the following set of equations:

$$y(0) = Cx(0)$$

$$y(1) = Cx(1) = C\Phi x(0)$$

$$\vdots$$

$$y(n-1) = C\Phi^{n-1}x(0)$$

Using vector notation gives

$$\begin{bmatrix} C \\ C\Phi \\ \vdots \\ C\Phi^{n-1} \end{bmatrix} x(0) = \begin{bmatrix} y(0) \\ y(1) \\ \vdots \\ y(n-1) \end{bmatrix} \tag{5.19}$$

The state $x(0)$ can be obtained from (5.19) if and only if the *observability matrix*

$$W_o = \begin{bmatrix} C \\ C\Phi \\ \vdots \\ C\Phi^{n-1} \end{bmatrix} \tag{5.20}$$

has rank n. The state $x(0)$ is unobservable if $x(0)$ is in the null space of W_o. If two states are unobservable, then any linear combination is also unobservable; that is, the unobservable states form a linear subspace.

Theorem 5.8. The system (5.1) is observable if and only if W_o has rank n.

□

The test of observability given by Theorem 5.8 is equivalent to that for continuous-time systems. It is straightforward to show that the observability matrix is independent of the coordinates in the same way as in the controllability matrix.

Example 5.8

Consider the system

$$x(k + 1) = \begin{bmatrix} 1.1 & -0.3 \\ 1 & 0 \end{bmatrix} x(k)$$

$$y(k) = [1 \ -0.5]x(k)$$

The observability matrix is

$$W_o = \begin{bmatrix} C \\ C\Phi \end{bmatrix} = \begin{bmatrix} 1 & -0.5 \\ 0.6 & -0.3 \end{bmatrix}$$

The rank of W_o is 1, and the unobservable states belong to the null space of W_o, i.e., $[0.5 \ 1]$. Figure 5.8 shows the output for four different initial states. All initial states that lie on a line parallel to $[0.5 \ 1]$ give the same output [see Fig. 5.8(b) and (d)].

□

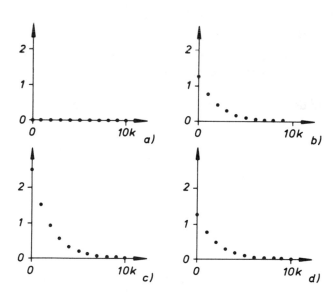

Figure 5.8 The output of the system in Example 5.8 for the initial states (a) $[0.5 \ 1]^T$ (b) $[1.5 \ 0.5]^T$, (c) $[2.5 \ 0]^T$, and (d) $[1 \ -0.5]^T$.

Kalman's Decomposition

The reachable and the unobservable parts of a system are two linear subspaces of the state space. Both subspaces are independent of the coordinates in the state space. Kalman showed that it is possible to introduce coordinates such that a system can be partitioned in the following way:

$$x(k + 1) = \begin{bmatrix} \Phi_{11} & \Phi_{12} & 0 & 0 \\ 0 & \Phi_{22} & 0 & 0 \\ \Phi_{31} & \Phi_{32} & \Phi_{33} & \Phi_{34} \\ 0 & \Phi_{42} & 0 & \Phi_{44} \end{bmatrix} x(k) + \begin{bmatrix} \Gamma_1 \\ 0 \\ \Gamma_3 \\ 0 \end{bmatrix} u(k)$$

$$y(k) = [C_1 \quad C_2 \quad 0 \quad 0]x(k)$$

where Φ_{ij}, Γ_i, and C_i are matrices of suitable orders. The state space is partitioned into four parts, which correspond to states that are reachable and observable, not reachable but observable, reachable and not observable, and neither reachable nor observable.

By simple algebraic manipulations, the pulse-transfer operator is given by

$$H(q) = C_1(qI - \Phi_{11})^{-1}\Gamma_1$$

The pulse-transfer operator is thus determined by the reachable *and* observable part of the system. The following theorem summarizes these results.

Theorem 5.9—Kalman's decomposition.　A linear system can be partitioned into four subsystems with the following properties:

S_{or}　Observable and reachable subsystem.

$S_{o\bar{r}}$　Observable but not reachable subsystem.

$S_{\bar{o}r}$　Not observable but reachable subsystem.

$S_{\bar{o}\bar{r}}$　Neither observable nor reachable subsystem.

Further, the pulse-transfer function of the system is uniquely determined by the subsystem that is both observable and reachable.　　□

A block diagram for the decomposition is given in Fig. 5.9, which shows

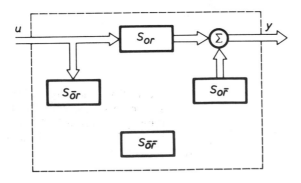

Figure 5.9 Block diagram of the Kalman decomposition.

how the subsystems are interconnected. The figure also shows that the input-output relationship is given only by the subsystem S_{or}.

Loss of Reachability and Observability Through Sampling

Sampling of a continuous-time system gives a discrete-time system with system matrices that depend on the sampling period. How will that influence the reachability and observability of the sample system? To get a reachable, discrete-time system, it is necessary that the continuous-time system also be reachable, because the allowable control signals for the sampled system—piecewise-constant signals—are a subset of the allowable control signals for the continuous-time system. However, it may happen that the reachability is lost for some sampling periods.

The conditions for unobservability are more restricted in the continuous-time case because the output has to be zero over a time interval, while the sampled-data output has to be zero only at the sampling instants. This means that the continuous output may oscillate between the sampling times and be zero at the sampling times. This condition is sometimes called *hidden oscillation*. The sampled-data system can thus be unobservable even if the corresponding continuous-time system is observable.

The harmonic oscillator can be used to illustrate the preceding discussion.

Example 5.9

The discrete-time model of the harmonic oscillator is given by (see Example A.3)

$$x(kh + h) = \begin{bmatrix} \cos \omega h & \sin \omega h \\ -\sin \omega h & \cos \omega h \end{bmatrix} x(kh) + \begin{bmatrix} 1 - \cos \omega h \\ \sin \omega h \end{bmatrix} u(kh)$$

$$y(kh) = [1 \quad 0]x(kh)$$

The determinants of the controllability and observability matrices are

$$\det W_c = -2 \sin \omega h(1 - \cos \omega h)$$

and

$$\det W_o = \sin \omega h$$

Both reachability and observability are lost for $\omega h = n\pi$, although the corresponding continuous-time system given by (A.7) is both controllable and observable. $\Box$

The example shows one obvious way to lose observability and/or reachability. If the sampling period is half the period time (or a multiple thereof) of the natural frequency of the system, then this frequency will not be seen in the output.

The rules of thumb for the choice of the sampling period given in Chapter 3 are such that this situation should not occur. The rules imply about twenty samples per period, not two.

Observability and/or reachability are lost when the pulse-transfer operator has common poles and zeros. The poles and zeros are functions of the sampling interval. This implies that there will be common factors only for isolated values of the sampling period. A change in sampling period will make the system observable and/or reachable again.

5.4 Analysis of Simple Feedback Loops

In this section the effect of feedback on stability, transient, and steady-state behavior is seen and simple feedback systems, as in Fig. 5.3, are discussed. Several advantages are obtained by using feedback in continous-time as well as in discrete-time systems. Feedback can, for instance, do the following:

Improve the transient behavior of the system.

Decrease the sensitivity to parameter changes in the open-loop system.

Eliminate steady-state errors if there are enough integrators in the open-loop system.

The stability of closed-loop systems can be investigated using the tools given in Sec. 5.2. The root-locus method is a suitable tool for analyzing simple feedback loops. Because feedback will change the poles of the system, it is important to understand the coupling between the discrete-time poles and the transient behavior of the system. This is treated in Sec. 3.6.

Steady-State Values

When analyzing control systems, it is important to calculate steady-state values of the output and of the error of the system. Assume a simple feedback system, as shown in Fig. 5.3. To generalize, it can be assumed that -1 in the feedback path is replaced by $-H(q)$. The error $e(k)$ is then given by

$$e(k) = \frac{1}{[1 + H_0(q)H(q)]} u_c(k) \tag{5.21}$$

The final-value theorem (Sec. 3.5, Table 3.2) can be used to calculate the steady-state value of $e(k)$, provided that it exists. If the input signal is a step, the steady-state error can be calculated simply by putting $q = 1$ in (5.21).

The number of integrators in the open-loop system determines the class of reference values that can be followed without steady-state errors. If the open-loop system has p integrators, then the error will be zero in steady state (provided that the closed-loop system is asymptotically stable) for reference signals that are polynomials in k of order less than or equal to $p - 1$.

Example 5.10

Consider the system

$$y(k) = H_0(q)u(k) = \frac{q - 0.5}{(q - 0.8)(q - 1)} u(k)$$

Closing the system, as in Fig. 5.3, gives

$$e(k) = \frac{(q - 0.8)(q - 1)}{(q - 0.8)(q - 1) + q - 0.5} u_c(k)$$

Assume that u_c is a unit step. Because the closed-loop system is stable, the final-value theorem can be used to show that the steady-state error is zero. This can be seen simply by putting $q = 1$. Another way is to observe that the open-loop system contains one integrator, i.e., a pole in $+1$.

If u_c is a unit ramp, use Table 3.3 in Sec. 3.5 to find the z-transform of the ramp. The steady-state error is then given by

$$\lim_{k \to \infty} e(k) = \lim_{z \to 1} \frac{(z - 0.8)(z - 1)}{(z - 0.8)(z - 1) + z - 0.5} \frac{z(1 - z^{-1})}{(z - 1)^2} = 0.4 \qquad \Box$$

To streamline the notation, it is often convenient to regard the input signals, or disturbances, as being generated by a dynamic system (see Fig. 5.10). It is assumed that the input to the dynamic system is a unit pulse δ_k, i.e.,

$$u_c(k) = H_c(q)\delta_k$$

In order to generate a step, use $H_c(q) = q/(q - 1)$; to generate a ramp, use $H_c(q) = q/(q - 1)^2$. The final value is then easily obtained from the final-value theorem.

Simulation

Simulation is a good way to investigate the behavior of dynamic systems—for example, the intersample behavior of computer-controlled systems. Computer simulation is a very good tool, but it should be remembered that simulation and analysis have to be used together. When making simulations, it is not always possible to investigate all combinations that are unfavorable, for instance, from the point of view of stability, observability, or reachability. These cases can be found through analysis.

It is important that the simulation program be so simple to use that the person primarily interested in the results can be involved in the simulation and in the evaluation of the simulation results.

In the beginning of the 1960s several digital simulation packages were developed. These packages were basically a digital implementation of analog simulation. The programming was done using block diagrams and fixed-operation modules. Later programs were developed in which the models were given directly as equations.

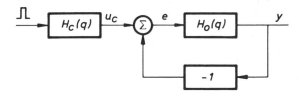

Figure 5.10 Generation of the reference value using a dynamic system with a pulse input.

It is important to have a good user-machine communication for simulations; the user should be able to change parameters and modify the model easily. Use of time-sharing operating systems makes it is possible to implement *interactive simulation programs,* which means that the user interacts with the computer on-line and decides the next step based on the results obtained so far. One way to implement interaction is to let the computer ask questions and the user select from predefined answers. This is called *menu-driven interaction.* Another possibility is *command-driven interaction,* which is like a high-level problem-solving language in which the user can choose freely from all commands available in the system. This is also a more flexible way of communicating with the computer, and it is very efficient for the experienced user, while a menu-driven program is easier to use for an inexperienced user.

In a simulation package, it is also important to have a flexible way of presenting the results, which are often curves. Finally, to be able to solve the type of problems of interest in this book, it is important to be able to mix continuous- and discrete-time systems.

Simnon is an interactive, command-driven simulation package developed at the Department of Automatic Control, Lund Institute of Technology. Most of the examples in this book have been generated using Simnon. A summary of the features of Simnon is found in Appendix C. (Also see the list of references.)

A simulation in Simnon is built up as a block diagram, where the blocks represent different subsystems. The blocks can be linear or nonlinear, time-continuous or discrete-time. The blocks are defined by their inputs and outputs. The relation between the blocks is specified in a connecting system.

The subsystems in Simnon are representations of the state equations for the subsystem. The state equations are given as systems of first-order differential of difference equations. The system can also contain parameters that can be changed by simple commands. The power of Simnon is illustrated below.

Control of the Double Integrator

The double integrator (Example A.1) will be used as the main example to show how the closed-loop behavior is changed with different controllers. The pulse-transfer operator of the double integrator for the sampling period $h = 1$ is

$$H_0(q) = \frac{0.5(q + 1)}{(q - 1)^2} \tag{5.22}$$

Assume that the purpose of the control is to make the output follow changes in the reference value. Also assume that the process is controlled by a computer using proportional feedback, i.e.,

$$u(k) = K[u_c(k) - y(k)] = Ke(k), \qquad K > 0 \tag{5.23}$$

where u_c is the reference value. The characteristic equation of the closed-loop system is

$$(q - 1)^2 + 0.5K(q + 1) = 0 \tag{5.24}$$

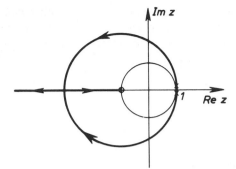

Figure 5.11 The root locus of (5.24) when $K > 0$.

A sketch of the root locus (see Fig. 5.11) shows that the closed-loop system is unstable for all values of the gain K. This can be seen from Jury's stability test also (compare with Example 5.2).

To get a stable system, the controller must be modified. It is known from continuous-time synthesis that derivative action improves stability, so proportional and derivative feedback can be tried for the discrete-time system; i.e.,

$$u(k) = K[e(k) - T_D \dot{y}(k)] \tag{5.25}$$

(see Fig. 5.12). The velocity $\dot{y}$ is thus also sampled and fed back. To find the input-output model of the closed-loop system with the controller (5.25), observe that

$$\frac{d\dot{y}}{dt} = u$$

Because u is constant over the sampling intervals,

$$\dot{y}(k + 1) - \dot{y}(k) = u(k)$$

or

$$\dot{y}(k) = \frac{1}{q - 1} u(k) \tag{5.26}$$

Equations (5.22), (5.25), and (5.26) give the closed-loop system

$$y(k) = \frac{0.5K(q + 1)}{(q - 1)(q - 1 + T_D K) + 0.5K(q + 1)} u_c(k) \tag{5.27}$$

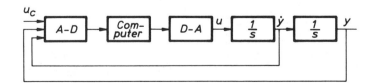

Figure 5.12 Discrete-time controller with feedback from position and velocity of the double integrator.

The system is of second order, and there are two free parameters, K and T_D, that can be used to select the closed-loop poles. The closed-loop system is stable if $K > 0$, $T_D > 0.5$, and $T_D K < 2$. The root locus with respect to K of the characteristic equation of (5.27) is shown in Fig. 5.13 when $T_D = 1.5$.

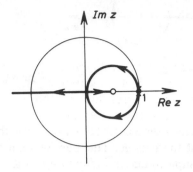

Figure 5.13 The root locus of the characteristic equation of the system in (5.27) when $T_D = 1.5$ with respect to the parameter K.

Let the reference signal be a step. Figure 5.14 shows the continuous-time output for four different values of K. The behavior of the closed-loop system varies from an oscillatory to a well-damped response. When $K = 1$, the output is equal to the reference value after two samples. This is called *deadbeat control* and is discussed further in Chapters 9 and 10. When $K > 1$ the output and the control signal oscillate because of the discrete-time pole on the negative real axis. The poles are inside the unit circle if $K < \frac{4}{3}$.

To determine the closed-loop response, it is important to understand the connection between the discrete-time poles and the response of the system. This is discussed in Sec. 3.6. From Fig. 3.8 it can be seen that $K = 0.75$ corresponds to a damping of $\zeta = 0.4$. The distance to the origin is a measure of the speed of the system.

Thus a good control system has been obtained by use of the discrete-time model. However, it is necessary to be careful. To demonstrate why, start with the model of (5.22), which can be written as the difference equation

$$y(k) = 2y(k-1) - y(k-2) + 0.5u(k-1) + 0.5u(k-2)$$

The purpose of the control is to follow the reference trajectory $u_c(k)$. If the control signal is chosen such that the right-hand side is equal to the reference value at

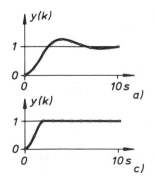

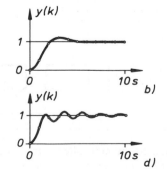

Figure 5.14 The continuous-time output of the system in Fig. 5.12 when $T_D = 1.5$ and $K = 0.5, 0.75, 1,$ and 1.25, respectively.

Analysis of Discrete-Time Systems   Chap. 5

time $k - 1$, the following causal controller is obtained:

$$u(k) = \frac{2q}{q + 1} u_c(k) - \frac{2(2q - 1)}{q + 1} y(k) \qquad (5.28)$$

The closed-loop system is given by

$$y(k) = u_c(k - 1)$$

The output is equal to the reference value after one step. Using the controller in (5.25) with $K = 1$ and $T_D = 1.5$, it took two steps. The step response and the control signal when using the control law (5.28) are shown in Fig. 5.15. At the sampling points, the system has the performance that was previously calculated, but there is an oscillation in the continuous-time output. It is thus important to simulate a system in order to investigate the behavior between the sampling points. It is, however, also possible to get an indication of the problem through analysis of the discrete-time system. That something is wrong is seen in the control signal. The control law of (5.28) and (5.22) gives the closed-loop system

$$y(k) = \frac{q(q + 1)}{(q + 1)[q^2 - 2q + 1 - (-2q + 1)]} u_c(k)$$

$$= \frac{q(q + 1)}{q^2(q + 1)} u_c(k) = u_c(k - 1)$$

The closed-loop system is of third order, the process has two modes, and the controller has one mode. The zero on the stability boundary is canceled by a pole. This mode is not observable in the closed-loop, discrete-time system. This means that observability of the closed-loop system has been lost by an improper choice of the controller.

The behavior of the double integrator with some simple controllers has been discussed; the results can be generalized to more complex systems. Also, the importance of analysis and simulation has been illustrated.

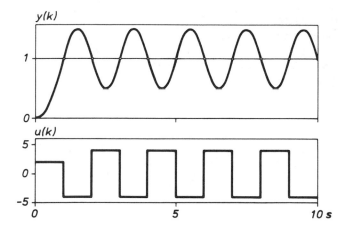

Figure 5.15 The step response and the control signal of the double integrator when the controller of (5.28) is used.

Example 5.11—Simulation using Simnon

The double integrator is defined by the state equation (A.2). The Simnon description of this is

```
CONTINUOUS SYSTEM dint
"System to simulate continuous time
"double integrator
STATE x1    x2        "Declare state variables
DER     dx1 dx2       "Declare corresponding derivatives
INPUT u               "Declare input
OUTPUT y              "Declare output
dx1 = x2              "Defines (A.2)
dx2 = u
y = x1
END
```

Reserved Simnon words are written in capitals. Text after a double quote mark ('') to the end of the line is comments. The discrete-time PD-controller (5.25) does not have any states and is defined as

```
DISCRETE SYSTEM reg
INPUT uc y yd         "Declare inputs
OUTPUT u              "Declare output
TIME t                "Declare present time
NSAMP ts              "Declare next sampling time
u = K*(uc-y-TD*yd)    "Defines (5.25)
ts = t + h            "Defines next sampling time
"Parameters
K:0.5                 "Gives parameter values
TD:1.5                "that can be changed
h:1                   "from outside the system
END
```

The systems dint and reg are connected together with

```
CONNECTING SYSTEM condint
"Connecting system for dint and reg
TIME t                              "Declares present time
ref = IF t < tstep THEN 0 ELSE step  "Defines ref signal
u[dint] = u[reg]                    "Define connections
uc[reg] = ref
y[reg] = y[dint]
yd[reg] = x2[dint]

tstep:0
step:1
END
```

We are now ready to simulate the system and generate Figure 5.14(a). This is done through the following comments:

```
SYST dint reg condint    "Defines and compiles the total system
SPLIT 2 1                "Splits screen into two areas
AXES H 0 10 V 0 1.5      "Draw axes with horizontal (H) and
                        "vertical (V) limits
PLOT y[dint] uc[reg]    "Define variable to be ploted
SIMU 0 10               "Simulate from t=0 to t=10
```

Figure 5.14(b) is now generated by the additional commands

```
AXES            "Draw axes with same length as before

PAR K:0.75      "Change the parameter K
SIMU            "Simulate over the same time
                "period as before
```

The system descriptions and commands are easy to understand. The possibility of mixing different types of subsystems makes Simnon very flexible and useful. □

Hidden Oscillations

Figures 5.14 and 5.15 show that the continuous-time output of the process may have oscillations that are not seen at the sampling points. These are called *hidden oscillations*, or *intersample ripple*. Simulation is an effective tool for finding hidden oscillations. The modified z-transform or (3.2) can also be used to calculate the continuous-time output between the sampling instants; however, it is also enlightening to do some analysis.

The intersample ripple is essentially determined by the open-loop dynamics because the system operates in open loop between the sampling points. Two cases can be distinguished.

Open- or closed-loop systems, where there is an oscillation in the continuous-time output which cannot be seen in the control signal.

Oscillations between the sampling points caused by an oscillation in the control signal.

The first case of intersample ripple may occur if observability of the open-loop system is lost due to sampling. The pulse-transfer function then has canceled poles and zeros. The effect of the canceled modes is then not seen at the sampling instants. There may then be hidden oscillations if the continuous-time open-loop system has oscillatory modes and if the sampling period matches the frequency of these modes. This type of hidden oscillation occurs only for certain values of the sampling period. A change in the sampling interval makes the system observable and the oscillation can be seen in the sampled output. To detect this type of intersample ripple, it is necessary to check the observability of the sampled-data system (compare with Example 5.9).

The second type of hidden oscillation occurs if there are poorly damped zeros in the open-loop system that are canceled by the controller. In this case, the oscillation can be seen in the control signal. This is the cause of the intersample ripple seen in Figs. 5.14 and 5.15. This type of hidden oscillation will not be detected if the sampling period is changed, provided that the design is still such that the process zeros are canceled.

To summarize, there are no hidden oscillations if the unobservable open-loop modes are not oscillatory and if unstable or poorly damped process zeros are not canceled by the regulator.

Example 5.12

Consider a continuous-time system with the transfer function

$$G(s) = \frac{1}{s+1} + \frac{\pi}{(s+0.02)^2 + \pi^2}$$

Sampling this system with $h = 2$ gives the pulse-transfer function

$$H(z) = \frac{1-a}{z-a}$$

where $a = e^{-2}$.

The discrete-time system is of first order while the continuous-time system is of third order. The cancellation of poles and zeros which are oscillatory is an indication that hidden oscillation may occur. Figure 5.16 shows the step response of the continuous-time system. The sampling points are indicated by small circles. The system behaves like a first-order system at the sampling points. □

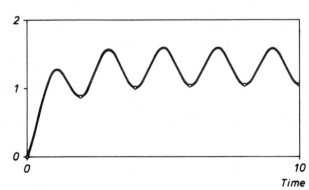

Figure 5.16 Continuous-time step response of the system in Example 5.12. The sampling instants are indicated by circles.

5.5 Problems

5.1. Determine if the following equations have all their roots inside the unit disc:
   (a) $z^2 - 1.5z + 0.9 = 0$
   (b) $z^3 - 3z^2 + 2z - 0.5 = 0$
   (c) $z^3 - 2z^2 + 2z - 0.5 = 0$
   (d) $z^3 + 5z^2 - 0.25z - 1.25 = 0$
   (e) $z^3 - 1.7z^2 + 1.7z - 0.7 = 0$

5.2. Consider the system in Fig. 5.3, and let

$$H_0(z) = \frac{K}{z(z - 0.2)(z - 0.4)} \qquad K > 0$$

Determine the values of K for which the closed-loop system is stable.

5.3. Consider the system in Fig. 4.20(a). Assume that the sampling is done periodically with the period h and that the D-A converter holds the control signal constant over the sampling interval. The control algorithm is assumed to be

$$u(kh) = K[u_c(kh - \tau) - y(kh - \tau)]$$

where $K > 0$ and τ is the computation time. The transfer function of the process is

$$G(s) = \frac{1}{s}$$

(a) How large are the values of the regulator gain, K, for which the closed-loop system is stable if $\tau = 0$ and $\tau = h$?

(b) Compare this system with the corresponding continuous-time systems, i.e., when there is a continuous-time proportional controller and a time delay in the process.

5.4. Determine the Nyquist curve for the system

$$H_0(z) = \frac{1}{z \quad 0.5}$$

5.5. Determine a Lyapunov function for the nonlinear equation in Example 1.5 and show that the solution is asymptotically stable.

5.6. From the system

$$x(k + 1) = \begin{bmatrix} 1 & 0 \\ 1 & 1 \end{bmatrix} x(k) + \begin{bmatrix} 1 \\ 0 \end{bmatrix} u(k)$$

$$y(k) = [0 \quad 1]x(k)$$

the following values are obtained

$$y(1) = 0 \qquad u(1) = 1$$
$$y(2) = 1 \qquad u(2) = -1$$

Determine the value of the state at $k = 3$.

5.7. Is the following system (a) observable, (b) reachable?

$$x(k + 1) = \begin{bmatrix} 0.5 & -0.5 \\ 0 & 0.25 \end{bmatrix} x(k) + \begin{bmatrix} 6 \\ 4 \end{bmatrix} u(k)$$

$$y(k) = [2 \quad -4]x(k)$$

5.8. Is the following system reachable?

$$x(k + 1) = \begin{bmatrix} 1 & 0 \\ 0 & 0.5 \end{bmatrix} x(k) + \begin{bmatrix} 1 & 1 \\ 1 & 0 \end{bmatrix} u(k)$$

Assume that a scalar input $u'(k)$ such that

$$u(k) = \begin{bmatrix} 1 \\ -1 \end{bmatrix} u'(k)$$

is introduced. Is the system reachable from $u'(k)$?

5.9. Given the system

$$x(k + 1) = \begin{bmatrix} 0 & 1 & 2 \\ 0 & 0 & 3 \\ 0 & 0 & 0 \end{bmatrix} x(k) + \begin{bmatrix} 0 \\ 1 \\ 0 \end{bmatrix} u(k)$$

(a) Determine a control sequence such that the system is taken from the initial state $x^T(0) = [1 \quad 1 \quad 1]$ to the origin.
(b) Which is the minimum number of steps that solve the problem in (a)?
(c) Explain why it is not possible to find a sequence such that the state $[1 \quad 1 \quad 1]^T$ is reached *from* the origin.

5.10. Verify the formula for W_c^{-1} given in Example 5.7. for an nth-order system.

5.11. The system

$$x(k + 1) = \Phi x(k) + \Gamma u(k)$$

has been obtained from the system

$$z(k + 1) = Fz(k) + Gu(k)$$

by a linear transformation

$$z = Tx$$

(a) Use the result in Section 5.3 to derive a formula for T when dim $(u) = 1$; dim $(u) = r$.
(b) Use the result to solve Problem 3.7.

5.12. Determine the stability and the stationary value of the output for the system described by Fig. 5.17 with

$$H_0(q) = \frac{1}{q(q - 0.5)}$$

when u_c is a step function and (a) $H_r(q) = K$ (proportional controller) $K > 0$ and (b) $H_r(q) = Kq/(q - 1)$ (integral controller) $K > 0$.

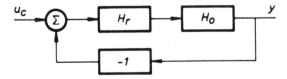

Figure 5.17

5.13. Consider the system in Problem 5.12. Determine the steady-state error between the command signal, u_c, and the output when u_c is a unit ramp, i.e., $u_c(k) = k$. Assume that H_r is (a) a proportional controller and (b) an integral controller.

5.14. Sample the system

$$G(s) = \frac{s + 1}{s^2 + 0.2s + 1}$$

and determine the sampling intervals for which the response of the system will have hidden oscillations. Verify by simulations.

5.15. Consider the tank system with the pulse-transfer operator given in Problem 3.10(b).

   (a) Introduce a controller as in Fig. 5.17. Let the command input be a step and determine the steady-state error when using a proportional controller K and an integral controller $K/(1 - q^{-1})$.

   (b) Simulate the system using the controllers in (a). Choose K such that the poles of the closed-loop system correspond to a damping of $\zeta = 0.7$.

5.16. Consider the system in Fig. 5.3. Derive a formula for the velocity *error coefficient*. That is an expression for the steady-state error when u_c is a unit ramp.

5.17. Determine the values of $K > 0$ for which the system

$$y(k) = K \frac{4q^{-1} + q^{-2}}{1 + q^{-1} + 0.16q^{-2}} u(k)$$

is stable under simple feedback.

5.18. Determine a coordinate transformation $z = Tx$ that transfers the system

$$x(k + 1) = \begin{pmatrix} 1 & 2 \\ 1 & 2 \end{pmatrix} x(k) + \begin{pmatrix} 3 \\ 4 \end{pmatrix} u(k)$$

$$y(k) = (5 \quad 6)x(k)$$

to controllable canonical form and to observable canonical form.

5.19. Assume that the continuous time system (CT)

$$\dot{x} = Ax + Bu \qquad\qquad (CT)$$
$$y = Cx$$

is sampled and gives the discrete-time system (DT)

$$x(kh + h) = \Phi x(kh) + \Gamma u(kh) \qquad\qquad (DT)$$
$$\Gamma(kh) = Cx(kh)$$

Consider the following statements:
   (a) CT stable $\Rightarrow$ DT stable
   (b) CT unstable $\Rightarrow$ DT unstable
   (c) CT stable inverse $\Rightarrow$ DT stable inverse
   (d) CT unstable inverse $\Rightarrow$ DT unstable inverse
   (e) CT controllable $\Rightarrow$ DT controllable
   (f) CT observable $\Rightarrow$ DT observable
   (g) CT pole excess $r \Rightarrow$ DT pole excess r
   Which statements are true for the cases
   (i) All sampling intervals $h > 0$.
   (ii) All $h > 0$ except for isolated values.
   (iii) Neither (i) nor (ii).
   Give short motivations.

5.20. Consider the system

$$x(k + 1) = \begin{pmatrix} 0 & -3 & 2 \\ 3 & -12 & 7 \\ 6 & -21 & 12 \end{pmatrix} x(k) + \begin{pmatrix} 0 \\ 1 \\ 2 \end{pmatrix} u(k)$$

Determine whether
   (a) The system is reachable.
   (b) The system is controllable.

5.21. Given the system

$$(q^2 + 0.4q)y(k) = u(k)$$

(a) For which values of K in the proportional controller

$$u(k) = K(u_c(k) - y(k))$$

is the closed-loop system stable?

(b) Determine the stationary error $u_c - y$ when u_c is a step and when $K = 0.5$ in the controller in (a).

5.22. Assume that the system

$$y(k) - 1.2y(k - 1) + 0.5y(k - 2) = 0.4u(k - 1) + 0.8u(k - 2)$$

is controlled by

$$u(k) = -Ky(k)$$

(a) Determine for which values of K the closed-loop system is stable.

(b) Assume that there is a computational delay in the controller, i.e.,

$$u(k) = -Ky(k - 1)$$

For which values of K is the closed-loop system now stable?

5.6 References

Original papers on tests for checking if a polynomial has all its poles inside the unit circle are

SCHUR, J. (1918): "Über Potenzreihen, die im Inneren des Einheitskreises beschänkt sind, II," *Zeitschrift für die reine und angewandte Matematik,* 148, 122–45.

COHN, A. (1922): "Über die Anzahl der Wurzeln einer algebraischen Gleichung in einem Kreise," *Matematische Zeitschrift,* 14, 110–48.

Jury's test is a simplification of the Schur-Cohn test and is found in

JURY, E. I. and J. BLANCHARD (1961): "A Stability Test for Linear Discrete Time Systems in Table Form," *Proc. IRE,* 49, 1947–48.

A simple proof of Jury's test is found in

ÅSTRÖM, K. J. (1970): *Introduction to Stochastic Control Theory.* New York: Academic Press.

The use of Lyapunov theory for discrete-time control systems is introduced in

KALMAN, R. E. and J. E. BERTRAM (1960): "Control System Analysis and Design via the Second Method of Lyapunov: II Discrete-Time Systems," *Trans. ASME Journal Basic Eng.,* Series D, no. 3, 394–400.

Controllability and observability are concepts introduced by Kalman in connection with analysis of optimal control systems. See

KALMAN, R. E. (1961): "On the General Theory of Control Systems," Proc. First IFAC Congress, Moscow: *Butterworths,* 1, 481–92.

KALMAN, R. E., Y. C. HO, and K. S. NARENDRA (1963): Controllability of Linear Dynamical Systems. In *Contributions to Differential Equations,* 1, 189–213, New York: John Wiley.

The concept of hidden oscillations and their cause are discussed in

JURY, E. I. (1957): "Hidden Oscillations in Sampled-Data Control Systems," *AIEE Trans.,* 75, pt. II, 391–95.

General aspects of digital simulation are discussed in

GORDON, G. (1969): *System Simulation.* Englewood Cliffs, N.J.: Prentice-Hall, Inc.

KHEIR, N. A. (1988): *Systems Modeling and Computer Simulation,* New York: Marcel Dekker, Inc.

The simulation package Simnon is described in

ELMQVIST, H. (1977): "Simnon: An Interactive Simulation Program for Nonlinear Systems," *Proc. International Symposium Simulation '77, Montreux.*

ELMQVIST, H., K. J. ÅSTRÖM, and TOMAS SCHÖNTHAL (1986): *Simnon—User's Guide for MS™-DOS Computers.* Department of Automatic Control, Lund Inst. of Techn.

ÅSTRÖM, K. J. (1983): "Computer-Aided Modeling, Analysis and Design of Control Systems: A perspective," *IEEE Control Systems Magazine, 3* no. 2, 4–16.

DISTURBANCE MODELS

6

GOAL

To Discuss Different Ways of Eliminating Disturbances and Their Effects. To Introduce Deterministic and Stochastic Models for Disturbances. To Analyze the Response of Linear Systems to Disturbances

6.1 Introduction

The presence of disturbances is one of the main reasons for using control. Without disturbances there is no need for feedback control. The character of the disturbances imposes fundamental limitations on the performance of a control system. Measurement noise in a servo system limits the achievable bandwidth of the closed-loop system. The nature of the disturbances determines the quality of regulation in a process-control system. Disturbances also convey important information about the properties of the system. By investigating the characteristics of the disturbances it is thus possible to detect the status of the process, including beginning process malfunctions.

Different ways to describe disturbances and to analyze their effect on a system are discussed in this chapter. An overview of different ways to eliminate disturbances is first given. This includes use of feedback, feedforward, and prediction. The discussion gives a reason for the different ways of describing disturbances.

The classic disturbance models, impulse, step, ramp, and sinusoid, are described in Sec. 6.3. All these disturbances can be thought of as generated by linear systems with suitable initial conditions. The problem of analyzing the effect of disturbances on a linear system can then be reduced to an initial-value problem. From the input-output point of view, a disturbance may also be modeled as an impulse response of a linear filter. The disturbance analysis is then reduced to a response calculation. This is particularly useful for disturbances that are steps or sinusoids. In all cases the disturbance analysis can be done with the tools developed in Chapter 5. When the response of a system to a specific disturbance needs to be known, it is often necessary to resort to simulation. This is easily done with a simulation program because the disturbance analysis is again reduced to an initial-value problem.

When disturbances can be neither eliminated at the source nor measured, it is necessary to resort to prediction. To do so it is necessary to have models of disturbances that lead to a reasonable formulation of a prediction problem. For this purpose the concept of piecewise deterministic disturbances is introduced in Sec. 6.4.

Another way to arrive at a prediction problem is to describe disturbances as random processes. This formulation is presented in Sec. 6.5. A simple version of the famous Wiener-Kolmogorov-Kalman prediction theory is also presented. As seen in Chapters 11 and 12, the prediction error expresses a fundamental limitation on regulation performance. Continous-time stochastic processes are discussed briefly in Sec. 6.6. Such models are required because of the desire to formulate models and specifications in continuous time. Sampling of continuous-time stochastic-state models is treated in Sec. 6.7.

6.2 Reduction of Effects of Disturbances

Before going into details of models for disturbances, it is useful to discuss how the effects of disturbances on a system can be reduced. Disturbances may be reduced at their source. The effects of disturbances can also be reduced by local feedback or by feedforward from measurable disturbances. Prediction may also be used to estimate unmeasurable disturbances. The predictable part of the disturbance can then be reduced by feedforward. These different approaches will be discussed in more detail.

Reduction at the Source

The most obvious way to reduce the effects of disturbances is to attempt to reduce the source of the disturbances. This approach is closely related to process design. The following are typical examples:

Reduce variations in composition by introducing a tank with efficient mixing.
Reduce friction forces in a servo by using better bearings.
Move a sensor to a position where there are smaller disturbances.
Modify sensor electronics so that less noise is obtained.
Replace a sensor with another having faster response.
Change the sampling procedure by spacing the samples better in time or space to obtain a better representation of the characteristics of the process.

These are just a few examples, but it is very important to keep these possibilities in mind.

Reduction by Local Feedback

If the disturbances cannot be reduced at the source, an attempt can be made to reduce them by local feedback. The generic principle of this approach is illustrated in Fig. 6.1. For this approach it is necessary that the disturbances enter the system locally in a well-defined way. It is also necessary to have access to a measured variable that is a result of the disturbance. It is also necessary to have access to a control variable that enters the system in the neighborhood of the disturbance. The effect of the disturbance can then be reduced by using local feedback. The dynamics relating the measured variable to the control variable should be such that a high-gain control loop can be used.

This use of feedback is often very simple and effective because it is not necessary to have detailed information about the characteristics of the process, provided that a high gain can be used in the loop. However, an extra feedback loop is required. The following are typical examples of local feedback:

> Reduce variations in supply pressure to valves, instruments, and regulators by introducing a pressure regulator.
>
> Reduce variations in temperature control by stabilizing the supply voltage.

Reduction by Feedforward

Measurable disturbances can also be reduced to feedforward. The generic principle is illustrated in Fig. 6.2. The disturbance is measured, and a control signal that attempts to counteract the disturbance is generated and applied to the process. If the transfer functions relating the output y to the disturbance w and the control u are H_w and H_p, respectively, it is easy to see that the transfer function H_{ff} of the feedforward compensator should ideally be

$$H_{ff} = -H_p^{-1}H_w$$

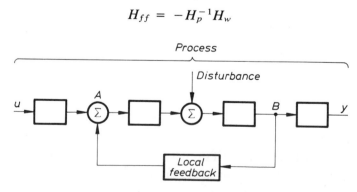

Figure 6.1 Reduction of disturbances by local feedback. The disturbance should enter the system between points A and B. The dynamics between A and B should be such that a high gain can be used in the loop.

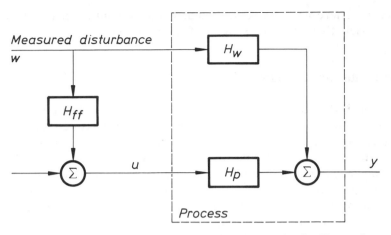

Figure 6.2 Reduction of disturbances by feedforward.

If this transfer function is unstable or nonrealizable, a suitable approximation is chosen instead. The design of the feedforward compensator is often based on a simple static model. The transfer function H_{ff} is then simply a static gain. Feedforward is particularly useful for disturbances generated by changes in the command or reference signals or for cascaded processes when disturbances downstream are generated by variations in processes upstream.

Reduction by Prediction

Reduction by prediction is an extension of the feedforward principle that may be used when the disturbance cannot be measured. The principle is very simple: The disturbance is predicted using measurable signals, and the feedforward signal is generated from the prediction.

It is important to observe that it is not necessary to predict the disturbance itself; it is sufficient to model a signal that represents the effect of the disturbance on the important process variables.

Goals for Modeling

Since different ways of reducing the effects of disturbances have been specified, it is now easy to see how models for disturbances will be used. To evaluate the needs for reduction of disturbances it is necessary to be able to estimate the influences of disturbances on important system variables, which is basically a problem of analyzing the response of a system to a given input. The models used for disturbances can be fairly simple, as long as they represent the major characteristics of true disturbances. Similarly simple models can also be used to estimate possible improvements obtained by local feedback and feedforward.

More accurate models of disturbances are needed if prediction is applied. In this case the performance obtained depends critically on the character of the disturbances. There are also some fundamental difficulties in formulating disturbance models that give a sensible prediction problem.

6.3 The Classic Disturbance Models

A discussion of the common disturbance models follows. The character of the disturbances is discussed first. The disturbances are classified as load disturbances, measurement errors, and parameter variations. Some simple disturbance models—impulse, step, ramp, and sinusoid—are then discussed. It is shown that these disturbances can be generated as outputs of simple linear systems. This observation suggests a general class of disturbances that are generated from arbitrary linear systems. The observation also makes it possible to simplify the analysis of effects of disturbances on a system.

Character of Disturbances

It is customary to distinguish among different types of disturbances, such as load disturbances, measurement errors, and parameter variations.

Load disturbances. Load disturbances influence the process variables. They may represent disturbance forces in a mechanical system—for example, wind gusts on a stabilized antenna, waves on a ship, load on a motor. In process control, load disturbances may be quality variations in a feed flow or variations in demanded flow. In thermal systems the load disturbances may be variations in surrounding temperature. Load disturbances typically vary slowly. They may also be periodic—for example, waves in ship control systems.

Measurement errors. Measurement errors enter in the sensors. There may be a steady-state error in some sensors due to errors in calibration. However, measurement errors typically have high-frequency components. There may also be dynamic errors because of sensor dynamics. A typical case is a thermocouple, which may have a time constant of 10–50 s depending on its encapsulation. There may also be complicated dynamic interaction between sensors and the process. Typical examples are gyroscopic measurements and measurement of liquid level in nuclear reactors. The character of the measurement errors often depends on the filtering in the instruments. It is often a good idea to look at the instrument and modify the filtering so that it fits the particular problem.

In some cases it is not possible to measure the controlled variable directly; the value of the variable is then inferred from indirect measurements of several other variables. The relationship between the controlled and the measured variables can be quite complex. It may be characterized as a nonlinear and time-varying dynamic system. A common situation is that one instrument gives a quick indication with large errors and another instrument gives an accurate measurement after a long delay. The accurate measurement can be obtained from a laboratory measurement, for example.

Parameter variations. Linear theory is used throughout this book. The load disturbance and the measurement noise then appear additively. Real systems are, however, often nonlinear. This means that disturbances can enter in a more complicated way. Because the linear models are obtained by linearizing the non-linear models, some disturbances then also appear as variations in the parameters of the linear model.

Simple Disturbance Models

There are four different types of disturbances—impulse, step, ramp, and sinusoid—that are commonly used in analyzing control systems. These disturbances are illustrated in Fig. 6.3 and a discussion of their properties follows.

The impulse and the pulse. The *impulse* and the *pulse* are simple idealizations of sudden disturbances of short duration. They can represent load disturbances as well as measurement errors. For continuous systems the disturbance is an impulse (a delta function); for sampled systems the disturbance is modeled as a pulse with unit amplitude and a duration of one sampling period.

The pulse and the impulse are also important for theoretical reasons because the response of a linear continuous-time system is completely specified by its impulse response, and a linear discrete-time system by its pulse response.

The step. The *step signal* is another prototype for a disturbance (see Fig. 6.3). It is typically used to represent a load disturbance or an offset in a measurement.

The ramp. The *ramp* is a signal that is zero for negative time and increases linearly for positive time (see Fig. 6.3). It is used to represent drifting measurement errors and disturbances that suddenly start to drift away. In practice, the disturbances are often bounded; however, the ramp is a useful idealization.

The sinusoid. The *sine wave* is the prototype for a periodic disturbance. Choice of the frequency makes it possible to represent low-frequency load disturbances, as well as high-frequency measurement noise.

It is convenient to view disturbances as being generated by dynamic systems. From an input-output viewpoint disturbances may be described as impulse responses. Disturbances may also be regarded as the responses of dynamic systems with zero inputs but nonzero initial conditions. In both cases the major characteristics of the disturbances are described by the dynamic systems that generate

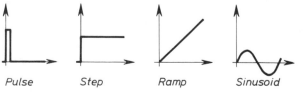

Pulse   Step   Ramp   Sinusoid

Figure 6.3 Idealized models of simple disturbances.

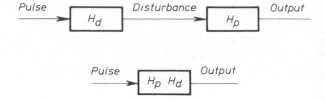

Figure 6.4 How analysis of the effects of disturbances can be reduced to the calculation of an impulse response (compare with Fig. 5.10).

them. The approach can, of course, be applied to continuous-time, as well as discrete-time, systems.

The step can be generated from an integrator, the ramp from a double integrator, and a sinusoid from a harmonic oscillator (compare Examples A.1 and A.3 in Appendix A).

Analysis of Effects of Disturbances

Because disturbances can be generated as outputs of linear dynamic systems, the effect of disturbances on a system can be analyzed in a straightforward manner.

State-space analysis. Because disturbances can be viewed as solutions to linear differential equations with initial conditions, it is easy to investigate the effects of disturbances on a linear system. The state variables of the system are simply augmented by the state variables used to describe the disturbances. The set of equations is then solved for the initial conditions, which correspond to the disturbances. Disturbance analysis can thus be dealt with using the tools for analysis and simulation presented in Chapter 5.

Input-output analysis. Because disturbances can also be thought of as generated by driving a linear system by an impulse, it is possible to analyze the effects of disturbance on a linear system by cascading it with the system, which generates the disturbance (see Fig. 6.4). This viewpoint is particularly useful when using the initial-value and final-value theorems or Bode-diagrams to make crude estimates of the effects of disturbances. It is also useful for simulation.

6.4 Piecewise Deterministic Disturbances

The disturbance models discussed in Sec. 6.3 are useful for analyzing the effects of disturbances on a system. Possible improvements by using local feedback and feedforward can also be investigated using the models. The disturbance models discussed are, however, not suitable for investigating disturbance reduction by prediction. To predict a signal it is necessary to have realistic models for the disturbances. It is a fundamental problem to construct models that allow a sensible formulation of a prediction problem. This fundamental issue is the first topic in this section. An investigation of the problem leads to introduction of the piecewise deterministic disturbances. Alternative models, which also permit formulation of a prediction problem, are discussed in Secs. 6.5 and 6.6.

A Fundamental Problem

It is not trivial to construct models for disturbances that permit a sensible formulation of a prediction problem.

Example 6.1—Predictor for a step signal

To predict the future value of a step signal, it seems natural to use the current value of the signal. For discrete-time signals, the predictor then becomes

$$\hat{y}[(k + m)h \mid kh] = y(kh)$$

The notation $\hat{y}(t \mid s)$ means the prediction of $y(t)$ based on data available at time s.

This predictor has a prediction error at times $t = 0, h, 2h, \ldots, (m - 1)h$, i.e., m steps after the step change in y. It then predicts the signal without error. ☐

Example 6.2—Predictor for a ramp signal

A predictor for a ramp can be constructed by calculating the slope from the past and the current observations and making a linear extrapolation, which can be expressed by the formula

$$\hat{y}[(k + m)h \mid kh] = y(kh) + m[y(kh) - y(kh - h)]$$

$$= (1 + m)y(kh) - my(kh - h)$$

This predictor has an initial error for $t = h, 2h, \ldots, mh$. After that it predicts the signal without error. ☐

The Basic Idea

These examples indicate that the prediction error will be zero except at a few points. This observation is not in close agreement with the practical experience that disturbances are hard to predict. The explanation is that the step and the ramp are not good models for prediction problems. It is in fact a nontrivial task to find signals that lead to sensible prediction problems. Analytic signals are useless because an analytic function is uniquely given by its values in an arbitrarily short interval. The step and the ramp are analytic everywhere except at the origin.

One possibility of constructing signals that are less regular is to introduce more points of irregularity. Thus signals can be introduced that are generated by linear dynamic systems with irregular inputs. Instead of having a pulse at the origin, inputs that are different from zero at several points can be introduced. An interesting class of signals is obtained if the pulses are assumed to be isolated and spread by at least n samples, where n is the order of the system. It is assumed, that it is not known a priori, when the pulses occur. The amplitudes of the pulses are also unknown. Such signals are called *piecewise deterministic signals*. The name comes from the fact that the signals are deterministic except at isolated points, where they change in an unpredictable way. An example of a piecewise deterministic signal is shown in Fig. 6.5.

Sensible prediction problems can be formulated for piecewise deterministic signals.

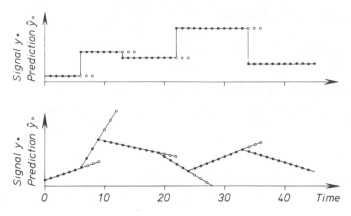

Figure 6.5 Piecewise constant and piecewise linear signals and their *m*-step predictions when $m = 3$.

State-Space Models

Let a signal be generated by the dynamic system

$$x(k + 1) = \Phi x(k) + v(k)$$
$$y(k) = Cx(k)$$

(6.1)

It is assumed that the output y is a scalar and that the system is completely observable. The input v is assumed to be zero except at isolated points. If the state of the system is known, it is straightforward to predict the state over any interval where the input is zero. However, when there is a pulse, the state can change in an arbitrary manner, but after a pulse there will always be an interval where the input is zero. Since the system is observable, the process state can then be calculated. Exact predictions can then be given until a new pulse occurs. This argument can be converted into mathematics in the following way. From the derivation of the condition for observability in Sec. 5.3, it is found that the state is given by

$$x(k - n + 1) = W_o^{-1}[y(k - n + 1) \cdots y(k)]^T$$

(6.2)

where W_o is the observability matrix given by Equation (5.20). The following predictor gives the state m steps ahead

$$\hat{x}(k + m \mid k) = \Phi^{m+n-1} W_o^{-1}[y(k - n + 1) \cdots y(k)]^T$$

(6.3)

The predictor for the signal is thus obtained from a linear combination of n values of the measured signal. The predictor can be expressed as

$$\hat{x}(k + m \mid k) = P^*(q^{-1})y(k)$$

where P is a polynomial of degree $n - 1$.

The predictor can also be represented by the recursive equation

$$\hat{x}(k \mid k) = \Phi \hat{x}(k - 1 \mid k - 1) + K[y(k) - C\Phi \hat{x}(k - 1 \mid k - 1)]$$

$$\hat{x}(k + m \mid k) = \Phi^m \hat{x}(k \mid k)$$

(6.4)

where the matrix K is chosen so that all eigenvalues of the matrix $(I - KC)\Phi$ are equal to zero.

Simple calculations for an integer and a double integrator give the same predictors as in Examples 6.1 and 6.2. This is a consequence of the fact that the important characteristics of the disturbances are captured by the dynamics of the systems that generate the disturbances. These dynamics determine the predictors uniquely; it does not matter if the systems are driven by a single pulse or by several pulses. The properties of the predictors are illustrated in Fig. 6.5.

Input-Output Models

Because the predictor for a piecewise deterministic signal becomes a polynomial, it seems natural to obtain it directly by polynomial calculations. For this purpose it is assumed that the signal is generated by the dynamic system

$$y(k) = \frac{C(q)}{A(q)} w(k)$$

(6.5)

where the input w is a signal that is zero except at isolated points, which are spaced more than deg $A + m$, and deg $C = $ deg A. Define $F(z)$ and $G(z)$ through the identity

$$z^{m-1}C(z) = A(z)F(z) + G(z)$$

It can be shown that the m-step predictor for y is given by the difference equation

$$C(q)\hat{y}(k + m \mid k) = qG(q)y(k)$$

(6.6)

A reference to the proof of this is given in Sec. 6.10.

Remark. Notice that the signals discussed in this section are similar to those in Sec. 6.3 in the sense that they are characterized by dynamic systems. The only difference between the signals is that the inputs to the systems are different. This idea is extended in the next section.

6.5 Stochastic Models of Disturbances

It is natural to use *stochastic,* or random, concepts to describe disturbances. By such an approach it is possible to describe a wide class of disturbances, which permits good formulation of prediction problems. The theory of random processes and the prediction theory were in fact developed under close interaction.

The general theory of stochastic processes is quite complex. For computer-control theory, it is fortunately often sufficient to work with a special case of the general theory, which requires much less sophistication. This theory is developed in this section. First, some elements of the theory of random processes are given, then the notion of discrete-time white noise is discussed. Disturbances are then modeled as outputs of dynamic systems with white-noise inputs. The disturbance models are thus similar to the models discussed in the previous sections; the only difference is the character of the input signals to the systems. Tools for analyzing the properties of the models are also given.

Stochastic Processes

The concept of a stochastic process is complex. It took brilliant researchers hundreds of years to find the right ideas. The concept matured in work done by the mathematician Kolmogorov around 1930. A simple presentation of the ideas is given here. Interested readers are strongly urged to consult the references.

A stochastic process (random process, random function) can be regarded as a family of stochastic variables $\{x(t), t \in T\}$. The stochastic variables are indexed with the parameter t, which belongs to the set T, called the *index set*. In stochastic control theory, the variable t is interpreted as time. The set T is then the real variables. When considering sampled-data systems, as in this book, the set T is the sampling instants, i.e., $T = \{\ldots, -h, 0, h, \ldots\}$, or $T = \{\ldots, -1, 0, 1, \ldots\}$ when the sampling period is chosen as the time unit.

A random process may be considered as a function $x(t, \omega)$ of two variables. For fixed $\omega = \omega_0$ the function $x(\cdot, \omega_0)$ is an ordinary time function called a *realization*. For fixed $t = t_0$, the function $x(t_0, \cdot)$ is a random variable. A random process can thus be viewed as generated from a random-signal generator. The argument ω is often suppressed in the notation.

Completely deterministic stochastic processes. One possibility of obtaining a random process is to pick the initial conditions of an ordinary differential equation as a random variable and to generate the time functions by solving the differential equations. These types of random processes are, however, not very interesting because they do not exhibit enough randomness. This is clearly seen by considering the stochastic process generated by an integrator with random initial conditions. Because the output of the integrator is constant it follows that

$$x(t, \omega) - x(t - h, \omega) = 0$$

for all t, h, and ω. A stochastic process with this property is called a *completely deterministic stochastic process,* because its future values can be predicted exactly from its past.

In general it will be said that a random process $x(t, \omega)$ is called *completely deterministic* if

$$\ell x(t, \omega) = 0, \qquad \text{for almost all } \omega$$

where ℓ is an arbitrary linear operator that is not identically zero. This means that completely deterministic random processes can be predicted exactly with linear predictors for almost all ω. (Almost all ω means all ω except for possibly a set of points with zero measure.)

The completely deterministic random processes are closely related to the signals discussed in Sec. 6.3. These signals will be completely deterministic random processes if the initial conditions to the dynamic systems are chosen as random processes. The completely deterministic processes are normally excluded because they are too regular to be of interest.

Concepts. Some important concepts for random processes follow.

The values of a random process at n distinct times are n-dimensional random variables. The function

$$F(\xi_1, \ldots, \xi_n; t_1, \ldots, t_n) = P\{x(t_1) \le \xi_1, \ldots, x(t_n) \le \xi_n\} \tag{6.7}$$

where P denotes probabilities, is called the *finite-dimensional distribution function* of the random process. An illustration is given in Fig. 6.6. A random process is called *Gaussian,* or *normal,* if all finite-dimensional distributions are normal.

The *mean-value function* of a random process x is defined by

$$m(t) = Ex(t) = \int_{-\infty}^{\infty} \xi \, dF(\xi, t) \tag{6.8}$$

The mean-value function is an ordinary time function. Higher moments are defined similarly.

The *covariance function* of a process is defined by

$$r_{xx}(s, t) = \text{cov}[x(s), x(t)]$$
$$= E[x(s) - m(s)][x(t) - m(t)]^T \tag{6.9}$$
$$= \iint [\xi_1 - m(s)][\xi_2 - m(t)]^T \, dF(\xi_1, \xi_2; s, t)$$

A Gaussian random process is completely characterized by its mean-value function and its covariance function.

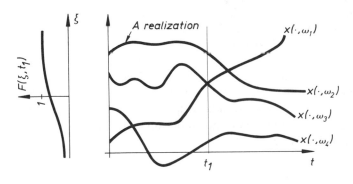

Figure 6.6 A stochastic process and a finite-dimensional distribution function.

The *cross-covariance function*

$$r_{xy}(s, t) = \text{cov}[x(s), y(t)]$$

of two stochastic processes is defined similarly.

A stochastic process is called *stationary* if the finite-dimensional distribution of $x(t_1)$, $x(t_2)$, . . . , $x(t_n)$ is identical to the distribution of $x(t_1 + \tau)$, $x(t_2 + \tau)$, . . . , $x(t_n + \tau)$ for all τ, n, t_1, . . . , t_n. The process is called *weakly stationary* if the first two moments of the distributions are the same for all τ.

The mean-value function of a (weakly) stationary process is constant. The cross-covariance function of weakly stationary processes is a function of the difference $s - t$ of the arguments only. With some abuse of function notation, write

$$r_{xy}(s, t) = r_{xy}(s - t) \tag{6.10}$$

The cross-covariance function of (weakly) stationary processes is a function of one argument only. Hence

$$r_{xy}(\tau) = \text{cov}[x(t + \tau), y(t)] \tag{6.11}$$

When x is scalar the function

$$r_x(\tau) = r_{xx}(\tau) = \text{cov}[x(t + \tau), x(t)] \tag{6.12}$$

is called the *autocovariance* function.

The *cross-spectral density* of (weakly) stationary processes is the Fourier transform of its covariance function. Hence,

$$\phi_{xy}(\omega) = \frac{1}{2\pi} \sum_{k=-\infty}^{\infty} r_{xy}(k)e^{-ik\omega} \tag{6.13}$$

and

$$r_{xy}(k) = \int_{-x}^{x} e^{ik\omega} \phi_{xy}(\omega) \, d\omega \tag{6.14}$$

It is also customary to refer to ϕ_{xx} and ϕ_{xy} as the *autospectral density* and the *cross-spectral density*. The autospectral density is also called *spectral density* for simplicity.

Interpretation of covariances and spectra. Stationary Gaussian processes are completely characterized by their mean-value functions and their covariance functions. In applications, it is useful to have a good intuitive understanding of how the properties of a stochastic process are reflected by these functions.

The mean-value function is almost self-explanatory. The value $r_x(0)$ of the covariance function at the origin is the variance of the process. It tells how large the fluctuations of the process are. The standard deviation of the variations is equal to the square root of $r_x(0)$. If the covariance function is normalized by $r_x(0)$, the *correlation function*, which is defined by

$$\rho_x(\tau) = \frac{r_x(\tau)}{r_x(0)}$$

is obtained. It follows from Schwartz's inequality that

$$| r_x(\tau) | \le r_x(0)$$

The correlation function is therefore less than one in magnitude.

The value $\rho_x(\tau)$ gives the correlation between values of the process with a spacing τ. Values close to one mean that there are strong correlations, zero values indicate no correlation, negative values indicate negative correlation. An investigation of the shape of the correlation function thus indicates the temporal interdependencies of the process.

It is very useful to study realizations of stochastic processes and their covariance functions to develop insight into their relationships. Some examples are shown in Fig. 6.7. All processes have unit variance.

The spectral-density function also has a good physical interpretation. The integral

$$2 \int_{\omega_1}^{\omega_2} \phi(\omega) \, d\omega \tag{6.15}$$

represents the power of the signal in the frequency band (ω_1, ω_2). The area under the spectral density curve thus represents the signal power in a certain frequency band. The presence of peaks in the spectrum indicates that there are almost periodic components. The total area under the curve is proportional to the total variance of the signal. In practical work it is useful to develop a good understanding of how signal properties are related to the spectrum (compare with Fig. 6.7).

Notice that the mean-value function, the covariance function, and the spectral density are characterized by the first two moments of the distribution only. Signals whose realizations are very different may have the same first moments. The random telegraph wave that switches between the values 0 and 1 thus has the same spectrum as the noise from a simple RC circuit.

Discrete-Time White Noise

A simple and useful random process is now introduced. Let time be the set of integers. Consider a stationary discrete-time stochastic process x such that $x(t)$ and $x(s)$ are independent if $t \ne s$. The stochastic process can thus be considered as a sequence $\{x(t, \omega), t = \ldots, -1, 0, 1, \ldots\}$ of independent, equally distributed random variables.

The covariance function is given by

$$r(\tau) = \begin{cases} \sigma^2 & \tau = 0 \\ 0 & \tau = \pm 1, \pm 2, \ldots \end{cases}$$

A process with this covariance function is called *discrete-time white noise*. It follows from (6.13) that the spectral density is given by

$$\phi(\omega) = \frac{\sigma^2}{2\pi}$$

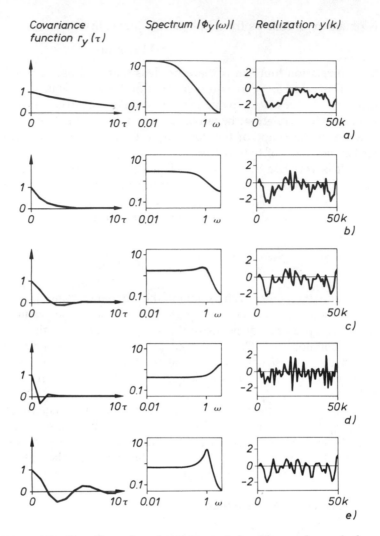

Figure 6.7 Covariance functions, spectral densities, and sample functions for some stationary random processes. All processes have unit variance.

The spectral density is thus constant for all frequencies. The analogy with the spectral properties of white light explains the name given to the process.

White noise plays an important role in stochastic control theory. All stochastic processes that are needed will be generated simply by filtering white noise. White noise is thus the equivalent of pulses for deterministic systems.

ARMA Processes

Large classes of stochastic processes can be generated by driving linear systems with white noise. Let $\{e(k), k = \ldots, -1, 0, 1, \ldots\}$ be discrete-time white noise.

The process generated by

$$y(k) = e(k) + b_1 e(k - 1) + \cdots + b_n e(k - n)$$

is called a *moving average*, or an MA process. The process generated by

$$y(k) + a_1 y(k + 1) + \cdots + a_n y(k - n) = e(k)$$

is called an *autoregression*, or an AR process. The process

$$y(k) + a_1 y(k - 1) + \cdots + a_n y(k - n) = e(k) + b_1 e(k - 1) + \cdots + b_n e(k - n)$$

is called an ARMA process.

The process

$$y(k) + a_1 y(k - 1) + \cdots + a_n y(k - n) = b_0 u(k - d) + \cdots$$
$$+ b_m u(k - d - m) + e(k) + c_1 e(k - 1) + \cdots + c_n e(k - n)$$

is called an ARMAX process, i.e., an ARMA process with an exogenous signal.

State-Space Models

The concept of state has its roots in cause and effect relationships in classical mechanics. The motion of a system of particles is uniquely determined for all future times by the present positions and moments of the particles and the future forces. How the present positions and moments were achieved is not important. The state is an abstraction of this property; it is the minimal information about the history of a system required to predict its future motion.

For stochastic systems, it cannot be required that the future motion be determined exactly. A natural extension of the notion of state for stochastic systems is to require that the probability distribution of future states be uniquely given by the current state. Stochastic processes with this property are called *Markov processes*. Markov processes are thus the stochastic equivalents of state-space models. They are formally defined as follows.

Definition 6.1—Markov process. Let t_i and t be elements of the index set T such that $t_1 < t_2 < \ldots < t_n < t$. A stochastic process $\{x(t), t \in T\}$ is called a Markov process if

$$P\{x(t) \le \xi \mid x(t_1), \ldots, x(t_n)\} = P\{x(t) \le \xi \mid x(t_n)\}$$

where $P\{\cdot \mid x(t_1), \ldots, x(t_n)\}$ denotes the conditional probability given $x(t_1), \ldots, x(t_n)$.

A Markov process is completely determined by the *initial probability distribution*

$$F(\xi; t_0) = P\{x(t_0) \le \xi\}$$

and the *transition probability distribution*

$$F(\xi_t, t \mid \xi_s, s) = P\{x(t) \le \xi_t \mid x(s) = \xi_s\}$$

All finite-dimensional distributions can then be generated from these distributions using the multiplication rule for conditional probabilities.

The Markov process is the natural concept to use when extending the notion of state model to the stochastic case.

Linear stochastic difference equations. Consider a discrete-time system where the sampling period is chosen as the time unit. Let the state at time k be given by $x(k)$. The probability distribution of the state at time $k + 1$ is then a function of $x(k)$. If the mean value is linear in $x(k)$ and the distribution around the mean is independent of $x(k)$, then $x(k + 1)$ can be represented as

$$x(k + 1) = \Phi x(k) + v(k) \qquad (6.16)$$

where $v(k)$ is a random variable with zero mean that is independent of $x(k)$ and independent of all past values of x. This implies that $v(k)$ also is independent of all past v's. The sequence $\{v(k), k = \ldots -1, 0, 1, \ldots\}$ is a sequence of independent, equally distributed random variables. The stochastic process $\{v(k)\}$ is thus discrete-time white noise.

Equation (6.16) is called a *linear stochastic difference equation*. To define the random process $\{x(k)\}$ completely, it is necessary to specify the initial conditions. It is assumed that initial state has the mean m_0 and the covariance matrix R_0. The covariance of the random variables v is denoted by R_1.

Properties of linear stochastic difference equations. The character of the random process defined by the linear stochastic difference equation of (6.16) will now be investigated and the first and second moments of the process will be calculated.

To obtain the mean-value function

$$m(k) = Ex(k)$$

simply take the mean values of both sides of (6.16). Because v has zero mean, the following difference equation is obtained:

$$m(k + 1) = \Phi m(k) \qquad (6.17)$$

The initial condition is

$$m(0) = m_0$$

The mean value will thus propagate in the same way as the unperturbed system.

To calculate the covariance function, introduce

$$P(k) = \text{cov}[x(k), x(k)] = E\tilde{x}(k)\tilde{x}^T(k)$$

where

$$\tilde{x} = x - m$$

It follows from Equations (6.16) and (6.17) that $\tilde{x}$ satisfies Equation (6.16) with initial condition having zero mean. To calculate the covariance, form the expression

$$\tilde{x}(k + 1)\tilde{x}^T(k + 1) = \{\Phi\tilde{x}(k) + v(k)\}\{\Phi\tilde{x}(k) + v(k)\}^T$$

$$= \Phi\tilde{x}(k)\tilde{x}^T(k)\Phi^T + \Phi\tilde{x}(k)v^T(k) + v(k)\tilde{x}^T(k)\Phi^T + v(k)v^T(k)$$

Taking mean values gives

$$P(k + 1) = \Phi P(k)\Phi^T + R_1$$

because $v(k)$ and $\tilde{x}(k)$ are independent. The initial conditions are

$$P(0) = R_0$$

The recursive equation for P tells how the covariance propagates.

To calculate the covariance function of the state, observe that

$$\tilde{x}(k + 1)\tilde{x}^T(k) = [\Phi\tilde{x}(k) + v(k)]\tilde{x}^T(k)$$

Because $v(k)$ and $\tilde{x}(k)$ are independent and $v(k)$ has zero mean,

$$r_{xx}(k + 1, k) = \text{cov}[x(k + 1), x(k)] = \Phi P(k)$$

Repeating this discussion,

$$r_{xx}(k + \tau, k) = \Phi^\tau P(k), \quad \tau \geq 0$$

The results obtained are so important that they deserve to be summarized.

Theorem 6.1. Consider a random process defined by the linear stochastic-difference equation (6.16), where $\{v(k)\}$ is a white-noise process with zero mean and covariance R_1. Let the initial state have mean m_0 and covariance R_0. The mean-value function of the process is then given by

$$m(k + 1) = \Phi m(k), \qquad m(0) = m_0 \tag{6.18}$$

and the covariance function by

$$r(k + \tau, k) = \Phi^\tau P(k), \qquad \tau \geq 0 \tag{6.19}$$

where $P(k) = \text{cov}[x(k), x(k)]$ is given by

$$P(k + 1) = \Phi P(k)\Phi^T + R_1, \qquad P(0) = R_0 \tag{6.20}$$

Remark 1. If the random variables are Gaussian, then the stochastic process is uniquely given by its mean-value function m and its covariance function r.

Remark 2. If the system has an output $y = Cx$, then the mean-value function of y is given by

$$m_y = Cm$$

and its covariance is given by

$$r_{yy} = Cr_{xx}C^T$$

The cross-covariance between y and x is given by

$$r_{yx} = Cr_{xx}$$

Remark 3. Notice that the difference equation in (6.20) for the matrix P is the same as Equation (5.15), which was used to calculate Lyapunov functions in Chapter 5.

Remark 4. The different terms of (6.20) have good physical interpretations. The covariance P may represent the uncertainty in the state, the term $\Phi P(k)\Phi^T$ tells how the uncertainty at time k propagates due to the system dynamics, and the term R_1 describes the increase of uncertainty due to the disturbance v.

Example 6.3

Consider the first-order system

$$x(k + 1) = ax(k) + v(k)$$

where v is a sequence of uncorrelated random variables with zero mean values and covariances r_1. Let the state at time k_0 have the mean m_0 and the covariance r_0. It follows from (6.18) that the mean value

$$m(k) = Ex(k)$$

is given by

$$m(k + 1) = am(k), \qquad m(k_0) = m_0$$

Hence

$$m(k) = a^{k-k_0}m_0$$

Equation (6.20) gives

$$P(k + 1) = a^2 P(k) + r_1, \qquad P(k_0) = r_0$$

Hence

$$P(k) = a^{2(k-k_0)}r_0 + \frac{1 - a^{2(k-k_0)}}{1 - a^2} r_1$$

Furthermore,

$$r_x(l, k) = a^{l-k}P(k), \qquad l \geq k$$

and

$$r_x(l, k) = a^{k-l}P(l), \qquad l < k$$

If $|a| < 1$ and $k_0 \to -\infty$, it follows that

$$m(k) \to 0$$

$$P(k) \to \frac{r_1}{1 - a^2}$$

$$r_x(k + \tau, k) \to \frac{r_1 a^{|\tau|}}{1 - a^2}$$

The process then becomes stationary because m is constant and the covariance function is a function of τ only.

If an output

$$y(k) = x(k) + e(k)$$

is introduced, where e is a sequence of uncorrelated random variables with zero mean and covariance r_2, it follows that the covariance function of y becomes

$$r_y(\tau) = \begin{cases} r_2 + \dfrac{r_1}{1 - a^2} & \tau = 0 \\[3mm] \dfrac{r_1 a^{|\tau|}}{1 - a^2} & \tau \neq 0 \end{cases}$$

The spectral density is obtained from (6.13). Hence

$$\phi_y(\omega) = \frac{1}{2\pi} \left[r_2 + \frac{r_1}{(e^{i\omega} - a)(e^{-i\omega} - a)} \right]$$

$$= \frac{1}{2\pi} \left[r_2 + \frac{r_1}{1 + a^2 - 2a \cos \omega} \right] \qquad\qquad \square$$

Input-Output Models

For additional insight an input-output description of signals generated by linear difference equations is given. Notice that the signal x given by (6.16) can be described as the output of a linear dynamic system driven by white noise. From this viewpoint it is then natural to investigate how the properties of stochastic processes change when they are filtered by dynamic systems.

Analysis. Consider a system as shown in Fig. 6.8. For simplicity it is assumed that the sampling period is chosen as the time unit. Assume that the input u is a stochastic process with a given mean value function m_u and a given covariance function r_u. Let the impulse response of the system be $\{h(k), k = 0, 1, \ldots\}$. Notice that h has also been used to denote the sampling period. It is, however, clear from the context what h should be. The input-output relationship is

$$y(k) = \sum_{n=-\infty}^{k} h(k - n)u(n) = \sum_{n=0}^{\infty} h(n)u(k - n) \qquad (6.21)$$

Taking mean values

$$m_y(k) = Ey(k) = E \sum_{n=0}^{\infty} h(n)u(k - n)$$

$$= \sum_{n=0}^{\infty} h(n)Eu(k - n) = \sum_{n=0}^{\infty} h(n)m_u(k - n) \qquad (6.22)$$

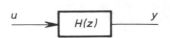

Figure 6.8 Generation of disturbances by driving dynamic sytems with white noise.

The mean value of the output is thus obtained by sending the mean value of the input through the system.

To determine the covariance, first observe that a subtraction of (6.22) from (6.21) gives

$$y(k) - m_y(k) = \sum_{n=0}^{\infty} h(n)[u(k - n) - m_u(k - n)]$$

The difference between the input signal and its mean value thus propagates through the system in the same way as the input signal itself. When calculating the covariance, it can be assumed that the mean values are zero. This simplifies the writing.

The definition of the covariance function gives

$$r_y(\tau) = Ey(k + \tau)y^T(k)$$

$$= E \sum_{n=0}^{\infty} h(n)u(k + \tau - n) \left[\sum_{l=0}^{\infty} h(l)u(k - l) \right]^T$$

$$= \sum_{n=0}^{\infty} \sum_{l=0}^{\infty} h(n)E[u(k + \tau - n)u^T(k - l)h^T(l)$$

$$= \sum_{n=0}^{\infty} \sum_{l=0}^{\infty} h(n)r_u(\tau + l - n)h^T(l)$$

(6.23)

A similar calculation gives the following formula for the cross-covariance of the input and the output:

$$r_{yu}(\tau) = Ey(k + \tau)u^T(k)$$

$$= E \sum_{n=0}^{\infty} h(n)u(k + \tau - n)u^T(k)$$

$$= \sum_{n=0}^{\infty} h(n)E[u(k + \tau - n)u^T(k)]$$

$$= \sum_{n=0}^{\infty} h(n)r_u(\tau - n)$$

(6.24)

Notice that it has been assumed that all infinite sums exist and that the operations of infinite summation and mathematical expectation have been freely exchanged in these calculations. This must of course be justified; it is easy to do in the sense of mean square convergence, if it is assumed that the fourth moment of the input signal is finite.

The relations expressed by Equations (6.23) and (6.24) can be expressed in a simpler form if spectral densities are introduced.

The definition of spectral density in (6.13) gives

$$\phi_y(\omega) = \phi_{yy}(\omega) = \frac{1}{2\pi} \sum_{n=-\infty}^{\infty} e^{in\omega} r_y(n)$$

Introducing r_y from (6.23) gives

$$\phi_y(\omega) = \frac{1}{2\pi} \sum_{n=-\infty}^{\infty} e^{-i\omega n} \sum_{k=0}^{\infty} \sum_{l=0}^{\infty} h(k) r_u(n + l - k) h^T(l)$$

$$= \frac{1}{2\pi} \sum_{k=0}^{\infty} \sum_{n=-\infty}^{\infty} \sum_{l=0}^{\infty} e^{-ik\omega} h(k) e^{-i(n+l-k)\omega} r_u(n + l - k) e^{il\omega} h^T(l)$$

$$= \frac{1}{2\pi} \sum_{k=0}^{\infty} e^{-ik\omega} h(k) \sum_{n=-\infty}^{\infty} e^{-in\omega} r_u(n) \sum_{l=0}^{\infty} e^{il\omega} h^T(l)$$

Introduce the pulse-transfer function H of the system. This is related to the impulse response h by

$$H(z) = \sum_{n=0}^{\infty} z^{-n} h(n) \tag{6.25}$$

The equation for the spectral density can then be written as

$$\phi_y(\omega) = H(e^{i\omega}) \phi_u(\omega) H^T(e^{-i\omega})$$

Similarly,

$$\phi_{yu}(\omega) = \frac{1}{2\pi} \sum_{n=-\infty}^{\infty} e^{-in\omega} r_{yu}(n)$$

$$= \frac{1}{2\pi} \sum_{n=-\infty}^{\infty} e^{-in\omega} \sum_{k=0}^{\infty} h(k) r_u(n - k)$$

$$= \frac{1}{2\pi} \sum_{k=0}^{\infty} e^{-ik\omega} h(k) \sum_{n=-\infty}^{\infty} e^{-in\omega} r_u(n)$$

$$= H(e^{i\omega}) \phi_u(\omega)$$

Main result.   To obtain the general result, the propagation of the mean value through the system must also be investigated.

Theorem 6.2—Filtering of stationary processes.   Consider a stationary discrete-time dynamic system with sampling period 1 and the pulse-transfer function H. Let the input signal be a stationary stochastic process with mean m_u and spectral density ϕ_u. If the system is stable, then the output is also a stationary process with the mean

$$m_y = H(1) m_u \tag{6.26}$$

and the spectral density

$$\phi_y(\omega) = H(e^{i\omega})\phi_u(\omega)H^T(e^{-i\omega}) \tag{6.27}$$

The cross-spectral density between the input and the output is given by

$$\phi_{yu}(\omega) = H(e^{i\omega})\phi_u(\omega) \tag{6.28}$$

Remark 1. The result has a simple physical interpretation. The number $|H(e^{i\omega})|$ is the steady-state amplitude of the response of the system to a sine wave with frequency ω. The value of the spectral density of the output is then the product of the power gain $|H(e^{i\omega})|^2$ and the spectral density of the input $\phi_u(\omega)$.

Remark 2. It follows from Equation (6.28) that the cross-spectral density is equal to the transfer function of the system if the input is white noise with unit spectral density. This fact can be used to determine the pulse-transfer function of a system.

The result is illustrated by an example.

Example 6.4

Consider the process $\{x(k)\}$ in Example 6.3. From the input-output point of view, the process can be thought of as generated by sending white noise through a filter with the pulse-transfer function

$$H(z) = \frac{1}{z - a}$$

Because the spectral density of $\{v(k)\}$ is

$$\phi_v(\omega) = \frac{r_1}{2\pi}$$

it follows from (6.27) that the spectral density of $\{x(k)\}$ is

$$\phi_x(\omega) = H(e^{i\omega})H(e^{-i\omega})\frac{r_1}{2\pi}$$

$$= \frac{r_1}{2\pi} \cdot \frac{1}{(e^{i\omega} - a)(e^{-i\omega} - a)} = \frac{r_1}{2\pi(1 + a^2 - 2a\cos\omega)}$$

The process

$$y(k) = x(k) + e(k)$$

then has the spectral density

$$\phi_y(\omega) = \frac{1}{2\pi}\left[r_2 + \frac{r_1}{1 + a^2 - 2a\cos\omega} \right]$$

(Compare with the calculation in Example 6.3.) ☐

Spectral Factorization

Theorem 6.2 gives the spectral density of a stochastic process obtained by filtering another stochastic process. The spectral density of a signal obtained by filtering white noise is obtained as a special case. The inverse problem is discussed next. A linear system that gives an output with a given spectral density when driven by white noise will be determined. This problem is important because it shows how a signal with a given spectral density can be generated by filtering white noise. The solution to the problem will also tell how general the model in (6.16) is. It follows from Theorem 6.2 that the random process generated from a linear system with a white-noise input has the spectral density given by (6.27). If the system is finite-dimensional, H is then a rational function in $\exp(i\omega)$ and the spectral density ϕ will also be rational in $\exp(i\omega)$ or equivalently in $\cos \omega$. With a slight abuse of language, such a spectral density is called *rational*. Introducing

$$z = e^{i\omega}$$

the right-hand side of (6.27) can be written as

$$F(z) = H(z)H^T(z^{-1})$$

If z_i is a zero of $H(z)$, then z_i^{-1} is a zero of $H(z^{-1})$. The zeros of the function F are thus symmetric with respect to the real axis and mirrored in the unit circle. If the coefficients of the rational function H are real, the zeros of the function F will also be symmetric with respect to the real axis. The same argument holds for the poles of H. The poles and zeros of F will thus have the pattern shown in Fig. 6.9.

It is now straightforward to find a function H that corresponds to a given rational spectral density as follows: First, determine the poles p_i and the zeros z_i of the function F associated with the spectral density. It follows from the symmetry of the poles and zeros, which has just been established, that the poles and zeros always appear in pairs such that

$$z_i z_j = 1$$

$$p_i p_j = 1$$

In each pair choose the pole or the zero that is less than or equal to one in magnitude; then form the desired transfer function from the chosen poles and

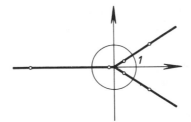

Figure 6.9 Symmetry of the poles and zeros of the spectral-density function.

zeros as

$$H(z) = K \frac{\prod (z - z_i)}{\prod (z - p_i)} = \frac{B(z)}{A(z)}$$

Since the stochastic process is stationary, the chosen poles p_i will all be strictly less than one in magnitude. There may, however, be zeros that have unit magnitude. The result is summarized as follows.

Theorem 6.3—Spectral factorization theorem. Given a spectral density $\phi(\omega)$, which is rational in $\cos \omega$, there exists a linear system with the pulse-transfer function

$$H(z) = \frac{B(z)}{A(z)} \tag{6.29}$$

such that the output obtained when the system is driven by white noise is a stationary random process with spectral density ϕ. The polynomial $A(z)$ has all its zeros inside the unit disc. The polynomial B has all its zeros inside the unit disc or on the unit circle.

Remark 1. The spectral factorization theorem is very important. It implies that all stationary random processes can be thought of as being generated by stable linear systems driven by white noise, i.e., an ARMA process of a special type. This means a considerable simplification both in theory and practice. It is sufficient to understand how systems behave when excited by white noise. It is sufficient to be able to simulate white noise. All other stationary processes with rational spectral density can then be formed by filtering.

Remark 2. Since a continuous function can be approximated uniformly arbitrarily well on a compact interval with a rational function, it follows that the models in (6.16) and (6.21) can give signals whose spectra are arbitrarily close to any continuous function. Notice, however, that there are models with nonrational spectral densities. In turbulence theory, for instance, there are spectral densities that decay as fractional powers of ω for large ω.

An important consequence of the spectral factorization theorem is that for systems with one output, it is always possible to represent the net effect of all disturbances with *one* equivalent disturbance. This disturbance is obtained by calculating the total spectral density of the output signal and applying the spectral factorization theorem.

Remark 3. It is often assumed that the polynomial $B(z)$ has all its zeros inside the unit disc. This means that the inverse of the system H is stable.

The results are illustrated by an example.

Example 6.5

Consider the process $\{y(k)\}$ of Examples 6.3 and 6.4. This process has the spectral density

$$\phi_y(\omega) = \frac{1}{2\pi}\left[r_2 + \frac{r_1}{(z - a)(z^{-1} - a)}\right]_{z=e^{i\omega}}$$

$$= \frac{1}{2\pi}\left[\frac{r_1 + r_2(1 + a^2) - r_2a(z + z^{-1})}{(z - a)(z^{-1} - a)}\right]_{z=e^{i\omega}}$$

The denominator is already in factored form. To factor the numerator, we observe that it can be written as

$$\lambda^2(z - b)(z^{-1} - b) \equiv r_1 + r_2(1 + a^2) - r_2a(z + z^{-1})$$

Identification of coefficients of equal powers of z gives

$$z^0: \quad \lambda^2(1 + b^2) = r_1 + r_2(1 + a^2)$$

$$z^1: \quad \lambda^2 b - r_2 u$$

Elimination of λ gives a second-order algebraic equation for b. This equation has the solution

$$b = \frac{r_1 + r_2(1 + a^2) - \sqrt{[r_1 + r_2(1 + a)^2][r_1 + r_2(1 - a)^2]}}{2ar_2}$$

The other root is discarded because it is outside the unit disc. Furthermore, the variable λ is given by

$$\lambda^2 = \frac{1}{2}\{r_1 + r_2(1 + a^2) + \sqrt{[r_1 + r_2(1 + a)^2][r_1 + r_2(1 - a)^2]}\}$$

$\square$

Innovation's Representations

Theorem 6.3 has some conceptually important consequences. It follows from the theorem that a process with rational spectral density can be represented as

$$y(k) = \sum_{n=-\infty}^{k} h(k - n)e(n) \tag{6.30}$$

where e is discrete-time white noise and h is the impulse response that corresponds to the pulse-transfer function (6.29). The system has a stable inverse if the polynomial $B(z)$ has all its zeros inside the unit disc. This means that

$$e(k) = \sum_{n=-\infty}^{k} g(k - n)y(n)$$

where g is the impulse response, which corresponds to the stable pulse-transfer function $A(z)/B(z)$. It thus follows that the sequences $y(k), y(k - 1), \ldots$ and $e(k), e(k - 1), \ldots$ are equivalent in the sense that one sequence can be calculated from the other.

Now consider

$$y(k + 1) = \sum_{n=-\infty}^{k+1} h(k + 1 - n)e(n)$$

$$= \sum_{n=-\infty}^{k} h(k + 1 - n)e(n) + h(0)e(k + 1)$$

$$= \sum_{n=-\infty}^{k} h(k + 1 - n) \sum_{l=-\infty}^{k} g(n - l)y(l) + h(0)e(k + 1)$$

The variable $y(k + 1)$ can be written as the sum of two terms: One term is a linear function of $y(k)$, $y(k - 1)$, . . . , and the other term is $h(0)e(k + 1)$. Thus $e(k + 1)$ can be interpreted as the part of $y(k + 1)$ that contains new information that is not available in the past values $y(k)$, $y(k - 1)$, ... The stochastic process $\{e(k)\}$ is therefore called the *innovations* of the process $\{y(k), k \in T\}$ and the representation in (6.30) is called the *innovation's representation* of the process. This representation is important in connection with filtering and prediction problems.

The term

$$\sum_{n=-\infty}^{k} h(k + 1 - n) \sum_{l=-\infty}^{n} g(n - l)y(l)$$

is in fact the best mean-square prediction of $y(k + 1)$ based on $y(k)$, $y(k - 1)$, ... This will be discussed in detail in Chapters 11 and 12.

Example 6.6

Consider the process $\{y(k)\}$ of Example 6.3. The process has the spectral density

$$\phi_y(\omega) = \frac{1}{2\pi} \left[r^2 + \frac{r_1}{1 + a^2 - 2a \cos \omega} \right]$$

It follows from Example 6.5 that the spectral density can be factored as

$$\phi_y(\omega) = \frac{\lambda^2}{2\pi} \frac{(z - b)(z^{-1} - b)}{(z - a)(z^{-1} - a)}$$

The process y can thus be generated by sending white noise through a system with the pulse-transfer function

$$H(z) = \frac{z - b}{z - a}$$

The input-output relation of such a system can be written as

$$y(k + 1) = ay(k) + e(k + 1) - be(k)$$

where $\{e(k)\}$ is white noise with variance λ^2. $\qquad\square$

Disturbance Models   Chap. 6

Calculation of Variances

The variance of a signal obtained by filtering white noise can be calculated from the recursive equation of (6.20), if the model is given in state-space form. For a system described by transfer functions it is possible to use the same equations, if the model is first transformed to state-space form. It is naturally convenient to have similar formulas when the system is given in input-output form. Such formulas will now be given.

Consider a signal generated by

$$y(k) = \frac{B(q)}{A(q)} e(k) \tag{6.31}$$

where e is white noise with unit variance. It follows from Theorem 6.2 that the spectral density of the signal y is given by

$$\phi(\omega) = \frac{1}{2\pi} \frac{B(z)B(z^{-1})}{A(z)A(z^{-1})}$$

where $z = \exp(i\omega)$. It also follows from Theorem 6.2 that the variance of the signal y is given by the complex integral

$$Ey^2 = \int_{-\pi}^{\pi} \phi(\omega)\, d\omega$$

$$= \frac{1}{i} \int_{-\pi}^{\pi} \phi(\omega) e^{-i\omega}\, d(e^{i\omega}) \tag{6.32}$$

$$= \frac{1}{2\pi i} \oint \frac{B(z)B(z^{-1})}{A(z)A(z^{-1})} \frac{dz}{z}$$

The evaluation of integrals of this form is closely related to Jury's stability test (compare with Sec. 5.2). To evaluate the integral, the following table is formed.

a_0	a_1	$\cdots$	a_{n-1}	a_n		b_0	b_1	$\cdots$	b_{n-1}	b_n	
a_n	a_{n-1}	$\cdots$	a_1	a_0	α_n	a_n	a_{n-1}	$\cdots$	a_1	a_0	β_n
a_0^{n-1}	a_1^{n-1}	$\cdots$	a_{n-1}^{n-1}			b_0^{n-1}	b_1^{n-1}	$\cdots$	b_{n-1}^{n-1}		
a_{n-1}^{n-1}	a_{n-2}^{n-1}	$\cdots$	a_0^{n-1}		$n-1$	a_{n-1}^{n-1}	a_{n-2}^{n-1}	$\cdots$	a_0^{n-1}		β_{n-1}
$\vdots$											
a_0^1	a_1^1					b_0^1	b_1^1				
a_1^1	a_0^1				α_1	a_1^1	a_0^1				β_1
a_0^0					1	b_0^0					β_0

where

$$\alpha_n = a_n/a_0 \qquad \beta_n = b_n/a_0$$

$$\alpha_k = a_k^k/a_0^k \qquad \beta_k = b_k^k/a_0^k$$

and

$$a_i^{k-1} = a_i^k - \alpha_k a_{k-i}^k$$

$$b_i^{k-1} = b_i^k - \beta_k a_{k-i}^k$$

The left half of the table is the same as Jury's stability test. The right half is built up in the same way with the exception that the even rows are taken from the left half of the table. The following theorem results.

Theorem 6.4. The integral (6.32) is given by

$$I_n = \frac{1}{a_0} \sum_{i=0}^{n} b_i^i \beta_i \tag{6.33}$$

Application of the theorem gives the following values of the integral for $n = 1$ and 2.

$$I_1 = \frac{(b_0^2 + b_1^2)a_0 - 2b_0b_1a_1}{a_0(a_0^2 - a_1^2)}$$

$$I_2 = \frac{B_0a_0e_1 - B_1a_0a_1 + B_2(a_1^2 - a_2e_1)}{a_0[(a_0^2 - a_2^2)e_1 - (a_0a_1 - a_1a_2)a_1]}$$

where

$$B_0 = b_0^2 + b_1^2 + b_2^2$$

$$B_1 = 2(b_0b_1 + b_1b_2)$$

$$B_2 = 2b_0b_2$$

$$e_1 = a_0 + a_2$$

6.6 Continuous-Time Stochastic Processes

It may be useful to formulate models and specifications in continuous time even if a computer is used to implement the control law. A brief account of continuous-time stochastic processes is therefore given.

Definitions

Continuous-time stochastic processes can be defined in the same way as discrete-time processes. The only difference is that the index set T is the set of real variables instead of a discrete set. Covariance functions and stationary processes are defined as for discrete-time processes using the finite-dimensional distribution functions. A spectral density can also be introduced as the Fourier transform of the covariance function. Equation (6.13) is then replaced by

$$\phi_{xy}(\omega) = \frac{1}{2\pi} \int_{-\infty}^{\infty} e^{-i\omega t} r_{xy}(t) \, dt \tag{6.34}$$

The inverse transform is given by

$$r_{xy}(t) = \int_{-\infty}^{\infty} e^{i\omega t} \phi_{xy}(\omega) \, d\omega \tag{6.35}$$

which replaces (6.14). The spectral density has the same interpretation as for discrete-time systems.

White Noise

White noise is defined as a stationary process with constant spectral density. If

$$\phi(\omega) = \frac{r_0}{2\pi} \tag{6.36}$$

it follows formally from (6.35) that the corresponding covariance is a delta function, i.e.,

$$r(t) = r_0 \delta(t) \tag{6.37}$$

Continuous-time white noise thus has the property that values of the signal at different times are uncorrelated as for discrete-time white noise. Continuous-time white noise has, however, infinite variance. This will cause some mathematical difficulties. Intuitively, continuous-time white noise is analogous to delta functions in the theory of linear systems.

Some of the difficulties with continuous-time white noise can be avoided by introducing a stochastic process that formally is the time integral

$$w(t) = \int_0^t e(s) \, ds \tag{6.38}$$

of white noise e. The stochastic process w has zero mean value. Its increments over disjoint intervals are uncorrelated. If the covariance function of e is

$$\operatorname{cov}[e(t), e(s)] = r_0 \delta(t - s)$$

then the variances of the increments of w are given by

$$E[w(t) - w(s)]^2 = |t - s| \cdot r_0$$

The stochastic process $\{w(t), t \in T\}$ is called a *Wiener process* if it also is Gaussian. The Wiener process is a model for random walk. The infinitesimal increment

$$dw = w(t + dt) - w(t)$$

has the variance

$$E(dw)^2 = r_0 \, dt$$

The increment dw thus has the magnitude $\sqrt{r_0\, dt}$ in the mean-square sense. The number $r_0\, dt$ is called the *incremental covariance* of the Wiener process.

State-Space Models

State models for continuous-time processes can be obtained by a formal generalization of (6.16) to

$$\frac{dx}{dt} = Ax + \dot{v}$$

where $\dot{v}$ is a vector whose elements are white-noise stochastic processes. Since $\dot{v}$ has infinite variance, it is customary to write the equation in terms of differentials as

$$dx = Ax\, dt + dv \tag{6.39}$$

where v is the integral of $\dot{v}$.

The signal v is thus assumed to have zero mean, uncorrelated increments, and the variance

$$\mathrm{cov}[v(t),\, v(t)] = R_1 t \tag{6.40}$$

It is also assumed that dv is uncorrelated with x.

A precise meaning can be given to (6.39) without any reference to white noise. This form is therefore common in mathematically oriented texts. The form is also useful as a reminder that dv has a magnitude proportional to $\sqrt{dt}$.

Equation (6.39) is called a *stochastic differential equation*. To specify it fully, it is also necessary to give the initial probability distribution of x at the starting time. The following continuous-time analog of Theorem 6.1 is then obtained.

Theorem 6.5. Consider a stochastic process defined by the linear stochastic differential Equation (6.39) where the process v has zero mean and incremental covariance $R_1\, dt$. Let the initial state have mean m_0 and covariance R_0. The mean-value function of the process x is then given by

$$\frac{dm(t)}{dt} = Am(t), \qquad m(0) = m_0 \tag{6.41}$$

and the covariance function is given by

$$\mathrm{cov}[x(s),\, x(t)] = e^{A(s-t)}P(t), \qquad s \geq t \tag{6.42}$$

where $P(t) = \mathrm{cov}[x(t),\, x(t]$ is given by

$$\frac{dP(t)}{dt} = AP(t) + P(t)A^T + R_1, \qquad P(0) = R_0 \tag{6.43}$$

Proof. The formula (6.41) for the mean value is obtained simply by taking the mean value of (6.39). Notice that dv has zero mean.

To obtain the differential equation in (6.43), notice that

$$d(xx^T) = (x + dx)(x + dx)^T - xx^T$$
$$= x\, dx^T + dx\, x^T + dx\, dx^T$$

Equation (6.39) then gives

$$d(xx^T) = x[Ax\, dt + dv]^T + [Ax\, dt + dv]x^T + [Ax\, dt + dv][Ax\, dt + dv]^T$$

Taking mean values gives

$$d(Exx^T) = (Exx^T)A^T\, dt + A(Exx^T)\, dt + E\, dv\, dv^T + A(Exx^T)A^T(dt)^2$$

because dv is uncorrelated with x. Furthermore, it follows from (6.40) that

$$E\, dv\, dv^T = R_1\, dt$$

Hence

$$dP = PA^T\, dt + AP\, dt + R_1\, dt + APA^T(dt)^2$$

Dividing by dt and taking the limit as dt goes to zero gives the differential equation in (6.43). To obtain Equation (6.42), let $s \geq t$ and integrate (6.39). Hence

$$x(s) = e^{A(s-t)}x(t) + \int_t^s e^{A(s-s')}\, dv(s')$$

Multiplying by $x^T(t)$ from the right and taking mathematical expectation gives (6.42). Notice that $dv(s')$ is uncorrelated with $x(t)$ if $s' \geq t$.   ⊔

Example 6.7

Consider the scalar stochastic differential equation

$$dx = -ax\, dt + dv$$
$$x(t_0) = m_0, \qquad \text{var}[x(t_0)] = r_0$$

where the process $\{v(t), t \in T\}$ has incremental covariance $r_1\, dt$. It follows from (6.41) that the mean-value function is given by

$$\frac{dm}{dt} = -am, \qquad m(t_0) = m_0$$

This equation has the solution

$$m(t) = m_0 e^{-a(t - t_0)}$$

The covariance function is given by

$$r(s, t) = \text{cov}[x(s), x(t)] = e^{-a(s-t)}P(t), \qquad s \geq t$$

and

$$r(s, t) = e^{-a(t-s)}P(s), \qquad s \leq t$$

Equation (6.43) gives the following differential equation for P.

$$\frac{dP}{dt} = -2aP + r_1, \qquad P(t_0) = r_0$$

This differential equation has the solution

$$P(t) = e^{-2a(t-t_0)}r_0 + \int_{t_0}^{t} e^{-2a(t-s)}r_1 \, ds$$

$$= e^{-2a(t-t_0)}r_0 + \frac{r_1}{2a}[1 - e^{-2a(t-t_0)}]$$

As $t_0 \to -\infty$, the mean value function goes to zero and the covariance function goes to

$$r(s, t) = \frac{r_1}{2a} e^{-a|t-s|}$$

Since the limiting covariance function depends only on the argument difference $s - t$, the limiting process is (weakly) stationary and its covariance function can be written as

$$r(\tau) = \frac{r_1}{2a} e^{-a|\tau|}$$

Equation (6.34) gives the corresponding spectral density

$$\phi(\omega) = \frac{r_1}{2\pi} \frac{1}{\omega^2 + a^2} \qquad \square$$

Filtering of Continuous-Time Processes

The analysis of linear systems with continuous-time stochastic processes as inputs is analogous to the corresponding analysis for discrete-time systems. Consider a time-invariant stable system with impulse response h. The input-output relationship is

$$y(t) = \int_{-\infty}^{t} h(t - s)u(s) \, ds = \int_{0}^{\infty} h(s)u(t - s) \, ds \qquad (6.44)$$

[compare with Equation (6.21)]. Let the input signal u be a stochastic process with mean-value function m_u and covariance function r_u.

The following result is analogous to Theorem 6.2 for discrete-time systems.

Theorem 6.6—Filtering of stationary processes. Consider a stationary linear system with the transfer function G. Let the input signal be a stationary continuous-time stochastic process with mean value m_u and spectral density ϕ_u. If the system is stable, then the output is also a stationary process with the mean value

$$m_y = G(0)m_u \qquad (6.45)$$

and the spectral density

$$\phi_y(\omega) = G(i\omega)\phi_u(\omega)G^T(-i\omega) \qquad (6.46)$$

The cross-spectral density between the input and the output is given by

$$\phi_{yu}(\omega) = G(i\omega)\phi_u(\omega) \tag{6.47}$$

The result may be interpreted in the same way as the corresponding result for discrete-time systems. Compare with Remarks 1 and 2 of Theorem 6.2.

Example 6.8

Consider the system in Example 6.7. The process x can be considered as the result of filtering white noise with the variance $r_1/2\pi$ through a system having the transfer function

$$G(s) = \frac{1}{s + a}$$

It follows from (6.46) that the spectral density is given by

$$\phi(\omega) = \frac{r_1}{2\pi} \cdot \frac{1}{i\omega + a} \cdot \frac{1}{-i\omega + a} = \frac{r_1}{2\pi} \cdot \frac{1}{\omega^2 + a^2} \qquad \square$$

Spectral Factorization

It follows from (6.46) that if the input is white noise with $\phi_u = 1$, then the spectral density of the output is given by

$$\phi_y(\omega) = G(i\omega)G^T(-i\omega) \tag{6.48}$$

This means that any disturbance whose spectral density can be written in this form may be generated by sending continuous-time white noise through a filter with the transfer function G.

Since finite linear dimensional systems have rational transfer functions, it follows that signals with arbitrary rational spectral densities can be generated from linear finite-dimensional systems. The covariance function is nonnegative and symmetric. It then follows from (6.34) that ϕ is also symmetric. If ϕ is rational, it then follows that its poles and zeros are symmetric with respect to the real and imaginary axes. The transfer function G in (6.48) can then be chosen so that all its poles are in the left half-plane and all its zeros in the left half-plane or on the imaginary axis. The following analog of Theorem 6.3 is thus obtained.

Theorem 6.7—Spectral factorization. Given a rational spectral density $\phi(\omega)$, there exists a finite-dimensional linear system with the rational transfer function

$$G(s) = \frac{B(s)}{A(s)}$$

such that the output obtained when the system is driven by white noise is a stationary stochastic process with the given spectral density. The polynomial A has all its zeros in the left half-plane. The polynomial B has no zeros in the right half-plane.

6.7 Sampling a Stochastic Differential Equation

If process models are presented in continuous time as stochastic differential equations, it is useful to sample these equations to obtain a discrete-time model.

Consider a process described by

$$dx = Ax \, dt + dv \tag{6.49}$$

where the process v has zero mean value and uncorrelated increments. The incremental covariance of v is $R_1 \, dt$. Let the sampling instants be $\{t_k; k = 0, 1, \ldots\}$. Integration of (6.49) over one sampling period gives

$$x(t_{k+1}) = e^{A(t_{k+1} - t_k)} x(t_k) + \int_{t_k}^{t_{k+1}} e^{A(t_{k+1} - s)} \, dv(s)$$

Consider the random variable

$$e(t_k) = \int_{t_k}^{t_{k+1}} e^{A(t_{k+1} - s)} \, dv(s)$$

This variable has zero mean because v has zero mean. The random variables $e(t_k)$ and $e(t_l)$ are also uncorrelated for $k \neq l$ because the increments of v over disjoint intervals are uncorrelated. The covariance of $e(t_k)$ is given by

$$E[e(t_k), e^T(t_k)] = E \iint_{t_k}^{t_{k+1}} e^{A(t_{k+1} - s)} \, dv(s) \, dv^T(t) \, e^{A^T(t_{k+1} - t)}$$

$$= \int_{t_k}^{t_{k+1}} e^{A(t_{k+1} - s)} R_1 e^{A^T(t_{k+1} - s)} \, ds \tag{6.50}$$

It is thus found that the random sequence $\{x(t_k), k = 0, 1, \ldots\}$ obtained by sampling the process $\{x(t)\}$ is described by the difference equation

$$x(t_{k+1}) = e^{A(t_{k+1} - t_k)} x(t_k) + e(t_k)$$

where $\{e(t_k)\}$ is a sequence of uncorrelated random variables with zero mean and covariance (6.50).

6.8 Conclusions

The main purpose of this chapter is to develop mathematical models for disturbances. The result is a uniform approach to models for a wide variety of signals. The signals are viewed as being generated from dynamic systems driven by a pulse, a sequence of pulses, or white noise. Equivalently, the signals may be considered as being generated by dynamic systems with initial conditions.

It is first shown in Sec. 6.3 that the simple disturbances like step, ramp, and sinusoid can be generated as outputs of linear systems driven by a pulse. More complicated disturbances may be viewed as pulse responses of more complicated systems.

Section 6.4 shows that the class of disturbances could be widened by driving the systems with signals composed of several pulses. This leads to the piecewise deterministic signals. A further extension is given in Sec. 6.5, where the input signal to the disturbance-generating system was chosen as white noise.

A unified way of modeling different types of disturbances is obtained. The disturbances are characterized by a dynamic system

$$y(k) = \frac{B(q)}{A(q)} \epsilon(k) \tag{6.51}$$

where the input ϵ is a pulse, several pulses, or white noise. The system is called *the disturbance generator*. The dynamic system can, of course, also be represented in state-space form.

The problem of prediction is important when controlling systems with disturbances that cannot be measured. The problem of predicting a signal given by (6.51) is in essence to compute ϵ from y. This is the same as inverting the dynamic system (6.51). To obtain a stable inverse, the polynomial $B(q)$ must then have all its zeros inside the unit disc. This will, in general, not be the case for deterministic disturbances. A consequence is that the performance of the prediction will deteriorate. It takes longer to obtain the prediction. For a stochastic system it follows from the spectral factorization theorem that the polynomial $B(q)$ has all its zeros inside the unit disc or on the unit circle.

The essential point of the discussion is that the predictors for the signals are uniquely given by the pulse-transfer function $H = B/A$. The predictors are thus the same for inputs that are pulses, pulse trains, or white noise. This means that predictors that are designed for deterministic disturbances can work very well also for stochastic disturbances if the disturbance generators for the signals are the same.

The unified approach to model disturbances also leads to a substantial simplification of the theory because it is sufficient to work with a few prototypes for disturbances only.

6.9 Problems

6.1. List situations in which it is possible to reduce the influence of disturbances by (a) reduction at the source, (b) local feedback, and (c) prediction.

6.2. Determine H_d in Fig. 6.4 such that the sampled output is (a) a sinusoid with frequency ω rad/s and (b) equal to $t \exp(-t)$ at the sampling points when the input is a pulse.

6.3. Show that the predictor (6.4) is equivalent to the predictor (6.3).

6.4. Determine the m-step predictor for the disturbance model

$$y(k) = \frac{C(q)}{A(q)} w(k)$$

where $w(k)$ is zero except at isolated points that are spaced more than deg A. Use the result to determine the signal and the prediction when $A(q) = q - 0.5$, $C(q) = q$, and $m = 3$ and when $w(k)$ is zero except for $k = 0$ and 5. The initial conditions are assumed to be equal to zero.

6.5. Use Theorem 6.1 to compute the stationary covariance function of the process

$$x(k + 1) = \begin{bmatrix} 0.4 & 0 \\ -0.6 & 0.2 \end{bmatrix} x(k) + v(k)$$

where v is a white-noise process with zero mean and the variance

$$R_1 = \begin{bmatrix} 1 & 0 \\ 0 & 2 \end{bmatrix}$$

6.6. Consider a stationary stochastic process generated by

$$x(k + 1) = \Phi x(k) + v(k)$$
$$y(k) = Cx(k)$$

where $v(k)$ is a sequence of equally distributed, independent, zero-mean, stochastic variables. Let Φ have the characteristic equation

$$z^n + a_1 z^{n-1} + \cdots + a_n = 0$$

Show that the autocovariance function of the output $r_y(\tau)$ satisfies

$$r_y(\tau) + a_1 r_y(\tau - 1) + \cdots + a_n r_y(\tau - n) = 0$$

for $\tau \geq n + 1$. (This equation is called the *Yule-Walker equation*.)

6.7. Consider the process

$$x(k + 1) = \begin{bmatrix} -a & 0 \\ 0 & -b \end{bmatrix} x(k) + v(k)$$

$$y(k) = \begin{bmatrix} 1 & 1 \end{bmatrix} x(k)$$

where $v(k)$ is white noise with zero mean and the covariance matrix

$$R_1 = \begin{bmatrix} \sigma_1^2 & 0 \\ 0 & \sigma_2^2 \end{bmatrix}$$

Show that $y(k)$ can be represented in the form

$$y(k) = \lambda \frac{q + c}{(q + a)(q + b)} e(k)$$

where $e(k)$ is white noise with zero mean and unit variance. Find the relationship from which λ and c can be determined.

6.8. A stochastic process $y(k)$ is described by

$$x(k + 1) = ax(k) + v(k)$$
$$y(k) = x(k) + e(k)$$

where v and e are normally distributed white-noise processes with the properties

$$Ev = Ee = 0$$
$$\text{Var } v = 1$$
$$\text{Var } e = r_2$$
$$Ev(k)e(j) = r_{12} \quad \text{when } k = j \text{ and } 0 \text{ otherwise}$$

Show that $y(k)$ can be represented as the output of a linear filter

$$y(k) = \lambda \frac{q - c}{q - a} \epsilon(k), \quad |c| \leq 1$$

where $\epsilon(k)$ is white noise with zero mean and unit variance. Determine λ and c.

6.9. Determine the covariance function, $r_y(\tau)$, and the spectrum, $\phi_y(\omega)$, of the process $y(k)$ when

$$y(k) - 0.7y(k - 1) = e(k) - 0.5e(k - 1)$$

where $e(k)$ is white noise with unit variance.

6.10. Determine the variance of the stochastic process $y(k)$ defined by

$$y(k) - 1.5y(k - 1) + 0.7y(k - 2) = e(k) + 0.2e(k - 1)$$

where e is white noise with unit variance.

6.11. Calculate the stationary covariance function $r_y(\tau)$, $\tau = 0, 1, 2, \ldots$ for the system

$$y(k) = e(k) - 2e(k - 1) + 3e(k - 2) - 4e(k - 3)$$

when e is zero-mean white noise with unit variance.

6.12. Assume that we want to generate a signal $y(k)$ with the spectral density

$$\phi_y(\omega) = \frac{1}{1.36 + 1.2 \cos \omega}$$

(a) Determine a stable filter $H(q)$ that gives the desired signal $y(k) = H(q)e(t)$ where e is white noise such that $e \in N(0, 1)$.
(b) What is the variance of y?

6.13. Consider the discrete-time system

$$x(k + 1) = \begin{pmatrix} 0.3 & 0.2 \\ 0 & 0.5 \end{pmatrix} x(k) + \begin{pmatrix} 0 \\ 1 \end{pmatrix} u(k) + v(k)$$

$$y(k) = (1 \quad 0)x(k) + e(k)$$

Assume that v and e are white noise processes that are uncorrelated and with the covariances

$$R_1 = \begin{pmatrix} 1 & 0 \\ 0 & 0.5 \end{pmatrix} \quad \text{and} \quad R_2 = 0.2$$

respectively. Assume that the initial value has the zero mean value and the covariance

$$\text{cov}[x(0), x(0)^T] = I$$

Compute the stationary value of the covariance of the state vector.

6.14. Assume that a white noise generator is available that gives a zero mean output with unit variance. Determine a filter that can be used to generate a stochastic signal with the spectral density

$$\phi(\omega) = \frac{3}{2\pi[5.43 - 5.40 \cos(\omega)]}$$

6.10 References

The principles for reducing disturbances by feedback and feedforward and the classic disturbance models are a key element of classic feedback theory. See

BROWN, G. S. and D. P. CAMPBELL (1948): *Principles of Servomechanisms*. New York: John Wiley.

CHESTNUT, H. and R. W. MAYER (1959): *Servomechanisms and Regulating System Design*. (vol. 1), New York: John Wiley.

GILLE, J. C., M. J. PELEGRIN, and P. DECAULNE (1959): *Feedback Control Systems*. New York: McGraw-Hill.

The notion of piecewise deterministic signals was introduced in

ÅSTRÖM, K. J. (1980): "Piece-wise Deterministic Signals," in *Time Series,* ed. O. D. Anderson. Amsterdam: North Holland.

where the formulas for prediction are proven and more details are given. Tables for the integrals for low values of n are given in

JURY, E. I. (1964): *Theory and Application of the z-Transform Method*. New York: John Wiley. Second printing: Krieger, 1973.

The ideas of representing disturbances as stochastic processes was also part of classic feedback theory. See

JAMES, H. M., N. B. NICHOLS, and R. S. PHILIPS (1947): *Theory of Servomechanisms*. New York: McGraw-Hill.

TSIEN, H. S. (1955): *Engineering Cybernetics*. New York: McGraw-Hill.

LANING, J. H. and R. H. BATTIN (1956): *Random Processes in Automatic Control*. New York: McGraw-Hill.

G. C. NEWTON, JR., L. A. GOULD, and J. F. KAISER (1957): *Analytical Design of Linear Feedback Controls*. New York: John Wiley.

A reasonably complete treatment of stochastic processes requires a full book. The books

KARLIN, S. (1966): *A First Course in Stochastic Processes*. New York: Academic Press.

PARZEN, E. (1962): *Stochastic Processes*. San Francisco: Holden-Day.

PAPOULIS, A. (1965): *Probability, Random Variables, and Stochastic Processes*. New York: McGraw-Hill.

CHUNG, K. L. (1974): *A Course in Probability Theory*. New York: Academic Press.

KUMAR, P. R. and P. VARAIYA (1986): *Stochastic Systems: Estimation, Identification and Adaptive Control*. Englewood Cliffs, NJ: Prentice Hall, Inc.

CAINES, P. E. (1988): *Linear Stochastic Systems*. New York: John Wiley.

give a good background. There are also shorter summaries in the books on stochastic control listed below.

Prediction theory originated in the papers

KOLMOGOROV, A. N. (1941): "Interpolation and Extrapolation of Stationary Random Sequences," *Bull. Moscow Univ.,* USSR, Ser. Math. 5.

WIENER, N. (1949): *The Extrapolation, Interpolation, and Smoothing of Stationary Time Series with Engineering Applications*. New York: John Wiley.

KALMAN, R. E. (1960): "A New Approach to Linear Filtering and Prediction Problems," *Trans. ASME, Ser. D., Journal Basic Eng.,* 82, 34–45.

KALMAN, R. E. and R. S. BUCY (1961): "New Results in Linear Filtering and Prediction Theory," *Trans ASME, Ser. D., Journal Basic Eng.,* 83, 95–107.

The papers by Wiener and Kolmogorov are not easy to read. A readable account of Kolmogorov's work is found in

WHITTLE, P. (1963): *Prediction and Regulation by Linear Least-Squares Methods*. London: English Universities Press.

Wiener's results were originally published as an MIT report in 1942. It became known as the "yellow peril" because of its yellow cover and its style of writing. The paper Kalman wrote (1960), which deals with discrete-time processes, is easy to read. There are also full books devoted to prediction, filtering theory, and stochastic control.

ÅSTRÖM, K. J. (1970): *Introduction to Stochastic Control Theory*. New York: Academic Press.

ANDERSON, B. D. O. and J. B. MOORE (1979): *Optimal Filtering*. Englewood Cliffs, N.J.: Prentice Hall, Inc.

BOX, G. E. P. and G. M. JENKINS (1970): *Time Series Analysis, Forecasting, and Control*. San Francisco: Holden-Day.

McGARTY, T. P. (1974): *Stochastic Systems and State Estimation*. New York: John Wiley.

DESIGN: AN OVERVIEW

7

GOAL

To Discuss Control-System Design and to Show How It Fits into Process Design and Operation. To Discuss Structuring and Selection of Control Principles. To Review Methods for Specifications and Design of Simple Loops.

7.1 Introduction

This chapter views the control problem in a wider perspective. In practice, more time is often spent formulating control problems than on solving them. It is therefore useful to be aware of these more general problems, although they are seldom discussed in current textbooks.

Most control problems arise from design of engineering systems. Such problems are typically large-scale and fuzzy. Typical tasks are design of power plants, chemical processes, industrial robots, aircraft, space vehicles, and biomedical systems. Control theory on the other hand deals with small-scale, well-defined problems. A typical problem is to design a feedback law for a given system, which is described by linear differential equations with constant coefficients, so that the closed-loop system has given poles.

A major difficulty in control-system design is to reconcile the large-scale, fuzzy, real problems with the simple, well-defined problems that control theory can handle. It is, however, in this intermediate area that a control engineer can effectively use creativity and ingenuity. This situation is not peculiar to control engineering. Similar situations are encountered in almost all fields of engineering design. Control is, however, one field of engineering where a comparatively sophisticated theory is needed to understand the problems.

It is useful to have some perspective on the design process and a feel for the role of theory in the design process. First, a good engineering design must satisfy a large number of specifications, and there often are many equally good solutions to a design problem. A good design is often a compromise based on reasonable trade-offs between cost and performance. Sadly enough, it is often true that the best is the worst enemy of the good. Consequently, when words like *optimal* are used in this context, they should be taken with a grain of salt.

Another aspect is that design is often arrived at by interaction between customer and vendor. Many subjective factors—such as pride, tradition, and ambition—enter into this interaction. This situation with regard to customer preference is particularly confused when technology is changing. Typical examples are discussions concerning pneumatic or electronic regulators or analog versus digital control, which are abundant in the trade journals.

What theory *can* contribute to the design process is to give insight and understanding. In particular, theory can often pinpoint fundamental limitations on control performance. There are also some idealized design problems, which can be solved theoretically. Such solutions can often give good insight into suitable structures and algorithms.

It is also useful to remember that control problems can be widely different in nature. They can range from design of a simple loop in a given system to design of an integrated control system for a complete process. The approach to design can also be widely different for mass-produced systems, and one-of-a-kind systems. For mass-produced systems, a substantial effort can be made to obtain a cheap system that will give good performance. For unique systems, it is often much better to install a flexible standard system and to tune it *in situ*.

The relation between process design and control design is also important. Control systems have traditionally been introduced into given processes to simplify or improve their operation. It has, however, become clear that much can be gained by considering process design and control design in one context. The availability of a control system always gives the designer an extra degree of freedom, which frequently can be used to improve performance or economy. Similarly, there are many situations where difficult control problems arise because of improper process design. Compare with the discussion of elimination of disturbances in Sec. 6.2.

Some operational aspects of control systems are first discussed in Sec. 7.2. This includes interfaces to the process, the operator, and the computer. Various aspects of design, commissioning, and process operation are also given. The problems of structuring are discussed in Sec. 7.3. The basic problem is to decompose a large, complicated problem into a set of smaller, simpler problems. This includes choice of control principles, selection, and pairing of control variables and measured variables. The common structuring principles—top-down, bottom-up, middle-out, and outside-in—are also discussed. The top-down approach is discussed in Sec. 7.4. This includes choice of control principles and selection and grouping of control signals and measurements. The bottom-up approach is discussed in Sec. 7.5, including a discussion of the elementary control structures, feedback, feedforward, prediction, estimation, optimization, and adaptation. Combinations of these concepts are also discussed. The design of simple loops is discussed in

Sec. 7.6. Specifications of servo and regulation performance are mentioned. Design methods for simple loops are also reviewed. A detailed treatment of these methods is given in the following chapters.

7.2 Operational Aspects

It is useful to understand how the control system interacts with its environment. This section discusses the interfaces between process and regulator design. Commissioning, operation, and modification of the system are also discussed.

Process and Regulator Design

In the early stages of automation, the control system was always designed when the process design was completed. This still happens in many cases. Since process design is largely based on static considerations, the process design can lead to a process that is difficult to control. For this reason, it is very useful to consider the control design jointly with the process design. The fact that a process will be controlled automatically also gives the process designers an additional degree of freedom, which can be used to make better trade-offs. The process and the regulator should therefore be designed together. An example illustrates the idea.

Example 7.1—Elimination of disturbances by mixing

Elimination of inhomogeneities in a product stream is one of the major problems in process control. One possibility for reducing the variations is to introduce large storage tanks and thus increase the material stored in the process. A system with large mixing tanks also has a slow response. It will take a long time to change product quality in such a system. One consequence is that the product may be off the specifications for a considerable time during a change in quality. Another possibility for eliminating inhomogeneities is to measure the product quality and to reduce the variations by feedback control. In this case, it is possible to use much smaller tanks and to get systems with a faster response. The control system does, however, become more complicated. Since the total system will always have a finite bandwidth, *small* mixing tanks must always be used to eliminate rapid variations. □

Stability Versus Controllability (Maneuverability)

It frequently happens that stability and controllability have contradictory requirements. This has been evident in the design of vehicles, for instance. The Wright brothers succeeded in the design of their aircraft because they decided to make a maneuverable, but unstable, aircraft, while their competitors were instead designing stable aircrafts. In ship design, a stable ship is commonly difficult to turn, but a ship that turns easily tends to be unstable. Traditionally, the tendency has been to emphasize stability. It is, however, interesting to see that if a control system is used, the basic system can instead be designed for controllability. The required stability can then be provided by the control system. An example from aircraft design is used to demonstrate that considerable savings can be obtained by this approach.

Example 7.2—Design of a supersonic aircraft

For a high-performance aircraft, which operates over a wide speed range, the center of pressure moves aft with increasing speed. For a modern supersonic fighter, the shift in center of pressure can be about 1 m. If the aircraft is designed so that it is statically stable at subsonic speeds, the center of mass will be a few decimeters in front of the center of pressure at low speed. At supersonic speeds, the distance between the center of mass and the center of pressure will then increase to over 1 m. Thus there will be a very strong stabilizing torque, which tends to keep the airplane on a straight course. The torque will be proportioned to the product of the thrust and the distance between the center of mass and the center of pressure. To maneuver the plane at high speeds, a large rudder is then necessary. A large rudder will, however, give a considerable drag.

There is a considerable advantage to changing the design so that the center of mass is in the middle of the range of variation of the center of pressure. A much smaller rudder can then be used, and the drag induced by the rudder is then decreased. The drag reduction can be over 10%. Such an airplane will, however, be statically unstable at low speeds—i.e., at takeoff and landing! The proper stability can, however, be obtained by using a control system. Such a control system must, of course, be absolutely reliable.

Current thinking in aircraft design is moving in the direction of designing an aircraft that is statically unstable at low speeds and providing sufficient stability by using a control system. Similar examples are common in the design of other vehicles. □

There are analogous cases also in the control of chemical processes. The following is a typical case.

Example 7.3—Exothermic chemical reactor

To obtain a high yield in an exothermic chemical reactor, it may be advantageous to run the reactor at operating conditions in which the reactor is open-loop unstable. Obviously, the safe operation then depends critically on the control system that stabilizes the reactor. □

Controllability, Observability, and Dynamics

When designing a process, it is very important to make sure that all the important process variables can be changed conveniently. The word *controllability* is often used in this context, although it is interpreted in a much wider sense than in the formal controllability concepts introduced in Sec. 5.3.

To obtain plants that are controllable in the wide sense, it is first necessary to have a sufficient number of actuators. If there are four important process variables that should be manipulated separately, there must be at least four actuators. Moreover, the system should be such that the static relationship between the process variables and the actuators is one-to-one. To achieve good control, the dynamic relationship between the actuators and the process variables should ideally be such that tight control is possible. This means that time delays and nonminimum phase relations should be avoided. Ideally the dynamic relations should be like an integrator or a first-order lag. It is, however, often difficult to obtain such processes. Nonminimum phase loops are therefore common in the dynamics of industrial processes.

Simple dynamic models are often very helpful in assessing system dynamics at the stage of process design. Actuators should be designed so that the process variables can be changed over a sufficient range with a good resolution. The relationships should also be such that the gain does not change too much over the whole operating range. A common mistake in flow systems is to choose a control valve that is too large. This leads to a very nonlinear relation between valve opening and flow. The flow changes very little when the valve opening is reduced until the valve is almost closed. There is then a drastic change in flow over a very small range of valve position.

The process must also have appropriate sensors, whose signals are closely related to the important process variables. Sensors need to be located to give signals that are representative for the important process variables. For example, care must be taken not to position sensors in pockets where the properties of the process fluid may not be typical. Time delays must also be avoided. Time lags can occur due to factors such as transportation or encapsulation of temperature sensors.

Simple dynamic models, combined with observability analysis, are very useful to assess suggested arrangements of sensors and actuators. It is also very useful for this purpose to estimate time constants from simple dynamic models.

Regulator Design, or On-Line Tuning

Another fact that drastically influences the regulator design is the effort that can be spent on the design. For systems that will be produced in large numbers, it may be possible to spend a lot of engineering effort to design a regulator. A regulator with fixed parameters not requiring any adjustments can then be designed. In many cases, however, it is not economically feasible to spend much effort on regulator design. For such applications it is common to use a standard, general-purpose regulator with adjustable parameters. The regulator is installed and appropriate parameters are found by tuning.

The possibilities for designing flexible, general-purpose regulators have increased drastically with computer control. When a regulator is implemented on a computer, it is also possible to provide the system with computer-aided tools, which simplify design and tuning.

In process control, the majority of the loops for control of liquid level, temperature, flow, and pressure are designed by rules of thumb and are tuned on-line. Systematic design techniques are, however, applied to control of composition and pH, as well as to control of multivariable, nonlinear, distributed systems like distillation columns.

Interaction Between Process, Regulator, and Operator

The regulator and the process must, of course, work well together. A regulator is normally designed for steady-state operation, which is one operating state. It is, however, necessary to make sure that the system will work well also during start-up and shutdown and under emergency conditions, such as drastic process failures. During normal conditions it is natural to design for maximum efficiency. At a failure, it may be much more important to recover and quickly return to a safe operating condition.

In process control it has been customary to use automatic regulation for steady-state operation. In other operating modes, the regulator is switched to manual and an operator takes over. With an increased level of automation, good control over more operating states will, however, be required.

7.3 Principles of Structuring

As mentioned earlier, real control problems are large and fuzzy, and control theory deals with small, well-defined problems. According to the dictionary, *structuring* can mean to construct a systematic framework for something. In this context, however, structuring is used to describe the process of bridging the gap between the real problems and the problems that control theory can handle.

The problems associated with structuring are very important for control-system design. Unfortunately, these problems can not yet be put into a complete systematic framework. For this reason they are often avoided both in textbooks and in research. Using an analogy, structuring can be said to have the same relation to control-system design as grammar has to composition. It is clearly impossible to write well without knowing grammar. It is also clear that a grammatically flawless essay is not necessarily a good essay. Structuring of control systems must, of course, be based on the scientific principles given by control theory. However, structuring also contains elements of creativity, ingenuity, and art. Perhaps the best way to introduce structuring is to teach it as a *craft*.

The problem of structuring occurs in many disciplines. Formal approaches have also been developed. The terminology used here is borrowed from the fields of computer science and problem solving, where structuring of large programs has been the subject of much work. There are two major approaches, called top-down and bottom-up.

The *top-down* approach starts with the problem definition. The problem is then divided into successively smaller pieces, adding more and more details. The procedure stops when all pieces correspond to well-known problems. It is a characteristic of the top-down approach that many details are left out in the beginning. More and more details are added as the problem is subdivided. The buzz word *successive refinement* is therefore often associated with the top-down approach.

The *bottom-up* approach starts instead with the small pieces, which represent known solutions for subproblems. These are then combined into larger and larger pieces, until a solution to the large problem is obtained.

The top-down approach is often considered to be more systematic and more logical. It is, of course, not possible to use such an approach unless the details of the system are known very well. Similarly, it is not easy to use the bottom-up approach unless the characteristics of the complete problem are known. In practice, it is common to use combinations of the approaches. This is sometimes called an *inside-out-outside-in* approach.

Structuring is an iterative procedure. It will be a long time before a fully systematic approach to structuring is obtained. It is hard to appreciate the structuring problems unless problems of reasonable size and complexity are considered. Therefore, most of the work on structuring is done in industry. It also appears that many industries have engineers who are very good at structuring. Students are therefore advised to learn what the "structuring masters" are doing, in the same way as painters have always learned from the grand masters.

7.4 A Top-Down Approach

This section describes a top-down approach to control-system design. This involves the selection of control principles, choice of control variables and measured variables, and pairing of these variables.

Control Principles

A control principle gives a broad indication of how a process should be controlled. The control principle thus tells how a process should respond to disturbances and command signals. The establishment of a control principle is the starting point for a top-down design. Some examples of control principles are given next.

Example 7.4—Flow control

When controlling a valve, it is possible to control the valve position, the flow, or both. It is simplest and cheapest to control the valve position. Since flow is, in general, a nonlinear function of the valve opening, this leads to a system in which the relationship between the control variable (valve position) and the physical variable (flow) is very nonlinear. The relationship will also change with such variables as changing pressure and wear of the valve. These difficulties are avoided if both valve position and flow are controlled. A system for flow control is, however, more complicated because it requires a flow meter. □

Example 7.5—Composition control

When controlling important product-quality variables, it is normally desired to keep them close to prescribed values. This can be done by minimizing the variance of product quality variations. If a flow is fed to a large storage tank with mixing, the quality variations in the mixing tank should be minimized. This is not necessarily the same as minimizing quality variations in the flow into the tank. □

Example 7.6—Control of a drum boiler

Consider a turbine and a generator, which are driven by a drum boiler. The control system can have different structures, as illustrated in Fig. 7.1, which shows three control modes: boiler follow, turbine follow, and sliding pressure control. The system has two key control variables, the steam valve and the oil flow. In boiler follow mode, the generator speed is controlled directly by feedback to the turbine valve, and the oil flow is controlled to maintain the steam pressure. In turbine follow mode, the generator speed is instead used to control the oil flow to the boiler, and the steam valve is used to control the drum pressure. In sliding pressure control, the turbine valve is fully open, and oil flow is controlled from the generator speed.

The boiler follow mode admits a very rapid control of generator speed and power output because it uses the stored energy in the boiler. There may, however, be rapid pressure and temperature variations that impose thermal strains on the turbine and the boiler. In turbine follow mode, steam pressure is kept constant and thermal stresses are thus much smaller. The response to power demand will, however, be much slower. The sliding pressure control mode may be regarded as a compromise between boiler follow and turbine follow. □

Example 7.7—Ship control

When designing an autopilot for a highly maneuverable ship, there are many alternatives for design. One possibility is to design the autopilot so that the captain can order a turn to a new course with a specified turning rate. Another possibility is to specify the turning radius instead of the turning speed. The advantage of specifying the turning radius is that the path of the ship will be independent of the ship's speed. Control of the turning radius leads to a more complicated system, since it is necessary to measure both turning rate and ship speed. □

Example 7.8—Material balance control

Many processes involve flow and storage of materials. Although the processes are very different, they all include material storage. The reason for introducing these is to smooth out variations in material flow. It is therefore not sensible to control these systems in such a way that the storages have constant mass. Instead the criteria should be to maintain the following:

Inventories between maximum and minimum limits.
An exact long-term material balance between input and output.
Smooth flow rates. □

Example 7.9—Constraint control

When designing systems, it is frequently necessary to consider several operating conditions. This means that constraints for safety or economical conditions may need to be considered. It may also be necessary to consider constraints during start-up and shutdown. The control during these situations is usually done with logical controllers. Today the logical control and the analog control are often done within the same equipment, programmable logic control (PLC) systems. This means that there are good possibilities to integrate different functions of the control system. □

The choice of a control principle is an important issue. A good control principle can often simplify the control problem. The selection often involves technical

Boiler follow

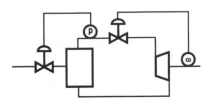

Turbine follow

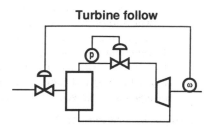

Sliding pressure

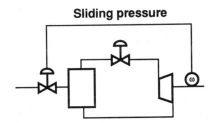

Figure 7.1. Control mode for a boiler-turbine unit.

and economical trade-offs. The selection of a control principle is often based on investigations of models of the process. The models used for this purpose are typically internal models derived from physical principles. It is therefore difficult to define general rules for finding control principles.

Choice of Control Variables

After the control principle has been chosen, the next logical step is to choose the control variables. The choice of control variables can often be limited for various practical reasons. Because the selection of control principle tells what physical variables should be controlled, it is natural to choose control variables that have

Design: An Overview   Chap. 7

a close relation to the variables given by the control principle. Since mathematical models are needed for the selection of control principles, these models can also be used for controllability studies when choosing control variables.

Choice of Measured Variables

When the control principle is chosen, the primary choice of measured variables is also given. If the variables used to express the control principle cannot be measured, it is natural to choose measured variables that are closely related to these control variables. Mathematical models and observability analysis can be very helpful in making this choice. Typical examples are found in chemical-process control, where temperatures—which are easy to measure—are used instead of compositions, which are difficult and costly to measure.

Pairing of Inputs and Outputs

A large system will typically have a large number of inputs and outputs. Even if a control principle, which involves only a few variables, is found initially, many variables typically must be considered once the variables that can be manipulated and measured are introduced. With a top-down approach, a system should be broken down into small subsystems. It is then desirable to group different inputs and outputs together, so that a collection of smaller systems is obtained. If possible, the grouping should be done so that (1) there are only weak couplings between the subsystems; and (2) each subsystem is dynamically well behaved; i.e., time constants are of the same magnitude and time delay, nonminimum phase, and severe variations in process dynamics are avoided.

There are no general rules for the grouping. Neither are there any good ways of deciding if it is possible to find a grouping with the desired properties. Trial and error, combined with analysis of models, is one possibility.

The following example illustrates the pairing problem.

Example 7.10—Material balance control

A system with material flow is shown in Fig. 7.2. The system consists of a series of tanks. The flows between the tanks are controlled by pumps. The figure illustrates two different control structures. In one structure, the flow *out of* each tank is controlled from the tank level. This is called control *in the direction of the flow*. To maintain balance between production and demand, it is necessary to control the flow into the first tank by feedback from the last tank level. In the other approach, the flow *into* each tank is controlled by the tank level. This is called control *in the direction opoposite to the flow*. This control mode is superior, because all control loops are simple first-order systems and there are no stability problems. With control in the direction of the flow, there may be instabilities due to the feedback around

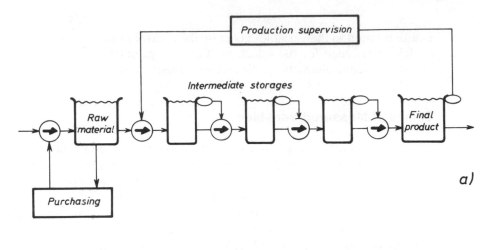

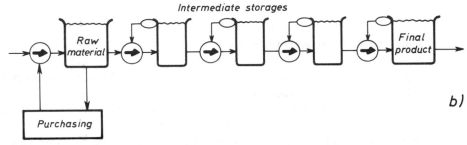

Figure 7.2 Material balance control in the direction of the flow (a) and in the direction opposite to the flow (b).

all tanks. It can also be shown that control in the direction opposite to the flow can be done by smaller storage tanks. □

7.5 A Bottom-Up Approach

In the bottom-up approach, a choice of control variables and measurements comes first. Different regulators are then introduced until a closed-loop system, with the desired properties, is obtained. The regulators used to build up the system are the standard types based on the ideas of feedback, feedforward, prediction and estimation, optimization, and adaptation. Since these techniques are familiar from elementary courses, they will be discussed only briefly.

Feedback

The feedback loops used include, for example, simple PID regulators and their cascade combinations. When digital computers are used to implement the regulators, it is also easy to use more sophisticated control, such as Smith-predictors for deadtime compensation, state feedback, and model reference control.

Feedback is used in the usual context. Its advantage is that sensitivity to disturbances and parameter variations can be reduced. Feedback is most effective when the process dynamics are such that a high bandwidth can be used. Design of feedback will be discussed in depth in the following chapters.

Many systems that are difficult to implement using analog techniques may be easy to implement using computer-control technology. The Smith-predictor is a typical example. A block diagram of this regulator is shown in Fig. 7.3, which shows that it is necessary to store signals. This is difficult to do in analog implementation, but easy with a computer. The Smith-predictor is discussed again in Chapter 8.

Feedforward

Feedforward is another control method. It is used, as described in Chapter 6, to eliminate disturbances that can be measured. The basic idea is to use the measured disturbance to anticipate the influence of the disturbance on the process variables and to introduce suitable compensating control actions. The advantage compared to feedback is that corrective actions may be taken before the disturbance has influenced the variables. Since feedforward is an open-loop compensation, it requires a good process model. With digital control, it is easy to incorporate a process model. Thus it can be anticipated that use of feedforward will increase with digital control.

The design of a feedforward compensator is in essence a calculation of the inverse of a dynamic system.

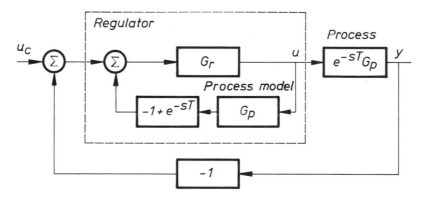

Figure 7.3 Block diagram of a Smith-predictor.

Selector Control

There are many cases where it is desirable to switch control modes, depending on the operating condition. This can be achieved by a combination of logic and feedback control. The same objective can, however, also be achieved with a combination of feedback regulators. A typical example is control of the air-to-fuel ratio in boiler control. In ship boilers it is essential to avoid smoke puffs when the ship is in the harbor. To do this it is essential that the air flow leads the oil flow when load is increased and that air flow lags oil flow when the load is decreased. This can be achieved with the system shown in Fig. 7.4, which has two selectors. The maximum selector gives an output signal which at each instant of time is the largest of the input signals, and the minimum selector chooses the smallest of the inputs. When the power demand is increased, the maximum selector chooses the demand signal as the input to the air flow controller, and the minimum selector chooses the air flow as the set point to the fuel-flow controller. The fuel will thus follow the actual air flow.

When the power demand is decreased, the maximum selector will choose the fuel flow as the set point to the air flow controller, and the minimum selector will choose the power demand as the set point to the fuel-flow controller. The air flow will thus lag the fuel flow.

Control using selectors is very common in industry. Selectors are very convenient for switching between different control modes.

Prediction and Estimation

State variables and parameters often cannot be measured directly. In such a case it is convenient to pretend that the quantities are known when designing a feedback. The unknown variables can then be replaced by estimates or predictions.

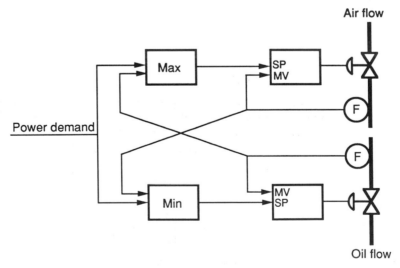

Figure 7.4. System with selectors for control of the air-to-fuel ratio in a boiler.

In some cases such a solution is in fact optimal. The notions of predictions and estimation are therefore important. Estimators for state variables in linear systems can easily be generated by analog techniques. They can also easily be implemented using a computer. Parameter estimators are more difficult to implement with analog methods. They can, however, easily be done with a computer. Prediction and estimation are thus easier to use with computer control.

Optimization

Some control problems can be conveniently expressed as optimization problems. With computer-control systems, it is possible to include optimization algorithms as elements of the control system.

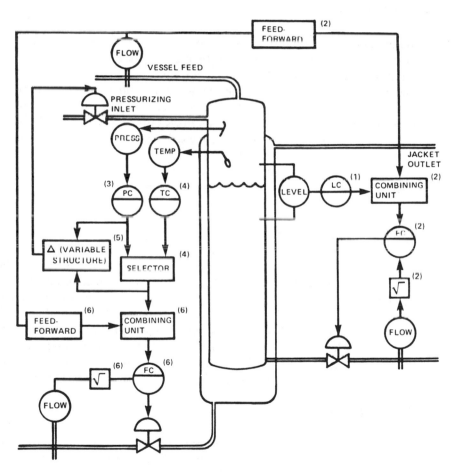

Figure 7.5. An example of a complicated control system built up from simple control structures. (Redrawn from Foxboro Company with permission.)

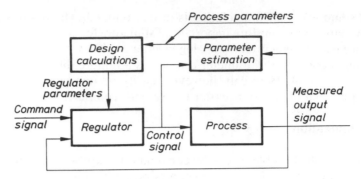

Figure 7.6. Block diagram of a self-tuning regulator obtained by combining a parameter estimator with a design calculation.

Combinations

When using a bottom-up approach, the basic control structures are combined to obtain a solution to the control problem. It is often convenient to make the combinations hierarchically. Many combinations, like cascade control, state feedback, and observers, are known from elementary control courses. Very complicated control systems can be built up by combining the simple structures. An example is shown in Fig. 7.5. This way of designing control using the bottom-up approach is in fact the technique predominantly used in process control. Its success depends largely on the experience and skill of the designer.

An adaptive system, which is obtained by combining a parameter estimator with a design procedure, is shown in Fig. 7.6.

7.6 Design of Simple Loops

If a top-down approach is used, the design procedure will end in the design of simple loops containing one or several controls, or measurements. If a bottom-up approach is used, the design will start with the design of simple loops. Therefore, the design of simple loops is an important step in both approaches. The design of simple loops is also one area in which there is a substantial theory available, which will be described in detail in the remaining chapters. To give some perspective, an overview of design methods for simple loops is given in this section. The prototype problems of regulator and servo design will be discussed.

Simple Criteria

A simple way to specify regulation performance is to give allowable errors for typical disturbances. For example, it can be required that a step disturbance give no steady-state error, and that the error due to a ramp disturbance be a fraction of the ramp velocity. These specifications are typically expressed in terms of the steady-state behavior, as discussed in Sec. 5.4. The error coefficients give requirements only on the low-frequency behavior. The bandwidth of the system should therefore be specified, in addition to the error coefficients.

Another more complete way to specify regulation performance is to give conditions on the transfer function from the disturbances to the process output.

Specifications for the Regulator Problem

The purpose of regulation is to keep process variables close to specified values in spite of process disturbances and variations in process dynamics.

Minimum variance control.   For regulation of important quality variables, it is often possible to state objective criteria for regulation performance. A common situation is illustrated in Fig. 7.7, which shows the distribution of the quality variables. It is often specified that a certain percentage of the production should be at a quality level above a given value. By reducing the quality variations, it is then possible to move the set point closer to the target. The improved performance can be expressed in terms of reduced consumption of energy or raw material or increased production. It is thus possible to express reductions in quality variations directly in economic terms.

For processes with a large production, reductions of a fraction of a percent can amount of a large sum of money. For example, a reduction in moisture variation of 1% in paper-machine control can amount to savings of $100,000 per year.

If the variations in quality can be expressed by Gaussian distributions, the criterion would simply be to minimize the variance of the quality variables. In these problems, the required control actions are irrelevant as long as they do not cause excessive wear or excessively large signals. A control strategy that minimizes the variance of the process output is called *minimum variance control.*

Optimal control.   Minimum variance control is a typical example of how a control problem can be specified as an optimization problem. In a more general case, it is not appropriate to minimize the variance of the output. Instead there

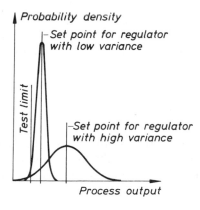

Figure 7.7. Expressing regulation performance in terms of variation in quality variables.

will be a criterion of the type

$$E \int_{t_0}^{t_1} g[x(s), u(s)] \, ds$$

where x is the state variable, u is the control variable, and E denotes the mean value. An example of such a criterion is given next.

Example 7.11—Ship steering

It can be shown that the relative increase in resistance due to deviations from a straight line course can be approximately expressed as

$$\frac{\Delta R}{R} = \frac{k}{T} \int_0^T [\psi^2(t) + \rho \delta^2(t)] \, dt$$

where ψ is the heading deviation, δ is the rudder angle, R is the resistance, and ρ is a parameter. Typical parameter values for a tanker are $k = 0.014$ and $\rho = 0.1$. ☐

Techniques for Regulator Design

There are many methods that may be used for estimating regulation performance and for designing regulators. Regulation problems are often solved by feedback, but feedforward techniques can be very useful to eliminate measurable disturbances.

If the specifications are given in terms of the transfer function, relating the output to the disturbance, it is natural to apply methods that admit control of this transfer function. One method is pole placement, which allows specification of the complete transfer function. This straightforward design technique is discussed in detail in Chapters 9 and 10. It is often too restrictive to specify the complete closed-loop transfer function, which is a drawback.

Another possibility is to use a frequency-response method, which admits control of the frequency response from the disturbance to the output. Such problems are most conveniently expressed in terms of continuous-time theory. The regulators obtained can then be translated to digital-control algorithms using the techniques described in Chapter 8.

If the criteria are expressed as optimization criteria, it is natural to use design techniques based on optimization. Techniques based on minimizing the variance of the process output and other types of quadratic criteria are discussed in Chapters 11 and 12.

Specifications for the Servo Problem

In the servo problem, the task is to make the process variables respond to changes in a command signal in a given way. Feedback techniques are useful for the servo problem. Since the command signal must be known, it is also natural to combine the feedback with feedforward.

Design: An Overview   Chap. 7

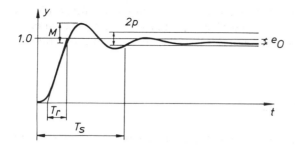

Figure 7.8. Expressing servo specifications in terms of requirements on the step response.

Model following. One way to express how the system should respond to a command signal is to give a model for the desired response. This can be done, for example, by specifying the desired transfer function from the command signal to the process variables. If it is not possible to get a perfect match, the deviation from the model response can be minimized.

Time and frequency domain specifications. Since a linear system is completely specified by its step response, it is possible to express servo response in terms of specifications on the desired closed-loop step response or the frequency response. These types of specifications include rise time, T_r, overshoot, M, settling time, T_s, steady-state errors, e_0, bandwidth, ω_B, and resonance peak, M_p. The settling time is defined as the time it takes before the step response is within $\pm p$ from its steady-state value. A common value of p is 5%. Compare Figs. 7.8 and 7.9.

There are also design parameters that can be determined from the open-loop Bode plot. Examples are the gain margin, A_m, the phase margin, φ_m, and the crossover frequency, ω_c. The gain and phase margins are related to the stability of the closed-loop system. The crossover frequency is essentially proportional to the bandwidth of the closed-loop system.

Techniques for Servo Design

There are many techniques that can be used for servo design. If all state variables can be measured, it is natural to use pole placement. If the system is controllable, a state feedback—which gives the desired closed-loop poles—can then be designed. If all state variables are not measured, they can be reconstructed by an

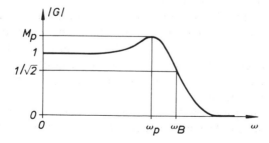

Figure 7.9. Bode plot for the closed-loop transfer function G_c; illustrates how servo specifications can be expressed in terms of requirements on the frequency response. The system in the figure has unity steady-state gain.

observer or by a Kalman filter. Design techniques based on these ideas are described in detail in Chapters 9, 10, and 11 where it is also shown that such a design procedure can be interpreted as a model following design that combines feedback and feedforward.

7.7 Key Issues in Control System Design

Control system can be quite complicated because design is a compromise between many different factors. The following issues must typically be considered:

- Command signals
- Load disturbances
- Measurement noise
- Model uncertainty
- Actuator saturation
- State constraint
- Regulator complexity

There are few design methods that consider all these factors. The design methods discussed in this book will typically focus on a few of the issues. In a good design it is often necessary to grasp all factors. To do this it is then necessary to investigate all factors by analysis or simulation.

7.8 Conclusions

This chapter present an overview of the design problems. There is a large step from the large and fuzzy problems of the real world to the small and well-defined problems that control theory can handle. Problems of structuring are discussed. The notion of control principle is introduced in order to apply the top-down approach. It is also shown how a bottom-up approach can be used to build complex systems from the simple control structures feedback, feedforward, estimation, and optimization. Finally, specifications and approaches to the design of simple loops are discussed.

A chemical process consists of many unit operations, such as reactors, mixers, and distillation columns. In a bottom-up approach to control-system design, control loops are first designed for the individual unit operations. Interconnections are then added to obtain a total system. In a top-down approach, control principles—such as composition control and material-balance control—are first postulated for the complete plant. In the decomposition, these principles are then applied to the individual units and loops.

In process control the majority of the loops for liquid level, flow, and pressure control are most frequently designed emperically and tuned on-line. However, control of composition and pH, as well as control of nonlinear distributed large systems with strong interaction, are often designed with care.

7.9 Problems

7.1. Consider the material balance problem shown in Fig. 7.2. Assume that each tank (storage) is an integrator and that each controller is a proportional controller. Discuss the influence on the two systems when there is a pulse disturbance out from the raw material storage.

7.2. Derive the transfer function of the closed-loop system in Fig. 7.3. Interpret the performance of the closed-loop system. What is the main feature of the Smith-predictor? (*Hint:* Compare with the case when there is no time delay and when the controller is G_r.)

7.3. Identify and discuss the use of (a) cascade control, (b) feedforward, and (c) nonlinear elements in Fig. 7.5.

7.10 References

The problem discussed in this chapter touches upon several aspects of problem solving. A reader with general interests may enjoy reading

POLYA, G. (1945): *How to Solve It*. Princeton, N.J.: Princeton University Press.

which takes problems from the mathematical domain, and

WIRTH, N. (1979): *Algorithms + Data Structures = Programs*. Englewood Cliffs, N.J.: Prentice Hall, Inc.

which applies to computer programming. There is some work on the structuring problem in the literature on process control; see, for instance,

BUCKLEY, P. S. (1964): *Techniques of Process Control*. New York: John Wiley.

SHINSKEY, F. G. (1988): *Process Control Systems* (3rd ed.). New York: McGraw-Hill.

BRISTOL, E. H. (1980): "Strategic Design: A Practical Chapter in a Textbook on Control," Paper WA4-A, Preprints JACC, San Francisco.

BALCHEN, J. G. and K. I. MUMMÉ (1988): *Process Control*. New York: Van Nostrand Reinhold.

The paper

BUCKLEY, P. S. (1978): "Distillation Column Design Using Multivariable Control, Part 1: Process and Control Design; Part 2: Economics, Energy, and Equipment," *Instrumentation Technology*, September, 115–22; October, 49–53.

contains much useful material of general interest although it deals with a very specific problem. The paper

FOSS, A. S. (1973): "Critique of Chemical Process Control Theory," *IEEE Trans. Autom. Control*, AC-18, 646–52.

is more general in scope.

There are only a few areas in which control design and process design have been considered jointly. Design of high-performance aircrafts is a notable example. See

BURNS, B. R. A. (1975): "Fly-by-wire and Control Configured Vehicles—Rewards and Risks," *Aeronautical Journal,* February.

BOUDREAU, J. A. (1976): "Integrated Flight Control System Design for CCV," *Proc. AIAA Conf. on Flight Mechanics, Guidance and Control.*

Specifications of regulator performance for simple loops are discussed in depth in standard texts on servomechanisms; see, for instance,

DORF, R. C. (1969): *Modern Control Systems* (5th ed.) Reading, Mass.: Addison-Wesley.

ELGERD, O. (1967): *Control System Theory.* New York: McGraw-Hill.

8

TRANSLATION OF
ANALOG DESIGN

GOAL

*To Show How Digital Regulators Can Be Obtained by Translating
and Modifying Analog Design, Including Digital PID-Controllers
and State-Feedback Redesign.*

8.1 Introduction

Sometimes an analog-control system is replaced by a computer-control system simply because the hardware of the latter is cheaper and more reliable. In such cases it is natural to look for methods of converting an analog system to a digital system with the same properties. A straightforward way to solve this problem is to use a short sampling interval and to make some discrete-time approximations of the continuous-time controller. This approach is illustrated in Example 1.3. Figure 1.7 contains a comparison of the continuous- and the approximating discrete-time controllers. Observe that the discrete-time control signal is delayed half a sample period, on the average, compared with the continuous-time control signal. This delay introduces a phase lag, and the closed-loop performance deteriorates. The choice of the sampling time is therefore important. In general, translation of continuous-time controllers using approximations will lead to quite short sampling periods. It is, however, possible to use digital-design methods, which allow longer sampling times (compare with Fig. 1.8). This is discussed in Chapters 9–11.

Section 8.2 discusses some ways to approximate continuous-time controllers. The discrete-time PID-controller is treated in Sec. 8.3. Different ways to implement PID-controllers are given, together with some operational aspects such as antireset windup and bumpless transfer. Section 8.4 shows how state-feedback continuous controllers can be modified in order to allow longer sampling intervals, compared with a straightforward approximation.

8.2 Different Approximations

This section assumes that a continuous-time controller is given as a transfer function, $G(s)$. It is desired to find an algorithm for a computer so that the digital system approximates the transfer function $G(s)$ (see Fig. 8.1). This problem is interesting for implementation of both analog controllers and digital filters. The approximation may be done in many different ways. Digital implementation includes a data reconstruction, which also can be made in different ways—e.g., zero- or first-order hold. A few simple common approximation methods are given in this section.

Differentiation and Tustin Approximations

A transfer function represents a differential equation. It is natural to obtain a difference equation by approximating the derivatives with a forward difference (Euler's method)

$$px(t) = \frac{dx(t)}{dt} \approx \frac{x(t + h) - x(t)}{h} = \frac{q - 1}{h} x(t)$$

or a backward difference

$$px(t) = \frac{dx(t)}{dt} \approx \frac{x(t) - x(t - h)}{h} = \frac{q - 1}{qh} x(t)$$

In the transform variables, this corresponds to replacing s by $(z - 1)/h$ or $(z - 1)/(zh)$. Section 3.6 shows that the variables z and s are related in some respects as $z = \exp(sh)$. The difference approximations correspond to the series

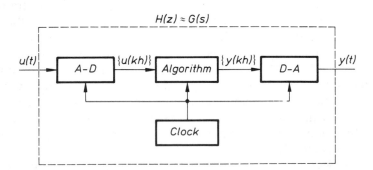

Figure 8.1 Approximating a continuous-time transfer function, $G(s)$, using a computer.

expansions

$$z = e^{sh} \approx 1 + sh \qquad \text{(Euler's method)} \qquad (8.1)$$

$$z = e^{sh} \approx \frac{1}{1 - sh} \qquad \text{(Backward difference)} \qquad (8.2)$$

Another approximation, which corresponds to the trapezoidal method for numerical integration, is

$$z = e^{sh} \approx \frac{1 + sh/2}{1 - sh/2} \qquad \text{(Trapezoidal method)} \qquad (8.3)$$

In digital control context, the approximation in (8.3) is often called *Tustin's approximation,* or *the bilinear transformation.* Compare this with the Möbius transformation in Sec. 5.2. Using the approximation methods above, the pulse-transfer function $H(z)$ is obtained by simply replacing the argument s in $G(s)$ by s', where

$$s' = \frac{z - 1}{h} \qquad \text{(Forward difference or Euler's method)} \qquad (8.4)$$

$$s' = \frac{z - 1}{zh} \qquad \text{(Backward difference)} \qquad (8.5)$$

$$s' = \frac{2}{h} \frac{z - 1}{z + 1} \qquad \text{(Tustin's approximation)} \qquad (8.6)$$

Hence

$$H(z) = G(s')$$

The methods are very easy to apply even for hand calculations. Figure 8.2 shows how the stability region Re $s < 0$ in the s-plane is mapped on the z-plane for the mappings (8.4), (8.5), and (8.6).

With the forward difference approximation it is thus possible that a stable continuous-time system is mapped into an unstable discrete-time system. When

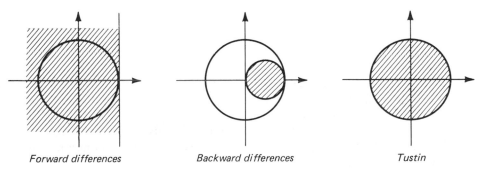

Forward differences   Backward differences   Tustin

Figure 8.2 Mapping of the stability region in the s-plane on the z-plane for the transformations (8.4), (8.5), and (8.6).

the backward approximation is used, a stable continuous-time system will always give a stable discrete-time system. There are, however, also unstable continuous-time systems that are transformed into stable discrete-time systems. Tustin's approximation has the advantage that the left half-s-plane is transformed into the unit disc. Stable continuous-time systems are therefore transformed into stable sampled systems, and unstable continuous-time systems are transformed into unstable discrete-time systems.

Frequency Prewarping

One problem with the approximations discussed earlier is that the frequency scale is distorted. For instance, if it is desired to design band-pass or notch filters, the digital filters obtained by the approximations may not give the correct frequencies for the band-pass or the notches. This effect is called *frequency warping*. Consider an approximation obtained by Tustin's approximation. The transmission of sinusoidals for the digital filter is given by

$$H(e^{i\omega h}) = \frac{1}{i\omega h} (1 - e^{-i\omega h}) G \left(\frac{2}{h} \frac{e^{i\omega h} - 1}{e^{i\omega h} + 1} \right)$$

The first two factors are due to the sample-and-hold operations; compare with (4.11). The argument of G is

$$\frac{2}{h} \frac{e^{i\omega h} - 1}{e^{i\omega h} + 1} = \frac{2}{h} \frac{e^{i\omega h/2} - e^{-i\omega h/2}}{e^{i\omega h/2} + e^{-i\omega h/2}} = \frac{2i}{h} \tan \left(\frac{\omega h}{2} \right)$$

The frequency scale is thus distorted. Assume for example that the continuous-time system blocks signals at the frequency ω'. Because of the frequency distorsion, the sampled system will instead block signal transmission at the frequency ω where

$$\omega' = \frac{2}{h} \tan \left(\frac{\omega h}{2} \right)$$

That is

$$\omega = \frac{2}{h} \tan^{-1} \left(\frac{\omega' h}{2} \right) \approx \omega' \left(1 - \frac{(\omega' h)^2}{12} \right) \tag{8.7}$$

This expression gives the distortion of the frequency scale (see Fig. 8.3.). It follows from (8.7) that there is no frequency distortion at $\omega = 0$ and that the distortion is small if ωh is small.

It is easy to introduce a transformation that eliminates the scale distortion at a specific frequency ω_1 by modifying Tustin's transformation from (8.6) to the transformation

$$s' = \frac{\omega_1}{\tan(\omega_1 h/2)} \frac{z - 1}{z + 1} \qquad \text{(Tustin with prewarping)} \tag{8.8}$$

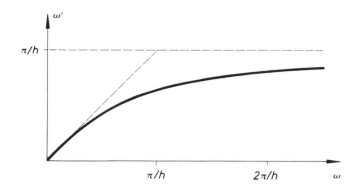

Figure 8.3 Frequency distortion (warping) obtained with Tustin's approximation.

From (8.8), it follows that

$$H[\exp(i\omega_1 h)] = G(i\omega_1)$$

i.e., the continuous-time filter and its approximation have the same value at the frequency ω_1. There is, however, still a distortion at other frequencies.

Example 8.1—Frequency prewarping

Assume that the integrator

$$G(s) = \frac{1}{s}$$

should be implemented as a digital filter. Using the transformation of (8.6) without prewarping gives

$$H_T(z) = \frac{1}{\dfrac{2}{h} \cdot \dfrac{z-1}{z+1}} = \frac{h}{2} \cdot \frac{z+1}{z-1}$$

Prewarping gives

$$H_P(z) = \frac{\tan(\omega_1 h/2)}{\omega_1} \cdot \frac{z+1}{z-1}$$

The frequency function of H_p is

$$H_P(e^{i\omega h}) = \frac{\tan(\omega_1 h/2)}{\omega_1} \cdot \frac{e^{i\omega h}+1}{e^{i\omega h}-1}$$

$$= \frac{\tan(\omega_1 h/2)}{i\omega_1} \cdot \frac{1}{\tan(\omega h/2)}$$

thus $G(i\omega)$ and $H_p[\exp(i\omega h)]$ are equal for $\omega = \omega_1$. $\square$

Step Invariance

Another way to generate approximations is to use the ideas developed in Chapter 3. In this way it is possible to obtain approximations that give correct values at the sampling instants for special classes of input signals. For example, if the input signal is constant over the sampling intervals, Table 3.1 or Equation (3.49) give an appropriate pulse-transfer function $H(z)$ for a given transfer function $G(s)$. Since this relation gives the correct values of the output when the input signal is a piecewise constant signal that changes at the sampling instants, it is called *step invariance*.

Ramp Invariance

The notion of step invariance is ideally suited to describe a system where the input signal is generated by a computer, because the input signal is then constant over the sampling period. The approximation is, however, not so good when dealing with input signals that are continuous. In this case it is much better to use an approximation where the input signal is assumed to vary linearly between the sampling instants. The approximation obtained is called *ramp invariance* because it gives the values of the output at the sampling instants exactly for ramp signals. The equations for ramp-invariant sampling are obtained in the same way as for step-invariant sampling. Compare with Section 3.2.

Consider a continuous-time system described by (3.1). Assume that the input signal is linear between the sampling instants. Integration of (3.1) over one sampling period gives

$$x(kh + h) = e^{Ah}x(kh) + \int_{kh}^{kh+h} e^{A(kh+h-s)}$$

$$\times B\left[u(kh) + \frac{s - kh}{h}(u(kh + h) - u(kh))\right] ds$$

Hence

$$x(kh + h) = \Phi x(kh) + \Gamma u(kh) + \frac{1}{h}\Gamma_1[u(kh + h) - u(kh)]$$

$$= \Phi x(kh) + \frac{1}{h}\Gamma_1 u(kh + h) + \left(\Gamma - \frac{1}{h}\Gamma_1\right)u(kh)$$

where

$$\Phi = e^{Ah}$$

$$\Gamma = \left(\int_0^h e^{As}\, ds\right)B \tag{8.9}$$

$$\Gamma_1 = \left(\int_0^h e^{As}(h - s)\, ds\right)B$$

The pulse-transfer function that corresponds to ramp invariant sampling thus becomes

$$H(z) = D + C[zI - \Phi]^{-1}\left[\frac{z}{h}\Gamma_1 + \Gamma - \frac{1}{h}\Gamma_1\right] \tag{8.10}$$

It follows from (8.9) that the matrices Φ, Γ, and Γ_1 satisfy the differential equations

$$\frac{d\Phi(t)}{dt} = \Phi(t)A$$

$$\frac{d\Gamma(t)}{dt} = \Phi(t)B$$

$$\frac{d\Gamma_1(t)}{dt} = \Gamma(t)$$

These equations can also be written as

$$\frac{d}{dt}\begin{pmatrix} \Phi(t) & \Gamma(t) & \Gamma_1(t) \\ 0 & I & It \\ 0 & 0 & I \end{pmatrix} = \begin{pmatrix} \Phi(t) & \Gamma(t) & \Gamma_1(t) \\ 0 & I & It \\ 0 & 0 & I \end{pmatrix}\begin{pmatrix} A & B & 0 \\ 0 & 0 & I \\ 0 & 0 & 0 \end{pmatrix}$$

This implies that the matrices Φ, Γ, and Γ_1 can be obtained as

$$(\Phi \quad \Gamma \quad \Gamma_1) = (I \quad 0 \quad 0)\exp\left\{\begin{bmatrix} A & B & 0 \\ 0 & 0 & I \\ 0 & 0 & 0 \end{bmatrix}h\right\} \qquad (8.11)$$

The calculation of ramp-invariant systems is illustrated by some examples.

Example 8.2—Ramp-invariant sampling of an integrator

Consider a system with the transfer function $G(s) = 1/s$. In this case we have $A = D = 0$ and $B = C = 1$. Using (8.11) we get

$$[\Phi \quad \Gamma \quad \Gamma_1] = [1 \quad 0 \quad 0]\exp\left\{\begin{bmatrix} 0 & 1 & 0 \\ 0 & 0 & 1 \\ 0 & 0 & 0 \end{bmatrix}h\right\}$$

$$= [1 \quad h \quad \tfrac{1}{2}h^2]$$

The pulse-transfer function becomes

$$H(z) = \frac{\tfrac{1}{2}zh + h - \tfrac{1}{2}h}{z - 1} = \frac{h}{2}\frac{z+1}{z-1}$$

This pulse-transfer function corresponds to the trapezoidal formula for computing an integral. Also notice that Tustin's transformation gives the same result in this case. $\quad\sqcap$

Example 8.3—Ramp-invariant sampling of a double integrator

Consider a system with the transfer function $G(s) = s^{-2}$. This system has the realization

$$\frac{dx}{dt} = \begin{bmatrix} 0 & 1 \\ 0 & 0 \end{bmatrix}x + \begin{bmatrix} 0 \\ 1 \end{bmatrix}u$$

$$y = [1 \quad 0]x$$

for the matrix

$$\tilde{A} = \begin{bmatrix} A & B & 0 \\ 0 & 0 & I \\ 0 & 0 & 0 \end{bmatrix} = \begin{bmatrix} 0 & 1 & 0 & 0 \\ 0 & 0 & 1 & 0 \\ 0 & 0 & 0 & 1 \\ 0 & 0 & 0 & 0 \end{bmatrix}$$

and its matrix exponential

$$e^{\tilde{A}h} = \begin{bmatrix} 1 & h & h^2/2 & h^3/6 \\ 0 & 1 & h & h^2/2 \\ 0 & 0 & 1 & h \\ 0 & 0 & 0 & 1 \end{bmatrix}$$

Hence from (8.11)

$$\Phi = \begin{bmatrix} 1 & h \\ 0 & 1 \end{bmatrix}, \quad \Gamma = \begin{bmatrix} h^2/2 \\ h \end{bmatrix}, \quad \Gamma_1 = \begin{bmatrix} h^3/6 \\ h^2/2 \end{bmatrix}$$

The pulse-transfer function is now obtained from (8.10), i.e.,

$$H(z) = \begin{bmatrix} 1 & 0 \end{bmatrix} \begin{bmatrix} z-1 & -h \\ 0 & z-1 \end{bmatrix}^{-1} \left\{ \begin{bmatrix} h^2/6 \\ h/2 \end{bmatrix} z + \begin{bmatrix} h^2/2 \\ h \end{bmatrix} - \begin{bmatrix} h^2/6 \\ h/2 \end{bmatrix} \right\}$$

$$= \frac{h^2}{6} \frac{z^2 + 4z + 1}{(z-1)^2} \qquad\qquad \square$$

A frequency domain approach. A formula relating the pulse-transfer function (8.10) directly to the transfer function $G(s)$ can also be derived. The key idea is to use the notion of ramp invariance directly. We thus proceed similarly as was done in Section 3.5.

1. Determine the ramp response of the transfer function $G(s)$.
2. Compute the z-transform of the ramp response.
3. Divide by the z-transform $hz(z-1)^{-2}$ of the ramp function.

This procedure will clearly give a pulse-transfer function that will give the same response as $G(s)$ for ramp functions. This explains the name ramp invariance. Carrying out the calculations, the following formula is obtained:

$$H(z) = \frac{(z-1)^2}{hz} \frac{1}{2\pi i} \int_{\gamma - i\omega}^{\gamma + i\omega} \frac{e^{sh}}{z - e^{sh}} \frac{G(s)}{s^2} \, ds \qquad (8.12)$$

A comparison with (3.49) shows that the pulse-transfer function associated with ramp invariance is obtained by computing the step-invariant pulse-transfer function for $G(s)/s$ and multiplying the result by $(z-1)/h$.

Example 8.4—Ramp-invariant approximation of s^{-1}

The step-invariant pulse-transfer function associated with s^{-2} is

$$\frac{h^2}{2} \frac{z+1}{(z-1)^2}$$

The desired pulse-transfer function thus becomes

$$H(z) = \frac{h}{2} \frac{z + 1}{z - 1}$$

This is the same result obtained in Example 8.2. $\qquad\square$

Example 8.5—Ramp invariance of s^{-2}

The step-invariant pulse-transfer function associated with s^{-3} is

$$\frac{h^3}{6} \frac{z^2 + 4z + 1}{(z - 1)^3}$$

The ramp-invariant pulse-transfer function is thus

$$H(z) = \frac{h^2}{6} \frac{z^2 + 4z + 1}{(z - 1)^2}$$

This agrees with the result obtained in Example 8.3. $\qquad\square$

Comparison of Approximations

The step-invariant method is not suitable for approximation of continuous-time transfer functions. The reason is that the approximation of the phase curve is unnecessarily poor. Both Tustin's method and the ramp-invariant method give better approximations. Tustin's method is a little simpler than the ramp-invariant method. The ramp-invariant method does give correct sampled poles. This is not the case for Tustin's method. This difference is particularly important when implementing notch filters where Tustin's method gives a frequency distortion. Another drawback with Tustin's method is that very fast poles of the continuous-time system appear in sampled poles close to $z = -1$, which will give rise to ringing in the digital filter. The different approximations are illustrated by an example.

Example 8.6—Sampled approximations of transfer function

Consider a continuous-time system with the transfer function

$$G(s) = \frac{(s + 1)^2(s^2 + 2s + 400)}{(s + 5)^2(s^2 + 2s + 100)(s^2 + 3s + 2500)}$$

Let $H(z)$ be the pulse-transfer function representing the algorithm in Fig. 8.1. The transmission properties of the digital filter in Fig. 8.1 depend on the nature of the D-A converter. If it is assumed that the converter keeps the output constant between the sampling periods, the transmission properties of the filter are described by

$$\hat{G}(s) = \frac{1}{sh} (1 - e^{-sh})H(e^{sh})$$

where the pulse-transfer function H depends on the approximation used. Figure 8.4 shows Bode diagrams of H for the different digital filters obtained by step equivalence, ramp equivalence, and Tustin's method. The sampling period is 0.03 s in all cases. This implies that the Nyquist frequency is 105 rad/s. All methods except

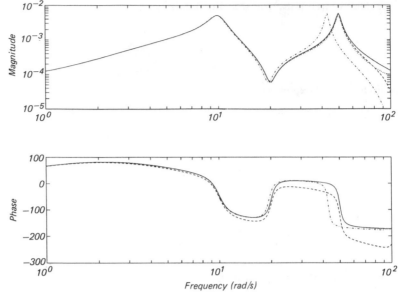

Figure 8.4 Bode diagrams of a continuous-time transfer functon $G(s)$ and different sampled approximations $H(e^{sh})$. The continuous transfer-function is the full line, ramp invariance is the dotted line, step invariance is the dashed line, and Tustin's approximation is the dot-dashed line.

Tustin's give a good approximation of the amplitude curve. The frequency distortion by Tustin's method is noticeable at the notch at 20 rad/s and very clear at the resonance at 50 rad/s. The step-equivalence method gives a noticeable phase error. This corresponds approximately to a time delay of half a sampling interval. Ramp equivalence gives a negligible phase error. The phase curve for Tustin's approximation also deviates because of the frequency warping. Notice that all approximations suffer from the time delay due to the sample and hold. Ramp equivalence gives the best approximation of both amplitude and phase. □

Antialiasing Filters

The consequences of aliasing and the importance of antialiasing filters was discussed in Section 2.5. Choice of sampling rate and antialiasing filters is important in digital systems that are based on translation of analog design. Some consequences of the selection of sampling rates have been discussed previously. The sampling rate must be so large that the errors due to sampling are negligible. The errors introduced by the antialiasing filters will now be discussed. Consider a controller for a system where the crossover frequency is ω_c. Assume that the prefilter has the transfer function

$$G(s) = \frac{\omega_f^2}{s^2 + 2\zeta\omega_f s + \omega_f^2}$$

with $\zeta = 0.707$. This filter gives an attenuation of

$$\beta = \frac{1}{|G(i\omega_N)|} \approx \left(\frac{\omega_f}{\omega_N}\right)^2$$

at the Nyquist frequency. The filter has a phase lag at ω_c of

$$\alpha = \text{atan} \frac{2\zeta\omega_c\omega_f}{\omega_f^2 - \omega_c^2} \approx 2\zeta\left(\frac{\omega_c}{\omega_f}\right)$$

Elimination of ω_f between these equations gives

$$\frac{\omega_s}{\omega_c} = 2\frac{\omega_N}{\omega_c} \approx \sqrt{\frac{8}{\alpha^2\beta}} \approx \frac{2.8}{\alpha\sqrt{\beta}}$$

To show the orders of magnitude involved, choose $\alpha = 0.1$ and $\beta = 0.01$. This means that the phase lag introduced by the prefilter decreases the phase margin by 0.1 rad $\approx 6°$ and that the filter has an attenuation of 0.01 at the Nyquist frequency. With these parameters we get $\omega_s = 280\omega_c$, i.e., $\omega_c h = 0.022$. The sampling rate should thus be much higher than the crossover frequency. The ratio ω_s/ω_c can be decreased by choosing an antialiasing filter of higher order. With a sixth-order Bessel filter we have

$$\frac{\omega_s}{\omega_c} \approx \frac{8.6}{\alpha\beta^{1/6}}$$

With $\alpha = 0.1$ and $\beta = 0.01$ we get $\omega_s = 185\omega_c$, i.e., $\omega_c h = 0.034$.

Selection of Sampling Interval

The choice of sampling period depends on many factors. One way to determine the sampling period is to use continuous-time arguments. The sampled system can be approximated by the hold circuit, followed by the continuous-time system. For small sampling periods, the transfer function of the hold circuit can be approximated as

$$\frac{1 - e^{-sh}}{sh} \approx \frac{1 - 1 + sh - (sh)^2/2 + \cdots}{sh} = 1 - \frac{sh}{2} + \cdots$$

The first two terms correspond to the series expansion of $\exp(-sh/2)$. That is, for small h, the hold can be approximated by a time delay of half a sampling interval. Assume that the phase marginal can be decreased by 5° to 15°. This gives the following rule of thumb:

$$h\omega_c \approx 0.15 - 0.5$$

where ω_c is the crossover frequency (in radians per second) of the continuous-time system. This rule gives quite short sampling periods. The Nyquist frequency will be about 5–20 times larger than the crossover frequency.

Example 8.7—Digital redesign of lead compensator

Consider the system in Example A.2, which is a normalized model of a motor. The closed-loop transfer function

$$G_c(s) = \frac{4}{s^2 + 2s + 4}$$ (8.13)

is obtained with the lead compensator

$$G_k(s) = 4\frac{s + 1}{s + 2}$$ (8.14)

The closed-loop system has a damping of $\zeta = 0.5$ and a natural frequency of $\omega_0 = 2$ rad/s. The objective is now to find $H(z)$ in Fig. 8.5, which approximates (8.14).
Euler's method gives the approximation

$$H_E(z) = 4\frac{z - 1 + h}{z - 1 + 2h} = 4\frac{z - (1 - h)}{z - (1 - 2h)}$$ (8.15)

while Tustin's approximation gives

$$H_T(z) = 4\frac{(2 + h)z - 2 + h}{(2 + 2h)z - 2 + 2h} = 4\frac{2 + h}{2 + 2h} \cdot \frac{z - (2 - h)/(2 + h)}{z - (1 - h)/(1 + h)}$$

Finally, zero-order hold sampling of (8.14) gives

$$H_{ZOH}(z) = \frac{4z - 2(1 + e^{-2h})}{z - e^{-2h}} = 4\frac{z - 0.5(1 + e^{-2h})}{z - e^{-2h}}$$

All approximations have the form

$$H(z) = \frac{b_0 z + b_1}{z + a_1}$$

The crossover frequency of the continuous-time process in cascade with the compensator (8.14) is $\omega_c = 1.6$ rad/s. The rule of thumb above gives a sampling period of about 0.1–0.3 s.

Figure 8.6 shows the control signal and the process output when Euler's approximation has been used for different sampling times. The other approximations give similar results. The closed-loop system has a satisfactory behavior for all compensators when the sampling time is short. The rule of thumb also gives reasonable values for the sampling period. The overshoot when $h = 0.5$ is about twice as large as for the continuous-time compensator. In the example, the change in u_c occurs at

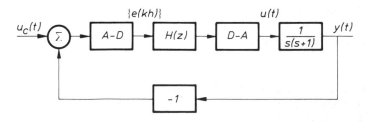

Figure 8.5 Digital control of the motor example.

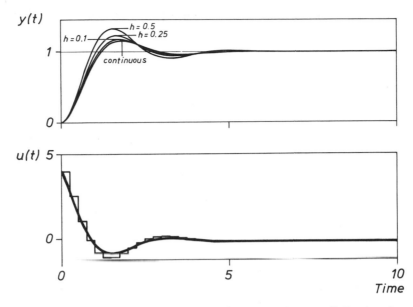

Figure 8.6 Process output, $y(t)$, when the motor is controlled using the compensator of (8.15) when $h = 0.1, 0.25$, and 0.5. The control signal is shown for $h = 0.25$. For comparison, the continuous-time signals are also shown.

a sampling instant. This is not true in practice, and there may be a delay in the response of at most one sampling period. ☐

8.3 Digital PID-Controllers

Many control problems can be solved using a PID controller. This is in fact the standard tool to solve process control problems. The "textbook" version of the PID-controller can be described by the equation

$$u(t) = K \left[e(t) + \frac{1}{T_i} \int^t e(s) \, ds + T_d \frac{de(t)}{dt} \right] \tag{8.16}$$

where error e is the difference between command signals u_c (the set point) and process output y (the measured variable). The PID-controller was originally implemented using analog technology that went through several development stages, i.e., pneumatic, relay and motors, transistors, and integrated circuits. In this development much know-how was accumulated that was imbedded into the analog design. Today virtually all PID-regulators are implemented digitally. Early implementations were often a pure translation of (8.16) which left out many of the extra features that were incorporated in the analog design. In this section we will discuss the digital PID-regulator in some detail. This is a good demonstration of the fact that a good controller is not just an implementation of a "textbook" algorithm. It is also a good way to introduce some of the implementation issues that will be discussed in depth in Chapter 15.

Modification of Linear Response

A pure derivative can, and should not be, implemented, because it will give a very large amplification of measurement noise. The gain of the derivative must thus be limited. This can be done by approximating the transfer function sT_d as follows:

$$sT_d \approx \frac{sT_d}{1 + sT_d/N}$$

The transfer function on the right approximates the derivative well at low frequencies but the gain is limited to N at high frequencies. N is typically in the range of 3–20.

In the work with analog controllers it was also found advantageous not to let the derivative act on the command signal. Later it was also found suitable to let only a fraction b of the command signal act on the proportional part. The PID-algorithm then becomes

$$U(s) = K \left[bU_c(s) - Y(s) + \frac{1}{sT_i} (U_c(s) - Y(s)) - \frac{sT_d}{1 + sT_d/N} Y(s) \right] \quad (8.17)$$

where U, U_c, and Y denote the Laplace transforms of u, u_c, and y. The idea of providing different signal paths for the process output and the command signal is a good way to separate command signal response from the response to measured signal. Alternatively it may be viewed as a way to position the closed-loop zeros. This is discussed further in Section 9.5. There are also several other variations of the PID-algorithm that are used in commercial systems. An extra first-order lag may be used in series with the controller to obtain a high-frequency roll-off. In some applications it has also been useful to include nonlinearities. The proportional term Ke can be replaced by $Ke \mid e \mid$ and a dead-zone can also be included.

Discretization

The regulator described by (8.17) can be discretized using any of the standard methods such as Tustin's approximation or ramp equivalence. Since the PID-regulator is so simple, there are some special methods that are used. The following is a popular approximation that is very easy to derive. The proportional part

$$P(t) = K(bu_c(t) - y(t)) \quad (8.18)$$

requires no approximation since it is a purely static part. The integral term

$$I(t) = \frac{K}{T_i} \int^t e(s) \, ds$$

is approximated by a rectangular approximation, i.e.,

$$I(kh + h) = I(kh) + \frac{Kh}{T_i} e(kh) \tag{8.19}$$

The derivative part given by

$$\frac{T_d}{N} \frac{dD}{dt} + D = -KT_d \frac{dy}{dt}$$

is approximated by taking backward differences. This gives

$$D(kh) = \frac{T_d}{T_d + Nh} D(kh - h) - \frac{KT_d N}{T_d + Nh} (y(kh) - y(kh - h)) \tag{8.20}$$

This approximation has the advantage that it is always stable and that the sampled pole goes to zero when T_d goes to zero. Tustin's approximation gives an approximation such that the pole instead goes to $z = -1$ as T_d goes to zero. The control signal is given as

$$u(kh) = P(kh) + I(kh) + D(kh) \tag{8.21}$$

This approximation has the pedagogical advantage that the proportional, integral, and derivative terms are obtained separately.

The other approximations give similar results. They can all be represented as

$$R(q)u(kh) = T(q)u_c(kh) - S(q)y(kh) \tag{8.22}$$

where the polynomials R, S, and T are of second order. The polynomial R has the form

$$R(q) = (q - 1)(q - a_d) \tag{8.23}$$

The number a_d and the coefficients of the polynomials S and T obtained for different approximation methods are given in Table 8.1.

Incremental Algorithms

Equation (8.21) is called a *position algorithm* or an *absolute algorithm*. In some cases it is advantageous to move the integral action outside the control algorithm. This is natural when a stepper motor is used. The output of the controller should then represent the increments of the control signal, and the motor implements the integrator. Another case is when an actuator with pulse-width control is used. In such a case the control algorithm is rewritten so that its output is the increment of the control signal. Since it follows from (8.23) that the polynomial R in (8.22) always has a factor $(q - 1)$ this is easy to do. Introducing

$$\Delta u(kh) = u(kh) - u(kh - h)$$

we get

$$(q - a_d)\Delta u(kh + h) = T(q)u_c(kh) - S(q)y(kh) \tag{8.24}$$

TABLE 8.1 Coefficients in different approximations of the continuous-time PID-regulator

	Special	Tustin	Ramp equivalence
s_0	$K(1 + b_d)$	$K(1 + b_i + b_d)$	
s_1	$-K(1 + a_d + 2b_d - b_i)$	$-K(1 + a_d + 2b_d - b_i(1 - a_d))$	
s_2	$K(a_d + b_d - b_i a_d)$	$K(a_d + b_d - b_i a_d)$	
t_0	Kb	$K(b + b_i)$	
t_1	$-K(b(1 + a_d) - b_i)$	$-K(b(1 + a_d) - b_i(1 - a_d))$	
t_2	$Ka_d(b - b_i)$	$Ka_d(b - b_i)$	
a_d	$\dfrac{T_d}{Nh + T_d}$	$\dfrac{2T_d - Nh}{2T_d + Nh}$	$\exp\left(-\dfrac{Nh}{T_d}\right)$
b_d	Na_d	$\dfrac{2NT_d}{2T_d + Nh}$	$\dfrac{T_d}{h}(1 - a_d)$
b_i	$\dfrac{h}{T_i}$	$\dfrac{h}{2T_i}$	$\dfrac{h}{2T_i}$

This is called the *incremental* form of the regulator. A drawback with the incremental algorithm is that it cannot be used for P- or PD-regulators. If this is attempted the regulator will be unable to keep the reference value, because an unstable mode $z - 1$ is cancelled.

Integrator Windup

A regulator with integral action combined with an actuator which becomes saturated can give some undesirable effects. If the control error is so large that the integrator saturates the actuator, the feedback path will be broken, because the

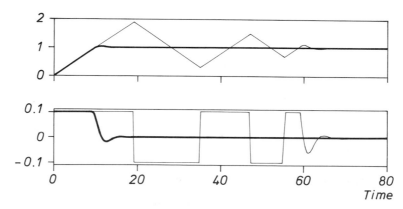

Figure 8.7 Illustration of integrator windup. The thin line shows response with an ordinary PID-regulator. The thick lines show the improvement with a regulator having antiwindup.

actuator will remain saturated even if the process output changes. The integrator, being an unstable system, may then integrate up to a very large value. When the error is finally reduced, the integral may be so large that it takes considerble time until the integral assumes a normal value again. This effect is called *integrator windup*. The effect is illustrated in Figure 8.7. There are several ways to avoid integrator windup. One possibility is to stop updating the integral when the actuator is saturated. Another method is illustrated by the block diagram in Figure 8.8(a). In this system an extra feedback path is provided by measuring the actuator output and forming an error signal (e_s) as the difference between the actuator output (u_c) and the controller output (v) and feeding this error back to the integrator through the gain $1/T_t$. The error signal e_s is zero when the actuator is not saturated. When the actuator is saturated the extra feedback path tries to make

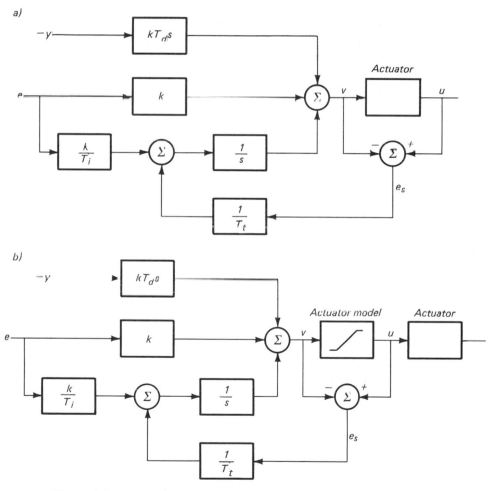

Figure 8.8 Regulator with antiwindup. A system where the actuator output is measured is shown in (a) and a system where the actuator output is estimated from a mathematical model is shown in (b).

the error signal e_s equal zero. This means that the integrator is reset so that the controller output is at the saturation limit. The integrator is thus reset to an appropriate value with the time constant T_t, which is called the tracking-time constant. The advantage with this scheme for antiwindup is that it can be applied to any actuator, i.e., not only a saturated actuator but also an actuator with arbitrary characteristics, such as a dead-zone or an hysteresis, as long as the actuator output is measured. If the actuator output is not measured, the actuator can be modeled and an equivalent signal can be generated from the model, as shown in Fig. 8.8(b).

Figure 8.7 shows the improved behavior with controllers having an antiwindup scheme.

Operational Aspects

Practically all PID-controllers can run in two modes: manual and automatic. In manual mode the regulator output is manipulated directly by the operator, typically by push buttons that increase or decrease the controller output. The controllers may also operate in combination with other controllers, as in a cascade or ratio connection, or with nonlinear elements such as multipliers and selectors. This gives rise to more operational modes. The regulators also have parameters which can be adjusted in operation. When there are changes of modes and parameters, it is essential to avoid switching transients. The way mode switchings and parameter changes are made depends on the structure chosen for the regulator.

Bumpless transfer. Because the regulator is a dynamic system, it is necessary to make sure that the state of the system is correct when switching the regulator between manual and automatic mode. When the system is in manual mode, the controller produces a control signal that may be different from the manually generated control signal. It is necessary to make sure that the value of the integrator is correct at the time of switching. This is called *bumpless transfer*. Bumpless switching is easy to obtain for a regulator in incremental form. This is shown in Figure 8.9(a). The integrator is provided with a switch so that the signals are either chosen from the manual or the automatic increments. Since the switching only influences the increments, there will not be any large transients. A related scheme for a position algorithm is shown in Fig. 8.9(b). In this case the integral action is realized as positive feedback around a first-order system. The transfer function from v to u is

$$\frac{1}{1 - \dfrac{1}{1 + sT_i'}} = \frac{1 + sT_i'}{sT_i'}$$

For simplicity the filters are shown in continuous-time form. In a digital system they are of course realized as sampled systems. The system can also be provided with an antiwindup protection as shown in Fig. 8.9(c). A drawback with this

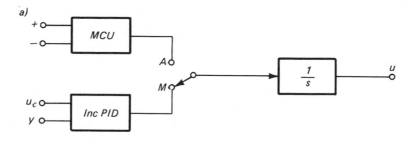

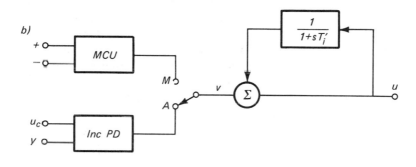

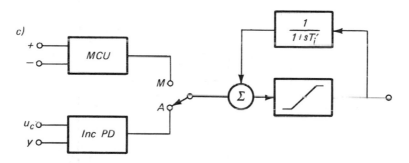

Figure 8.9 Regulators with bumpless transfer from manual to automatic mode. The regulator in (a) is incremental. The regulators in (b) and (c) are special forms of position algorithms. The regulator in (c) has anti-windup (MCU = Manual Control Unit).

scheme is that the PID-controller must be of the form

$$G(s) = K' \frac{(1 + sT_i')(1 + sT_d')}{sT_i'} \tag{8.25}$$

which is less general than (8.16). Moreover the reset-time constant is equal to T_i. More elaborate schemes have to be used for general PID-algorithms on position form. Such a regulator is built up of a manual control module and a PID-module, each having an integrator. See Fig. 8.10.

Bumpless Parameter Changes

A regulator is a dynamic system. A change of the parameters of a dynamic system will naturally result in changes of its output even if the input is kept constant. Changes in the output can in some cases be avoided by a simultaneous change of the state of the system. The changes in the output will also depend on the chosen realization. With a PID-regulator it is natural to require that there be no drastic changes in the output if the parameters are changed when the error is zero. This will obviously hold for all incremental algorithms, because the output of an incremental algorithm is zero when the input is zero irrespective of the parameter values. It also holds for a position algorithm with the structure shown in Fig. 8.9(b) and (c). For a position algorithm it depends, however, on the implementation. Assume, for example, that the state is chosen as

$$x_I = \int^t e(s) \, ds$$

when implementing the algorithm. The integral term is then

$$I = \frac{K}{T_i} x_I$$

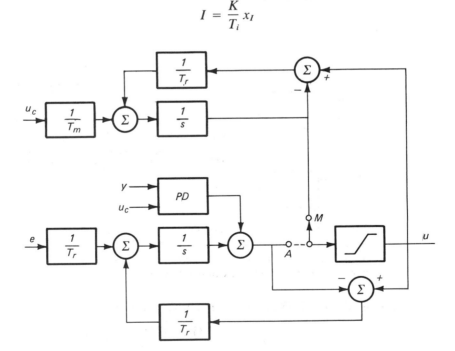

Figure 8.10 PID-regulator with bumpless switching between manual and automatic control.

Any change of K or T_i will then result in a change of I. To avoid bumps when the parameters are changed it is therefore essential that the state be chosen as

$$x_I = \frac{K}{T_i} \int^t e(s) \, ds$$

when implementing the integral term.

Computer Code

A typical computer code for a discrete PID-regulator is given in Listing 8.1. The code is given in Simnon format. The discretization of the integral term is made using a forward difference. The derivative term is approximated using a backward difference.

The calculation given between the labels INITIAL and SORT are made initially only. This saves computing time. In a real system these calculations have to be made each time the parameters are changed. The code given admits bumpless parameter changes if $b = 1$. When $b \neq 1$ the proportional term (P) is different from zero in steady state. To ensure bumpless parameter changes it is necessary that the quantity $P + I$ is invariant to parameter changes. This means that the state I has to be changed as follows:

$$I_{new} = I_{old} + K_{old}(b_{old}u_c - y) - K_{new}(b_{new}u_c - y) \tag{8.26}$$

Word length and integration offset. The integral part in digital PID-controllers is approximated as a sum. Computational problems, such as *integration offset*, may then occur due to the finite precision in the representation in the computer. Assume that there is an error, $e(kh)$. The integrator term is then increased at each sampling time with

$$\frac{Kh}{T_i} e(kh)$$

(see (8.19)). Assume that the gain is small and that the reset time is large compared to the sampling time. The change in the output may then be smaller than the quantization step in the D-A converter. For instance, a 12-bit D-A converter (i.e., a resolution of 1/4096) should give sufficiently good resolution for control purposes. Yet if $K = h = 1$ and $T_i = 3600$, then any error less than 90% of the span of the A-D converter gives a calculated change in the integral part less than the quantization step. If the integral part is stored in the same precision as that of the D-A converter, then there will be an offset in the output. One way to avoid this is to use higher precision in the internal calculations. The results then have an error that is less than the quantization level of the output. Frequently at least

```
DISCRETE SYSTEM reg

"PID regulator based on Tustin discretization

INPUT uc        "Set point
INPUT y         "Measured variable
OUTPUT u        "Regulator output
STATE I         "Integral part
STATE D         "Derivative part
STATE yold      "Delayed measured variable
NEW   nI nD nyold
TIME t
TSAMP ts

INITIAL
bi = K*h/Ti
ar = h/Tt
bd = K*N*Td/(Td + N*h)
ad = Td/(Td + N*h)

SORT
v = P + I + nD
u = IF v<ulow THEN ulow ELSE IF v<uhigh THEN v ELSE uhigh

"Proportional part
P = K*(b*R - y)

"Integral part
nI = I + bi*(uc - y) + ar*(u - v)

"Derivative part
nD = ad*D - bd*(y - yold)
nyold = y

"Update sampling time
ts = t + h

"Parameters
    K:1            "Regulator gain
    Ti:4           "Integral time
    Td:1           "Derivative time
    Tt:0.1         "Reset time
    N:10           "Maximum derivative gain
    b:1            "Fraction of setpoint in proportional term
    ulow: -1       "Low output limit
    uhigh:1        "High output limit
    h:0.02         "Sampling period

END
```

Listing 8.1 PID-regulator based on Tustin discretization.

24 bits are used to implement the integral part in a computer, in order to avoid integration offset.

Tuning

A PID-regulator has parameters K, T_i, T_d, T_t, b, N, u_{low}, and u_{high} that must be chosen. The primary parameters are K, T_i, and T_d. Parameter N can often be given a fixed default value, e.g., $N = 10$. The tracking-time constant (T_t) is often related to the integration time (T_i). In some implementations it has to be equal to T_i; in other cases it can be chosen as 0.1–0.5 times T_i. The parameters u_{low} and u_{high} should be chosen close to the true saturation limits.

If the process dynamics and the disturbances are known parameters, then K, T_i, and T_d can be computed using the design methods of Chapters 9, 10, 11, and 12. Some special methods have, however, been developed to tune the PID-parameters experimentally. The behavior of the discrete-time PID-controller is very close to the analog PID-controller if the sampling interval is short. The traditional tuning rules for continuous-time controllers can thus be used. There are two classical heuristic rules due to Ziegler and Nichols (1942) that can be used to determine the regulator parameters: the step-response method and the ultimate-period method.

The step-response method. In this method the unit step response of the open loop process is determined experimentally. The technique can be applied to processes whose step response is monotone or essentially monotone apart from an initial nonminimum phase characteristic. To use the method the tangent to the step response that has the steepest slope is drawn and the intersections of the tangent with the axes are determined. See Fig. 8.11. The controller parameters are then obtained from Table 8.2. The Ziegler-Nichols rule was designed to give good response to load disturbances. It does, however, give fairly low damping of the dominant poles.

Parameter L is called the apparent dead-time. For stable processes parameter T, which is called the apparent-time constant, can also be determined. The suitability of Ziegler-Nichols tuning and regulator selection can be based on the ratio L/T. Ziegler-Nichols tuning will work well for $0.15 < L/T < 0.6$.

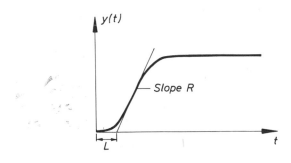

Figure 8.11 Determination of parameters $a = RL$ and L from the unit step response.

TABLE 8.2 PID parameters obtained from the Ziegler-Nichols step-response method

Regulator type	K	T_i	T_d
P	$1/a$		
PI	$0.9/a$	$3L$	
PID	$1.2/a$	$2L$	$0.5L$

The Ultimate-Sensitivity Method

In this method the key idea is to determine the point where the Nyquist curve of the open-loop system intersects the negative real axis. This is done by connecting the controller to the process and setting the parameters so that pure proportional control is obtained. The gain of the controller is then increased until the closed-loop system reaches the stability limit. The gain (K_u) when this occurs and the period of the oscillation (T_u) are determined. These parameters are called ultimate gain and ultimate period. The regulator parameters are then given by Table 8.3. Let K_p be the static process gain. The number $K_p K_u$, which is dimension-free, can be used to choose the regulator type and to judge if PID-control is appropriate. Ziegler-Nichols tuning will work well for $2 < K_p K_u < 20$.

The tuning rules above should be used only as a first approximation. The final tuning usually has to be done manually. There are also several other methods for tuning digital PID-controllers. Some involve a compensation for the length of the sampling interval (see the references).

Selection of Sampling Interval

When DDC-control was first introduced, computers were not as powerful as they are today. Long sampling intervals were needed to handle many loops. The following recommendations for the most common process variables are given for DDC-control.

Type of variable	Sampling time s
Flow	1–3
Level	5–10
Pressure	1–5
Temperature	10–20

TABLE 8.3 PID parameters obtained from Ziegler-Nichols ultimate-period method

Regulator type	K	T_i	T_d
P	$0.5K_u$		
PI	$0.45K_u$	$T_u/1.2$	
PD	$0.6K_u$	$T_u/2$	$T_u/8$

Commercial digital controllers for few loops often have a short fixed-sampling interval on the order of 200 ms. This implies that the controllers can be regarded as continuous-time controllers, and the continuous-time tuning rules may be used.

Several rules of thumb for choosing the sampling period for a digital PID-regulator are given in the literature. There is a significant difference between PI- and PID-regulators. For PI-regulators the sampling period is related to the integration time. A typical rule of thumb is

$$\frac{h}{T_i} \approx 0.1\text{--}0.3$$

When Ziegler-Nichols tuning is used this implies

$$\frac{h}{L} \approx 0.3\text{--}1$$

or

$$\frac{h}{T_u} \sim 0.1 \quad 0.3$$

With PID-control the critical issue is that the sampling period must be so short that the phase lead is not adversely affected by the sampling. This implies that the sampling period should be chosen so that the number hN/T_d is in the range of 0.2 to 0.6. With $N = 10$ this means that for Ziegler-Nichols tuning the ratio h/L is between 0.01 and 0.06. This gives

$$\frac{hN}{T_d} \approx 0.2 - 0.6$$

Significantly shorter sampling periods are thus required for controllers with derivative action. If computer time is at a premium, it is advantageous to use the design methods discussed in Chapters 9–12.

Smith-Predictor

The Smith-predictor is introduced in Sec. 7.5. One drawback of the continuous-time version of the Smith-predictor is that the predictor contains a time delay. This will not cause any problems when using digital implementation, because the time delay is implemented as a vector that is shifted at each sampling instant. A simple example illustrates the design of a Smith-predictor.

Example 8.8—Smith-predictor

A time-delay process is described in Example A.4. The process can, for instance, represent a paper machine. Assume that the process in (A.9) has a delay of 2 time units and that the sampling time is $h = 1$. The system is then described by the model

$$y(k + 1) = 0.37y(k) + 0.63u(k - 2)$$

(see Example 3.6). If there were no time delays, a PI-regulator with gain 0.4 and integration time $T_i = 0.4$ would give good control (see Fig. 8.12). The set point is changed at $t = 0$ and a step disturbance is introduced in the output at $t = 20$. This

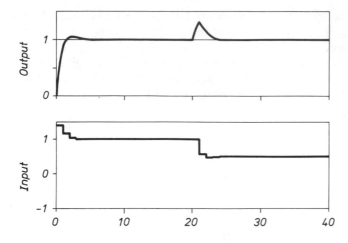

Figure 8.12 PI-control of the process in Example 8.8 without time delay.

PI-regulator will not give good control if the process has a time delay. To obtain a good PI-regulation, it is necessary to have a gain of 0.1 and $T_i = 0.5$. The response of this regulator is illustrated in Fig. 8.13. Comparison with Fig. 8.12 shows that the response is slower because the gain is smaller.

A Smith-predictor based on a first order process model and a PI-regulator can be described by the following Simnon program:

```
DISCRETE SYSTEM Smith
"Smith-predictor for first-order system with delay
INPUT y uc
OUTPUT u
STATE i u1 ym ym1 ym2
NEW ni nu1 nym nym1 nym2
TIME t
TSAMP ts


e = uc − y + ym2 − ym
u = K*(e +  e*h/Ti + i)


ni = i + e*h/Ti
nym = am*ym + bm1*u + bm2*u1 "simulates process model
nu1 = u
num1 = ym
nym2 = ym1
ts = t + h


h:1          "sampling time
Ti:0.4       "integration time
K:0.4        "gain
am:0.37      ""
bm1:0.63     "process parameters
bm2:0        ""


END
```

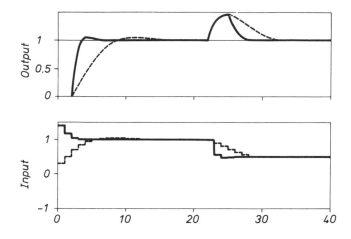

Figure 8.13 PI-control (dashed line) and Smith-predictor control (thick line) of the process in Example 8.8 with time delay.

Notice how easy it is to implement the predictor with computer control. The response of the closed-loop system obtained with the Smith-predictor is shown in Fig. 8.13, which shows the improvement in comparison with PI-control. The control signal at the initial step for the Smith-predictor can be observed to be identical to the control signal with PI-control for a process without delay. □

8.4 State-Feedback Redesign

In this section state-feedback, continuous-time controllers are translated into discrete-time controllers. State-feedback controllers can be regarded as generalized P-controllers. The formulation of the problem assumes that the process is described by the equations

$$\dot{x} = Ax + Bu \tag{8.27}$$
$$y = Cx$$

where all the states are assumed to be measurable. Using a controller of the form

$$u(t) = Mu_c(t) - Lx(t) \tag{8.28}$$

it is possible to place the poles of the closed-loop system arbitrarily if the system is controllable. The controller in (8.28) can be implemented in digital form by sampling the states and holding the control signal constant over the sampling intervals. This is how the control is done in Example 1.3. If the sampling period is increased, then the behavior of the closed-loop system starts to deteriorate. It is, however, possible to modify the controller in order to improve the performance of the closed-loop system. Assume that the discrete-time controller is

$$u(kh) = \tilde{M}u_c(kh) - \tilde{L}x(kh) \tag{8.29}$$

One way to solve the problem is to design the controller in (8.29) using sampled-data theory. This is done in Chapter 9. Here, an approximate method is used to translate the controller in (8.28) into discrete-time form.

Controlling (8.27) with the continuous-time controller in (8.28) gives the closed-loop system

$$\dot{x} = (A - BL)x + BMu_c = A_c x + BMu_c$$

$$y = Cx$$

If $u_c(t)$ is constant over one sampling period, then this equation can be integrated; this gives

$$x(kh + h) = \Phi_c x(kh) + \Gamma_c M u_c(kh) \qquad (8.30)$$

where

$$\Phi_c = \exp(A_c h)$$

$$\Gamma_c = \int_0^h \exp(A_c s)\, ds\, B$$

If the discrete-time controller in (8.29) is used to control (8.27), then

$$x(kh + h) = (\Phi - \Gamma\tilde{L})x(kh) + \Gamma\tilde{M}u_c(kh) \qquad (8.31)$$

where Φ and Γ are the system matrices obtained when (8.27) is sampled. It is in general not possible to choose $\tilde{L}$ such that

$$\Phi_c = \Phi - \Gamma\tilde{L}$$

However, we can make a series expansion and equate terms of different powers of h. Assume that

$$\tilde{L} = L_0 + L_1 h/2$$

then

$$\Phi_c \approx I + (A - BL)h + [A^2 - BLA - ABL + (BL)^2]h^2/2 + \cdots$$

and

$$\Phi - \Gamma\tilde{L} \approx I + (A - BL_0)h + (A^2 - ABL_0 - 2BL_1)h + \cdots$$

The systems (8.30) and (8.31) have the same poles up to and including order h^2 when

$$\tilde{L} = L[I + (A - BL)h/2] \qquad (8.32)$$

Without modification of L the poles are the same up to and including order h.

The modification of M is determined by assuming that the steady-state values are the same for (8.30) and (8.31). Let the reference value be constant and assume that the steady-state value of the state is x^0. This gives the relations

$$(I - \Phi_c)x^0 = \Gamma_c M u_c$$

and

$$[I - (\Phi - \Gamma\tilde{L})]x^0 = \Gamma\tilde{M}u_c$$

The series expansions of left-hand sides of the relations above are equal for

Translation of Analog Design   Chap. 8

powers of h up to and including h^2. Now determine $\tilde{M}$ such that the series expansions of the right-hand sides are the same for h and h^2. Assume that

$$\tilde{M} = M_0 + M_1 h/2$$

then

$$\Gamma_c M \approx BMh + (A - BL)BMh^2/2 + \cdots$$

and

$$\Gamma \tilde{M} \approx BM_0 h + (BM_1 + ABM_0)h^2/2 + \cdots$$

This gives

$$\tilde{M} = (I - LBh/2)M \tag{8.33}$$

The modifications (8.32) and (8.33) are easily computed from the continuous-time system and the continuous-time controller.

Example 8.9

The system in Example 1.3 is the double integrator; i.e., the system is defined by the matrices

$$A = \begin{bmatrix} 0 & 1 \\ 0 & 0 \end{bmatrix}, \quad B = \begin{bmatrix} 0 \\ 1 \end{bmatrix}, \quad \text{and} \quad C = [1 \quad 0]$$

Let the continuous-time controller be

$$u(t) = u_c(t) - [1 \quad 1]x(t)$$

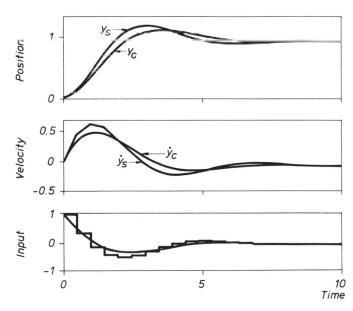

Figure 8.14 Digital control of the double integrator using the control law in (8.34) when $h = 0.5$; y_c is the continuous-time response.

Sec. 8.4   State-Feedback Redesign

237

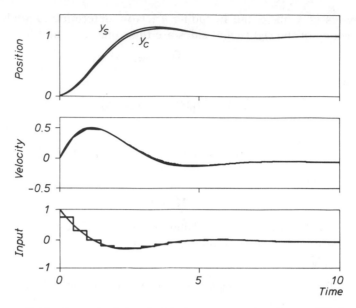

Figure 8.15 Control of the double integrator using the modified controller in (8.35) when $h = 0.5$; y_c is the continuous-time response.

Fig. 8.14 shows the behavior when the sampled controller

$$u(kh) = u_c(kh) - [1 \quad 1]x(kh) \tag{8.34}$$

is used when $h = 0.5$. Using the modifications in (8.32) and (8.33), we get

$$\begin{aligned} \tilde{L} &= [1 - 0.5h \quad 1] \\ \tilde{M} &= 1 - 0.5h \end{aligned} \tag{8.35}$$

Figure 8.15 shows the behavior when the modified controller is used for $h = 0.5$; there is an improvement compared with the unmodified controller. However, the sampling period cannot be increased much further before the closed-loop behavior starts to deteriorate, even when the modified controller is used. □

8.5 Frequency-Response Design Methods

This chapter has so far shown how continuous-time controllers can be translated into discrete-time forms. This section discusses how continuous-time frequency design methods can be used to design discrete-time controllers.

Frequency design methods based on Bode and Nichols plots are useful for designing compensators for systems described by transfer functions. The usefulness of the methods depends on the simplicity of drawing the Bode plots and on rules of thumb for choosing the compensators. The Bode plots are easy to draw since the transfer functions are in general rational functions in $i\omega$, except for pure time delays.

Frequency curves for discrete-time systems are more difficult to draw since

Translation of Analog Design   Chap. 8

the pulse-transfer functions are not rational functions in $i\omega$, but in $\exp(i\omega h)$. The *w-transform method* is one way to circumvent this difficulty. The method can be summarized into the following steps:

1. Sample the continuous-time system that should be controlled using a zero-order-hold circuit. This gives $H(z)$.
2. Introduce the variable

$$w = \frac{2}{h} \frac{z - 1}{z + 1}$$

[compare (8.6)]. Transform the pulse-transfer function of the process into the w-plane giving

$$H'(w) = H(z) \Big|_{z = \frac{1 + wh/2}{1 - wh/2}}$$

For $z = \exp(i\omega h)$ then

$$w = i\frac{2}{h} \tan(\omega h/2) = iv$$

(compare frequency prewarping in Sec. 8.2). The transformed transfer function $H'(iv)$ is a rational function in iv.

3. Draw the Bode plot of $H'(iv)$ and use conventional methods to design a compensator $H'_r(iv)$ that gives desired frequency domain properties. The distortion of the frequency scale between v and ω must be taken into account when deciding, for instance, crossover frequency and bandwidth.
4. Transform the compensator back into the z-plane and implement $H_r(z)$ as a discrete-time system.

The advantage with the w-transform method is that conventional Bode diagram techniques can be used to make the design. One difficulty is to handle the distortion of the frequency scale and to choose the sampling interval.

8.6 Conclusions

Different ways of translating an analog-control system to a digital-control system are necessary to bridge the gap between the analog and digital worlds. The problem is also of substantial interest if an analog design is available, and a digital solution needs to be found. Several methods to compute a pulse-transfer function that corresponds to the continuous-time transfer function have been discussed, based on step invariance, ramp invariance, and Tustin's approximation. The method based on ramp invariance is a good choice. Tustin's approximation is commonly used because of its simplicity. It does, however, distort the frequency scale of the filter. Digital systems designed in this way are always (slightly) inferior to analog systems because of the inherent time delay caused by the hold circuit. This time delay is approximately $h/2$.

The translation methods work well if the sampling period is short. A good way to choose the sampling period is to observe that the extra time delay decreases the phase margin by $\omega_c h/2$ radians or by $180\omega_c/\omega_s$ degrees, where ω_c is the crossover frequency.

There are possibilities of creating better designs than those discussed in this chapter, as discussed in the following chapters.

8.7 Problems

8.1. Find how the left half-s-plane is transformed into the z-plane when using the mappings in (8.4)–(8.6).

8.2. Make an approximation of the transfer function

$$G(s) = \frac{a}{s + a}$$

using (a) Euler's method, (b) Tustin's approximation, and (c) Tustin's approximation with prewarping if the warping frequency is $\omega_1 = a$ rad/s.

8.3. The lead network given in (8.14) gives about 20° phase advance at $\omega_c = 1.6$ rad/s. Approximate the network for $h = 0.25$ using (a) Euler's method, (b) backward differences, (c) Tustin's approximation, (d) Tustin's approximation with prewarping using $\omega_1 = \omega_c$ as the warping frequency, and (e) zero-order hold sampling.
Compute the phase of the approximated networks at $z = \exp(i\omega_c h)$.

8.4. Verify the calculations leading to the rule of thumb for the choice of the sampling interval given in Sec. 8.2.

8.5. Show that (8.21) is obtained from (8.17) by approximating the integral part using Euler's method and backward difference for the derivative part. Discuss advantages and disadvantages in each of the following cases.
(a) The integral part is approximated using backward difference.
(b) The derivative part is approximated using Euler's method. (*Hint:* Consider the case when T_d is small.)

8.6. A continuous-time PI-controller is given by the transfer function

$$K \left(1 + \frac{1}{T_i s} \right)$$

Use the bilinear approximation to find a discrete-time controller. Find the relation between the continuous-time parameters K and T_i and the corresponding discrete-time parameters in (8.21).

8.7. Consider the tank system in Problem 3.10. Assume the following specifications for the closed-loop system:

1. The steady-state error after a step in the reference value is zero.
2. The crossover frequency of the compensated system is 0.025 rad/s.
3. The phase margin is about 50°.

(a) Design a PI-controller such that the specifications are fulfilled.

(b) Determine the poles and the zero of the closed-loop system. What is the damping corresponding to the complex poles?

(c) Choose a suitable sampling interval and approximate the continuous-time controller using Tustin's method with warping. Use the crossover frequency as the warping frequency.

(d) Simulate the system when the sampled-data controller is used. Compare with the desired response, i.e., when the continuous-time controller is used.

8.8 Make an approximation, analogous to (8.32) and (8.33) such that the modifications are valid for terms up to and including h^3.

8.9. The normalized motor has a state-space representation given by (A.5). The control law

$$u(t) = Mu_c(t) - Lx(t)$$

with $M = 4$ and $L = [2 \quad 4]$ gives the continuous-time transfer function

$$\frac{4}{s^2 + 3s + 4}$$

from u_c to y. This corresponds to $\zeta = 0.75$ and $\omega_0 = 2$.

(a) Make a sampled-data implementation of the controller above.

(b) Modify the control law using (8.32) and (8.33).

(c) Simulate the controllers in (a) and (b) for different sampling periods and compare with the continuous-time controller.

8.10. Given the continuous-time system

$$\frac{dx}{dt} = \begin{bmatrix} -3 & 1 \\ 0 & -2 \end{bmatrix} x + \begin{bmatrix} 0 \\ 1 \end{bmatrix} u$$

$$y = [1 \quad 0]x$$

(a) Determine a continuous-time state-feedback controller

$$u(t) = -Lx(t)$$

such that the characteristic polynomial of the closed-loop system is

$$s^2 + 8s + 32$$

A computer is then used to implement the controller as

$$u(kh) = -Lx(kh)$$

(b) Modify the controller using (8.32).

(c) Simulate the controllers in (a) and (b) and decide suitable sampling intervals. Assume that $x(0) = [1 \quad 0]^T$.

8.11. Use the w-plane method to design a compensator for the motor in Example 8.7 when $h = 0.25$. Design the compensator such that the transformed system has a crossover frequency corresponding to 1.4 rad/s and a phase margin of 50 degrees. Compare with the continuous-time design and the discrete-time approximations in Example 8.7. Investigate how long a sampling interval can be used for the w-plane method.

8.12. Consider the continuous-time double integrator described by (3.11). Assume that a time-continuous design has been made giving the controller

$$u(t) = 2u_c(t) - (1 \quad 2)\hat{x}(t)$$

$$\frac{d\hat{x}(t)}{dt} = A\hat{x}(t) + Bu(t) + K(y(t) - C\hat{x}(t))$$

with $K^T = (1 \quad 1)$.

   (a) Assume that the controller should be implemented using a computer. Modify the controller (not the observer part) for the sampling interval $h = 0.2$ using (8.32) and (8.33).
   (b) Approximate the observer using backward-difference approximation.
   (c) Simulate the continuous-time controller and the discrete-time approximation. Let the initial values be $x(0) = (1 \quad 1)^T$ and $\hat{x}(0) = (0 \quad 0)^T$.

8.13. Derive ramp-invariant approximations of the transfer function

$$G(s) = \frac{1}{s + a}$$

and

$$G(s) = \frac{s}{s + a}$$

8.14. Derive the ramp-invariant equivalent of the PID-controller.

8.15. There are many different ways to sample a continuous time system. The key difference is the assumption made on the behavior of the control signal over the sampling interval. So far we have discussed step invariance and ramp invariance. Derive formula for impulse invariant sampling of the system (3.1) when the continuous time signal is assumed to be a sequence of impulses that occur just after the sampling instants.

8.16. Derive the impulse invariant approximations of the transfer functions in Problem 8.13.

8.17. The frequency prewarping in Section 8.2 gives the correct transformation at one frequency along the imaginary axis. Derive the necessary warping transformation such that one point at an arbitrary ray through the origin is transformed correctly.

8.8 References

The problem of designing digital filters that implement analog-transfer functions approximately is discussed in the digital filtering literature

RABINER, L. R. and B. GOLD (1975): *Theory and Application of Digital Signal Processing.* Englewood Cliffs, N.J.: Prentice-Hall, Inc.

OPPENHEIM, A. V. and R. W. SCHAFER (1989): *Discrete-Time Signal Processing.* Englewood Cliffs, N.J.: Prentice Hall, Inc.

ANTONIOU, A. (1979): *Digital Filters: Analysis and Design.* New York: McGraw-Hill.

Interesting views on similarities and differences between digital signal processing and control theory are presented in

Willsky, A. S. (1979): *Digital Signal Processing and Control and Estimation Theory*. Cambridge, Mass.: MIT Press.

A more control-oriented presentation of different approximations is found in

Franklin, G. F. and J. D. Powell (1989): *Digital Control of Dynamic Systems* (2nd ed.). Reading, Mass.: Addison-Wesley.

Digital PID-controllers and their operational aspects are discussed in

Goff, K. W. (1966): "A Systemic Approach to DDC Design," *ISA Journal,* December, 44–54.

Bristol, E. H. (1977): "Design and Programming Control Algorithms for DDC Systems," *Control Engineering,* January, 24–26.

Shinskey, F. G. (1988): *Process Control Systems* (3rd ed.). New York: McGraw-Hill.

The basic reference for tuning PID-controllers is

Ziegler, J. G. and N. B. Nichols (1942): "Optimum Settings for Automatic Controllers," *Trans. ASME,* 64, 759–68.

A modification of the rules by Ziegler and Nichols which takes the sampling interval into account is given in

Takahashi, Y., C. S. Chan, and D. M. Auslander (1971): "Parametereinstellung bei linearen DDC-Algorithmen," *Regelungstechnik und Prozess-Datenverarbeitung,* 19, 237–44.

A detailed treatment of PID-control is given in

Åström, K. J. and T. Hägglund (1988): Automatic Tuning of PID Controllers. Research Triangle Park: Instrument Society of America.

The Smith-predictor is first described in

Smith, O. J. M. (1957): "Closer Control of Loops with Deadtime," *Chem. Eng. Progr.,* 53, 217–19.

Redesign of state feedback is discussed in more detail in

Kuo, B. C. (1980): *Digital Control Systems*. Tokyo: Holt-Saunders.

STATE-SPACE DESIGN METHODS

GOAL

To Develop Design Techniques Based on the State-Space Approach.

9.1 Introduction

This chapter presents design methods based on internal models of the system. In the light of the discussion in Chapter 7, the methods developed in this chapter can be viewed as solutions to specific, idealized control problems. The solutions give insight into the nature of control problems. This chapter also shows that many of the concepts introduced earlier are useful. When applying the methods, subjective judgments will always be involved. This is expressed formally with so-called design parameters which have to be chosen by the designer.

State-feedback is discussed in Sec. 9.2. How to recover state variables that are not measured directly is discussed in Sec. 9.3. This leads to the introduction of observers. The output-feedback problem is solved in Sec. 9.4 by combining state-feedback and observers. Section 9.5 deals with the servo problem, i.e., how to introduce command signals. A design example is given in Sec. 9.6.

9.2 Regulation Based on Pole Placement by State Feedback

A simple regulation problem is discussed in this section. One of the fundamental design methods, pole placement, is developed. This design method can be viewed as an extension of the classical root-locus method. The purpose is to arrange a feedback so that all poles of the closed-loop system assume prescribed values. In this section, the problem is solved under the restrictive assumption that all state variables can be measured directly. This restriction is relaxed in Sec. 9.3.

Problem Formulation

To specify the design problem, the process, the disturbances, the criterion, and the admissible control signals must be given.

The process. It is assumed that the process to be controlled can be described by the model

$$\frac{dx}{dt} = Ax + Bu \tag{9.1}$$

where u represents the control variables, x represents the state vector, and A and B are constant matrices. Further, only the single-input–single-output case will be discussed. Because computer control is considered, the control signals will be constant over sampling periods of constant length. Sampling the system of (9.1) gives the discrete-time system

$$x(kh + h) = \Phi x(kh) + \Gamma u(kh) \tag{9.2}$$

where the matrices Φ and Γ are given by

$$\Phi = e^{Ah}$$

$$\Gamma = \left(\int_0^h e^{As} \, ds \right) B$$

(see Sec. 3.2).

The disturbances. It is assumed that the disturbances are impulses that occur irregularly, but that the spacing between the impulses is so large that the system can settle down from one disturbance before the next impulse occurs. Because the effect of an impulse is simply to change the process state, the disturbance can be represented by an initial state.

The criterion. The problem is a regulation problem, in which it is attempted to bring the state to zero after a perturbation in the initial condition. In the pole-placement formulation, the rate of decay of the state is given indirectly by specifying the poles of the closed-loop system.

Admissible controls. Because feedback solutions are desired, it is necessary to specify the information available for generating the control signal. Because the properties of the system are specified by the closed-loop poles, the closed-loop system must be linear. The feedback must then also be linear. It is assumed that all state variables can be measured directly, so the admissible controls can be expressed as a linear feedback:

$$u(kh) = -Lx(kh) \qquad (9.3)$$

Design parameters. In the formal specification of the problem, the design parameters are the sampling period and the desired closed-loop poles. It is rare that a user of a control system can give specifications in terms of these parameters. Therefore, the designer must be able to relate the design parameters to quantities that are more meaningful to the user. For this purpose, it is often useful to consider the time histories of the state variables and the control variables. It is particularly useful to discuss the trade-off between the magnitude of the control signals and the speed at which the system recovers from a disturbance (see the discussion Sec. 7.6).

An Example

To introduce the design method and to illustrate the influence of the design parameters, a special case is first discussed.

Example 9.1—Pole placement for the double-integrator plant

The sampled double-integrator plant is described by

$$x(kh + h) = \begin{bmatrix} 1 & h \\ 0 & 1 \end{bmatrix} x(kh) + \begin{bmatrix} h^2/2 \\ h \end{bmatrix} u(kh) \qquad (9.4)$$

A general linear feedback can be described by

$$u = -l_1 x_1 - l_2 x_2 \qquad (9.5)$$

With this feedback, the closed-loop system becomes

$$x(kh + h) = \begin{bmatrix} 1 - l_1 h^2/2 & h - l_2 h^2/2 \\ -l_1 h & 1 - l_2 h \end{bmatrix} x(kh) \qquad (9.6)$$

The characteristic equation of the closed-loop system is

$$z^2 + \left(\frac{l_1 h^2}{2} + l_2 h - 2 \right) z + \left(\frac{l_1 h^2}{2} - l_2 h + 1 \right) = 0$$

Assume that the desired characteristic equation is

$$z^2 + p_1 z + p_2 = 0 \qquad (9.7)$$

This leads to the following linear equations for l_1 and l_2:

$$\frac{l_1 h^2}{2} + l_2 h - 2 = p_1$$

$$\frac{l_1 h^2}{2} - l_2 h + 1 = p_2$$

These equations have the solution

$$\begin{cases} l_1 = \dfrac{1}{h^2}(1 + p_1 + p_2) \\[2ex] l_2 = \dfrac{1}{2h}(3 + p_1 - p_2) \end{cases} \tag{9.8}$$

In this example it is always possible to find the parameters in the controller (9.5) such that the desired characteristic equation (9.7) is obtained. The linear system of equations for l_1 and l_2 has a solution for all values of p_1 and p_2. $\qquad\square$

The General Case

The solution of the pole-placement problem is now given in the general case for systems with one input signal.

Let the system be described by (9.2), where u is a scalar. Assume that the sampled system is reachable. This implies that the system (9.2) can be transformed to controllable canonical form:

$$z(kh + h) = \begin{bmatrix} -a_1 & -a_2 & \cdots & -a_{n-1} & -a_n \\ 1 & 0 & \cdots & 0 & 0 \\ \vdots & \vdots & & \vdots & \vdots \\ 0 & 0 & \cdots & 1 & 0 \end{bmatrix} z(kh) + \begin{bmatrix} 1 \\ 0 \\ \vdots \\ 0 \end{bmatrix} u(kh) \tag{9.9}$$

where

$$z^n + a_1 z^{n-1} + \cdots + a_n$$

is the characteristic polynomial of the matrix Φ. It is easy to solve the pole-placement problem when the system equations are in this form. Assume that it is desired to obtain a closed-loop system with the characteristic polynomial

$$P(z) = z^n + p_1 z^{n-1} + \cdots + p_n \tag{9.10}$$

It follows from (9.9) that the feedback law

$$u = -\tilde{L}z \tag{9.11}$$

where

$$\tilde{L} = [p_1 - a_1 \quad p_2 - a_2 \quad \cdots \quad p_n - a_n] \tag{9.12}$$

gives a closed-loop system with the desired characteristic polynomial. The feedback law can be written as

$$u = -\tilde{L}z = -\tilde{L}Tx = -Lx$$

where T defines the transformation from x to z. It remains to determine the transformation matrix T. Let W_c and $\tilde{W}_c$ be the controllability matrices of (9.2) and (9.9). These are related through

$$\tilde{W}_c = TW_c$$

Further,

$$\tilde{W}_c^{-1} = \begin{bmatrix} 1 & a_1 & \cdots & a_{n-1} \\ 0 & 1 & \cdots & a_{n-2} \\ \vdots & \vdots & & \vdots \\ 0 & 0 & \cdots & 1 \end{bmatrix} \tag{9.13}$$

(compare Equation 5.17 and Example 5.7). The transformation matrix T can be expressed as

$$T = \tilde{W}_c W_c^{-1} \tag{9.14}$$

Using (9.10) the matrix polynomial

$$P(\Phi) = \tilde{\Phi}^n + p_1 \tilde{\Phi}^{n-1} + \cdots + p_n I$$
$$= (p_1 - a_1)\tilde{\Phi}^{n-1} + \cdots + (p_n - a_n)I$$

can be obtained, where $\tilde{\Phi}$ is the system matrix of the transformed system (9.9). The second equality is obtained by using the Cayley-Hamilton theorem. Further, the last row of $\tilde{\Phi}^k$ is zero except in position $n - k$, which is 1; hence, from (9.12),

$$\tilde{L} = [0 \ \ldots \ 0 \ \ 1]P(\tilde{\Phi})$$

But

$$\tilde{\Phi} = T\Phi T^{-1}$$

and thus

$$L = \tilde{L}T = [0 \ \ldots \ 0 \ \ 1]P(T\Phi T^{-1})T$$
$$= [0 \ \ldots \ 0 \ \ 1]TP(\Phi)$$

The matrix T is, however, given by (9.14). Furthermore, from (9.13)

$$[0 \ \ldots \ 0 \ \ 1]\tilde{W}_c = [0 \ \ldots \ 0 \ \ 1]$$

so it follows that

$$L = \tilde{L}\tilde{W}_c W_c^{-1} = [0 \ \ldots \ 0 \ \ 1]W_c^{-1}P(\Phi)$$

Equation (9.15) is sometimes called *Ackermann's formula*. The result is summarized in Theorem 9.1

Theorem 9.1. Consider the system of (9.2) with one input. There exists a feedback that gives a closed-loop system with poles specified by $P(z) = 0$ if and only if the sampled system is reachable. The feedback is given by

$$u(kh) = -Lx(kh)$$

[Equation (9.3)], where

$$L = \tilde{L}\tilde{W}_c W_c^{-1} = [0 \ \ldots \ 0 \ \ 1]W_c^{-1}P(\Phi) \tag{9.15}$$

$\tilde{L}$ is given by (9.12), and W_c and $\tilde{W}_c$ are the controllability matrices of the systems in (9.2) and (9.9), respectively.

Remark 1. Constructing a counterexample shows that reachability is a necessary condition.

Remark 2. Notice that the pole-placement problem can be formulated as the following abstract problem. Given matrices Φ and Γ, find a matrix L such that the matrix $\Phi - \Gamma L$ has prescribed eigenvalues.

Remark 3. Notice that it follows from (9.13), (9.14), and the definition of W_c that

$$T^{-1} = [\Gamma \quad \Phi\Gamma + a_1\Gamma \quad \ldots \quad \Phi^{n-1}\Gamma + a_1\Phi^{n-2} + \cdots + a_{n-1}\Gamma]$$

The theorem is illustrated by an example.

Example 9.2

Consider the double-integrator plant in Example 9.1. Assume that the desired characteristic polynomial is given by (9.7). We have

$$W_c = [\Gamma \quad \Phi\Gamma] = \begin{bmatrix} h^2/2 & 3h^2/2 \\ h & h \end{bmatrix}$$

and the characteristic polynomial of Φ is

$$z^2 - 2z + 1$$

Hence

$$W_c^{-1} = \begin{bmatrix} -1/h^2 & 1.5/h \\ 1/h^2 & -0.5/h \end{bmatrix}$$

We have

$$P(\Phi) = \Phi^2 + p_1\Phi + p_2 I = \begin{bmatrix} 1 + p_1 + p_2 & 2h + p_1 h \\ 0 & 1 + p_1 + p_2 \end{bmatrix}$$

Formula (9.15) now gives

$$L = [0 \quad 1] W_c^{-1} P(\Phi) = [1/h^2 \quad -0.5/h] P(\Phi)$$

$$= \begin{bmatrix} \dfrac{1 + p_1 + p_2}{h^2} & \dfrac{3 + p_1 - p_2}{2h} \end{bmatrix}$$

which is the same result obtained by the direct calculation in Example 9.1 [compare with (9.8)]. $\square$

Example 9.3

The system

$$x(k + 1) = \begin{bmatrix} 0.3 & 0 \\ 0 & 0.5 \end{bmatrix} x(k) + \begin{bmatrix} 0 \\ 1 \end{bmatrix} u(k)$$

is not reachable since

$$\det W_c = \det \begin{bmatrix} 0 & 0 \\ 1 & 0.5 \end{bmatrix} = 0$$

Let the controller be given by (9.5). Then the closed-loop system has the characteristic equation

$$(z - 0.5 + l_2)(z - 0.3) = 0$$

With the parameter l_2 the open-loop pole in 0.5 can be moved to arbitrary positions. The second pole, which corresponds to the nonreachable state, cannot be changed.

$\square$

Practical Aspects

It is easy to solve the pole-placement design problem explicitly. Notice that reachability is a necessary and sufficient condition for solving the problem. To apply the pole-placement design method in practice, it is necessary to understand how the properties of the closed-loop system are influenced by the design parameters— i.e., the closed-loop poles and the sampling period. This is illustrated by an example.

Example 9.4

Consider the double-integrator plant. Instead of using the parameters p_1 and p_2 in (9.7), two other parameters—which admit a more direct physical interpretation— will be introduced. If the desired discrete-time system is obtained by sampling a second-order system, we find that

$$p_1 = -2e^{-\zeta\omega h} \cos(\omega h \sqrt{1 - \zeta^2})$$

$$p_2 = e^{-2\zeta\omega h}$$

where ω is the natural frequency and ζ is the relative damping (compare with Example 3.16). The parameter ζ influences the relative damping of the response and ω influences the response speed.

To discuss the magnitude of the control signal, it is assumed that the system has an initial position x_0 and an initial velocity v_0. The initial value of the control signal is then

$$u(0) = -l_1 x_0 - l_2 v_0$$

If the sampling period is short, then the expressions for p_1 and p_2 can be approximated using series expansion. The following approximation:

$$u(0) \approx -\omega^2 x_0 + 2\zeta\omega v_0$$

The expression clearly shows that the magnitude of the control signal increases with increasing ω. Thus an increase in the speed of the response of the system will require an increase in the control signals. If the bounds on the control signal and typical disturbances are known, it is possible to determine reasonable values of ω. The consequences of different choices of ω when $x_0 = 1$ and $v_0 = 1$ are illustrated in Fig. 9.1. A larger ω gives a faster system but also larger control signals.

If the parameters ζ and ω are fixed, a sampling period must still be selected. The choice of the sampling period will influence how quickly the disturbances are detected. The sampling period will also influence the response curves. Section 3.7 discusses how to select the sampling period for an open-loop system. The same arguments can be used for the closed-loop system. The sampling period can be cho-

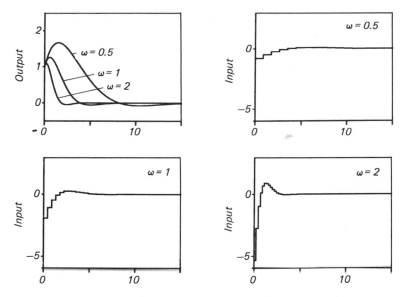

Figure 9.1 Response of the closed-loop system in Example 9.4 obtained for different values of ω with $\omega h = 0.44$ and $\zeta - 0.707$.

sen such that

$$N_r \approx 4\text{–}10$$

where N_r is the number of samples per rise time of the *closed-loop system*. This means that the sampling period should be chosen in relation to the desired behavior of the closed-loop system. The sampling period can also be related to the damped frequency of the closed-loop system. It is convenient to introduce the parameter N defined by

$$N = \frac{2\pi}{\omega h \sqrt{1 - \zeta^2}} \tag{9.16}$$

This parameter is the ratio of the damped period and the sampling period. Figure 9.2 illustrates the consequences of different choices of N in the special case. It is clear that as long as $N > 10$, there are very small differences in the responses. The responses obtained for $N > 50$ are indistinguishable in the graph.

It is important to consider how quickly disturbances are detected. Assume that an impulse hits the process just after the sampling at time $t = 0$. This implies that the effect of the disturbance is not detected until after a full sampling interval. Figure 9.3 shows the response of the system when the system is disturbed such that $x(0+) = [1 \quad 1]^T$, i.e., when the control signal is zero over the first sampling interval. Before any control action is taken, the system moves further away from equilibrium when the sampling interval is long compared to the case when a short sampling value is used.

The selection of sampling interval is now more important, and it can be reasonable to choose $N \approx 25\text{–}75$. This corresponds to $\omega h \approx 0.12\text{–}0.36$ for $\zeta = 0.707$. $N_r \approx 4\text{–}10$ corresponds to $N \approx 15\text{–}45$ for the damping $\zeta = 0.707$. ☐

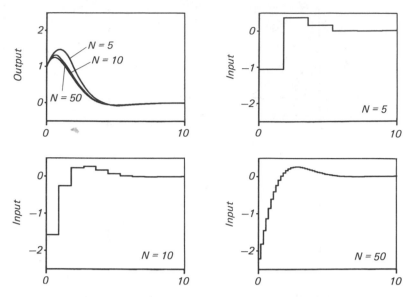

Figure 9.2 Response for different choices of sampling periods for the system in Example 9.4 when $\omega = 1$ and $\zeta = 0.707$.

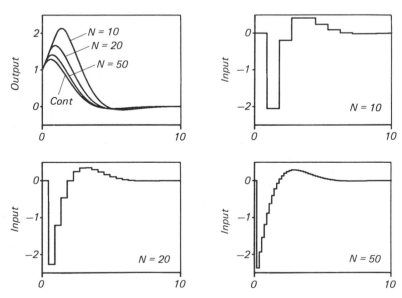

Figure 9.3 Response of the closed-loop system in Example 9.4 for different sampling periods when $\omega = 1$, $\zeta = 0.707$, and $x^T(0+) = [1 \quad 1]$.

The example above indicates that it can be necessary to sample fast if the main disturbances are load disturbances that are not synchronized with the sampling. As a rule of thumb we suggest that the sampling period be chosen as

$$\omega h = 0.2\text{--}0.6$$

where ω is the desired natural frequency of the *closed-loop system*. The choice depends on the nature of the disturbances acting on the system.

Deadbeat Control

If the desired poles in the pole-placement problem are all chosen to be at the origin, the characteristic equation for the closed-loop system becomes

$$z^n = 0$$

It then follows from the Cayley-Hamilton theorem that the system matrix $\Phi_c = \Phi - \Gamma L$ of the closed-loop system satisfies

$$\Phi_c^n = 0$$

This strategy has the property that it will drive all the states to zero in at most n steps after an impulse disturbance in the process state. The control strategy is called *deadbeat control*.

In deadbeat control there is only one design parameter—the sampling period. Since the error goes to zero in at most n sampling periods, the settling time is at most nh. However, the sampling period will drastically influence the magnitude of the control signal. When using deadbeat control, the magnitude of the control signal normally increases drastically with a decreasing sampling period. This fact has given the deadbeat control an undeservedly bad reputation. Thus it is important to choose the sampling period carefully when using deadbeat control.

The deadbeat strategy is unique to sampled-data systems. There is no corresponding feature for continuous-time systems. An example demonstrates the properties of deadbeat control.

Example 9.5—Deadbeat control of a double integrator

Consider a double-integrator plant. It follows from (9.8) that the deadbeat control strategy is given by (9.5) with

$$\begin{cases} l_1 = \dfrac{1}{h^2} \\[2ex] l_2 = \dfrac{3}{2h} \end{cases}$$

If the process has the initial state $x(0) = \text{col}[x_0, v_0]$, it follows that

$$u(0) = -\frac{x_0}{h^2} - \frac{3v_0}{2h}$$

$$u(h) = \frac{x_0}{h^2} + \frac{v_0}{2h}$$

TABLE 9.1 Control signals for deadbeat control of a double integrator with $x(0) = \mathrm{col}$ [1, 1] and different sampling periods.

h	100	10	1	0.1	0.01
$u(0)$	-0.0151	-0.16	-2.5	-115	$-10{,}150$
$u(h)$	0.0051	0.06	1.5	105	$10{,}050$

The control signals obtained for $x_0 = 1$ and $v_0 = 1$ are listed in Table 9.1. The output and the control signals are shown in Fig. 9.4. ☐

More General Disturbances

There are many ways to generalize the pole-placement design problem. In practice it is important to be able to handle disturbances that are more general than impulses. One way to do this is to use the idea described in Chapter 6. There, a disturbance is considered to be generated by sending an impulse through a linear system.

For example, assume that the system is described by

$$\frac{dx}{dt} = Ax + Bu + v$$

where v is a disturbance described by

$$\frac{d\xi}{dt} = A_v\xi$$

$$v = C_v\xi$$

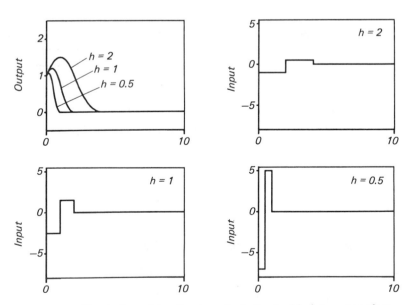

Figure 9.4 Simulation of deadbeat control of a double-integrator plant. The initial condition is $x^T(0) = [1 \quad 1]$.

State-Space Design Methods   Chap. 9

with given initial conditions. It is also assumed that ξ can be measured. By introducing the augmented state vector

$$z = \begin{bmatrix} x \\ \xi \end{bmatrix}$$

the system can be described by

$$\frac{d}{dt}\begin{bmatrix} x \\ \xi \end{bmatrix} = \begin{bmatrix} A & C_v \\ 0 & A_v \end{bmatrix}\begin{bmatrix} x \\ \xi \end{bmatrix} + \begin{bmatrix} B \\ 0 \end{bmatrix} u \qquad (9.17)$$

Thus we have a problem of the same form as the basic pole-placement problem. Notice, however, that there is one important difference: The system of (9.17) is not completely reachable. The poles associated with the description of the disturbance—i.e., the eigenvalues of A_v—cannot be influenced by the feedback. This is very natural. Compare Example 9.3.

When the system is sampled, the following discrete-time system is obtained:

$$\begin{bmatrix} x(kh + h) \\ \xi(kh + h) \end{bmatrix} = \begin{bmatrix} \Phi & \Phi_{xv} \\ 0 & \Phi_v \end{bmatrix}\begin{bmatrix} x(kh) \\ \xi(kh) \end{bmatrix} + \begin{bmatrix} \Gamma \\ 0 \end{bmatrix} u(kh)$$

The general, linear-state feedback is given by

$$u(kh) = -Lx(kh) - L_v\xi(kh) \qquad (9.18)$$

The closed-loop system is then described by

$$x(kh + h) = [\Phi - \Gamma L]x(kh) + [\Phi_{xv} - \Gamma L_v]\xi(kh)$$

$$\xi(kh + h) = \Phi_v\xi(kh)$$

If the pair (Φ, Γ) is reachable, the matrix L can be chosen so that the matrix $\Phi - \Gamma L$ has prescribed eigenvalues. The matrix Φ_v cannot be influenced by feedback. To reduce the effect of disturbances, it is useful to choose L_v so that the matrix $\Phi_{xv} - \Gamma L_v$ is small. This makes the coupling between the disturbances weak.

Notice that the feedback law in (9.18) can be interpreted as a combination of a feedback term Lx and a feedforward term $L_v\xi$. The feedforward term arises from the disturbance model. The solution clearly requires both x and ξ to be directly measurable.

Computational Aspects

For simple, low-order systems, it is often easiest to calculate the state feedback, as was done in Example 9.1. The general procedure is to introduce a general state feedback with unknown coefficients, determine the characteristic polynomial, and equate this to the desired characteristic polynomial. A set of linear equations for the feedback coefficients is then obtained. The equations can always be solved if the system is reachable.

Of course, it is also possible to use the general formula in (9.15) to calculate the state feedback. It is easy to write a computer code for the problem. However, notice that the formula is not well suited for very precise numerical calculations. As a rule, any method using computation of powers of matrices should be avoided. There are other ways to compute the feedback matrix L that are better for numerical calculations. See the references at the end of the chapter.

9.3 Observers

It is often highly unrealistic to assume that all states of the system and the disturbances can be measured. If a mathematical model of a system is available, one can attempt to compute the state from measured inputs and outputs. Some different ways to do this are discussed in this section.

It is assumed that the system is described by the sampled model

$$\begin{cases} x(k + 1) = \Phi x(k) + \Gamma u(k) \\ \quad\quad y(k) = Cx(k) \end{cases} \tag{9.19}$$

The problem of calculating the state $x(k)$ from input and output sequences $y(k)$, $y(k - 1), \ldots, u(k), u(k - 1), \ldots$ is considered next.

To determine the state of the system from the inputs and outputs, the system must be completely observable (compare with Sec. 5.3). There are many ways to determine the state and its predicted values. The alternatives of direct calculation, using dynamic models and Luenberger observers, is discussed in this section. The method best suited for a particular system depends on the nature of the disturbances and the measurement noise. If models for these two factors are known, it is possible to design optimal observers and predictors. This is discussed in Chapter 11.

Direct Calculation of the State Variables

Calculating the state vector directly from inputs and outputs is attempted first. For the general case with a scalar output, it follows from (9.19) that

$$y(k - n + 1) = Cx(k - n + 1)$$

$$y(k - n + 2) = C\Phi x(k - n + 1) + C\Gamma u(k - n + 1)$$

$$\vdots$$

$$y(k) = C\Phi^{n-1}x(k - n + 1) + C\Phi^{n-2}\Gamma u(k - n + 1)$$

$$+ \cdots + C\Gamma u(k - 1)$$

These equations can be written as

$$\begin{bmatrix} y(k - n + 1) \\ y(k - n + 2) \\ \vdots \\ y(k) \end{bmatrix} = W_o x(k - n + 1) + \Omega \begin{bmatrix} u(k - n + 1) \\ u(k - n + 2) \\ \vdots \\ u(k - 1) \end{bmatrix}$$

where W_o is the observability matrix

$$W_o = \begin{bmatrix} C \\ C\Phi \\ \vdots \\ C\Phi^{n-1} \end{bmatrix} \tag{9.20}$$

and

$$\Omega = \begin{bmatrix} 0 & 0 & \cdots & 0 \\ C\Gamma & 0 & \cdots & 0 \\ \vdots & & & \\ C\Phi^{n-2}\Gamma & C\Phi^{n-3}\Gamma & \cdots & C\Gamma \end{bmatrix}$$

If the system of (9.19) is observable, then the state vector $x(k - n + 1)$ can be determined from

$$\hat{x}(k - n + 1) = W_o^{-1} \begin{bmatrix} y(k - n + 1) \\ y(k - n + 2) \\ \vdots \\ y(k) \end{bmatrix} - W_o^{-1}\Omega \begin{bmatrix} u(k - n + 1) \\ u(k - n + 2) \\ \vdots \\ u(k - 1) \end{bmatrix}$$

Repeated use of (9.19) gives

$$\hat{x}(k) = \Phi^{n-1}W_o^{-1} \begin{bmatrix} y(k - n + 1) \\ y(k - n + 2) \\ \vdots \\ y(k) \end{bmatrix} + \Psi \begin{bmatrix} u(k - n + 1) \\ u(k - n + 2) \\ \vdots \\ u(k - 1) \end{bmatrix} \tag{9.21}$$

where

$$\Psi = [\Phi^{n-2}\Gamma \quad \Phi^{n-3}\Gamma \quad \cdots \quad \Gamma] - \Phi^{n-1}W_o^{-1}\Omega$$

Thus the state vector is given as a linear combination of $y(k)$, $y(k - 1)$, ..., $y(k - n + 1)$ and $u(k - 1)$, $u(k - 2)$, ..., $u(k - n + 1)$. Equation (9.21) can be written as

$$\hat{x}(k) = F^*(q^{-1})y(k) + G^*(q^{-1})u(k - 1) \tag{9.22}$$
$$= \frac{F(q)}{z^{n-1}} y(k) + \frac{G(q)}{z^{n-1}} u(k)$$

where $F(q)$ and $G(q)$ are vector polynomials of order $n - 1$ and $n - 2$ respectively. Equation (9.22) can be regarded as an observer which has the characteristic equation z^{n-1}, i.e., all the poles are at the origin.

Equations (9.21) and (9.22) show that after a disturbance, the correct estimate will be obtained after at most n measurements of inputs and outputs. The observer of (9.21) or (9.22) may be called a deadbeat observer. The result is summarized in Theorem 9.2.

Theorem 9.2. Consider the system of (9.19). Assume that it is completely observable. Then the state vector can be calculated from (9.21).

The general formula is illustrated by an example.

Example 9.6—Double integrator

For the double integrator, we have

$$\Phi = \begin{bmatrix} 1 & h \\ 0 & 1 \end{bmatrix}, \quad \Gamma = \begin{bmatrix} h^2/2 \\ h \end{bmatrix}, \quad C = \begin{bmatrix} 1 & 0 \end{bmatrix}$$

The observability matrix is given by

$$W_o = \begin{bmatrix} 1 & 0 \\ 1 & h \end{bmatrix} \quad \text{and} \quad W_o^{-1} = \frac{1}{h}\begin{bmatrix} h & 0 \\ -1 & 1 \end{bmatrix} = \begin{bmatrix} 1 & 0 \\ -1/h & 1/h \end{bmatrix}$$

This gives

$$\Phi^{n-1}W_o^{-1} = \begin{bmatrix} 1 & h \\ 0 & 1 \end{bmatrix}\begin{bmatrix} 1 & 0 \\ -1/h & 1/h \end{bmatrix} = \begin{bmatrix} 0 & 1 \\ -1/h & 1/h \end{bmatrix}$$

Furthermore,

$$W_o^{-1}\begin{bmatrix} 0 \\ C\Gamma \end{bmatrix} = \begin{bmatrix} 1 & 0 \\ -1/h & 1/h \end{bmatrix}\begin{bmatrix} 0 \\ h^2/2 \end{bmatrix} = \begin{bmatrix} 0 \\ h/2 \end{bmatrix}$$

and

$$\Psi = \Gamma - \Phi W_o^{-1}\begin{bmatrix} 0 \\ C\Gamma \end{bmatrix} = \begin{bmatrix} h^2/2 \\ h \end{bmatrix} - \begin{bmatrix} 1 & h \\ 0 & 1 \end{bmatrix}\begin{bmatrix} 0 \\ h/2 \end{bmatrix} = \begin{bmatrix} 0 \\ h/2 \end{bmatrix}$$

Equation (9.21) thus yields

$$x(kh) = \begin{bmatrix} 0 & 1 \\ -1/h & 1/h \end{bmatrix}\begin{bmatrix} y(kh-h) \\ y(kh) \end{bmatrix} + \begin{bmatrix} 0 \\ h/2 \end{bmatrix}u(kh-h)$$

$$= \begin{bmatrix} 1 \\ 1/h \end{bmatrix}y(kh) - \begin{bmatrix} 0 \\ 1/h \end{bmatrix}y(kh-h) + \begin{bmatrix} 0 \\ h/2 \end{bmatrix}u(kh-h)$$

The first component is measured directly while the second is obtained through

$$x_2(kh) = \frac{y(kh) - y(kh-h)}{h} + \frac{h}{2} \cdot u(kh-h)$$

As expected, $x_2 \to \dot{y}$ when $h \to 0$. However, direct calculation is much simpler in this case than using the general formula. $\square$

Reconstruction Using a Dynamic System

The direct calculation of the state vector has the advantage that the state variable is obtained after at most n measurements. The disadvantage of the method is that it may be sensitive to disturbances. It is therefore useful to have other alternatives.

Consider the system of (9.19). Assume that the state x is to be approximated by the state $\hat{x}$ of the model

$$\hat{x}(k + 1) = \Phi\hat{x}(k) + \Gamma u(k) \qquad (9.23)$$

which has the same input as the system of (9.19). If the model in (9.23) is perfect in the sense that the parameters are identical to those of the system in (9.19) and if the initial conditions of (9.19) and (9.23) are the same, then the state $\hat{x}$ of the model of (9.23) will be identical to the state x of the true system in (9.19). If the initial conditions of (9.19) and (9.23) are different, then $\hat{x}$ will converge to x only if the system (9.19) is asymptotically stable.

Notice that the reconstruction in (9.23) does not make use of the measured output. The reconstruction in (9.23) can be improved by introducing the difference between the measured and estimated outputs, $y - C\hat{x}$, as a feedback to obtain

$$\hat{x}(k + 1 \mid k) = \Phi\hat{x}(k \mid k - 1) + \Gamma u(k) + K[y(k) - C\hat{x}(k \mid k - 1)] \quad (9.24)$$

Here, K is a matrix, which has to be chosen in a suitable way. The notation $\hat{x}(k + 1 \mid k)$ is used to indicate that it is an estimate, or prediction, of $x(k + 1)$ based on measurements available at time k. Introduce the reconstruction error

$$\tilde{x} = x - \hat{x} \qquad (9.25)$$

Subtraction of (9.24) from (9.19) gives

$$\begin{aligned} \tilde{x}(k + 1 \mid k) &= \Phi\tilde{x}(k \mid k - 1) - K[y(k) - C\hat{x}(k \mid k - 1)] \\ &= [\Phi - KC]\tilde{x}(k \mid k - 1) \end{aligned} \qquad (9.26)$$

Hence if K is chosen so that the system in (9.26) is asymptotically stable, the reconstruction error will always converge to zero. Hence, by introducing a feedback from the measurements in the reconstruction, it is possible to make the error go to zero even if the system of (9.19) is unstable. The system in (9.24) is called an *observer* for the system of (9.19) because it produces the state of the system from measurements of inputs and outputs.

There are many variations of the observer in (9.24). The observer has a delay, because $\hat{x}(k \mid k - 1)$ depends only on measurements up to time $k - 1$. To avoid this, the following observer can be used:

$$\begin{aligned} \hat{x}(k \mid k) &= \Phi\hat{x}(k - 1 \mid k - 1) + \Gamma u(k - 1) \\ &\quad + K[y(k) - C\{\Phi\hat{x}(k - 1 \mid k - 1) + \Gamma u(k - 1)\}] \\ &= [I - KC][\Phi\hat{x}(k - 1 \mid k - 1) + \Gamma u(k - 1)] + Ky(k) \end{aligned} \qquad (9.27)$$

The reconstruction error when using (9.27) is given by

$$\tilde{x}(k \mid k) \doteq x(k) - \hat{x}(k \mid k) = (\Phi - KC\Phi)\tilde{x}(k - 1 \mid k - 1)$$

This equation is similar to (9.26), and from the definition of the observability matrix W_o, it is found that the pair $(\Phi, C\Phi)$ is detectable if the pair (Φ, C) is detectable. This implies that $\Phi - KC\Phi$ can be given arbitrary eigenvalues by selecting K. Further,

$$y(k) - C\hat{x}(k \mid k) = C\tilde{x}(k \mid k)$$
$$= (C\Phi - CKC\Phi)\tilde{x}(k - 1 \mid k - 1)$$
$$= (I - CK)C\Phi\tilde{x}(k - 1 \mid k - 1)$$

If the system has p outputs, then $I - CK$ is a $p \times p$ matrix; K may be chosen such that $CK = I$ if rank $(C) = p$. This implies that $C\hat{x}(k \mid k) = y(k)$—i.e., the outputs of the system are estimated without error. This will make it possible to eliminate p equations from (9.27), and the order of the observer will be reduced. Reduced-order observers of this type are called *Luenberger observers*.

Deciding When a Suitable Gain Matrix *K* Can Be Found

It now remains to find a suitable way to choose the matrix K so that the system (9.26) is stable. Given the matrices Φ and C, find a matrix K such that the matrix $\Phi - KC$ has prescribed eigenvalues. Since a matrix and its transpose have the same eigenvalues, the problem is the same as finding a matrix K^T such that $\Phi^T - C^T K^T$ has prescribed eigenvalues. However, this problem is solved in Sec. 9.2 (remark 2 of Theorem 9.1) in connection with the pole-placement problem. If those results are translated, then the problem can be solved if the matrix

$$W_o^T = [C^T \quad \Phi^T C^T \quad \ldots \quad (\Phi^{n-1})^T C^T]$$

has full rank. Notice that the matrix is the transpose of the observability matrix for the system of (9.19). The result is summarized in Theorem 9.3.

Theorem 9.3. Assume that the system of (9.19) is completely observable. Then it is possible to find a matrix K such that the observer in (9.24) has the characteristic polynomial $P(z)$ defined by (9.10). $\square$

Determining *K*

The determination of the matrix K in the observer (9.24) is found to be the same mathematical problem as the problem of determining the feedback matrix L in the pole-placement problem. The practical aspects are also closely related. The determination of the observer poles is a compromise between sensitivity to measurement errors and rapid recovery of initial errors. A fast observer will converge quickly, but it will also be sensitive to measurement errors.

A normal procedure for determining the matrix K is to decide on suitable observer poles and then to find a matrix K that gives the desired poles. For simple problems it is easiest to introduce the matrix K and to determine the equations for $\Phi - KC$ to have prescribed eigenvalues.

It is shown earlier that selection of K is related to that of selection of L for pole placement. The choice of K can be determined from Theorem 9.1 by using the translations

$$L \to K^T \qquad W_c \to W_o^T \qquad \Phi \to \Phi^T$$

From (9.15) it follows that K is given by

$$K^T = [0 \ \ldots \ 0 \ \ 1](W_o^T)^{-1}P(\Phi^T)$$

or

$$K = P(\Phi)W_o^{-1}[0 \ \ldots \ 0 \ \ 1]^T \tag{9.28}$$

The characteristic polynomial of $\Phi - KC$ is then $P(z)$, as defined by (9.10). The duality with pole placement also implies that K is especially simple to determine if the system is in observable form.

Example 9.7—Full-order observer for the double integrator

Consider a double-integrator plant. The matrix $\Phi - KC$ is given by

$$\Phi_o = \Phi - KC = \begin{bmatrix} 1 & h \\ 0 & 1 \end{bmatrix} - \begin{bmatrix} k_1 \\ k_2 \end{bmatrix}[1 \ \ 0] = \begin{bmatrix} 1 - k_1 & h \\ -k_2 & 1 \end{bmatrix}$$

Thus the characteristic equation is given by

$$z^2 - (2 - k_1)z + 1 - k_1 + k_2 h = 0$$

If the observer (9.24) must have the characteristic equation

$$z^2 + p_1 z + p_2 = 0$$

the following equations are obtained:

$$2 - k_1 = -p_1$$

$$1 - k_1 + k_2 h = p_2$$

These linear equations give

$$k_1 = 2 + p_1$$

$$k_2 = (1 + p_1 + p_2)/h \qquad\qquad \Box$$

Example 9.8—Reduced-order observer for double integrator

The observer of (9.27) applied to the double integrator gives the equations

$$\hat{x}(kh \mid kh) = \begin{bmatrix} 1 - k_1 & h(1 - k_1) \\ -k_2 & 1 - hk_2 \end{bmatrix} \hat{x}(kh - h \mid kh - h)$$

$$+ \begin{bmatrix} (1 - k_1)h^2/2 \\ h(1 - hk_2/2) \end{bmatrix} u(kh - h) + \begin{bmatrix} k_1 \\ k_2 \end{bmatrix} y(kh)$$

If $I - CK = 0$—i.e., if $k_1 = 1$—then the first equation is reduced to

$$\hat{x}_1(kh \mid kh) = y(kh)$$

The reduced-order observer is now given by the second equation, which can be simplified to

$$\hat{x}_2(kh \mid kh) = (1 - hk_2)\hat{x}_2(kh - h \mid kh - h)$$

$$+ k_2[y(kh) - y(kh - h)] + h(1 - hk_2/2)u(kh - h)$$

By choosing k_2, the reduced-order observer can be given an arbitrary eigenvalue. For instance, if $k_2 = 1/h$, the deadbeat response, then the same result is obtained as when making the direct calculation in Example 9.6. $\qquad \Box$

9.4 Output Feedback

A solution to the pole-placement problem for output feedback can be obtained by combining the results of Secs. 9.2 and 9.3. Let the system be described by

$$x(k + 1) = \Phi x(k) + \Gamma u(k) \tag{9.29}$$
$$y(k) = Cx(k)$$

A linear feedback law relating u to y such that the closed-loop system has given poles is desired. The disturbances are assumed to be impulses or equivalently unknown initial states.

The admissible control law is such that $u(k)$ is a function of $y(k - 1)$, $y(k - 2), \dots , u(k - 1), u(k - 2), \dots$. If all state variables are measured, it is shown in Sec. 9.2 that the feedback

$$u(k) = -Lx(k)$$

gives the desired poles. When the state cannot be measured, it seems intuitively reasonable to use the control law

$$u(k) = -L\hat{x}(k \mid k - 1) \tag{9.30}$$

where $\hat{x}$ is obtained from the observer

$$\hat{x}(k + 1 \mid k) = \Phi \hat{x}(k \mid k - 1) + \Gamma u(k) + K[y(k) - C\hat{x}(k \mid k - 1)] \tag{9.31}$$

Thus the feedback is a dynamic system of order n. Notice that the dynamics are due to the dynamics of the observer. A block diagram of the feedback is shown in Fig. 9.5. The dynamics of the observer described by (9.30) and (9.31) can be represented by the nth-order pulse-transfer function from y to u:

$$G_r(z) = -L[zI - \Phi + \Gamma L + KC]^{-1}K$$

Analysis of the Closed-Loop System

The closed-loop system has desirable properties. To show this, introduce

$$\tilde{x} = x - \hat{x}$$

Using (9.29), (9.30), and (9.31), the closed-loop system can be described by the equations

$$\begin{cases} x(k + 1) = [\Phi - \Gamma L]x(k) + \Gamma L\tilde{x}(k \mid k - 1) \\ \tilde{x}(k + 1 \mid k) = [\Phi - KC]\tilde{x}(k \mid k - 1) \end{cases} \tag{9.32}$$

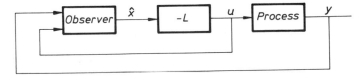

Figure 9.5 Block diagram of a regulator obtained by combining state feedback with an observer.

The closed-loop system has order $2n$. The eigenvalues of the closed-loop system are the eigenvalues of the matrices $\Phi - \Gamma L$ and $\Phi - KC$. Notice that the eigenvalues of $\Phi - \Gamma L$ are the desired closed-loop poles obtained in Sec. 9.2 and the eigenvalues of $\Phi - KC$ are the poles of the observer given in Sec. 9.3.

Example 9.9—Output feedback of double integrator

Consider the double integrator plant. Assume that the feedback vector L is determined as in Examples 9.2 and 9.4 with the closed-loop natural frequency $\omega = 1$, the damping $\zeta = 0.7$, and $h = 0.44$. This gives $L = [0.73, \quad 1.21]$. First assume that the observer is designed as in Example 9.7 with the poles of the observer in $z = 0.75$. Figure 9.6(a) shows the true and the estimated states when the estimated states are used in the control law. Figure 9.6(b) shows the states and the estimate of the second state when the reduced-order observer in Example 9.8 is used. The observer pole is in $z = 0.75$. □

Extensions

The problem can be extended to other disturbances by introducing models for the process disturbances and the measurement noise. The model can then be represented as follows (compare with Sec. 9.2):

$$x(k + 1) = \Phi x(k) + \Phi_{xv}\xi(k) + \Gamma u(k)$$

$$\xi(k + 1) = \Phi_v\xi(k) \tag{9.33}$$

$$\eta(k + 1) = \Phi_w\eta(k)$$

$$y(k) = Cx(k) + C_w\eta(k)$$

The control law is given by

$$u(k) = -L\hat{x}(k) - L_v\hat{\xi}(k) - L_w\hat{\eta}(k) \tag{9.34}$$

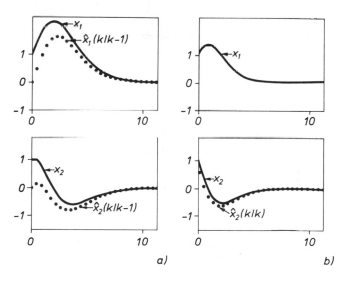

Figure 9.6 Control of the double-integrator plant using estimated states. The states and the estimated states are shown for:
a. Second-order observer.
b. Reduced-order observer.

a)

b)

where

$$
\begin{bmatrix} \hat{x}(k+1) \\ \hat{\xi}(k+1) \\ \hat{\eta}(k+1) \end{bmatrix} = \begin{bmatrix} \Phi & \Phi_{xv} & 0 \\ 0 & \Phi_v & 0 \\ 0 & 0 & \Phi_w \end{bmatrix} \begin{bmatrix} \hat{x}(k) \\ \hat{\xi}(k) \\ \hat{\eta}(k) \end{bmatrix} + \begin{bmatrix} \Gamma \\ 0 \\ 0 \end{bmatrix} u(k) + \begin{bmatrix} K \\ K_v \\ K_w \end{bmatrix} \epsilon(k)
$$

$$
\epsilon(k) = y(k) - C\hat{x}(k) - C_w\hat{\eta}(k)
$$

(9.35)

Notice that (9.34) is a typical example of combination of feedback and feedforward from estimated disturbances.

9.5 The Servo Problem

Only the regulator problem has been discussed so far. The criterion has been to eliminate impulse disturbances and to drive the states of the system to zero. The servo problem is another important prototype problem. There, the objective is to make the states and the outputs of the system follow desired trajectories. Practical control systems often have specifications that involve both servo and regulation properties. This is traditionally solved using a two-degree-of-freedom structure as shown in Figure 9.7. This configuration has the advantage that the servo and regulation problems become separated. The feedback controller G_{fb} is first designed to obtain a closed-loop system that is insensitive to load disturbances and to plant uncertainty. The design typically involves compromises that ensure that measurement errors do not generate excessive fluctuations in the control signal. The feedforward compensator is then designed to obtain the desired servo properties.

Pole-Placement Design

The primary concern in pole-placement design is to make sure that errors in the initial state of a system will decay in a specified way. This is achieved by positioning the closed-loop poles appropriately. The pole-placement method with perfect-state information therefore is primarily concerned with load disturbances.

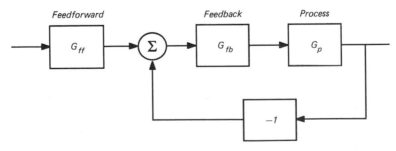

Figure 9.7 Block diagram of a feedback system with a two-degree-of-freedom structure.

When all states are not measured exactly, measurement errors also enter in the design of the observer. Because of the structure in the controller there is a separation. The observer dynamics are specified by the polynomial

$$A_o(z) = \det(zI - \Phi + KC)$$

whose zeros determine how quickly the observer can find the disturbed state. The dynamics given by the polynomial

$$A_r(z) = \det(zI - \Phi + \Gamma L)$$

tell how quickly the state will go to zero. Finding proper closed-loop poles involves a compromise between the influences of load disturbances and measurement errors.

Servo Properties

Since servo properties do not enter the pole-placement problem directly, they have to be introduced separately. To do this, first notice that the pole-placement design gives the control law

$$u(k) = -L\hat{x}(k)$$

where $\hat{x}$ is the estimated state. When it is no longer desired to drive the state to zero, it is natural to introduce the feedback law

$$u(k) = L(x_m(k) - \hat{x}(k)) + u_m(k) \tag{9.36}$$

where $x_m(k)$ specifies the desired state at time k, and $u_m(k)$ is the nominal value of the control signal at time k. The term $L(x_m - \hat{x})$ represents the feedback and u_m the feedforward compensation.

Equation (9.36) has a good physical interpretation. The feedforward signal u_m will ideally produce the desired time variation in the process state. If the estimated process state $\hat{x}$ equals the desired state x_m, the feedback signal $L(x_m - \hat{x})$ is zero. If there is a difference between $\hat{x}$ and x_m the feedback, however, will generate corrections.

There are many ways to generate x_m and u_m. In a robotics problem x_m can, for example, represent the values of the joint angles and their derivatives and u_m could be proportional to the joint acceleration evaluated at the nominal path.

In many cases x_m can be generated from a dynamic system, which tells how the state ideally should respond to the command signal. If the plant dynamics has a stable inverse, then the system can be represented as shown in Figure 9.8. If the plant dynamics do not have a stable inverse, the block labeled 'Inverse process model' has to be replaced by an approximate inverse. The block diagram in Figure 9.8 shows the conceptual solution. In practice the model and the inverse process model are generated as one dynamic system with u_c as input and u_m as the output. Notice that the model can be a nonlinear system.

The servo problem is closely related to properties of the closed-loop zeros. Before giving a formal solution to the problem, we will therefore give an interpretation of poles and zeros.

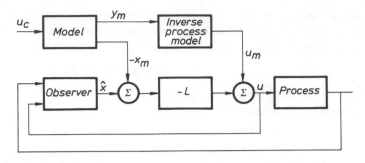

Figure 9.8 The introduction of command signals in a system with output feedback.

An Interpretation of Poles and Zeros

Before discussing how command signals can be introduced, a digression gives an interpretation of poles and zeros. The poles, or the eigenvalues, of a linear system reflect the internal couplings in the system. They tell how the system behaves when it is left to itself. The zeros tell how the system is coupled to its environment.

Basically, to introduce command signals in a system means to couple the system to its environment. It is clear that this involves manipulation of the zeros of the system. To discuss how the zeros of a system are influenced by how command signals are introduced, consider the signal-flow diagrams of a single-input–single-output system shown in Fig. 9.9. The coefficients of the characteristic polynomial, which determines the poles of the system, are labeled a_i. The coefficients of the polynomial that determines the zeros of the system are labeled b_i.

It is immediately clear from Fig. 9.9(b) that the poles of the system can be changed by state-feedback. In the same way it is seen from Fig. 9.9(a) that all the zeros of the system can be influenced by changing the feedforward to all the state variables.

It is in general very difficult to modify the feedforward connections to the state variables of a process. The process can be influenced only via the control variable. New actuators are required to influence the full-process state directly. However, the state variables associated with the compensator can be influenced freely to modify some of the zeros.

Specifications

Generation of the feedforward signals will now be discussed in detail for single-input–single-output systems. Let the process be described by the nth order system

$$x(k + 1) = \Phi x(k) + \Gamma u(k) \tag{9.37}$$
$$y(k) = Cx(k)$$

It is assumed that the system is reachable and observable. The input-output prop-

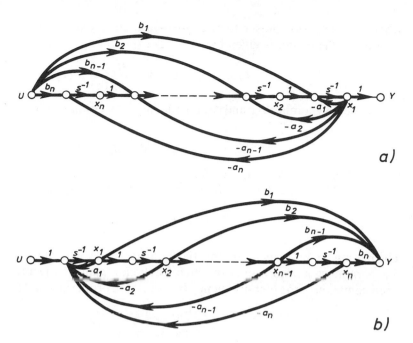

Figure 9.9 Signal-flow diagrams of a system with the transfer function

$$G(s) = \frac{b_1 s^{n-1} + b_2 s^{n-2} + \cdots + b_n}{s^n + a_1 s^{n-1} + \cdots + a_n}$$

a) Observable canonical form   b) Controllable canonical form

erties of the system are defined by the pulse-transfer operator

$$y(k) = C(qI - \Phi)^{-1}\Gamma u(k)$$

$$= \frac{B(q)}{A(q)} u(k) \tag{9.38}$$

$$= H(q)u(k)$$

where $\deg A = n$ and $d = \deg A - \deg B$. The polynomials A and B are defined
as

$$A(q) = q^n + a_1 q^{n-1} + \cdots + a_n$$

$$B(q) = b_0 q^{n-d} + b_1 q^{n-d-1} + \cdots + b_{n-d}$$

where d is the pole excess, i.e., the time delay of the process. The desired response
of the system to command signal is given by

$$x_m(k + 1) = \Phi_m x_m(k) + \Gamma_m u_c(k) \tag{9.39}$$

$$y_m(k) = C_m x_m(k)$$

where the states are chosen to be equivalent to the model states. The input-output relation of the model is given by

$$y_m(k) = H_m(q)u_c(k) = \frac{B_m(q)}{A_m(q)} u_c(k) \tag{9.40}$$

where the polynomials B_m and A_m do not have any common factors. It is assumed that

$$d_m = \deg A_m - \deg B_m \geq d \tag{9.41}$$

Different restrictions on A_m and B_m are discussed in connection with the specific solutions. The model transfer function thus has the property

$$\frac{B_m(q)}{A_m(q)} = b_{m0}q^{-d_m} + \cdots$$

The parameter b_{m0} thus represents the high-frequency gain of the model. Similarly the parameter b_0 represents the high-frequency gain of the process. A simple approximation of the high-frequency transmission properties of $AB_m/(BA_m)$ is thus

$$\frac{b_{m0}}{b_0} q^{d-d_m}$$

With this approximation the feedback law becomes

$$u(k) = L(x_m(k) - \hat{x}(k)) + \frac{b_{m0}}{b_0} u_c(t + d - d_m)$$

which is realizable if $d_m \geq d$. A more accurate solution is obtained by using more terms in the series expansion. Ideal model following is obtained by implementing the complete transfer function $AB_m/(BA_m)$. To do this, however, we must require that the polynomial B is stable.

A Direct Implementation of Model-Following

In some cases it is inconvenient to implement a model like (9.39), whose states correspond directly to the states of the model. The servo problem can then be solved in a straightforward manner using a two-degree-of-freedom configuration. Compare with Figure 9.7. If the plant is controlled with the feedback law

$$u(k) = -L\hat{x}(k) + \tilde{u}_c(k) \tag{9.42}$$

the closed-loop system is described by

$$y(k) = C(qI - (\Phi - \Gamma L))^{-1}\Gamma\tilde{u}_c(k) = \frac{B(q)}{A_r(q)} \tilde{u}_k(k) \tag{9.43}$$

Notice that the observer dynamics does not appear, because when the command signal is introduced, as in (9.42), changes in $\tilde{u}_c$ will not generate observer errors.

A straightforward implementation of a two-degree-of-freedom system now shows that to obtain perfect model-following, the feedforward compensator should be described by

$$\bar{u}_c(k) = \frac{B_m(q)A_r(q)}{B(q)A_m(q)} u_c(k) \tag{9.44}$$

Notice that there may be factors that are common to the numerator and the denominator. Such factors should be canceled before implementing the feedforward compensator. The polynomials A_r and A_m are stable. Unstable factors of B can thus only be canceled by corresponding factors in B_m. We thus find that perfect model-following can be achieved only if B is stable or if the unstable factors of B are also factors of B_m. If this is not the case the model-following can only be realized approximately, e.g., by replacing B in (9.44) with a B^+ such that A_r/B^+ is an approximate inverse of B/A_r.

Analysis

Assume that a realization of the feedforward system (9.44) is

$$x_f(k + 1) = \Phi_f x_f(k) + \Gamma_f u_c(k) \tag{9.45}$$
$$\bar{u}_c(k) = -L_f x_f(k) + l_c u_c(k)$$

Introducing an observer of the form (9.24) gives the closed-loop system described by

$$x(k + 1) = \Phi x(k) + \Gamma u(k)$$
$$x_f(k + 1) = \Phi_f x_f(k) + \Gamma_f u_c(k)$$
$$\hat{x}(k + 1) = \Phi\hat{x}(k) + \Gamma u(k) + K(y(k) - C\hat{x}(k))$$
$$y(k) = Cx(k)$$

A general linear controller that uses the states of the observer and the feedforward system is

$$u(k) = L\hat{x}(k) \quad L_f x_f(k) \quad l_c u_c(k) \tag{9.46}$$

Introducing $\bar{x}(k) = x(k) - \hat{x}(k)$ gives

$$\begin{bmatrix} x(k + 1) \\ x_f(k + 1) \\ \bar{x}(k + 1) \end{bmatrix} = \begin{bmatrix} \Phi - \Gamma L & -\Gamma L_f & \Gamma L \\ 0 & \Phi_f & 0 \\ 0 & 0 & \Phi - KC \end{bmatrix} \begin{bmatrix} x(k) \\ x_f(k) \\ \bar{x}(k) \end{bmatrix} + \begin{bmatrix} \Gamma l_c \\ \Gamma_f \\ 0 \end{bmatrix} u_c \tag{9.47}$$

$$y(k) = \begin{bmatrix} C & 0 & 0 \end{bmatrix} \begin{bmatrix} x(k) \\ x_f(k) \\ \bar{x}(k) \end{bmatrix}$$

It follows from (9.47) that the command signals will not influence observer errors. This is natural because it would be unwise to introduce the command signals in such a way that they would generate estimation errors. As a result the observer poles will not appear in the transfer function, which relate the output to the command signal. The close-loop system is in block triangular form. The dynamics are then determined by the eigenvalues of the matrices $\Phi - \Gamma L$, Φ_f, $\Phi - KC$. These parts correspond to the regulator problem dynamics, A_r, the feedforward dynamics, $A_m B$, and the observer dynamics A_o.

In the case of scalar measurements the controller can be considered as a system with two inputs, the command signal u_c and the measurement y. The controller (9.46) can then be written in input-output form as

$$R(q)u(k) = -S(q)y(k) + T(q)u_c(k) \qquad (9.48)$$

where R, S, and T are polynomials in the shift operator. The input-output relation is obtained from (9.47)

$$y(k) = C(qI - (\Phi - \Gamma L))^{-1}\Gamma(l_c - L_f(qI - \Phi_f)^{-1}\Gamma_f)u_c(k) \qquad (9.49)$$

or

$$y(k) = \frac{B(q)}{A_r(q)} \cdot \frac{B_f(q)}{A_f(q)} u_c(k)$$

The polynomial A_f is assumed to be monic. The first part of the pulse-transfer function is the pole shifting due to the solution of the regulator problem. The denominator of the second part is the characteristic polynomial of Φ_f in (9.45). The numerator of the second part has the same order as the denominator. The controller has the structure shown in Fig. 9.10. The polynomials A_r, A_f, and B_f can be arbitrarily chosen through L, L_f, and l_c. Perfect model-following is obtained if

$$B_f(q) = A_r(q)B_m(q)/b_0$$

$$A_f(q) = A_m(q)B(q)/b_0$$

The assumption (9.41) implies that B_f/A_f is causal. If $n_m = \deg A_m$ then the order

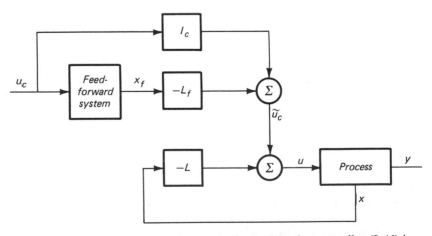

Figure 9.10 The control scheme obtained when the controller (9.46) is used.

of the controller is

$$n + n_m - d$$

which is the same as that of (9.44). Notice that it is possible to rewrite (9.44) as

$$\bar{u}_c(k) = \frac{B_m(q)A_r(q)}{B(q)A_m(q)} u_c(k)$$

$$= \frac{B_m(q)}{A_m(q)} u_c(k) + \frac{B_m(q)(A_r(q) - B(q))}{A_m(q)B(q)} u_c(k) \qquad (9.50)$$

The first term is y_m, see (9.40), and contains only known parts, while the second term also contains parts that are dependent upon the possibly uncertain process. If A_m and B do not have any common factors it is easy to show that the feedforward system (9.45) can be implemented as a block diagonal system of order $n + n_m - d$ with two output signals that correspond to the two terms in (9.50). The importance of this separation will be apparent below when integrators are introduced.

In the next case it is assumed that the process (9.37) and the model (9.39) are 'compatible.' Without giving a formal definition this means that

- The orders of the system and the model are the same and the states have the same physical interpretation.
- The model dynamics can be obtained from the process through feedback, i.e.,

$$\Phi_m = \Phi - \Gamma L$$

It is then meaningful to look at the difference between the states of the process and the model. Further assume that

$$A_r(q) = A_m(q)$$

The system can now be implemented as shown in Fig. 9.8 where the observer also is included. Notice that the system contains an inverse of the process, which may be noncausal. The model together with the inverse can, however, be implemented as a causal system (see above). The advantage with the implementation is that some nonlinearities of the process can be taken care of.

Introduction of Integrators

The methods discussed in this chapter utilize all available information about the system. This means that all disturbances are modeled and that the models are exact. However, it is often desirable for the system to be robust against minor errors in the process model. For instance, the outputs in steady state should be equal to given desired values. One way to eliminate load disturbances is to estimate them and use feedforward from the estimated disturbance (compare with Sec. 9.4). In classical design, steady-state errors are eliminated by introducing integrators. This can also be done in the pole-placement formulation discussed in this chapter.

Assume that the system is given by (9.29). For simplicity it is assumed that the input and output are scalars. Let u_c be the desired steady-state level for y. One way to introduce an integrator is to introduce a new state that integrates the error $u_c - y$, i.e.,

$$x_{n+1}(k+1) = x_{n+1}(k) + u_c(k) - Cx(k)$$

The system can be augmented with the new state, which gives

$$\begin{bmatrix} x(k+1) \\ x_{n+1}(k+1) \end{bmatrix} = \begin{bmatrix} \Phi & 0 \\ -C & 1 \end{bmatrix} \begin{bmatrix} x(k) \\ x_{n+1}(k) \end{bmatrix} + \begin{bmatrix} \Gamma \\ 0 \end{bmatrix} u(k) + \begin{bmatrix} 0 \\ 1 \end{bmatrix} u_c(k)$$

The pole-placement design can be used for the new system. If this is reachable, then $x_{n+1} \to 0$ if u_c is constant. The control law has the structure

$$u(k) = -Lx(k) - l_{n+1}x_{n+1}(k) + l_c u_c(k) \tag{9.51}$$

where the last term can be interpreted as a feedforward from a measurable disturbance (see Fig. 9.11). If the poles of the augmented closed-loop system are stable, then—in steady state—$y = u_c$. Because the eigenvalues of a matrix depend continuously on the elements, the steady-state error will be zero as long as the closed-loop system is stable, even if there are minor errors in the process model.

The Servo Problem with Integrator

The servo problem discussed above can now be extended with an integrator. Let the regulator be a combination of (9.45) and (9.51):

$$u(k) = -Lx(k) - l_{n+1}x_{n+1}(k) - L_f x_f(k) + l_c u_c(k)$$

$$= -Lx(k) - l_{n+1}x_{n+1}(k) + \tilde{u}_c(k)$$

The closed-loop system is given by

$$x(k+1) = (\Phi - \Gamma L)x(k) + \Gamma \tilde{u}_c(k) - \Gamma \frac{l_{n+1}}{q-1}(y_m(k) - y(k))$$

where

$$y_m(k) = \frac{B_m}{A_m} u_c(k)$$

is the reference value we want the output to follow. The output is now given by

$$y(k) = C(qI - (\Phi - \Gamma L))^{-1}\Gamma \left(\tilde{u}_c(k) - \frac{l_{n+1}}{q-1}(y_m(k) - y(k)) \right)$$

$$= \frac{B}{A_r} \left(\tilde{u}_c(k) - \frac{l_{n+1}}{q-1}\frac{B_m}{A_m} u_c(k) + \frac{l_{n+1}}{q-1} y(k) \right)$$

or

$$y(k) = \frac{B((q-1)A_m\tilde{u}_c(k) - l_{n+1}B_m u_c(k))}{A_m(A_r(q-1) - l_{n+1}B)}$$

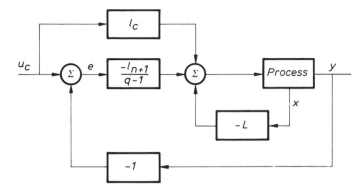

Figure 9.11. Block diagram showing the introduction of an integrator in pole-placement design.

where

$$\det(qI - \Phi + \Gamma L) = A_r(q)$$

Let

$$\bar{u}_c(k) = \frac{A_r B_m}{BA_m} u_c(k)$$

This gives

$$y(k) = \frac{B_m}{A_m} \cdot \frac{B(A_r(q - 1) - l_{n+1}B)}{B(A_r(q - 1) - l_{n+1}B)} u_c(k) = \frac{B_m}{A_m} u_c(k) \qquad (9.52)$$

To summarize the control law

$$u(k) = -Lx(k) - l_{n+1}x_{n+1}(k) + \bar{u}_c(k) \qquad (9.53)$$

$$\bar{u}_k(k) = \frac{A_r B_m}{BA_m} u_c(k)$$

Where L and l_{n+1} are chosen such that

$$A_r(q - 1) - l_{n+1}B - 0$$

get arbitrary roots. This is possible since the open-loop system is controllable and observable. The dynamics that are not observable from u_c to y are given by $B(A_r (q - 1) - l_{n+1}B)$, which is assumed to be stable.

Example 9.10—Introduction of integrator

Consider the system

$$x(k + 1) = \varphi x(k) + \gamma u(k)$$

$$y(k) = x(k)$$

where $\varphi = 0.9$ and $\gamma = 1$. The desired input-output, pulse-transfer operator is assumed to be

$$H_m(q) = \frac{\gamma_m}{q - \varphi_m}$$

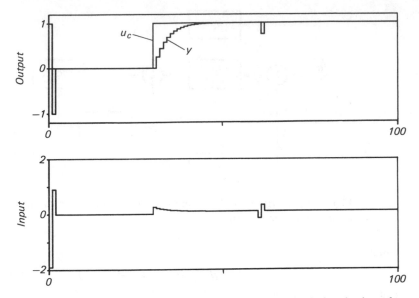

Figure 9.12. Output, reference value, and the control signal when the controller (9.53) is used to control the process in Example 9.10. Input load disturbance at $t = 60$.

with $\varphi_m = 0.75$ and $\gamma_m = 0.25$. Using (9.53) gives the closed-loop system (9.52)

$$y(k) = \frac{\gamma_m}{q - \varphi_m} \frac{(q - \varphi + \gamma l)(q - 1) - \gamma l_{n+1}}{(q - \varphi + \gamma l)(q - 1) - \gamma l_{n+1}} u_c(k)$$

Placing the cancelled poles at the origin gives

$$l = (1 + \varphi)/\gamma = 1.9$$

$$l_{n+1} = -1/\gamma = -1$$

$$\bar{u}_c(k) = \frac{0.25(q + 1)}{q - 0.75} u_c(k)$$

Figure 9.12 shows the output and the control signal when (9.53) is used. The initial value is such that $y(0) = 1$. The controller eliminates the input disturbance, and the dynamics from the reference value u_c are different from the dynamics from the input disturbance. □

9.6 A Design Example

To illustrate different design methods a more complex example will be used.

Process

Consider a motor with current constant k_I that drives a load consisting of two masses coupled with a spring with spring constant, k (see Fig. 9.13). The input signal is the motor current I. The angular velocities and the angles of the masses are ω_1, ω_2, φ_1, and φ_2, the moments of inertia are J_1 and J_2. It is assumed that there is a damping, d, in the spring and that the first mass may be disturbed by a torque, v. Finally the output of the process is the angular velocity ω_2.

Introduce the states

$$x_1 = \varphi_1 - \varphi_2$$

$$x_2 = \omega_1/\omega_0$$

$$x_3 = \omega_2/\omega_0$$

where

$$\omega_0 = \sqrt{k(J_1 + J_2)/(J_1 J_2)}$$

The process is then described by

$$\frac{dx}{dt} = \omega_0 \begin{bmatrix} 0 & 1 & -1 \\ \alpha - 1 & -\beta_1 & \beta_1 \\ \alpha & \beta_2 & -\beta_2 \end{bmatrix} x + \begin{bmatrix} 0 \\ \gamma \\ 0 \end{bmatrix} u + \begin{bmatrix} 0 \\ \delta \\ 0 \end{bmatrix} v$$

$$y = \begin{bmatrix} 0 & 0 & \omega_0 \end{bmatrix} x$$

(9.54)

where

$$\alpha = J_1/(J_1 + J_2)$$

$$\beta_1 = d/(J_1 \omega_0)$$

$$\beta_2 = d/(J_2 \omega_0)$$

$$\gamma = k_I/(J_1 \omega_0)$$

$$\delta = 1/(J_1 \omega_0)$$

The following values have been used in the example: $\omega_0 = 1$, $J_1 = 10/9$, $J_2 = 10$, $k = 1$, $d = 0.1$, and $k_I = 1$. With these values the process (9.54) has three poles $p_1 = 0$, and $p_{23} = -0.05 \pm 0.999i$ and one zero $z_1 = -10$. The complex poles have a damping of $\zeta_p = 0.05$ and a natural frequency $\omega_p = 1$ rad/s. The Bode plot of the process is shown in Fig. 9.14 and the impulse response in Fig. 9.15.

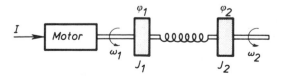

Figure 9.13 Robot mechanism process.

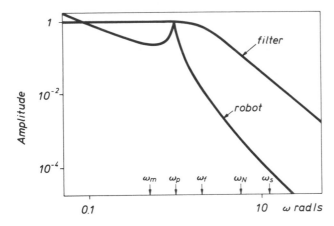

Figure 9.14. Bode plot of the process and the antialiasing filter. Only the magnitudes are shown.

Specifications

It is desired that the closed-loop system has a response from the reference signal such that the dominating modes have a natural frequency $\omega_m = 0.5$ rad/s and a damping $\zeta_m = 0.7$.

Design

Choice of sampling interval. The desired model has a natural frequency $\omega_m = 0.5$ rad/s. Using the rule of thumb in Sec. 9.2 gives $h = 0.5$ s as a reasonable choice for the sampling interval. This gives a Nyquist frequency of $\omega_N = \pi/h \approx 6$ rad/s.

In practice an antialiasing filter is necessary to avoid frequency folding of disturbances. In this first design the disturbances are disregarded and the design is done for the plant only.

State feedback design. It is assumed that all the states in (9.54) are measured. The system is of third order which implies that three poles can be placed using the controller

$$u(k) = -Lx(k) + l_c u_c(k) \tag{9.55}$$

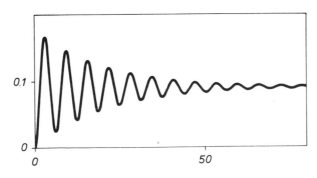

Figure 9.15. Impulse response of the process.

State-Space Design Methods   Chap. 9

Let the desired poles be specified by

$$(s^2 + 2\zeta_m\omega_m s + \omega_m^2)(s + \alpha_1\omega_m) = 0 \qquad (9.56)$$

This characteristic equation is transferred to sampled form with $h = 0.5$. The parameter l_c is determined such that the steady-state gain from u_c to y is unity, i.e., no integrator is introduced in the controller. Figure 9.16 shows the behavior of the closed-loop system when the state-feedback controller (9.55) is used when $\alpha_1 = 2$. The reference signal is a step at $t = 0$ and the disturbance v is a pulse at $t = 25$ of height -10 and a duration of 0.1 time units.

Observer design. It is now assumed that only the output can be measured. The other states are reconstructed using a full-state observer of the form (9.24). The eigenvalues of $\Phi - KC$ are chosen in the same pattern as the closed-loop poles but a factor α_0 further away from the origin, i.e., in continuous time we assume that we have

$$(s^2 + 2\zeta_m\alpha_0\omega_m s + (\alpha_0\omega_m)^2)(s + \alpha_0\alpha_1\omega_m) = 0$$

This characteristic equation is transferred to sampled-data form using $h = 0.5$. Figure 9.17 shows the same as Fig. 9.16 when an observer is used. The output is shown for $\alpha_0 = 2$. The continuous-time equivalence of the fastest pole of the closed-loop system when using the observer is $-\alpha_0\alpha_1\omega_m$. For $\alpha_0 = 2$, $\alpha_1 = 2$, and $\omega_m = 0.5$ we get the pole -2. This implies that the used sampling interval ($h = 0.5$) is a little too long. There is, however, no significant difference in the response when h is decreased to 0.25.

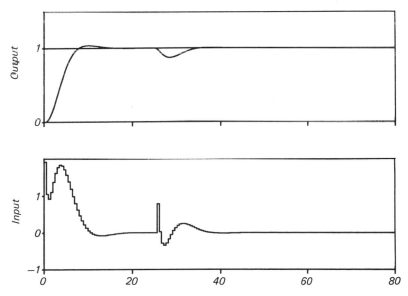

Figure 9.16. Output and input when the reference signal u_c is a step and the disturbance v a short pulse.

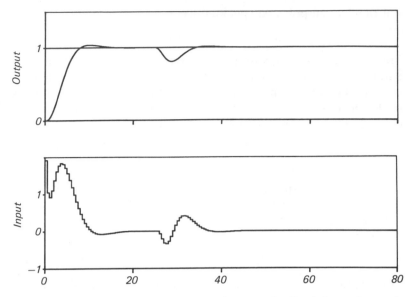

Figure 9.17. Same as Fig. 9.16, but using state-feedback from observed states when $\alpha_0 = 2$.

Summary

The example shows the design using state-feedback and observer. The response to the reference value change is the same, since the system and the observer have the same initial values. The response to the disturbance deteriorates slightly when the observer is used compared to direct-state feedback. The observer is twice as fast as the desired closed-loop response.

9.7 Conclusions

The chapter shows how the regulator and servo design problems can be solved using pole placement and observers. The solution has three major components: the feedback matrix L, the observer, and the response model. The feedback matrix L is chosen in such a way that load disturbances decay properly using the techniques discussed in Sec. 9.2. The observer is designed by considering the load disturbances and the measurement noise, as discussed in Sec. 9.3. The major trade-off is between quick convergence and sensitivity to measurement errors. The regulation properties are taken care of by the matrix L and the observer. The response model and the inverse process model are then chosen to obtain the desired servo performance in response to command signals.

    The pole-placement design is done here for the single-input–single-output case. With n parameters in the state-feedback vector, it is possible to place n poles arbitrarily, if the system is reachable. In the multivariable case, there are more degrees of freedom. This makes it possible to determine not only the poles, but also some eigenvectors of the closed-loop system. Further details can be found in the references.

9.8 Problems

9.1. A general second-order, discrete-time system can be written as

$$x(k + 1) = \begin{bmatrix} a_{11} & a_{12} \\ a_{21} & a_{22} \end{bmatrix} x(k) + \begin{bmatrix} b_1 \\ b_2 \end{bmatrix} u(k)$$

$$y(k) = [c_1 \quad c_2]x(k)$$

Determine a state-feedback controller of the form

$$u(k) = -Lx(k)$$

such that the characteristic equation of the closed-loop system is

$$z^2 + p_1 z + p_2 = 0$$

Use the result to verify the deadbeat controller for the double integrator given in Example 9.5.

9.2. Given the system

$$x(k + 1) = \begin{bmatrix} 1.0 & 0.1 \\ 0.5 & 0.1 \end{bmatrix} x(k) + \begin{bmatrix} 1 \\ 0 \end{bmatrix} u(k)$$

$$y(k) = [1 \quad 1]x(k)$$

Determine a linear, state-feedback controller

$$u(k) = -Lx(k)$$

such that the closed-loop poles are in 0.1 and 0.25.

9.3. Determine the deadbeat controller for the normalized motor in Example A.2. Assume that $x(0) = [1 \quad 1]^T$. Determine the sample interval such that the control signal is less than one in magnitude. It can be assumed that the maximum value of $u(kh)$ is at $k = 0$.

9.4. Consider the system in Problem 8.10. Sampling the system with $h = 0.2$ gives

$$x(kh + h) = \begin{bmatrix} 0.55 & 0.12 \\ 0 & 0.67 \end{bmatrix} x(kh) + \begin{bmatrix} 0.01 \\ 0.16 \end{bmatrix} u(kh)$$

(a) Determine a state-feedback control law such that the closed-loop characteristic polynomial is

$$z^2 - 0.63z + 0.21$$

This corresponds to the specifications given in Problem 8.10.
(b) Simulate the closed-loop system and compare with the previous results.

9.5. The system

$$x(k + 1) = \begin{bmatrix} 0.78 & 0 \\ 0.22 & 1 \end{bmatrix} x(k) + \begin{bmatrix} 0.22 \\ 0.03 \end{bmatrix} u(k)$$

$$y(k) = [0 \quad 1]x(k)$$

represents the normalized motor for the sampling interval $h = 0.25$. Determine observers for the state based on the output by using each of the following.

(a) Direct calculation using (9.21).

(b) A dynamic system that gives $\hat{x}(k + 1 \mid k)$ using (9.24).

(c) The reduced-order observer in (9.27).

Let the observer be of deadbeat type; i.e., the poles of the observer should be in the origin.

9.6. Determine the full-state observer based on (9.24) for the tank system in Problem 3.10. Choose the observer gain such that the observer is twice as fast as the open-loop system.

9.7. Consider the observer of (9.27) and let the control law be given by

$$u(k) = -L\hat{x}(k \mid k)$$

Show that the resulting controller can be written as

$$\xi(k + 1) = \Phi_o \xi(k) + \Gamma_o y(k)$$
$$u(k) = C_o \xi(k) + D_o y(k)$$

where

$$\Phi_o = (I - KC)(\Phi - \Gamma L) \qquad \Gamma_o = (I - KC)(\Phi - \Gamma L)K$$

$$C_o = -L \qquad\qquad\qquad D_o = -LK$$

9.8. Given the discrete-time system

$$x(k + 1) = \begin{bmatrix} 0.5 & 1 \\ 0.5 & 0.7 \end{bmatrix} x(k) + \begin{bmatrix} 0.2 \\ 0.1 \end{bmatrix} u(k) + \begin{bmatrix} 1 \\ 0 \end{bmatrix} v(k)$$

$$y(k) = [1 \quad 0]x(k)$$

where v is a constant disturbance. Determine controllers such that the influence of v can be eliminated in steady state in each case.

(a) the state and v can be measured.

(b) the state can be measured.

(c) only the output can be measured.

9.9. Consider the two-tank system in Problem 3.10 for $h = 12$ s.

(a) Determine a state-feedback controller such that the closed-loop poles are given by the characteristic equation

$$z^2 - 1.55z + 0.64 = 0$$

This corresponds to $\zeta = 0.7$ and $\omega = 0.027$ rad/s.

(b) Introduce a command signal and determine a controller such that the steady-state error between the command signal and the output is zero in steady state; i.e., introduce an integrator in the system.

(c) Simulate the system using the regulators in (a) and (b). Compare the results with those obtained in Problem 8.7.

9.10. Consider the double integrator with a load disturbance acting on the process input. The disturbance can be described as a sinusoidal with frequency ω_0, but with unknown amplitude and phase. Design a state-feedback controller and an observer such that there is no steady-state error due to the sinusoidal perturbation.

9.11. Show that (9.48) can be obtained from (9.46).

9.12. Consider the discrete-time process

$$x(k + 1) = \begin{bmatrix} 0.9 & 0 \\ 1 & 0.7 \end{bmatrix} x(k) + \begin{bmatrix} 1 \\ 0 \end{bmatrix} u(k)$$

$$y(k) = [0 \quad 1]x(k)$$

(a) Determine a state deadbeat controller that gives unit static gain, i.e., determine m and L in the controller

$$u(k) = mu_c(k) - Lx(k)$$

(b) Determine the stability range for the parameters in L, i.e., use the controller from (a) and determine how much the other parameters may change before the closed-loop system becomes unstable.

9.13. Consider the system

$$x(k + 1) = \begin{bmatrix} 0.25 & 0.5 \\ 1 & 2 \end{bmatrix} x(k) + \begin{bmatrix} 1 \\ 4 \end{bmatrix} u(k)$$

$$y = [1 \quad 0]x(k)$$

(a) Determine the state-feedback controller $u(k) = mu_c(k) - Lx(k)$ such that the states are brought to the origin in two sampling intervals.
(b) Is it possible to determine a state-feedback controller that can take the system from the origin to $x(k) = [2 \quad 8]^T$?
(c) Determine an observer that estimates the state such that the estimation error decreases as $p(k)0.2^k$.

9.9 References

Pole placement was one of the first applications of the state-space approach. One of the first to solve the problem was J. Bertram in 1959. The first published solution is given in

RISSANEN, J. (1960): "Control System Synthesis by Analogue Computer Based on 'Generalized Linear Feedback' Concept," Proceedings of the Symposium on Analog Computation Applied to the Study of Chemical Processes, Brussels, November 21–23, 1–13.

Treatment of the multivariable case of pole-placement can be found, for instance, in

ROSENBROCK, H. H. (1970): *State-Space and Multivariable Theory*. London: Nelson.

WOLOWICH, W. A. (1974): *Linear Multivariable Systems*. New York: Springer-Verlag.

KAILATH, T. (1980): *Linear Systems*. Englewood Cliffs, N.J.: Prentice Hall, Inc.

Observers are also described in the books above. The reduced-order observer was first described in a Ph.D. thesis by Luenberger. Easier available references are

LUENBERGER, D. G. (1964): "Observing the State of a Linear System," *IEEE Trans. Mil. Electron.*, MIL-8, 74–80.

LUENBERGER, D. G. (1971): "An Introduction to Observers," *IEEE Trans. Autom. Control,* AC-16, 596–603.

The servo problem and introduction of reference values are discussed in

WITTENMARK, B. (1985): "Design of Digital Controllers—The Servo Problem," in S. G. Tzafestas, ed., *Applied Digital Control,* Elsevier Science Publishers, Amsterdam.

Numerical aspects of computing the state feedback and the observer gain are discussed in

MIMINIS, G. S. and C. C. PAIGE (1982): "An Algorithm for Pole Assignment of Time Invariant Linear Systems," *Int. Journal Control,* 35, No. 2, 341–54.

PETKOV, P. HR., N. D. CHRISTOV, and M. M. KONSTANTINOV (1984): "A Computational Algorithm for Pole Assignment of Linear Input Systems," *Preprints 23rd IEEE Conference on Decision and Control,* 1770–73.

MIMINUS, G. S. and C. C. PAIGE (1988): "A Direct Algorithm for Pole Assignment of Time-Invariant Multi-Input Systems Using State Feedback," *Automatica,* 24, 343–56.

POLE-PLACEMENT DESIGN BASED ON INPUT-OUTPUT MODELS

GOAL

To Solve the Pole-Placement Problem Using Output Feedback and Polynomial Design.

10.1 Introduction

In this chapter the servo problem is solved using input-output models. The design is based on polynomial manipulations. The main idea is to start with a general linear regulator and to determine its parameters so that the closed-loop system has desired properties. The problem is formulated as a pole-placement problem; the design method is simple to use and to understand. It is easy to introduce various constraints, such as high-loop gain at certain frequencies and low-loop gain at other frequencies.

The problem is formulated in Sec. 10.2, and the solution is given in Sec. 10.3. The solution leads to a linear polynomial equation, which has to be solved. Sec. 10.4 gives conditions for the existence of solutions to the design problem. A design procedure is given in Sec. 10.5, where some examples also are given. The design is usually done based on approximate models. Section 10.6 treats the sensitivity to modeling errors. Relations to other design methods are given in Sec. 10.7. Special emphasis is made in Sec. 10.8 to relate the design procedure to natural physical specifications—for instance, sensitivity to disturbances is analyzed. Some examples showing the properties of the design method are given in Sections 10.9–10.11.

10.2 Problem Formulation

The design problem is specified in the usual way by giving the model, the criterion, and the admissible control laws. The design considers a servo problem, which is formulated as a model-following problem. Thus it is desired to find a control law such that the appropriate response to command inputs is obtained. This formulation includes pole-placement design as a special case. Therefore, an alternative solution to the design problem in Sec. 9.5 is obtained.

Process Dynamics

It is assumed that the process has one input, u, and one measured output, y. The relation between u and y is given by the pulse-transfer function

$$H(z) = \frac{B(z)}{A(z)} \tag{10.1}$$

where $A(z)$ and $B(z)$ are polynomials. Notice that the pulse-transfer function represents the dynamics of the process, including hold circuit, actuator, sensor, and antialiasing filter. It is assumed that the polynomials A and B do not have any common factors. Recall from Sec. 3.2 that the model of (10.1) may represent a discrete-time model of a continuous-time system with a rational transfer function and an arbitrary time delay.

The Criterion

The servo specifications are expressed in terms of a model that gives the desired response to command signals. The desired closed-loop pulse-transfer function is given by

$$H_m(z) = \frac{B_m(z)}{A_m(z)} \tag{10.2}$$

where B_m and A_m do not have any common factors. In general, it is not sufficient to specify H_m. With output feedback, there will be additional dynamics that are not excited by the command signal. The results of Sec. 9.4 show that it is also necessary to specify the observer dynamics. This is done by specifying the characteristic polynomial A_o of the observer.

Disturbances

There are three major types of disturbances: load disturbances, measurement errors, and plant uncertainties, as discussed in Sec. 6.3. In a servo problem it may also be useful to specify the nature of the command signals.

Specific disturbance models are not used in the pole-placement design method. The disturbances are instead considered indirectly by introducing constraints on the model H_m, the observer polynomial A_o, and the admissible control law.

The sensitivity of the closed-loop system to modeling errors and to high-frequently measurement noise is influenced by the model-transfer function H_m. The sensitivity requirements conflict with requirements on rapid model-following. The observer polynomial also influences the sensitivity to load disturbances and measurement noise.

Admissible Control Laws

The regulator has one output, u, and two inputs, the command signal, u_c, and the measured output, y. A general linear structure for a regulator with these inputs and outputs may be represented by

$$u(k) = \frac{T_1(q)}{R_1(q)} u_c(k) - \frac{S_1(q)}{R_2(q)} y(k) \tag{10.3}$$

where R_1, R_2, T_1, and S_1 are polynomials in the forward-shift operator q. The control law of (10.3) can be written as

$$R(q)u(k) = T(q)u_c(k) - S(q)y(k) \tag{10.4}$$

where $R = R_1 R_2$, $T = T_1 R_2$, and $S = S_1 R_1$. It is assumed that R is monic; i.e., the coefficient of the highest power in R is unity. The control law of (10.4) represents a combination of a feedforward from the command signal u_c with the pulse-transfer function

$$H_{ff}(z) = \frac{T(z)}{R(z)} \tag{10.5}$$

and a feedback from the measured output y with the pulse-transfer function

$$H_{fb}(z) = \frac{S(z)}{R(z)} \tag{10.6}$$

Some of the requirements on the control law are discussed next.

Causality. The conditions

$$\deg R \geq \deg T \tag{10.7}$$

$$\deg R \geq \deg S \tag{10.8}$$

ensure that the feedback and the feedforward transfer functions are causal. If the time to calculate the control signal in the computer is only a small fraction of the sampling period, then it is natural to require that

$$\deg R = \deg T = \deg S \tag{10.9}$$

This means that there is no time delay in the regulator.

If the computation time is close to one sampling period, the corresponding constraint is instead

$$\deg R = 1 + \deg T = 1 + \deg S \tag{10.10}$$

This means that there is a time delay in the control law of one sampling period.

The constraint of (10.9) is normally used as a standard case. Possible computational delays can be included in the process model instead of in the regulator (compare with Sec. 3.2).

Disturbances and plant uncertainties.   Disturbances are handled by introducing additional specifications on the admissible control law. The contraints are conveniently expressed in terms of the loop gain

$$H_{lg}(z) = \frac{S(z)B(z)}{R(z)A(z)} \qquad (10.11)$$

To make sure that low-frequency disturbances give small errors, the loop gain $H_{lg}(\exp i\omega h)$ must be large for low frequencies. This may be achieved by requiring that the polynomial R has the form

$$R(z) = (z - 1)^l R_1'(z) \qquad (10.12)$$

with a suitable l. This is the classical principle of integral control. The condition also ensures that the closed-loop system is insensitive to modeling uncertainties that influence the low-frequency signal transmission.

The influence of uncertainties in the high-frequency dynamics of the process and of high-frequency measurement noise may be reduced by choosing S and R so that the loop gain falls off rapidly for high frequencies (see Theorem 5.4). Notice that sampling is very useful in this respect because the combination of the anti-aliasing filter and the hold circuit reduces the signal transmission effectively for frequencies above the Nyquist frequency. Thus the selection of the sampling period is very important for the sensitivity to high-frequency disturbances and modeling errors.

There are also situations in which the loop gain must be low for some frequencies, which are lower than the bandwidth of the servo. This may happen when there are very large periodic disturbances that cannot be reduced by the servo or when there are rapid phase changes in the signal transmission. Typical examples are disturbances due to waves in station-keeping of offshore drilling ships and low-frequency bending modes in aerospace vehicles. These types of disturbances are classically handled by introducing notch filters, which block the signal transmission for certain frequencies. In the pole-placement design, the same effect can be obtained by introducing a factor in the polynomial S.

10.3 Solution

The design problem formulated in the previous section may be summarized as follows: A process model specified by the input-output relation

$$A(q)y(k) = B(q)u(k) \qquad (10.13)$$

where u is the control signal and y is the measured output signal. Find an admissible control law of the form of (10.4) such that the closed-loop system has the input-output relation given by the pulse-transfer function of (10.2), and an observer with the characteristic polynomial $A_o(z)$.

A block diagram of the closed-loop system is shown in Fig. 10.1. The disturbances are also shown in the figure. To simplify the writing in the analysis that follows, the arguments of polynomials and time functions are suppressed. The input-output relationship for the closed-loop system is obtained by eliminating u between Equations (10.4) and (10.13). Hence

$$y = \frac{BT}{AR + BS} u_c + \frac{BR}{AR + BS} v + \frac{AR}{AR + BS} e \qquad (10.14)$$

Requiring that this input-output relation is equivalent to (10.2) gives

$$\frac{BT}{AR + BS} = \frac{B_m}{A_m} \qquad (10.15)$$

The design problem is thus reduced to the problem of finding the polynomials R, S, and T that satisfy (10.15) and the additional requirements on the admissible controls.

Examples

There are many possibilities to choose the polynomials R, S, and T. Two examples are given as illustrations.

Example 10.1—A pure feedforward solution

It is easily verified that the polynomials

$$R = BA_m \qquad S = 0 \qquad T = AB_m$$

satisfy (10.15). With this choice of polynomials, it follows that the feedback, pulse-transfer function is

$$H_{fb} = \frac{S}{R} = 0$$

and that the feedforward, pulse-transfer function is

$$H_{ff} = \frac{T}{R} = \frac{B_m A}{A_m B}$$

Thus the controller is a pure feedforward compensator without feedback. The compensator simply cancels the process dynamics and adds the desired dynamics. Although the solution satisfies (10.15), it does not satisfy the other constraints on the

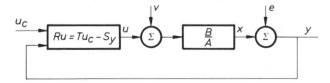

Figure 10.1 Block diagram of the closed-loop system.

control law. Because the feedback is zero, the condition that the loop gain is large for low frequencies is obviously not satisfied. All poles and zeros of the process are canceled, so the system will be unstable if the process has poles or zeros outside the unit disc. □

Example 10.2—Error feedback

If the polynomials S and T are equal, the control law (10.4) can be written as

$$Ru = S(u_c - y) = Se$$

where e is the control error. This means that the control law is based on feedback from the error only. It is easy to verify that the polynomials

$$R = B(A_m - B_m) \qquad S = T = AB_m$$

satisfy (10.15). This choice represents a feedback solution. However, there is no guarantee that the feedback has high gain at low frequencies. Notice that all poles and zeros of the process are canceled. The closed-loop system will be unstable if there are poles or zeros outside the unit disc. The necessity of canceling poles and zeros is due to the restriction imposed by error feedback. □

The examples clearly show that there are many possible choices of R, S, and T that satisfy (10.15). Rational ways of making the proper choices that satisfy (10.15) and the constraints on admissible controls are given next.

Pole-Zero Cancellations

It follows from (10.15) that the poles of the closed-loop system are the solutions to the characteristic equation

$$AR + BS = 0 \tag{10.16}$$

The zeros of the closed-loop system are the zeros of the polynomials B and T. The order of the closed-loop system is usually higher than the order of the model. In order to satisfy the condition in (10.15), there must then be cancellation of poles and zeros.

First consider the open-loop zeros, i.e., the zeros of the polynomial B. If a factor of B is not a factor of B_m then it must be a factor of $AR + BS$, so it must be canceled by a closed-loop pole. Since the closed-loop system must be stable, it follows that only stable zeros may be canceled. Hence factor B as

$$B = B^+ B^- \tag{10.17}$$

where B^- has all its zeros outside the unit disc and B^+ has all its zeros inside the unit disc. To get a unique factorization the coefficient of the highest power in B^+ is fixed to unity. The polynomial B^+ is then said to be monic.

Since B^- cannot be a factor of $AR + BS$, it follows that it must divide B_m, i.e.,

$$B_m = B^- B_m' \tag{10.18}$$

This implies that unstable process zeros cannot be changed, but must be included in B_m. Since B^+ is a factor of $AR + BS$, it follows that it is also a factor of R. Hence

$$R = B^+ R' \tag{10.19}$$

Equation (10.15) can then be written as

$$\frac{B^+ B^- T}{B^+ (AR' + B^- S)} = \frac{B^- B'_m}{A_m}$$

Cancellation of common factors give

$$\frac{T}{AR' + B^- S} = \frac{B'_m}{A_m}$$

and it follows that A_m is a factor of $AR' + B^- S$. Furthermore, it follows from Sec. 9.5 that the observer polynomial is canceled in the transfer function from the reference signal to the output. Therefore, it follows that A_o is a factor of $AR + BS$. Hence the following conditions are obtained:

$$AR' + B^- S = A_o A_m \tag{10.20}$$

and

$$T = B'_m A_o \tag{10.21}$$

The closed-loop characteristic equation becomes

$$AR + BS = B^+ A_o A_m \tag{10.22}$$

Thus the closed-loop poles are the canceled stable-process zeros, B^+, the model poles, A_m, and the observer poles, A_o.

Remark. It may also be desirable to cancel stable-process poles. These can then be included in A_o. The polynomial A_o then loses its interpretation as observer polynomial. From (10.20) and (10.21) it is seen that the cancelled poles are a factor of S and T.

Interpretation of the Closed-Loop Poles

The characteristic polynomial of the closed-loop system is $B^+ A_o A_m$. When comparing the polynomial approach with the solution obtained by state feedback and observer it is natural to consider the polynomials as

$$A_m(z) = \det(zI - \Phi + \Gamma L)$$

and

$$A_o(z) = \det(zI - \Phi + KC)$$

Compare (9.32). From this point of view it is natural to call A_o the observer polynomial. This is also the reason for making the polynomial T divisible by A_o. This implies that the command signal does not generate observer errors. When making a polynomial design we can, however, be more flexible. It is of course

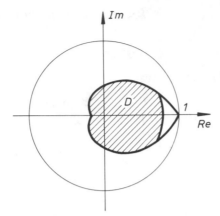

Figure 10.2 A region D such that points in the region have a minimum relative damping and a minimum absolute damping.

not necessary to view the closed-loop characteristic polynomial as split up in two polynomials of the same degree. We can simply consider $A_o A_m$ as the desired closed-loop characteristic polynomial, where A_o represents the modes that are not excited by the command signal. Also it is possible to give different responses to the reference value changes and to load disturbances. This is further discussed in Sections 10.7 and 10.11.

Practical Constraints

In practice it is useful to have more stringent requirements on allowable cancellations. Sometimes cancellation may not be desirable at all. In other cases it may be reasonable to cancel zeros that are sufficiently well damped. One way to express this formally is to introduce a region D in the complex plane that corresponds to modes with sufficient relative and absolute damping. Only zeros inside D may be canceled. An example of a region is shown in Fig. 10.2. From Sec. 3.6, recall that lines with constant relative damping are logarithmic spirals in the z-plane and that lines with constant absolute damping are circles.

Summary

In summary, the pole-placement problem is solved by finding R' and S that satisfy (10.20) and such that B is factored as $B^+ B^-$. The polynomial B^+ is monic and has all its zeros inside D; B^- has all its zeros outside D. Further, B^-, must be a factor of B_m. Finally, the polynomials R and T should be chosen by (10.19) and (10.21).

10.4 An Algebraic Problem

The discussion in the previous section gave some insight into the choice of the polynomials R, S, and T. There still remains one problem, namely, to determine two polynomials R' and S that satisfy (10.20) and the other requirements on the admissible feedback. The basic mathematical problem is to determine two polynomials X and Y that satisfy a linear polynomial equation

$$AX + BY = C \qquad (10.23)$$

where A, B, and C are known polynomials. This is a well-known problem in elementary algebra. The identity (10.23) is a called a *Diophantine equation*. Another name is the *Aryabhatta's identity*. When $C = q^m$ it is also called the *Bezout identity*.

A Digression

Equation (10.23) looks strange at first because two unknowns have to be determined from one equation. A simpler, but related, problem from high school algebra gives insight.

Example 10.3—The Diophantine equation

Consider the equation

$$3x + 2y = 5 \qquad (10.24)$$

where x and y are integers. This is called a *Diophantine equation* after Diophantus ($\approx$ A.D. 300), who was one of the original inventors of algebra.

It is obvious that $x = 1$ and $y = 1$ is a solution. Another solution may be found by increasing x by 2 and decreasing y by 3. Hence if x_0 and y_0 satisfy (10.24), then another solution is given by

$$\begin{aligned} x &= x_0 + 2n \\ y &= y_0 - 3n \end{aligned} \qquad (10.25)$$

where n is an integer. A few solutions are listed below

x:	-5	-3	-1	1	3	5	7
y:	10	7	4	1	-2	-5	-8

It follows from (10.25) that if a solution x_0, y_0 is known, it is possible to add or subtract 2 from x_0 until a unique solution with

$$0 \le x < 2$$

is obtained. Similarly, there is also a unique solution such that

$$0 \le y < 3 \qquad \square$$

Equation (10.24) is closely related to Equation (10.23) because the integers and the polynomials with real coefficients obey the same algebraic rules. Integers and polynomials may be multiplied and added with the usual rules. However, division of two integers (or polynomials) does not necessarily result in an integer (or a polynomial). In algebraic terminology this is expressed by saying that integers and polynomials with real coefficients are *rings*.

Another example shows that there may not be a solution to an equation such as (10.24).

Example 10.4

Consider the equation

$$4x + 6y = 1$$

where x and y are integers. Since the left-hand side is an even number and the right-hand side an odd number, it is clear that the equation does not have a solution. The difficulty in finding a solution is due to the fact that the numbers 4 and 6 have 2 as a common factor, while the right-hand side does not. □

Main Result

Examples 10.3 and 10.4 essentially reveal the important issues about Equation (10.23). It is now simply a matter of giving a formal analysis of the equation. The solutions are characterized by the following result.

Theorem 10.1. Let A, B, and C be polynomials with real coefficients. Then Equation (10.23) has a solution if and only if the greatest common factor of A and B divides C.

A proof of the theorem is given in Appendix D.
It follows from (10.23) that if X_0, Y_0 is a solution, then

$$X = X_0 + QB$$
$$Y = Y_0 - QA$$
(10.26)

is also a solution, where Q is an arbitrary polynomial. If a solution exists, it is possible to find infinitely many solutions by adding or subtracting a suitable multiple of A and B.

Corollary. There are unique solutions to (10.23) such that

$$\deg X < \deg B \tag{10.27}$$

or

$$\deg Y < \deg A \tag{10.28}$$

10.5 A Design Procedure

The development of an understanding for the algebraic properties of (10.23) leads to a return to the pole-placement design problem. The feedback in (10.4) is uniquely given by the polynomials R, S, and T. Section 10.3 shows that T is given by (10.21). Also, R' and S should satisfy (10.20). Because the polynomials A and B do not have any common factors, it follows from Theorem 10.1 that there are polynomials R' and S that satisfy (10.20). Also, there are infinitely many polynomials that satisfy the equation. All solutions give the same pulse-transfer function from the command signal to the output signal. Next, the question of whether solutions can be found that also satisfy the other constraints on the admissible control laws is investigated.

Causality

The control law of (10.4) is causal if Inequalities (10.7) and (10.8) are satisfied. The consequences of these inequalities are summarized in Theorem 10.2.

Theorem 10.2. There exists a causal solution to the pole-placement design if

$$\deg A_m - \deg B_m \geq \deg A - \deg B \tag{10.29}$$

and

$$\deg A_o \geq 2\deg A - \deg A_m - \deg B^+ - 1 \tag{10.30}$$

Proof. Equations (10.19) and (10.20) give

$$AR + BS = B^+ A_o A_m$$

Since $\deg S \leq \deg R$ and $\deg B < \deg A$, it follows that

$$\deg AR = \deg (AR + BS) = \deg B^+ A_o A_m$$

Hence

$$\deg R = \deg A_o + \deg A_m + \deg B^+ - \deg A$$

Furthermore, it follows from (10.21) that

$$\deg T = \deg A_o + \deg B'_m$$

Now investigate the consequences of (10.7), which implies that

$$\deg A_o + \deg A_m + \deg B^+ - \deg A \geq \deg A_o + \deg B'_m$$

Hence,

$$\deg A_m - \deg B'_m \geq \deg A - \deg B^+$$

Subtraction of $\deg B^-$ from both sides and noting (10.18) gives (10.29). It follows from the corollary of Theorem 10.1 that there exists a solution to (10.20) such that

$$\deg S < \deg A$$

The choice of $\deg S = \deg A - 1$ and Inequality (10.8) then imply that

$$\deg A_o + \deg A_m + \deg B^+ - \deg A \geq \deg A - 1$$

Rearrangement of the terms gives (10.30).

Remark 1. It is easy to understand the condition in (10.29) intuitively. It implies that the delay in the model H_m must be at least as large as the delay in the system H.

Remark 2. The condition in (10.30) implies that the degree of the observer polynomial A_o must be sufficiently high in order to obtain a causal control law.

High Gain at Low Frequencies

It is useful to have a high feedback gain at low frequencies in order to get a system that is insensitive to low-frequency modeling errors and low-frequency disturbances. This can be achieved by requiring that $(z - 1)^l$ be a factor of R. Introducing

$$R = (z - 1)^l R_1 \qquad (10.31)$$

the design equation of (10.20) becomes

$$A(z - 1)^l R_1' + B^- S = A_o A_m \qquad (10.32)$$

Simple calculations show that the causality conditions in (10.7) and (10.8) are satisfied if (10.29) holds and if

$$\deg A_o \geq 2\deg A - \deg A_m - \deg B^+ + l - 1 \qquad (10.33)$$

The degree of the observer polynomial must thus be increased by l in comparison with (10.30).

A Design Algorithm

The results obtained can be summarized in a straightforward procedure for solving the pole-placement design problem.

Algorithm 10.1—Pole-placement design

Data. A process model specified by the pulse-transfer function B/A, an observer polynomial A_o, and specifications in terms of the desired closed-loop pulse-transfer function B_m/A_m. Further, the stability region D is specified.

Conditions. It is assumed that the data satisfy (10.18), (10.29), and (10.33).

Step 1. Factor B and B_m as

$$B = B^- B^+ \qquad B_m = B^- B_m'$$

where B^+ is monic and has all its zeros inside the stability region D, and B^- has all its zeros outside D.

Step 2. Solve the equation

$$(z - 1)^l A R_1' + B^- S = A_o A_m$$

with respect to R_1' and S. Choose a solution such that

$$\deg S < l + \deg A$$

and

$$\deg R_1' = \deg A_o + \deg A_m - \deg A - l$$

Step 3. The control law is then

$$Ru = Tu_c - Sy$$

where

$$R = B^+ R' \qquad T = B'_m A_o \qquad R' = (z - 1)^l R'_1.$$

Remark. From the polynomial identity it follows that R' is monic. This implies that R is also monic.

Solution of Linear Polynomial Equations

To carry out the design, it is necessary to solve a linear polynomial equation. There are several ways to do this.

One possibility is to introduce polynomials R' and S with unknown coefficients and given orders and identify the coefficients of equal powers of z. A linear system of equations is then obtained. From Theorem 10.1, it follows that the linear equation can always be solved if A and B do not have any common factors. The equations obtained may be simplified if A and B are given in factored form. This method is illustrated in the examples that follow.

By equating coefficients of equal order, the Diophantine equation given by Eq. (10.23) can be written as a set of linear equations:

$$
\begin{bmatrix}
1 & 0 & \cdots & 0 & b_0 & 0 & \cdots & 0 \\
a_1 & 1 & & \vdots & b_1 & b_0 & & \vdots \\
a_2 & a_1 & & 0 & b_2 & b_1 & & 0 \\
\vdots & \vdots & & 1 & \vdots & \vdots & & b_0 \\
a_n & \vdots & & a_1 & b_n & \vdots & & b_1 \\
0 & a_n & & \vdots & 0 & b_n & & \vdots \\
\vdots & & & & \vdots & & & \\
0 & \cdots & 0 & a_n & 0 & \cdots & 0 & b_n
\end{bmatrix}
\begin{bmatrix}
x_1 \\
\vdots \\
\\
x_k \\
y_0 \\
\vdots \\
\\
y_l
\end{bmatrix}
=
\begin{bmatrix}
c_1 - a_1 \\
\vdots \\
\\
c_n - a_n \\
c_{n+1} \\
\vdots \\
\\
c_{k+l+1}
\end{bmatrix}
$$

$$\underbrace{\qquad\qquad}_{k \text{ columns}} \qquad \underbrace{\qquad\qquad}_{l + 1 \text{ columns}}$$

The matrix on the left-hand side is called the *Sylvester matrix* and occurs frequently in applied mathematics. It has the property that it is nonsingular if and only if the polynomials A and B do not have any common factors. If there are no common factors, there exists a unique solution to Eq. (10.23). Notice, however, the nonuniqueness with respect to the orders of X and Y. Different choices of k and l will give different X and Y, as discussed above.

The solution to the linear equation above can be obtained by Gaussian elimination. This method does not use the special structure of the Sylvester matrix. There are also special polynomial methods for solving the system of equations directly. The pole-placement procedure with all process zeros or all poles canceled has the advantage that the Sylvester matrix is triangular. This means that the coefficients of the X and Y polynomials can be obtained one at a time.

It is also possible to solve the Diophantine equation using polynomial calculations. This has the advantage that possible common factors in the polynomials

Sec. 10.5   A Design Procedure

A and *B* can be canceled. The method is based on a classical algorithm due to Euclid. See Appendix D and the references at the end of the chapter.

Examples

A few examples will be given to illustrate control-system design with the pole-placement algorithm.

Example 10.5—Motor with cancellation of process zero

The pulse-transfer function of a DC motor can be written as

$$H(z) = \frac{K(z - b)}{(z - 1)(z - a)} \tag{10.34}$$

(see Example A.2) where

$$K = e^{-h} - 1 + h$$

$$a = e^{-h}$$

$$b = 1 - \frac{h(1 - e^{-h})}{e^{-h} - 1 + h}$$

Notice that $b < 0$; i.e., the zero is on the negative real axis. It is first assumed that the desired closed-loop system is characterized by the pulse-transfer function

$$H_m(z) = \frac{z(1 + p_1 + p_2)}{z^2 + p_1 z + p_2} \tag{10.35}$$

The pulse-transfer function H has a zero $z = b$ which is not included in H_m. With the given specifications, it is necessary to cancel the zero $z = b$. Factor B as

$$B^+ = z - b$$

$$B^- = K$$

Then

$$B'_m = \frac{B_m}{K} = \frac{z(1 + p_1 + p_2)}{K}$$

The degree of the observer polynomial is given by (10.30), which gives deg $A_o \geq 0$. Choose

$$A_o(z) = 1$$

The degree of the polynomials R' and S are given by

$$\deg R' = \deg A_o + \deg A_m - \deg A = 0$$

$$\deg S = \deg A - 1 = 1$$

Introduce R' as a zero-order polynomial and S as a first-order polynomial in the design equation. The following polynomial identity is then obtained.

$$(z - 1)(z - a)r_0 + K(s_0 z + s_1) = z^2 + p_1 z + p_2$$

Equating coefficients of equal powers of z gives the equations

$$r_0 = 1$$

$$-(1 + a)r_0 + Ks_0 = p_1$$

$$ar_0 + Ks_1 = p_2$$

Hence,

$$r_0 = 1$$

$$s_0 = \frac{1 + a + p_1}{K}$$

$$s_1 = \frac{p_2 - a}{K}$$

Equation (10.21) gives

$$T(z) = A_o(z)B'_m(z) = \frac{z(1 + p_1 + p_2)}{K} = t_0z$$

The control law (see step 3 of Algorithm 10.1) can be written as

$$u(k) = t_0u_c(k) - s_0y(k) - s_1y(k - 1) + bu(k - 1) \qquad (10.36)$$

A simulation of the step response of the system is shown in Fig. 10.3. Notice the "ringing," or the "ripple," in the control signal, which is caused by the cancellation of the zero on the negative real axis. The ripple is not noticeable in the output signal at the sampling instants. It is, however, seen as a ripple in the output between the sampling instants. The amplitude of the ripple in the output depends on the sampling period. It goes down rapidly as the sampling period is decreased. □

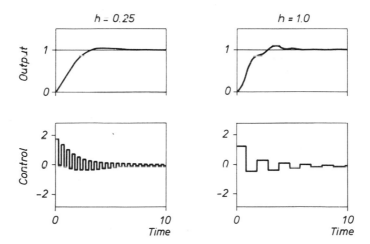

Figure 10.3 Step response for a motor with pole-placement control. The specifications are $\zeta = 0.7$ and $\omega = 1$. The sampling periods are $h = 0.25$ and $h = 1.0$. The process zero is canceled.

Example 10.6—Motor with no cancellation of process zero

Consider the same motor as in Example 10.5, but assume that the desired closed-loop transfer function is

$$H_m(z) = \frac{1 + p_1 + p_2}{1 - b} \frac{z - b}{z^2 + p_1 z + p_2} \qquad (10.37)$$

Notice that the process zero on the negative real axis is now also a zero of the desired closed-loop transfer function. This means that the zero does not have to be canceled by the regulator. Factor B as

$$B^+ = 1$$

$$B^- = K(z - b)$$

Hence,

$$B'_m = \frac{1 + p_1 + p_2}{K(1 - b)}$$

The degree of the observer polynomial is

$$\deg A_o \geq 2 \deg A - \deg A_m - \deg B^+ - 1 = 1$$

Therefore, the observer polynomial should be of at least first degree. A deadbeat observer is chosen:

$$A_o(z) = z$$

The minimal degrees of the polynomials R and S are then given by

$$\deg R = \deg A_m + \deg A_o - \deg A = 1$$

$$\deg S = \deg A - 1 = 1$$

The design equation of (10.20) can then be written as

$$(z - 1)(z - a)(z + r_1) + K(z - b)(s_0 z + s_1) = z^3 + p_1 z^2 + p_2 z \qquad (10.38)$$

To determine r_1, put $z = b$ in (10.38). Hence,

$$(b - 1)(b - a)(b + r_1) = b^3 + p_1 b^2 + p_2 b$$

which gives

$$r_1 = -b + \frac{b(b^2 + p_1 b + p_2)}{(b - 1)(b - a)}$$

Now put $z = 1$ and $z = a$ in (10.38). This gives

$$K(1 - b)(s_0 + s_1) = 1 + p_1 + p_2$$

$$K(a - b)(s_0 a + s_1) = a^3 + p_1 a^2 + p_2 a$$

from which s_0 and s_1 can be determined. Equation (10.21) gives

$$T(z) = A_o B'_m = z \frac{1 + p_1 + p_2}{K(1 - b)} = t_0 z$$

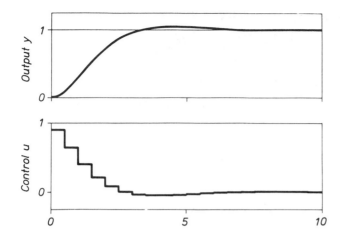

Figure 10.4   Step response of a motor with pole-placement control for $h = 0.5$. The specifications are $\zeta = 0.7$ and $\omega = 1$. The process zero is not canceled.

The control law is then

$$u(k) = t_0 u_c(k) - s_0 y(k) - s_1 y(k-1) - r_1 u(k-1)$$

Notice that this feedback law is of the same form as (10.36). However, the coefficients are different. A simulation of the step response of the system is shown in Fig. 10.4. A comparison with Fig. 10.3 shows that the control signal is much smoother, there is no ringing. The response start is also a little slower, since A_o is of higher degree than in Example 10.5.   □

10.6 Sensitivity to Modeling Errors

It is highly unrealistic to assume that the process model used in a control design is accurate. Therefore, it is important to understand how modeling errors will influence the closed-loop properties, which is discussed in this section. It is assumed that the design is based on the model $H = B/A$, while the true model is $H^0 = B^0/A^0$.

Stability

The following result describes the influence of modeling errors on the stability of the closed-loop system.

Theorem 10.3.   Consider a pole-placement design based on an approximate model

$$H = B/A$$

Let H^0 be the pulse-transfer function of the system to be controlled. Assume that H and H^0 have the same number of poles outside the unit disc and that H_m

is stable. Then the closed-loop system related to $H^0(z)$ is stable if

$$| H(z) - H^0(z) | < \left| \frac{H(z)T(z)}{H_m(z)S(z)} \right| = \left| \frac{H(z)}{H_m(z)} \right| \left| \frac{H_{ff}(z)}{H_{fb}(z)} \right| \tag{10.39}$$

for $| z | = 1$, where H_{ff} and H_{fb} are defined by (10.5) and (10.6).

Proof. The proof is based on the general sensitivity results of Sec. 5.2. Introduce the loop gain

$$H_{lg} = \frac{BS}{AR} \tag{10.40}$$

It follows from (10.22) that

$$1 + H_{lg} = 1 + \frac{BS}{AR} = \frac{AR + BS}{AR} = \frac{B^+ A_o A_m}{AR}$$

From (10.21) it follows that

$$A_o = \frac{T}{B_m'} = \frac{TB^-}{B_m' B^-} = \frac{TB^-}{B_m}$$

Hence,

$$1 + H_{lg} = \frac{B^+ B^- TA_m}{ARB_m} = \frac{BA_m T}{AB_m R} = \frac{HT}{H_m R}$$

After multiplication by S/R, the condition given by (10.39) can be written as

$$\left| \frac{S}{R} H - \frac{S}{R} H^0 \right| < \left| \frac{HT}{H_m R} \right| = | 1 + H_{lg} |$$

or

$$| H_{lg} - H_{lg}^0 | < | 1 + H_{lg} |$$

The result now follows from Theorem 5.4.

Remark 1. It is easy to apply this theorem. When a design is performed, the right-hand side of (10.39) can easily be calculated for $z = \exp(i\omega h)$, and it does not depend on the true pulse-transfer function. Then the requirements on model precision can be expressed in terms of frequency domain conditions.

Remark 2. Notice that the inequality in (10.39) is automatically satisfied if

$$| H_{lg} | = \left| \frac{SB}{RA} \right| < \frac{1}{3}$$

and

$$| H_{lg}^0 | = \left| \frac{SB^0}{RA^0} \right| < \frac{1}{3}$$

It is thus sufficient to check (10.39) for frequencies where the loop gain is larger than $\frac{1}{3}$.

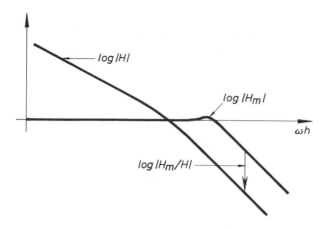

Figure 10.5 Bode diagrams for H and H_m. The ratio H/H_m, which appears in (10.39), is easily found in the figure.

Remark 3. Notice that the relative accuracy is

$$\frac{|H(z) - H^0(z)|}{|H(z)|} \leq \frac{1}{|H_m(z)|} \left|\frac{H_{ff}(z)}{H_{fb}(z)}\right|$$

The condition given in (10.39) has good physical interpretation. Consider first the ratio H/H_m. The pulse-transfer function H of the process is typically large for low frequencies and decreases for high frequencies (see Fig. 10.5). The desired pulse-transfer function H_m of the closed-loop system is typically unity for low frequencies. There is a small increase around the crossover frequency and H_m decreases for high frequencies (compare with Fig. 7.9). The frequency response of H_m is also shown in Fig. 10.5. The ratio H/H_m is easy to obtain from the figure. It is clear from the figure that it is sufficient to have good model precision only in certain frequency ranges. The consequences of changing the desired bandwidth of the closed-loop system can also be determined. The requirements on the model accuracy are relaxed if the closed-loop bandwidth is decreased. A more-precise model will be needed if the desired bandwidth is increased. The ratio

$$\left|\frac{H_{ff}}{H_{fb}}\right| = \left|\frac{T}{S}\right|$$

is the quotient between the feedforward- and the feedback-transfer functions. The quotient is unity if the design procedure gives feedback from the error only. The requirements on model precision are smaller for frequencies where the feedforward gain is larger than the feedback gain.

Accuracy of the Closed-Loop Transfer Function

Some calculations give the pulse-transfer function of the closed-loop system as

$$H_{cl} = \frac{B^0 T}{A^0 R + B^0 S} = \frac{T/R}{A^0/B^0 + S/R} = H_m \frac{1}{1 + (RB^-/A_o A_m)(1/H^0 - 1/H)} \tag{10.41}$$

where A_o is the observer polynomial. This shows how errors in the model are reflected in errors in the closed-loop pulse-transfer function. It is clear from the expression that the errors are small when the open-loop pulse-transfer functions H and H^0 are large.

10.7 Relationships to Other Design Methods

Pole placement is a general approach to design of single-input–single-output systems. Many other design methods may be interpreted as pole placement design.

Root Locus

The root-locus method is a classical technique for design of control systems. The method is based on the idea of attempting to place the closed-loop poles in desired positions. Thus it is closely related to pole placement. In this method, polynomials R and S are first chosen as $R = 1$ and $S = K$, which correspond to proportional control. The gain K is then changed and the roots of the characteristic equation

$$A + KB = 0 \qquad (10.42)$$

are investigated. The roots of (10.42) can easily be sketched for varying K. If a reasonable pole-placement cannot be obtained, the orders of the polynomials R and S are increased using heuristic rules. The procedure is then repeated.

It is clear that the pole-placement design procedure given in Sec. 10.5 is easier to use than the root-locus design. All poles are positioned in one operation. On the other hand, it is also clear that the root-locus method will give a simple regulator if there is one that solves the problem. However, the general design method will give a regulator whose complexity is determined by the complexity of the process model used in the design.

Dipole Compensation

In classical servo design with error feedback, it is customary to introduce a dipole in the desired closed-loop transfer function— i.e., a pole and a zero that are close together and also close to the origin. This dipole has only a marginal influence on the transmission of high-frequency signals, although it does influence the transmission of low-frequency signals. It is not intuitively clear why it is useful to introduce a dynamic mode in the regulator, which is almost canceled, i.e., almost unreachable, or unobservable. The dipole compensation may be interpreted in light of the pole-placement design, and the dynamics introduced by the dipole may be interpreted as observer dynamics.

When only feedback from the error is used, it is not possible to make the observer error unreachable from the command signal. Thus the observer dynamics will appear in the pulse-transfer function from command to output signal. The design method used in this chapter makes it totally possible to avoid excitation of the modes associated with the observer by introducing a different feedforward pulse-transfer function. The dipole will then disappear completely in the pulse-transfer function that relates the process output to the command signal.

Smith-Predictor

The Smith-predictor is mentioned in Sec. 7.5. It is a particular control law that has been proposed for systems with time delays. Smith's predictor can be described as follows.

Consider a process with the pulse-transfer function

$$H(z) = \frac{1}{z^d} H'(z) = \frac{B'(z)}{z^d A'(z)} \tag{10.43}$$

where deg A' = deg B'. The number d thus represents the time delay of the system. First design a feedback

$$H_{fb}(z) = \frac{S'(z)}{R'(z)}$$

which gives the desired closed-loop properties if there is no delay in the system. For a system with time delay, the Smith-predictor gives the control law

$$u(k) = H_{fb}(u_c(k) - y(k)) - H' H_{fb}(1 - q^{-d}) u(k) \tag{10.44}$$

The closed-loop pulse-transfer function is then given by

$$\frac{B_m}{A_m} = \frac{B'S'}{z^d(A'R' + B'S')} \tag{10.45}$$

The Smith-predictor can easily be derived using the pole-placement design. Let the process be given by (10.43) and the desired model by (10.45). Further, let the regulator use error feedback only. Observe that the process zeros are preserved in the model. With this constraint, it follows from Example 10.2 that the control law is given by

$$u(k) = \frac{S}{R} [u_c(k) - y(k)] = \frac{AB_m}{B(A_m - B_m)} [u_c(k) - y(k)]$$

The controller is thus

$$\frac{S}{R} = \frac{z^d A'S'}{z^d(A'R' + B'S') - B'S'} = \frac{H_{fb}}{1 + H' H_{fb}(1 - z^{-d})}$$

or

$$u(k) = H_{fb}(u_c(k) - y(k)) - H' H_{fb}(1 - q^{-d}) u(k)$$

which is the Smith-predictor of (10.44). It is necessary to cancel all the process poles because only error feedback is used. Notice that it is not apparent from (10.44) that the process poles are canceled. Of course, it is possible to derive controls that are similar to the Smith-predictor by using the more general feedback-feedforward structure of (10.4). The cancellation of process poles may then be avoided.

Model Following

The control law obtained from the pole-placement design can be interpreted as a model-following controller. To see this the design equation of (10.20) is used to rewrite the control law. It follows from (10.21) that

$$\frac{T}{R} = \frac{A_o B'_m}{B^+ R'} = \frac{A_o A_m B'_m}{A_m B^+ R'} = \frac{(AR' + B^- S)B'_m}{A_m B^+ R'}$$

$$= \frac{AB'_m}{B^+ A_m} + \frac{B^- B'_m S}{A_m B^+ R'} = \frac{AB'_m B^-}{B^- B^+ A_m} + \frac{B^- B'_m S}{A_m B^+ R'}$$

$$= \frac{AB_m}{BA_m} + \frac{B_m S}{A_m R}$$

The feedback law in (10.4) now becomes

$$u = \frac{T}{R} u_c - \frac{S}{R} y = \frac{AB_m}{BA_m} u_c + \frac{SB_m}{RA_m} u_c - \frac{S}{R} y$$

$$= \frac{AB_m}{BA_m} u_c - \frac{S}{R} (y - y_m)$$

$$(10.46)$$

The first term in the control law may be interpreted as a feedforward term and the second as a feedback term. The feedforward term is a combination of the desired response and an inverse process model. The feedback term has the pulse-transfer function S/R. The feedback signal is obtained by forming the difference between the actual output and the desired output y_m, where

$$y_m = H_m u_c = \frac{B_m}{A_m} u_c$$

A block diagram representation of the feedback law of (10.46) is given in Fig. 10.6. Notice the similarity to Fig. 9.8.

Notice that it is possible to choose

$$AR + BS = A_r$$

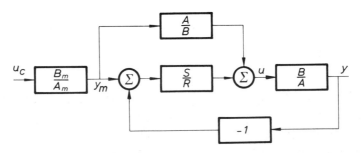

Figure 10.6 Block diagram of the control law of (10.46).

This implies that the disturbance rejection is determined by A_r while the model following is determined by A_m. Compare Section 9.5.

The Dahlin-Higham Algorithm

The Dahlin-Higham design method was popular in early digital process control design because the calculations required are so simple. It is assumed that the process dynamics can be described by (10.43). The polynomials A' and B' are first canceled by the regulator. After the cancellation, the process dynamics are a pure time delay

$$H''(z) = z^{-d}$$

To reduce steady-state errors, it is necessary to have integral action. This is ensured with $l = 1$ in (10.31). The desired model for the design is

$$H_m(z) = \frac{1 - a}{z^{d-1}(z - a)}$$

The controller is determined as in Example 10.2, i.e., using error feedback. The controller obtained can also be interpreted as a controller with a Smith-predictor. Since the algorithm is based on cancellation of all poles and zeros of the process, no poles or zeros can be allowed outside the unit disc. There will also be problems with ringing due to cancellation of stable but poorly damped zeros; compare to Fig. 10.3.

Model Algorithmic Control

Model algorithmic control is a particular design method that has also been used in process control. This method can be interpreted as a pole-placement design with

$$H(z) = z^{-d}B(z)$$

and

$$H_m(z) = \frac{1 - a}{z^r(z - a)}$$

The process is thus modeled as a finite pulse response.

10.8 Practical Aspects

The pole-placement design algorithm is a straightforward design procedure if the model, H, the desired closed-loop pulse-transfer function, H_m, and the observer polynomial, A_o, are given. It is, however, not necessarily easy to determine the design parameters H_m and A_o such that physical specifications are fulfilled.

To provide some guidance in the practical use of the pole-placement design method, influence of specifications on the magnitude of the control signal, sensitivity to load disturbances, measurement errors, and modeling errors will be discussed.

Desired Servo Performance

The desired servo performance is specified by the closed-loop pulse-transfer function H_m. It follows from the discussion in Sec. 10.3 that unstable or poorly damped process zeros must correspond to factors of B_m. Failure to realize this results in ringing, as Fig. 10.3 and the Dahlin-Higham algorithm illustrate. Thus it is very important to know the unstable process zeros. Rapid sampling of continuous-time systems with pole excess larger than two always results in unstable zeros (see Sec. 3.6). Hence it is necessary to know the time delay and the unstable zeros of the system to be controlled. Notice that high-order dynamics can often be approximated by time delay.

Specification of all poles and zeros for a high-order system requires many parameters. It rarely makes sense in practice to give so much data. It would be much more desirable to give some global characteristics of the desired dynamics, such as bandwidth and resonance peak (see Fig. 7.9). For discrete-time systems, this may be done by specifying the dominant poles. For oscillatory systems, it may be done by specifying a second-order factor

$$P_1(z) = z^2 + p_1 z + p_2 \tag{10.47}$$

where

$$p_1 = -2e^{-\zeta \omega h} \cos(\omega h \sqrt{1 - \zeta^2})$$

$$p_2 = e^{-2\zeta \omega h}$$

Also, ζ corresponds to the desired relative damping and ω is the natural frequency.

For systems without overshoot, the dominant dynamics may instead be given by a first-order factor

$$P_1(z) = z - a \tag{10.48}$$

where

$$a = \exp(-h/T)$$

and T is the desired time constant.

Equations (10.47) and (10.48) only specify some of the desired closed-loop poles. Further at least d poles should be in the origin, where d is the time delay of the process. The desired closed-loop characteristic equation can now be chosen as

$$A_m(z) = z^d P_1(z) P_2(z) = z^d P(z)$$

where P_1 is given by (10.47) or (10.48). The polynomial P_2 can have the same pattern as P_1 but with faster dynamics. Thus the desired closed-loop dynamics will typically be of the form

$$H_m(z) = \frac{P(1)}{B^-(1)} \frac{B^-(z)}{z^d P(z)} \qquad (10.49)$$

where B^- corresponds to the unstable or poorly damped zeros and d is a number that accounts for time delay of the open-loop process dynamics.

It should be noticed that the open loop system often has fast modes. These should usually not be changed. A typical situation is the extra dynamics introduced by the antialiasing filter. The position of these poles should not be changed by the controller.

Magnitude of Control Signals

It is intuitively clear that large control signals are needed if a high closed-loop bandwidth is specified. Some analysis gives a quantitative understanding. Neglecting all disturbances, it follows from (10.1) and (10.2) that

$$y = \frac{B_m}{A_m} u_c = H_m u_c$$

and

$$y = \frac{B}{A} u = Hu$$

Hence

$$u = \frac{H_m}{H} u_c \qquad (10.50)$$

This formula tells precisely the control signals that result from specific command signals. Notice that the ratio H_m/H also appeared in the robustness analysis. Compare this with Theorem 10.3.

The orders of magnitude are easily estimated from Bode diagrams of the pulse-transfer functions H and H_m. The analysis in Sec. 4.4 indicates the use of the corresponding continuous-time Bode diagrams, possibly including the hold circuit, which are good approximations for high sampling rates, at least. As Fig. 10.5 shows, the ratio $|H_m/H|$ is easily obtained from the diagram. The effect on the magnitude of the control signal of changes in the desired closed-loop bandwidth may also be estimated from the figure. If the slope of the curves is -20 dlog/decade for high frequencies, a doubling of the desired bandwidth means that control signals are quadrupled. The magnitude of the control signal may thus have a high sensitivity to changes in the bandwidth.

Sensitivity to Disturbances

A block diagram of the system, including disturbances and measurement errors, is shown in Fig. 10.1. Simple calculations give

$$x = \frac{BT}{AR + BS} u_c + \frac{BR}{AR + BS} v - \frac{BS}{AR + BS} e$$

$$= \frac{B_m}{A_m} u_c + \frac{H_{lg}}{1 + H_{lg}} \frac{1}{H_{fb}} v - \frac{H_{lg}}{1 + H_{lg}} e \qquad (10.51)$$

$$= \frac{B_m}{A_m} u_c + \frac{BR'}{A_m A_o} v - \frac{B^- S}{A_m A_o} e$$

where x is the process output and H_{lg} is the loop gain defined by (10.11). Notice the similarity in the pulse-transfer functions relating x to u_c, v, and e. The only difference is the presence of the polynomials T, R, and S, respectively, in the numerator. All polynomials R and S that satisfy the design equation (10.20) will have the same effect on the signal transmission from u_c to x. The arbitrary solution of (10.20) may be used to reduce the influence of load disturbances and measurement errors.

It follows from (10.51) that low-frequency measurement errors may propagate through the system with unit gain, because the errors are picked up in the measured signal y. Through the feedback mechanism, they will then generate variations in the process output. High-frequency measurement errors will, however, not propagate through the system because they will be attenuated by the regulator dynamics. It also follows from (10.51) that the signal transmission may be blocked at certain frequencies by requiring that $S(\exp i\omega h)$ be small for these frequencies. For instance, an ideal notch filter is obtained by requiring that

$$Q(z) = z^2 - 2z \cos \omega h + 1 \qquad (10.52)$$

is a factor of S.

It also follows from (10.51) that low-frequency load disturbances will be attenuated for those frequencies where the feedback gain H_{fb} is high or where the polynomial R is small. A typical example is to eliminate steady-state errors due to constant load disturbances by requiring that $R(1) = 0$, which is equivalent to integral compensation. To eliminate a load disturbance with frequency ω, it may be necessary that $Q(z)$—given by (10.52)—be a factor of R.

When it is required that R and S have specific factors to achieve notch filtering or integral control, the degree of the observer polynomial A_o must be increased. The choice of the observer polynomial will also influence the signal transmission because R and S are related to A_o through (10.20).

Influence of the Observer Polynomial

The effect of the observer polynomial on the transmission of disturbances is illustrated in two examples.

Example 10.7

Consider a system with the pulse-transfer function

$$H(z) = \frac{0.1}{z - 1} \qquad (10.53)$$

Assume that the desired pulse-transfer function from command to output is given by

$$H_m(z) = \frac{0.2}{z - 0.8}$$

It is easily verified that $A_o = 1$, and that the proportional feedback

$$u(k) = 2(u_c(k) - y(k))$$

gives the desired closed-loop transfer function. The following expression is then obtained for the process output.

$$x = \frac{0.2}{z - 0.8} u_c + \frac{0.1}{z - 0.8} v - \frac{0.2}{z - 0.8} e$$

The Bode diagrams for the transmission of load disturbances and measurement errors are shown in Fig. 10.7. □

Figure 10.7 shows that the proportional feedback gives a closed-loop system that is sensitive to load disturbances. A less-sensitive system can be obtained by introducing an observer polynomial of higher degree and constraints on the polynomial R, as shown in the next example.

Example 10.8

Consider the same system and the same desired closed-loop response as in Example 10.7. Let the observer polynomial be

$$A_o(z) = z - a$$

The design equation of (10.20) becomes

$$(z - 1)(z + r_1) + 0.1(s_0 z + s_1) = (z - a)(z - 0.8)$$

Hence the following conditions are obtained.

$$-1 + r_1 + 0.1 s_0 = -a - 0.8$$

$$-r_1 + 0.1 s_1 = 0.8a$$

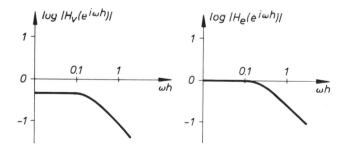

Figure 10.7 Bode diagrams for the transmission of load disturbances and measurement noise for the system in Example 10.7.

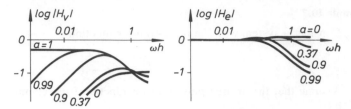

Figure 10.8 Bode diagram for the signal transmission from load disturbances and measurement error to process output for different observer polynomials for the controller in Example 10.8.

Because there are two linear equations and three unknowns, one extra condition may be introduced. Choose $r_1 = -1$ to make sure that $R(1) = 0$ (integral action). Hence

$$s_0 = 12 - 10a$$

$$s_1 = 8a - 10$$

The following expression is obtained for the process output.

$$x = \frac{0.2}{z - 0.8} u_c + \frac{0.1(z - 1)}{(z - a)(z - 0.8)} v - \frac{(1.2 - a)z - 1 + 0.8a}{(z - a)(z - 0.8)} e$$

The Bode diagrams for the signal transmission from load disturbances and measurement errors are shown in Fig. 10.8; compare this with Fig. 10.7. ☐

Selection of Sampling Interval

The choice of sampling interval is discussed in Sec. 9.2 for the pole-placement design based on state feedback. The same arguments can be used for the method given in this chapter. This means that the sampling interval should be chosen in relation to the desired closed-loop behavior. Notice, however, that the A_o polynomial must also be taken into consideration. This is further discussed in the following sections. One rule of thumb is to have 4 to 10 samples per rise time of the closed-loop system, or 15 to 45 samples per period.

10.9 Pole-Placement Design—The Double Integrator

There are many issues to consider in control-system design, as was mentioned in Section 7.7. In pole-placement design we are primarily choosing the polynomials A_m and A_o, whose zeros are the closed-loop poles, and the sampling period. To make proper choices it is important to understand how they influence response to command signals, load disturbances, measurement noise, and sensitivity to modelling errors. The double integrator will now be used to illustrate these relations.

Process Model

Consider a process with the transfer function

$$G(s) = \frac{k}{s^2}$$

where $k = 1$. This could for example be a simplified model of the arm servo of a compact disc player. The sampled pulse-transfer function is

$$H(q) = \frac{h^2}{2} \frac{q + 1}{(q - 1)^2}$$

Specifications

The properties of the closed-loop system are specified indirectly by requiring that the response to a command signal have the characteristic polynomial

$$s^2 + 2\zeta\omega s + \omega^2$$

The equivalent discrete-time polynomial is

$$A_m(q) = q^2 - 2qe^{-\zeta\omega h} \cos (\omega h \sqrt{1 - \zeta^2}) + e^{-2\zeta\omega h}$$

$$= q^2 + a_{m1}q + a_{m2}$$

It is also required that the observer have two poles at $s = -\alpha$ or equivalently at $z = e^{-\alpha h}$ in discrete time. The discrete-time observer polynomial is thus

$$A_o(q) = (q \quad e^{-\alpha h})^2$$

$$= q^2 + a_{o1}q + a_{o2}$$

It is also required that the controller have integral action.

Controller Design

The controller design is now straightforward. The Diophantine equation is

$$(q - 1)^2 R(q) + \frac{h^2}{2} (q + 1)S(q) = A_o(q)A_m(q)$$

where it is required that R have a zero at $q = 1$. The minimum-degree solution with no controller delay is such that both R and S are second-order polynomials, i.e.,

$$S(q) = s_0 q^2 + s_1 q + s_2$$

$$R(q) = q^2 + r_1 q + r_2 = (q + r)(q - 1)$$

Straightforward calculations give

$$s_0 = \frac{P(1) - 2P'(1) + 2P''(1)}{4h^2}$$

$$s_1 = \frac{P(1) + 4P'(1) - 4P''(1)}{4h^2}$$

$$s_2 = \frac{P(1) - P'(1) + P''(1)}{2h^2}$$

$$r_1 = \frac{P(-1)}{8}$$

$$r_2 = 1 - \frac{P(-1)}{8}$$

where $P(q) = A_o(q)A_m(q)$. Since $R(1) = 0$ the polynomial T is given by

$$T(q) = \frac{A_m(1)A_o(q)}{h^2}$$

Nominal Design

The design parameters that the user has to choose are the polynomials A_m and A_o, which specify the closed-loop poles, and the sampling period h. The closed-loop poles are parameterized in terms of ζ, ω, and α, i.e., in terms of continuous-time equivalents. The nominal parameter values are chosen as $\zeta = 0.707$, $\omega = 0.2$, $\alpha = 2$, and $h = 1$. This choice means that the observer poles are an order of magnitude faster than the dominant poles. The sampling rate is chosen so that $\omega h = 0.2$ according to the recommendation in Section 10.8. This choice does not, however, take the observer dynamics into account. With the chosen sampling period we have $\exp(-\alpha h) = 0.135$. The sampled observer poles are thus close to the origin. A simulation experiment is performed to illustrate the properties of the nominal design. The experiment is chosen to show responses to command signals, load disturbances, and measurement noise. A unit-step command signal is first applied to the process. A load disturbance in the form of a negative step with amplitude 0.05 at the plant input is then applied at time 50. Finally, a high-frequency sinusoid $n(t) = 0.01 \sin 2t$ is introduced at time 100 to show the response to high-frequency measurement noise. The results are shown in Figure 10.9. Notice that frequency folding is clearly noticeable in the control signal. The Nyquist frequency is 0.5 Hz $= \pi$ rad/s and the measurement noise has the frequency 2 rad/s. In a practical case it would thus be important to use a proper prefilter. Consequences of changing design parameters will be investigated.

Changing ω and ζ

The polynomial A_m determines the response to command signals. It also influences the response to load disturbances and measurement errors. Figure 10.10 illustrates the consequences of changing ω. The results are as we can expect. The response

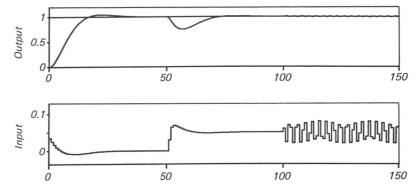

Figure 10.9 Simulation of the nominal design which has parameters ω = 0.2, ζ = 0.707, α = 2, and h = 1.

time and the error due to load disturbances decrease inversely proportional to the bandwidth. When the bandwidth is increased, the control signals also increase. The initial control signal is approximately proportional to the square of bandwidth. Saturation of the control signal thus limits the admissible bandwidth.

Different choices of ω have only moderate effect on the response to measurement noise. The fluctuations in the control signal are increased a little when the bandwidth is increased. The effects of changing damping ζ are also as can be expected. A command response without overshoot is obtained for ζ = 1.

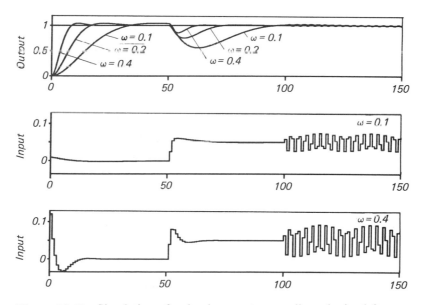

Figure 10.10 Simulation of pole-placement controllers obtained for ω = 0.10, 0.20, and 0.40. The control signal is only shown for ω = 0.10 and ω = 0.40.

Changing Observer Poles

The observer has two poles at $z = e^{-\alpha h}$ or equivalently at $s = -\alpha$ in the continuous-time representation. Figure 10.11 shows the effect of changing α from its nominal value $\alpha = 2$. The figure shows that the observer poles influence the response to load disturbances and measurement noise. The response to command signals is, however, the same for all observer polynomials as can be expected. The response to load disturbances is improved when the observer is made faster ($\alpha = 10$). The reason for this is that the disturbance is observed faster which implies that the control signal responds faster to counteract the disturbance. Compare the control signals for $\alpha = 0.5$ and $\alpha = 10$ in Figure 10.11. Also notice that the improvement in increasing α from 2 to 10 is marginal. The reason for this is that with the chosen sampling period an observer with $\alpha = 2$ is close to a deadbeat observer. The response to load disturbances is essentially detemined by the delay in observing the disturbance due to the sampling. The influence of measurement noise is decreased by making the observer slower as is also clearly seen in Figure 10.11. Selection of the observer polynomial is thus a compromise between response to load disturbances and measurement noise.

Changing the Sampling Period

The sampling period was chosen so that $\omega h = 0.2$ where ω represented the dominating (slowest) closed-loop poles. Figure 10.12 shows the response of the system when the sampling period is changed. The figure is obtained by sampling the system and calculating the control laws for different sampling periods. Figure

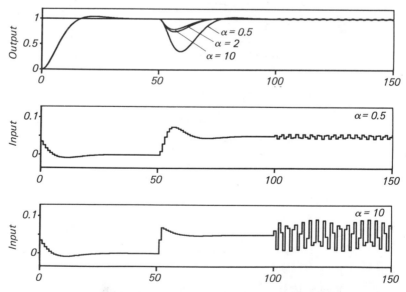

Figure 10.11   Response of the pole-placement controller for $\alpha = 0.5$, 2, and 10.

10.12 shows clearly that the sampling period has a significant influence on the response to load disturbances. The error due to load disturbances increases with increasing sampling period and decreases with decreasing sampling period. The reason is that with a sampled system there is always a delay in observing and reacting to a disturbance. This is clearly noticeable in the control signal in Figure 10.12. The disturbance is a step in the load applied shortly before the sampling instant at $t = 50$. With a sampling period $h = 2$ the control system first reacts at time $t = 52$ when the disturbance has generated a large error. With a sampling interval $h = 0.1$ the control signal reacts much quicker before a control error is built up. The result is that the overshoot in the control signal is also much smaller. The benefits in making the sampling period shorter than 0.2 s are marginal. The reason is that the observer poles are at $\alpha = -2$. With $h = 0.2$ the disturbance response is essentially determined by α. It is necessary to reduce h and increase α to further improve the response to load disturbance.

Notice that a reasonable choice of sampling period is $\alpha h \approx 0.2$. We can thus draw the important conclusion that to choose the sampling period properly it is necessary to consider all closed-loop poles, not just the corresponding poles to A_m.

Robustness

In control-system design it is important to assess the required accuracy of the mathematical model of the process used in the design. Figure 10.13 illustrates the consequences of changing the process gain for the nominal design with $k = 1$.

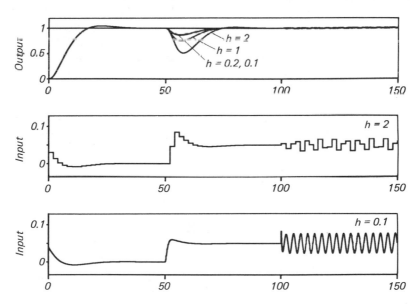

Figure 10.12 Response of the pole-placement controller for different sampling periods.

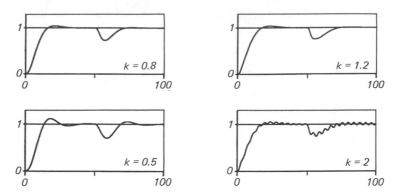

Figure 10.13 Outputs of the system with the nominal controller when the process gain is changed.

The figure shows that a gain change of 20% has little effect on the system but that an increase or decrease with a factor of 2 is not acceptable.

An interesting observation is that the sensitivity to process changes is reduced by using a shorter sampling period. This is illustrated in Figure 10.14. The simulation results in Figure 10.13, and Figure 10.14 can be well understood from the loop gain

$$H_p H_c = \frac{BS}{AR}$$

of the process. Figure 10.15 shows the Bode diagram of the loop-transfer functions $H_{lg} = H_p H_c$ for systems with sampling periods 0.2 and 1. Figure 10.15 gives a good insight into the behavior of the system. The loop-transfer function has a phase lag of $-270°C$ at low frequencies. The controllers provide a significant phase lead to obtain a stable closed-loop system. Notice that the phase curves of the two systems are almost the same for low frequencies. The phase of the system

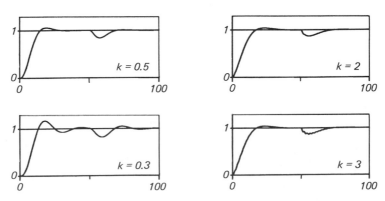

Figure 10.14 Outputs of the system with the controller obtained for $h = 0.2$ when the process gain is changed.

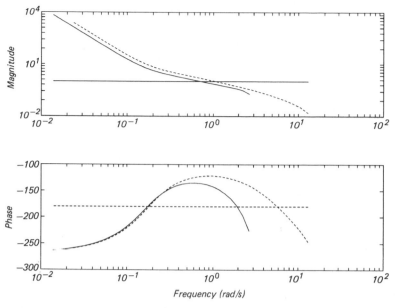

Figure 10.15 Bode diagram for the loop-transfer functions $H_p H_c$ of the systems with sampling periods $h = 0.2$ (dashed line) and $h = 1$ (full line).

with sampling period $h = 1$ does, however, decrease more rapidly after the maximum. With $h = 1$ the phase margin is $\varphi = 42°$; the amplitude margins are $a_{ml} = 0.33$, $a_{mh} = 1.9$. For $h = 0.2$ the corresponding values are $\varphi_m = 57°$, $a_{ml} = 0.25$, $a_{mh} = 4$, which explains the differences in robustness. Further insight into the robustness problems is obtained from Theorem 5.4, which says that a design based on a model with the relative error $\Delta H_p / H_p$ is stable provided that

$$\left| \frac{\Delta H_p}{H_p} \right| < \left| \frac{1 + H_{lg}}{H_{lg}} \right|$$

where H_{lg} is the loop-transfer function. The right-hand side of the above inequality is shown in Fig. 10.16. The figure indicates that with the shorter sampling period it is necessary to know the model accurately over a wider frequency range. The figure gives good insight into the robustness problem. Even if the bandwidth of the system measured from the command signal to the output is 0.2 rad/s it is necessary that the process model is reasonably accurate for frequencies up to 1 rad/s. The model precision required in the frequency range 0.1 to 1 rad/s is a little higher if the sampling period $h = 1$ is used. The precision at frequencies higher than 1 rad/s is, however, less for the system with slow sampling. If there are considerable unmodeled dynamics at frequencies higher than 2 rad/s, the design with $h = 1$ may thus be preferable. Also notice that in a properly designed system there will be antialiasing filters that influence the sensitivity. This will be discussed in the next section.

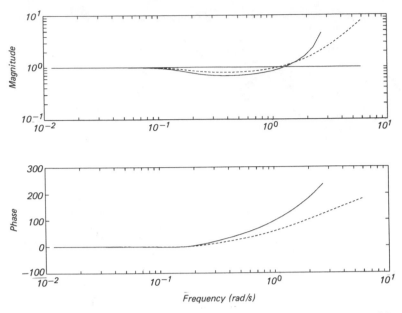

Figure 10.16 Relative model precision required for the systems with sampling periods $h = 0.2$ (dashed) and $h = 1$ (full line).

10.10 Pole-Placement Design—The Harmonic Oscillator

The discussion of pole-placement design based on polynomial methods will be continued in this section. The process considered is the harmonic oscillator. Particular emphasis is given to the influence of the antialiasing filter.

Process Model

Let the process be the harmonic oscillator with the transfer function (see Example A.3)

$$G(s) = \frac{\omega_0^2}{s^2 + \omega_0^2} \qquad \omega_0 = 1$$

The sampled pulse-transfer operator is

$$H(q) = \frac{(1 - \beta)(q + 1)}{q^2 - 2\beta q + 1} = \frac{B(q)}{A(q)} \qquad \beta = \cos(\omega_0 h)$$

Specifications

The desired response is characterized by the continuous-time characteristic equation

$$s^2 + 2\zeta\omega s + \omega^2 = 0$$

The sampled-data form of this polynomial is $A_m(q)$. Since the pulse-transfer function has a zero at -1 no zero cancellation is allowed. This implies that $B^+ = 1$ and $B_m = B^- = B$. It is specified that the regulator should have integral action. This implies that the observer polynomial at least should be of second order.

$$s^2 + 2\zeta_{obs}\omega_{obs}s + \omega_{obs}^2$$

This polynomial is also transformed to sampled-data form $A_o(q)$.

Controller Design

The Diophantine equation is

$$(q^2 - 2\beta q + 1)R(q) + (1 - \beta)(q + 1)S(q) = A_o(q)A_m(q) = P(q)$$

where

$$R(q) = (q + r)(q - 1)$$
$$S(q) = s_0 q^2 + s_1 q + s_2$$
$$P(q) = A_o(q)A_m(q) = q^4 + p_1 q^3 + p_2 q^2 + p_3 q + p_4$$

The controller parameters are given by

$$r = 1 - \frac{P(-1)}{4(1 + \beta)}$$

$$s_0 = \frac{p_1 - r + 1 + 2\beta}{1 - \beta}$$

$$s_1 = \frac{p_2 - p_1 - 2\beta + 2(1 + \beta)(r - 1)}{1 - \beta}$$

$$s_2 = \frac{p_4 + r}{1 - \beta}$$

The polynomial T is given by

$$T(q) = \frac{A_m(1)A_o(q)}{B(1)}$$

Nominal Design

The nominal parameter values are chosen as $\zeta = 0.7$, $\omega = 1.5$, $\zeta_{obs} = 0.7$, $\omega_{obs} = 3$, and $h = 0.2$. These specifications imply that significant damping is introduced and that the response speed is increased compared with the open-loop system. The choice of sampling rate implies that $\omega h = 0.3$ and $\omega_{obs} = 0.6$. Recall the rate of thumb $0.2 \leq \omega h \leq 0.6$. Figure 10.17 shows the output and input when the reference signal is a step at $t = 0$, a step disturbance at the input at $t = 15$, and discrete-time white measurement noise with standard deviation 0.01 at $t = 30$.

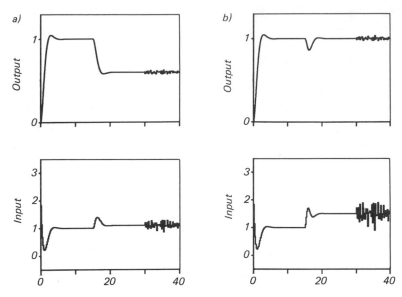

Figure 10.17 Simulation of the nominal design for the harmonic oscillator when $\omega = 1.5$, $\omega_{obs} = 3$, $\zeta = \zeta_{obs} = 0.7$ and $h = 0.2$. a. Without integrator. b. With integrator in the regulator.

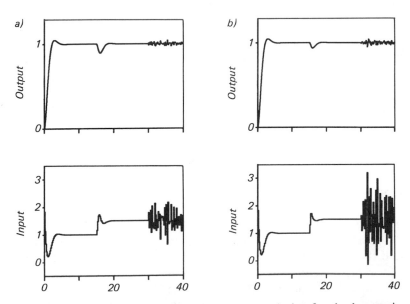

Figure 10.18 Response of the pole-placement design for the harmonic oscillator for different observer dynamics. a. $\omega_{obs} = 4$. b. $\omega_{obs} = 8$.

Changing Observer Poles

Figure 10.18 shows the response when the observer poles are changed to ω_{obs} = 4 and 8. The load disturbance is eliminated faster with a faster observer dynamics, but the noise sensitivity also increases.

Changing the Sampling Period

The sampling period in the nominal design was chosen such that

$$\omega_{obs}h = 0.6$$

which is according to the upper limit of the rule of thumb. Figure 10.19 shows the responses when the sampling period is changed, $h = 0.1$ and 1. As for the double integrator in the previous section the controller will respond faster after load disturbances when the sampling interval is decreased. A too-long sampling interval will increase the deviation after the load disturbance. Also, the aliasing effect is seen when there is measurement noise, since no antialiasing filter is used. The example shows that the rule of thumb for the choice of sampling interval gives a sensible result.

Influence of Antialiasing Filter

The influence of the measurement noise in Figure 10.17 indicates that an anti-aliasing filter should be used to reduce the effect of the disturbance. The filter will, however, also influence the closed-loop response unless the bandwidth of

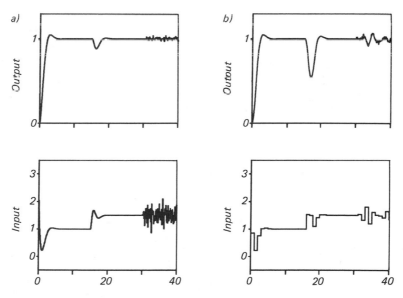

Figure 10.19 Response of the pole-placement design for the harmonic oscillator for sampling intervals a. $h = 0.1$ and b. $h = 1$.

the filter is an order of magnitude higher than the desired bandwidth of the closed-loop system. The closed-loop system will be unstable when the nominal controller is used on the system with an antialiasing filter. The filter introduces too much phase lag at the crossover frequency. The filter dynamics should thus be considered when designing the controller. The results obtained in Section 2.5 show that the Bessel filter can be well approximated by a delay. This simplifies the design and reduces the order of the controller. A sixth-order Bessel filter is approximated as a delay $\tau = 2.7/\omega_B$. The filter bandwidth is chosen $\omega_B = 2\pi$, which gives $\tau = 0.43$. To incorporate the delay it is necessary to increase the order of the regulator such that deg R = deg S = deg T = 5. Figure 10.20 shows the response with an antialiasing filter when the design is made by approximating the filter by a delay. The measurement noise, which is a discrete-time white-noise sequence with a sampling period of 0.01, starts at $t = 30$. The filter will significantly reduce the influence of the measurement noise. The filter will, however, also increase the deviation after the load disturbance. Compare with Figure 10.17b. The simulation also shows that the filter must be taken into account in the design. The filter can be disregarded provided the filter bandwidth is more than ten times the desired closed-loop bandwidth. This also implies, however, that the sampling frequency must be increased by a magnitude.

Robustness

The consequences of added unmodeled dynamics are illustrated next. It is assumed that the plant is

$$G'(s) = \frac{\omega_0^2}{s^2 + \omega_0^2} \cdot \frac{k(a_d - s)}{a_d + s}$$

i.e., that the unmodeled part is nonminimum phase. The disturbance is thus mainly a phase lag compared to the nominal system. Figure 10.21 shows the influence of unmodeled dynamics when $a_d = 15$ and 10 and when the nominal controller is used. The figure also shows the sensitivity to gain variations. Figure 10.22 shows the Bode diagram for the loop-transfer function H_{lg} when the pole-placement controller is designed for the nominal case. For comparison, the loop-transfer

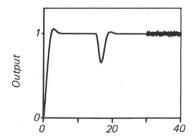
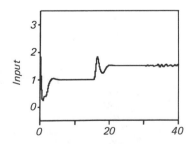

Figure 10.20 Response of pole-placement design when using an antialiasing filter. Output and input when the filter is approximated with a delay in the design.

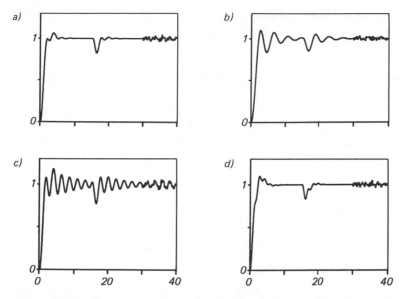

Figure 10.21 Responses when using the nominal controller when a. a_d = 15 and k = 1, b. a_d = 25 and k = 0.5, c. a_d = 10 and k = 1, and d. a_d = 25 and k = 1.5.

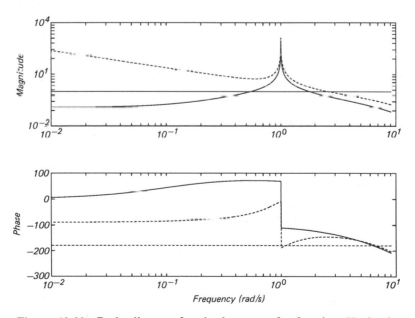

Figure 10.22 Bode diagram for the loop-transfer function H_{lg} in the nominal case when the controller is designed with (dashed line) and without (full line) integral action.

function is shown for the case when no integral action is required in the controller. The enforced integral action implies that less model precision is needed for all frequencies.

When Can Dynamics of Antialiasing Filters Be Neglected?

The effect of antialiasing filters was discussed in Section 8.2, where the phase lag introduced by the antialiasing filter was calculated. It was shown that the phase lag at the frequency ω_0 introduced by a second-order Butterworth filter is

$$\alpha \approx \frac{2.6}{\sqrt{\beta}} \frac{\omega_0}{\omega_s}$$

where ω_s is the sampling frequency and β is the attenuation of the filter at the Nyquist frequency. For a Bessel filter of sixth order the relation is

$$\alpha \approx \frac{8.6}{\beta^{1/6}} \frac{\omega_0}{\omega_s}$$

Our rules for selecting the sampling rates in digital systems require that $\omega_0 h$ is in the range of 0.2 to 0.6. With $\omega_0 h = 0.2$ the above equation implies that the phase lag of the second-order antialiasing filter is

$$\alpha = \frac{0.083}{\sqrt{\beta}}$$

With $\beta = 0.1$ we get $\alpha = 0.26$ rad or 15°. With the sixth-order Bessel filter as antialiasing filter and $\beta = 0.1$, we get $\alpha = 0.4$ rad or 23°. These calculations show that it is necessary to take the dynamics of the antialiasing filter into account for practically all the digital designs. Approximating the filter by a delay is a convenient way of doing that.

10.11 Pole-Placement Design—Robot Mechanism

A more complex design example will give further insights into the designers' choices in polynomial design. The process is the robot mechanism presented in Section 9.6. The process is of third order. In Section 9.6 it was remarked that an antialiasing filter should be used to avoid problems with frequency folding. A second-order filter of the form

$$\frac{\omega_f^2}{s^2 + 1.4\omega_f s + \omega_f^2}$$

is chosen with $\omega_f = 2$ rad/s. The filter has a gain of about 0.1 at the Nyquist frequency $\omega_N \approx 6$ rad/s.

Sampling the process. The frequencies ω_p and ω_f are of the same magnitude. It is then necessary to take the antialiasing filter into account in the design.

Sampling the process and the filter with $h = 0.5$ gives the discrete-time model

$$y(k) = \frac{B(q)}{A(q)} u(k)$$

with

$$A(q) = \underbrace{(q^2 - 0.7505q + 0.2466)}_{\text{filter}} \underbrace{(q^2 - 1.7124q + 0.9512)(q - 1)}_{\text{process}}$$

$$B(q) = 0.1420 \cdot 10^{-3}(q + 12.1314)(q + 1.3422)(q + 0.2234)(q - 0.0023)$$

The poles and zeros of the sampled system are shown in Fig. 10.23.

Design specifications. The design specifications were discussed in Section 9.6. The system is now of fifth order, because of the antialiasing filter. Three of the poles are chosen as the sampled-data correspondence to (9.56). The filter poles are quite fast compared to the desired poles given by (9.56) and therefore we keep them unchanged. This specifies the five poles of A_m.

Notch filter design. The frequency associated with the mechanical resonance $\omega_p = 1$ is close to the desired closed-loop frequency $\omega_m = 0.5$. It is then necessary to take the mechanical resonance into account when designing the control loop. One classic method for doing this is to introduce a compensating network that avoids excitation of the oscillatory process poles. The filter that accomplishes this is called a *notch filter* because its Bode diagram has a notch at the frequency of the undesired modes. This approach ensures that the oscillatory modes will not be excited by the command signals or the control action. However, it does not introduce any damping of the oscillatory modes. This means that the system will respond to excitation of the oscillatory modes in the same way as the open-loop system. The notch-filter compensator can be obtained using the polynomial design approach.

Consider the characteristic polynomial of the closed loop system. It is, of

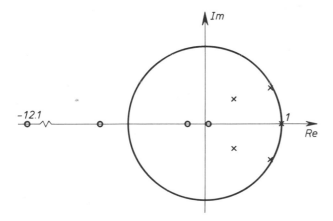

Figure 10.23 Pole-zero diagram for the process and the filter sampled with $h = 0.5$.

course, desired that this polynomial has A_m as a factor. It must also contain the polynomial A_n, which corresponds to the oscillatory modes, as a factor because the oscillatory modes remain in the closed-loop system and are not influenced by the feedback. The observer polynomial A_o must also be a factor of the closed-loop characteristic polynomial. It is further assumed that no process zeros are cancelled. The polynomial identity (10.22) will then be

$$AR + BS = A_m A_n A_o \qquad (10.54)$$

Since A_n divides A but not B, it follows that it also divides S. The regulator will thus have zeros at the oscillatory poles. The observer is assumed to have the same poles as the antialiasing filter.

From Theorem 10.2 it follows that the design requires $\deg R = 4$, $\deg S = 4$, and $\deg A_o = 2$. The identity (10.22) is of ninth order and gives the regulator polynomials R and S. The closed-loop pulse transfer function is

$$\frac{BT}{AR + BS} = t_0 \frac{B}{A_m}$$

where

$$t_0 = \frac{A_m(1)}{B(1)}$$

The polynomial T is thus

$$T = t_0 A_n A_o$$

The steady-state gain of the closed-loop system is unity with this choice of T. Notice that both S and T have A_n as a factor. The response of the closed-loop system when using the notch-design regulator is shown in Fig. 10.24. The reference signal is a step at $t = 0$, and the disturbance v is a pulse at $t = 25$ of height -10 and a duration of 0.1 s. The response of the system is according to the specifications. Compare Fig. 9.17.

There is no excitation of the weakly damped modes by the reference signal or by the control signal. However, the pulse disturbance excites these modes and causes the oscillation in the response. Note that the oscillation does not introduce any control actions because of the notch filter.

Active damping of oscillatory modes. With the notch-filter design the regulator makes no attempt to damp the oscillatory modes. A new design will now be done such that the servo performance is the same but the oscillations are also damped. Assume that the damping of the oscillatory modes should be changed from the open-loop damping $\zeta_p = 0.05$ to 0.707. Further assume that the damped frequency should be the same as before. This corresponds to the continuous-time poles

$$p_{12} = -0.707 \pm 0.707i$$

Let the corresponding discrete-time polynomial be denoted A_d. The design

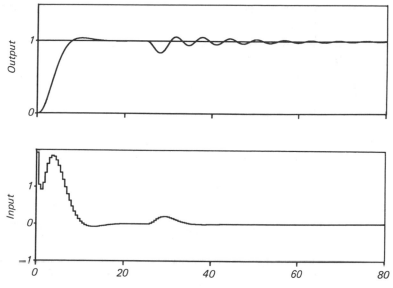

Figure 10.24 Response of the closed-loop system using the notch-design regulator.

is now made in the same way as above but with the identity

$$AR + BS = A_m A_d A_o \qquad (10.55)$$

where A_m and A_o are the same as above. The response of the closed-loop system is shown in Fig. 10.25. Compare Fig. 10.24. The servo performance is the same as before and the oscillatory modes are now damped by the regulator.

Comparison

The notch-filter design and the active damping design are two ways to solve the design problem using the polynomial approach. The design example can be elaborated further by introducing an integrator and by studying the effect of different observer polynomials.

Equation (10.51) can be used to see the difference between the two designs. The relationships between the variable x in Fig. 10.1 and the reference signal and the disturbances are given by

$$x(k) = \frac{BT}{AR + BS} u_c(k) + \frac{BR}{AR + BS} v(k) - \frac{BS}{AR + BS} e(k)$$

This shows that the notch-filter design does not damp the oscillatory modes. The polynomial A_n is a factor of the denominator of the pulse-transfer operator from the disturbance v but not from e. In the design with active damping, A_d is instead a factor in the denominator and the oscillations are getting better damped. The

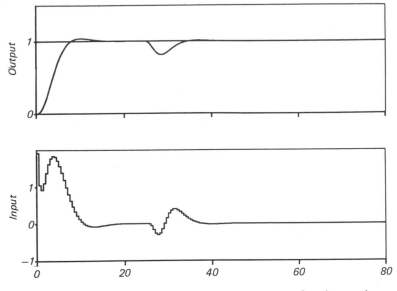

Figure 10.25 Response of the closed-loop system using the regulator designed for active damping.

damping of disturbances is also seen by looking at the return difference which is

$$1 + \frac{BS}{AR} = \frac{AR + BS}{AR}$$

From (10.54) and (10.55) it follows that the return difference is

$$\frac{A_m(z)A_n(z)A_o(z)}{A(z)R(z)}$$

for the notch design and

$$\frac{A_m(z)A_d(z)A_o(z)}{A(z)R(z)}$$

for the active damping design. The magnitudes of the return differences for $|z| = 1$ are shown in Fig. 10.26. The return difference indicates the attenuation of disturbances like measurement noise at the process output. Compare (10.51). It is seen that the active damping design gives a better damping of the oscillatory modes.

Theorem 10.3 can be used to evaluate the precision of the model needed for the design. The relative precision is given by

$$\frac{1}{|H_m|} \left| \frac{H_{ff}}{H_{fb}} \right| = \left| \frac{A_m(z)T(z)}{t_0 B_m(z)S(z)} \right|$$

This function is plotted for $|z| = 1$ in Fig. 10.27. The figure shows that the notch design is more insensitive than the active damping design to modeling errors for high frequencies.

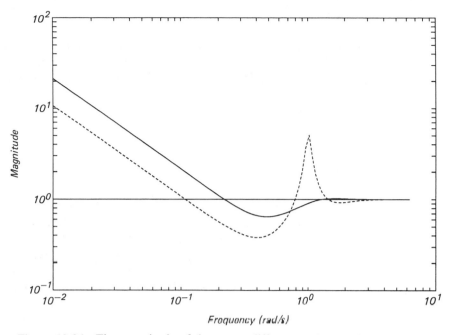

Figure 10.26 The magnitude of the return differences. Notch design (fall line) and active damping (dashed line).

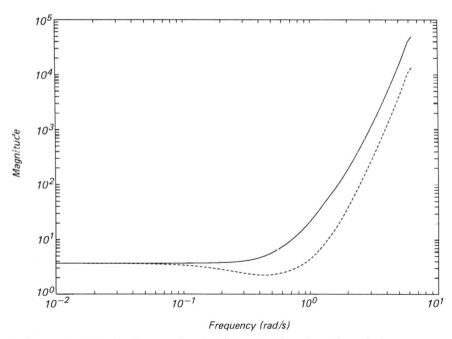

Figure 10.27 Bode diagram for $A_m T/(t_0 B_m S)$ that gives the relative model precision needed for a stable design. Notch design (fall line) and active damping (dashed line).

Summary

The example shows that the polynomial design method is a flexible and useful way to design controllers. However, some care must be taken in order to avoid numerical problems if the process is of high order (fourth order or higher). Numerical problems may arise in the solution of the polynomial identity and in the implementation of the regulator.

10.12 Conclusions

This chapter deals with a pole-placement procedure based on polynomial manipulations. The design starts with a process model, H, a desired model, H_m, and an observer polynomial, A_o. Algorithm 10.1 leads to a regulator that includes a feedback term from the measured output and a feedforward term from the command signal.

The choices of the desired model and the observer polynomial are carefully related to physical specifications.

10.13 Problems

10.1. Use Euclid's algorithm in Appendix D to determine the largest common factor of the polynomials

$$B(z) = z^3 - 2z^2 + 1.45z - 0.35$$

$$A(z) = z^4 - 2.6z^3 + 2.25z^2 - 0.8z + 0.1$$

10.2. Given the pulse-transfer operator

$$H(q) = \frac{1}{q + a}$$

and let the desired system be given by

$$H_m(q) = \frac{1 + \alpha}{q + \alpha}$$

(a) Determine a controller of the form (10.4) using Algorithm 10.1.
(b) Determine the characteristic polynomial of the closed-loop system.

10.3. Consider the system given by the pulse-transfer function

$$H(z) = \frac{z + 0.7}{z^2 - 1.8z + 0.81}$$

Use polynomial design to determine a controller such that the closed-loop system has the characteristic polynomial

$$z^2 - 1.5z + 0.7$$

Let the observer polynomial have as low order as possible and place all observer poles in the origin. Consider the following two cases:

The process zero is canceled.

The process zero is not canceled.

Simulate the two cases and discuss the differences between the two controllers. Which one should be preferred?

10.4. For the system in Problem 10.2, assume that the feedback can be made only from the error. Thus the controller has the form

$$u(k) = \frac{S}{R}[u_c(k) - y(k)]$$

(a) Determine S/R such that the desired closed-loop system is obtained.
(b) Determine the characteristic equation of the closed-loop system and compare it with Problem 10.2. Consider, for instance, the case when $|a| > 1$.

10.5. Consider the system in Problem 10.2 and assume that the closed-loop system should be able to eliminate step disturbances at the input of the process. This means that v in Fig. 10.1 is a step.
(a) Analyze what happens when the controller derived in Problem 10.2 is used and when v is a step.
(b) Redesign the controller such that the specifications will be fulfilled.

10.6. Show that (10.41) is correct.

10.7. Consider the system in Problem 10.2 and assume that $a = -0.9$ and $\alpha = -0.5$.
(a) Use straightforward calculations to determine the influence of modeling errors. Assume that the design is made for $a = -0.9$ and determine the stability of the closed-loop system if the true process has a pole in a^o.
(b) Use Theorem 10.3 to determine the influence of modeling errors. What happens when α is decreased?

10.8. Consider the system in Problem 10.2. Use (10.50) to determine the maximum value of the control signal as a function of a and α when the command signal is a step.

10.9. A polynomial design for the normalized motor is given in Example 10.6. Simulate the system and investigate the sensitivity of the design method with respect to the choice of the sampling interval. Assume that the closed-loop specifications correspond to a second-order, continuous-time system with damping $\zeta = 0.7$ and natural frequency $\omega = 1$ rad/s.

10.10. Consider the system described by

$$A_1(q)x(k) = B_1(q)u(k)$$

$$A_2(q)y(k) = B_2(q)x(k)$$

Assume that the variable to be controlled is $x(k)$, but that the measured variable is $y(k)$. Further assume that A_2 has its roots inside the unit disc.

Derive a controller of the form (10.4) such that the closed-loop system is

$$A_m(q)x(k) = B_m(q)u_c(k)$$

What are the restrictions that have to be imposed? How will uncertainties in A_2 and B_2 influence the pulse-transfer function of the closed-loop system?

10.11. Consider the two-tank system in Problem 3.10 for $h = 12$ s.

(a) Use polynomial methods to design a controller with an integrator. Assume that the desired closed-loop characteristic equation is

$$z^2 - 1.55z + 0.64 = 0$$

This corresponds to $\zeta = 0.7$ and $\omega = 0.027$ rad/s.

(b) Redesign the controller for different values of ω and study how the magnitude of the control signal varies with ω.

10.12. Consider the control of the normalized motor in Example A.2. Show that velocity feedback can be designed using pole-placement design (*Hint:* First design a feedback law with position feedback only. Show then that the control law can be rewritten as a combination of position and velocity feedback.)

10.13. Generalize the results in Problem 10.12 to a general process with several outputs.

10.14. Assume that the desired closed-loop system is given as the continuous-time model

$$G_m(s) = \frac{0.01}{s^2 + 0.14s + 0.01}$$

(a) Choose an appropriate sampling interval.

(b) Determine the corresponding discrete-time transfer operator. Sketch the singularity diagram for the continuous and the discrete time systems respectively.

10.15. Assume that the process has the pulse-transfer operator

$$H(q) = \frac{0.4q + 0.3}{q^2 - 1.6q + 0.65}$$

Use pole placement to design a controller satisfying the following specifications:

- Static gain $= 1$
- Minimal degree of the observer polynomial
- Cancellation of process zero
- No integrator
- Desired characteristic polynomial

$$A_m = q^2 - 0.7q + 0.25$$

10.16. Consider the process and specifications in the previous problem. Redo the design under the assumption that the controller has an integrator.

10.17. Consider the system

$$H(z) = \frac{z}{(z - 1)(z - 2)}$$

Determine an error-feedback controller that places both poles in the origin, i.e., use the controller

$$Ru(k) = -Sy(k) + Tu_c(k)$$

with $S = T$. Show by using the Diophantine equation that there is more than one causal controller that solves the problem. Assume that the observer poles are placed at the origin. Determine two controllers that fulfill the specifications, and determine the closed-loop zeros.

10.14 References

The polynomial approach for pole placement is treated in

KUČERA, V. (1979): *Discrete Linear Control.* Prague: Academia.

WOLOWICH, W. A. (1974): *Linear Multivariable Systems.* New York: Springer-Verlag.

PERNEBO, L. (1981): "An Algebraic Theory for the Design of Controllers for Multivariable Systems—Part I: Structure Matrices and Feedforward Design and Part II: Feedback Realizations and Feedback Design," *IEEE Trans. Autom. Control,* AC-26, 171–82 and 183–94.

The method discussed in this chapter has been used in connection with adaptive pole-placement algorithms, as in

ÅSTRÖM, K. J. and B. WITTENMARK (1980): "Self-tuning Controllers Based on Pole-Zero Placement," *Proc. IEE,* Part D, 127, 120–30.

The Dahlin-Higham algorithm was derived independently in

DAHLIN, E. B. (1968): "Designing and Tuning Digital Controllers," *Instruments & Control Systems,* 41, No. 6, 77–83.

HIGHAM, J. D. (1968): "'Single-Term' Control of First- and Second-Order Processes with Dead Time," *Control,* February, 136–40.

The model algorithmic controller is described in

RICHALET, J., A. RAULT, J. L. TESTUD, and J. PAPON (1978): "Model Predictive Heuristic Control: Applications to Industrial Processes," *Automatica,* 14, 413–28.

More about the Sylvester matrix can be found in

BARNETT, S. (1971): *Matrices in Control Theory.* New York: Van Nostrand Reinhold.

BARNETT, S. (1983): *Polynomials and Linear Control Systems.* New York: Marcel Dekker.

Solution of the Diophantine equation is discussed in

BLANKINSHIP, W. A. (1963): "A New Version of the Euclidean Algorithm," *American Mathematics Monthly,* 70, 742–45.

KUČERA, V. (1979): *Discrete Linear Control—The Polynomial Equation Approach.* New York: John Wiley.

JEŽEK, J. (1982): "New Algorithm for Minimal Solution of Linear Polynomial Equations," *Kybernetica,* 18, 505–16.

OPTIMAL DESIGN METHODS: STATE-SPACE APPROACH

11

GOAL

To Develop Optimal Design Techniques and to Solve the Prediction and Filtering Problems for Systems Described by Linear State-Space Models and Quadratic Criteria.

11.1 Introduction

In the previous two chapters the synthesis problem is solved using pole-placement techniques. The main design parameters have been the locations of the closed-loop poles, and the presentations have been limited to single-input–single-output systems. In this chapter a more general control problem is discussed. The process is still assumed to be linear, but it may be time-varying and have several inputs and outputs. Further process and measurement noise are introduced in the models. The synthesis problem is formulated to minimize a criterion, which is a quadratic function of the states and the control signals. The resulting optimal controller is linear. The problem, which is stated formally below, is called the *Linear Quadratic (LQ) control problem,* or the *Linear Quadratic Gaussian (LQG) control problem* if Gaussian stochastic disturbances are allowed in the process models. The stationary solution to the LQ-problem for time-invariant systems leads to a control law of the same structure as the state-feedback controller in Chapter 9. The LQ-controller can then also be interpreted as a pole-placement controller. The degrees of freedom of the multivariable version of the controller in Chapter 9 is resolved by the minimization of a loss function instead of specifying only the closed-loop poles and the closed-loop eigenvectors.

LQ-control is a large topic treated in many books. In this chapter, only a brief review of the main ideas and results is given. The problem is stated and some useful results are given in this section. The solution of the LQ-control problem, if all the states are available, is given in Sec. 11.2, where the properties of LQ-controllers are also discussed. If all the states are not measurable, they have to be estimated, which can be done using a dynamic system as in Sec. 9.3. For the case with Gaussian disturbances, it is possible to determine the optimal estimator, which minimizes the variance of the estimation error. This is called the Kalman filter. The estimator has the same structure as in (9.24). However, the gain matrix, K, is determined differently and is in general time-varying. Kalman filters are discussed in Sec. 11.3. The LQG-problem is solved in Sec. 11.4, where the states are estimated using a Kalman filter. The solution is based on the *separation theorem* or the *certainty equivalence principle*. This implies that the optimal control strategy can be separated into two parts: one state estimator, which gives the best estimates of the states from the observed outputs, and one linear-feedback law from the estimated states. The linear controller used is the same as the one used if there are no disturbances acting on the system. Some practical aspects are discussed in Sec. 11.5.

Problem Formulation

The design problem is specified by giving the process, the criterion, and the admissible control signals.

The process. It is assumed that the process to be controlled is described by the continuous-time model

$$dx = Ax\, dt + Bu\, dt + dv_c \tag{11.1}$$

where A and B may be time-varying matrices. The process v_c has mean value of zero and uncorrelated increments. The incremental covariance of v_c is $R_{1c}\, dt$ (compare with Sec. 6.6). The model in (11.1) can be sampled as in Sec. 6.7. Some modifications must be made because the system is allowed to be time-varying. The input $u(t)$ is constant over the sampling period, for the noise-free case the solution of (11.1) can be written as

$$x(t) = \Phi(t, kh)x(kh) + \Gamma(t, kh)u(kh) \tag{11.2}$$

where $\Phi(t, kh)$ is the fundamental matrix of (11.1) satisfying

$$\frac{d}{dt}\Phi(t, kh) = A(t)\Phi(t, kh); \qquad \Phi(kh, kh) = I \tag{11.3}$$

and

$$\Gamma(t, kh) = \int_{kh}^{t} \Phi(t, s)B(s)\, ds \tag{11.4}$$

Omitting the time arguments of the matrices, the sampled model can be written as

$$x(kh + h) = \Phi x(kh) + \Gamma u(kh) + v(kh) \tag{11.5}$$
$$y(kh) = Cx(kh) + e(kh)$$

where v and e are discrete-time Gaussian white-noise processes with zero mean valued and

$$Ev(kh)v^T(kh) = R_1$$

$$Ev(kh)e^T(kh) = R_{12}$$

$$Ee(kh)e^T(kh) = R_2$$

The expressions for the covariance matrices are given in Sec. 6.7. Further, it is assumed that the initial state $x(0)$ is Gaussian distributed with

$$Ex(0) = m_0 \quad \text{and} \quad \text{cov}(x(0)) = R_0$$

The matrices R_0, R_1, and R_2 are positive semidefinite. The covariance matrices may be time-varying. It is assumed that the model (11.5) is reachable and observable.

As discussed in Chapter 9, it is possible to include different types of disturbances and effects from the environment by augmenting the state vector of the process.

The criterion.   The purpose of the control is to minimize the loss function

$$J = E \left\{ \int_0^{Nh} [x^T(t)Q_{1c}x(t) + 2x^T(t)Q_{12c}u(t) \right.$$

$$\left. + u^T(t)Q_{2c}u(t)]\, dt + x^T(Nh)Q_{0c}x(Nh) \right\} \quad (11.6)$$

where the matrices Q_{0c}, Q_{1c} and Q_{2c} are symmetric and positive definite. The matrices in the loss function may depend on time.

Admissible control laws.   It is important to specify the data available for determining the control signal. The first assumption is that periodic sampling is used and that the control signal is constant over the sampling periods.

If C equals the unit matrix and if $e(kh) = 0$ in (11.5), then the full state vector is available. The control signal is then allowed to be a function of the state up to and including time kh. This is called *complete state information*. In most cases the state variables are not known exactly. This is called *incomplete state information*. In this case the control signal at time kh is allowed to be a function of the outputs and inputs up to and including time $kh - h$.

The problem.   The optimal control problem is now defined to be finding the admissible control signal that minimizes the loss function of (11.6) when the process is described by the model of (11.1) or the equivalent model of (11.5). The design parameters are the matrices in the loss function and the sampling period.

Sampling of the Loss Function

The loss function in (11.6) is expressed in continuous time. It is first transformed into a discrete-loss function. Integrating (11.6) over intervals of lengths h gives

$$J = E \left\{ \sum_{k=0}^{N-1} J(k) + x^T(Nh)Q_{0c}x(Nh) \right\} \qquad (11.7)$$

where

$$J(k) = \int_{kh}^{kh+h} [x^T(t)Q_{1c}x(t) + 2x^T(t)Q_{12c}u(t) + u^T(t)Q_{2c}u(t)] \, dt \qquad (11.8)$$

Using (11.2) in (11.8) and the fact that $u(t)$ is constant over the sampling period gives

$$J(k) = x^T(kh)Q_1x(kh) + 2x^T(kh)Q_{12}u(kh) + u^T(kh)Q_2u(kh)$$

where

$$Q_1 = \int_{kh}^{kh+h} \Phi^T(s, kh)Q_{1c}\Phi(s, kh) \, ds \qquad (11.9)$$

$$Q_{12} = \int_{kh}^{kh+h} \Phi^T(s, kh)[Q_{1c}\Gamma(s, kh) + Q_{12c}] \, ds \qquad (11.10)$$

$$Q_2 = \int_{kh}^{kh+h} [\Gamma^T(s, kh)Q_{1c}\Gamma(s, kh) + 2\Gamma^T(s, kh)Q_{12c} + Q_{2c}] \, ds \qquad (11.11)$$

Minimizing the loss function of (11.6) when $u(t)$ is constant over the sampling period is thus the same as minimizing the discrete-time loss function

$$J = E \left\{ \sum_{k=0}^{N-1} [x^T(kh)Q_1x(kh) + 2x^T(kh)Q_{12}u(kh) \right.$$

$$\left. + u^T(kh)Q_2u(kh)] + x^T(Nh)Q_0x(Nh) \right\} \qquad (11.12)$$

The matrices Q_1, Q_{12}, and Q_2 are given by (11.9)–(11.11), respectively, and $Q_0 = Q_{0c}$. In the following it is assumed that Q_1 is positive semidefinite and that Q_2 is positive definite. Further, it is assumed that $Q_{12} = 0$. How Q_{12} can be eliminated is shown below. Notice that the sampled loss function (11.12) will have a cross-coupling term Q_{12} even if $Q_{12c} = 0$.

When the stochastic case is considered, one additional term depending on the noise is obtained in (11.12). However, this term is independent of the control signal and can thus be disregarded when performing the minimization.

The optimal control problem has now been transformed into the discrete-time problem of minimizing the loss function (11.12) when the process is described by (11.5). To facilitate the writing in the sequel, it is assumed that the sampling period is used as time unit, i.e., $h = 1$.

Tranformation of the Loss Function

A transformation is done to simplify the writing. Introduce a new control signal

$$\tilde{u} = u + M^T x$$

where

$$M = Q_{12} Q_2^{-1}$$

The system of (11.5) will be transformed to

$$x(k + 1) = \tilde{\Phi} x(k) + \Gamma \tilde{u}(k) + v(k)$$

$$y(k) = C x(k) + e(k)$$

where

$$\tilde{\Phi} = \Phi - \Gamma M^T$$

Further, the loss function of (11.12) is transformed to

$$J = E \left\{ \sum_{k=0}^{N-1} [x^T(k) \tilde{Q}_1 x(k) + \tilde{u}^T(k) Q_2 \tilde{u}(k)] + x^T(N) Q_0 x(N) \right\} \quad (11.13)$$

where

$$\tilde{Q}_1 = Q_1 - Q_{12} Q_2^{-1} Q_{12}^T$$

For the transformed system, the cross-product between states and control is eliminated. Thus the case where $Q_{12} = 0$ can be discussed.

Mean Value of a Quadratic Form

In the following, expressions of the form

$$E x^T S x$$

will be evaluated, where x is a Gaussian random variable with mean m and covariance matrix R. We have

$$E x^T S x = E(x - m)^T S(x - m) + E m^T S x + E x^T S m - E m^T S m$$

$$= E(x - m)^T S(x - m) + m^T S m$$

Further,

$$E(x - m)^T S(x - m) = E tr(x - m)^T S(x - m) = E tr S(x - m)(x - m)^T$$

$$= tr S E(x - m)(x - m)^T = tr S R$$

Thus

$$E x^T S x = m^T S m + tr S R \quad (11.14)$$

Completing the Squares

Quadratic functions will be minimized several times in the sequel. One way to find the minimum is to use the method of completing the squares. Consider the function

$$F(u) = u^T S u + r^T u + u^T r$$

where S is a symmetric, positive, definite matrix of order $n \times n$, and u and r are n vectors. The minimum of $F(u)$ can be obtained by rewriting the function as follows:

$$
\begin{aligned}
F(u) &= u^T S u + r^T u + u^T r \\
&= u^T S u + r^T u + u^T r + r^T S^{-1} r - r^T S^{-1} r \\
&= (u + S^{-1} r)^T S (u + S^{-1} r) - r^T S^{-1} r
\end{aligned}
$$

The first term is always nonnegative; thus the minimum is obtained for

$$u = -S^{-1} r \tag{11.15}$$

and the minimum is

$$F_{\min} = -r^T S^{-1} r \tag{11.16}$$

11.2 Linear Quadratic Control

The LQ-control problem will now be solved for the case of complete state information. The solution is obtained by using dynamic programming.

The Deterministic Case

The deterministic case where $v(k) = 0$ and $e(k) = 0$ in (11.5) is first considered. The system is thus described by

$$x(k + 1) = \Phi x(k) + \Gamma u(k) \tag{11.17}$$

where $x(0)$ is given. The problem is now to determine the sequence $u(0)$, $u(1)$, ..., $u(N - 1)$ such that the loss function in (11.12) is minimized. The solution is given by the following theorem.

Theorem 11.1. Consider the system of (11.17). Allow $u(k)$ to be a function of $x(k)$, $x(k - 1)$, ... Introduce

$$
\begin{aligned}
S(k) &= \Phi^T S(k + 1)\Phi + Q_1 - L^T(k)(Q_2 + \Gamma^T S(k + 1)\Gamma)L(k) \\
&= [\Phi - \Gamma L]^T S(k + 1)\Phi + Q_1 \\
&= [\Phi - \Gamma L]^T S(k + 1)[\Phi - \Gamma L] + Q_1 + L^T Q_2 L
\end{aligned} \tag{11.18}
$$

where the matrix L is defined by

$$L(k) = (Q_2 + \Gamma^T S(k + 1)\Gamma)^{-1}\Gamma^T S(k + 1)\Phi \qquad (11.19)$$

and with end condition $S(N) = Q_0$. Assume that $S(k)$ has a solution that is positive semidefinite and that $Q_2 = \Gamma^T S(k)\Gamma$ is positive definite. Then there exists a unique, admissible, control strategy

$$u(k) = -L(k)x(k) \qquad (11.20)$$

that minimizes the loss (11.12) when $Q_{12} = 0$. The minimal value of the loss is

$$\min J = V_0 = x^T(0)S(0)x(0) \qquad (11.21)$$

Proof. To prove the theorem, dynamic programming will be used. Introduce

$$V_k = \min_{u(k)...u(N-1)} E\left\{\sum_{i=k}^{N-1} [x^T(i)Q_1x(i) + u^T(i)Q_2u(i)] + x^T(N)Q_0x(N)\right\}$$

For $k = N$ we have

$$V_N = x^T(N)S(N)x(N)$$

where

$$S(N) = Q_0$$

For $k = N - 1$,

$$V_{N-1} = \min_{u(N-1)} \{x^T(N - 1)Q_1x(N - 1) + u^T(N - 1)Q_2u(N - 1) + V_N\} \quad (11.22)$$

Using (11.5) gives

$$\begin{aligned}
V_{N-1} &= \min_{u(N-1)} \{x^T(N - 1)Q_1x(N - 1) + u^T(N - 1)Q_2u(N - 1) \\
&\quad + [\Phi x(N - 1) + \Gamma u(N - 1)]^T S(N)[\Phi x(N - 1) + \Gamma u(N - 1)]\} \\
&= \min_{u(N-1)} \{x^T(N - 1)[Q_1 + \Phi^T S(N)\Phi]x(N - 1) \\
&\quad + x^T(N - 1)\Phi^T S(N)\Gamma u(N - 1) + u^T(N - 1)\Gamma^T S(N)\Phi x(N - 1) \\
&\quad + u^T(N - 1)[\Gamma^T S(N)\Gamma + Q_2]u(N - 1)\}
\end{aligned}$$

Using (11.15) and (11.16), the control law

$$u(N - 1) = -L(N - 1)x(N - 1)$$

gives the minimum loss

$$V_{N-1} = x^T(N - 1)S(N - 1)x(N - 1)$$

where

$$S(N-1) = \Phi^T S(N)\Phi + Q_1 - L^T(N-1)(Q_2 + \Gamma^T S(N)\Gamma)L(N-1)$$

and

$$L(N-1) = (Q_2 + \Gamma^T S(N)\Gamma)^{-1}\Gamma^T S(N)\Phi$$

The same arguments give

$$V_{N-2} = \min_{u(N-2)} \{x^T(N-2)Q_1 x(N-2) + u^T(N-2)Q_2 u(N-2) + V_{N-1}\}$$

This is the same as (11.22), but with the time arguments shifted one step. The procedure can now be repeated, and V_0, which is the minimum of J, is obtained by iterating backward in time.   □

Remark 1. Notice that it is not necessary that Q_2 be positive definite; it is sufficient that $Q_2 + \Gamma^T S(k)\Gamma$ be positive definite.

Remark 2. If there is a cross-product term in the loss function, as in (11.12), then the gain matrix is changed to

$$L(k) = (Q_2 + \Gamma^T S(k+1)\Gamma)^{-1}(\Gamma^T S(k+1)\Phi + Q_{12}^T)$$

and Equation (11.18) is changed to

$$\begin{aligned}
S(k) &= \Phi^T S(k+1)\Phi + Q_1 - L^T(Q_2 + \Gamma^T S\Gamma)L \\
&= (\Phi - \Gamma L)^T S(k+1)\Phi + Q_1 - L^T Q_{12}^T \\
&= (\Phi - \Gamma L)^T S(k+1)(\Phi - \Gamma L) + Q_1 - L^T Q_{12}^T - Q_{12}L + L^T Q_2 L
\end{aligned}$$

Remark 3. The calculations needed to determine the LQ-controller can be made by hand only for very simple examples. In practice it is necessary to have access to interactive programs, which can compute the control law and simulate the systems.

The Riccati Equation

Equation (11.18) is called the discrete-time Riccati equation. The equation can be written as

$$S(k) = [\Phi - \Gamma L(k)]^T S(k+1)[\Phi - \Gamma L(k)] + Q_1 + L^T(k)Q_2 L(k) \quad (11.23)$$

The matrices $S(N) = Q_0$ and Q_1 are assumed to be symmetric and positive semidefinite. It then follows that $S(k)$ is symmetric and positive semidefinite.

It is possible to use the Riccati equation to rewrite the loss function of (11.12) when $Q_{12} = 0$, which gives the following theorem.

Theorem 11.2. Assume that the Riccati equation of (11.18) has a solution that is nonnegative definite for $0 \leq k \leq N$; then

$$x^T(N)Q_0x(N) + \sum_{k=0}^{N-1} [x^T(k)Q_1x(k) + u^T(k)Q_2u(k)] = x^T(0)S(0)x(0)$$

$$+ \sum_{k=0}^{N-1} [u(k) + L(k)x(k)]^T$$

$$\times [\Gamma^T S(k+1)\Gamma + Q_2][u(k) + L(k)x(k)] \qquad (11.24)$$

$$+ \sum_{k=0}^{N-1} \{v^T(k)S(k+1)[\Phi x(k) + \Gamma u(k)]$$

$$+ [\Phi x(k) + \Gamma u(k)]^T S(k+1)v(k)\}$$

$$+ \sum_{k=0}^{N-1} v^T(k)S(k+1)v(k)$$

where $x(k+1)$ is given by (11.5).

Proof. We have the identity

$$x^T(N)Q_0x(N) = x^T(N)S(N)x(N)$$

$$= x^T(0)S(0)x(0) + \sum_{k=0}^{N-1} [x^T(k+1)S(k+1)x(k+1) \qquad (11.25)$$

$$- x^T(k)S(k)x(k)]$$

Consider the different terms in the sum and use (11.5) and (11.18). Then

$$x^T(k+1)S(k+1)x(k+1)$$

$$= (\Phi x(k) + \Gamma u(k) + v(k))^T S(k+1)(\Phi x(k) + \Gamma u(k) + v(k)) \qquad (11.26)$$

and

$$x^T(k)S(k)x(k) = x^T(k)[\Phi^T S(k+1)\Phi + Q_1$$

$$- L^T(k)(\Gamma^T S(k+1)\Gamma + Q_2)L(k)]x(k) \qquad (11.27)$$

Introducing (11.26) and (11.27) in (11.25) gives

$$x^T(N)Q_0x(N) = x^T(0)S(0)x(0) + \sum \{[\Phi x(k) + \Gamma u(k)]^T S(k+1)v(k)$$

$$+ v^T(k)S(k+1)[\Phi x(k) + \Gamma u(k)] + v^T(k)S(k+1)v(k)\}$$

$$+ \sum \{u^T(k)(\Gamma^T S(k+1)\Gamma + Q_2)u(k)$$

$$+ u^T(k)\Gamma^T S(k+1)\Phi x(k) + x^T(k)\Phi^T S(k+1)\Gamma u(k)$$

$$+ x^T(k)L^T(k)(\Gamma^T S(k+1)\Gamma + Q_2)L(k)x(k)$$

$$- x^T(k)Q_1x(k) - u^T(k)Q_2u(k)\}$$

where the term $u^T Q_2 u$ has been added and subtracted in the last sum. Rearrangement of the terms completes the proof.

Complete State Information

Assume that $v(k) \equiv 0$ in (11.5) but that the initial state is uncertain. Theorem 11.2 gives

$$J = E \left\{ \sum_{k=0}^{N-1} [x^T(k)Q_1 x(k) + u^T(k)Q_2 u(k)] + x^T(N)Q_0 x(N) \right\}$$

$$= E\{x^T(0)S(0)x(0)\} + E \left\{ \sum_{k=0}^{N-1} [u(k) + L(k)x(k)]^T [\Gamma^T S(k+1)\Gamma + Q_2] \right.$$

$$\times [u(k) + L(k)x(k)] \Big\}$$

Since $S(k)$ is positive semidefinite, the second term is nonnegative. Further, $S(k)$ is independent of $u(k)$, and it follows that

$$J \geq E x^T(0)S(0)x(0) = m_0^T S(0) m_0 + tr S(0) R_0 \tag{11.28}$$

where (11.14) has been used. Equality is obtained for the control law of (11.20). Theorem 11.2 and (11.28) give an alternative way to prove Theorem 11.1.

Now assume that there are stochastic disturbances acting on the system and that the full state is still measurable. Using Theorem 11.2, and the fact that $v(k)$ is independent of $u(k)$ and $x(k)$, gives

$$J = E \left\{ x^T(0)S(0)x(0) + \sum_{k=0}^{N-1} v^T(k)S(k+1)v(k) \right.$$

$$+ \sum_{k=0}^{N-1} [u(k) + L(k)x(k)]^T [\Gamma^T S(k+1)\Gamma + Q_2][u(k) + L(k)x(k)] \Big\} \tag{11.29}$$

Using (11.14) gives the relationship

$$J \geq m_0^T S(0) m_0 + tr S(0) R_0 + \sum_{k=0}^{N-1} tr S(k+1) R_1 \tag{11.30}$$

Equality is obtained for the control law of (11.20), which is an admissible control law. The difference in the optimal costs of (11.28) and (11.30) is due to the disturbance $v(k)$. The control law of (11.20) thus minimizes the loss for the complete state information case.

The solution to the LQ-problem gives a time-varying controller. The feedback matrix does not depend on x and can be precomputed from $k = N$ to $k = 0$ and stored in the computer. Usually only the stationary controller—the constant controller obtained when the time horizon increases—is used. For time-invariant processes and loss functions, $S(k)$ will—under quite general assumptions—converge to a constant matrix as the time horizon increases. In general there exist several solutions. If the system of (11.17) is reachable and if (Φ, U) is an observable pair where

$$Q_1 = U^T U \tag{11.31}$$

then there exists a unique, symmetric, nonnegative definite solution to the Riccati equation. If there is a cross-product term in the loss function, $Q_{12} \neq 0$, the matrices $\tilde{\Phi}$ and $\tilde{Q}_1$ should be considered instead.

Example 11.1—LQ-control of the double integrator

Consider the double integrator (see Example A.1) and use the sampling period $h = 1$. Let the weighting matrices in (11.12) be

$$Q_1 = \begin{bmatrix} 1 & 0 \\ 0 & 0 \end{bmatrix} \quad \text{and} \quad Q_2 = [\rho]$$

The influence of the weighting can now be investigated. The stationary feedback vector has been calculated for different values of ρ. Figure 11.1 shows the states and the control signal for some values. When $\rho = 0$, which means there is penalty only on the output, then the resulting controller is the same as the deadbeat controller in Sec. 5.4. When ρ is increased, then the magnitude of the control signal is decreased.

$\square$

Properties of the LQ-Controller

The pole-placement controller in Sec. 9.2 and the stationary LQ-controller have the same structure. However, they are obtained differently, so there are some differences in their properties.

The linear state-feedback controller of (11.20) has n parameters. It is, in

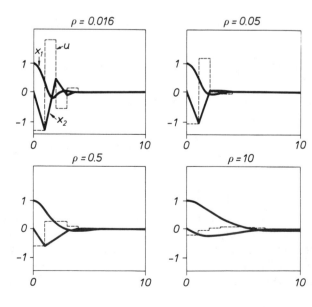

Figure 11.1 Linear quadratic control of the double integrator plant for different weightings, ρ, on the control signal. The initial value of the state (full line) is $x(0)^T = [1 \quad 0]$. The control signal is shown as a dashed line.

general, difficult to tune the parameters directly such that a good performance of the closed-loop system is obtained. Instead, the tuning procedure can be to choose the n eigenvalues of the closed-loop system and use the design procedure in Sec. 9.2. This procedure is well suited for single-input–single-output systems. It is, however, difficult to compromise between the speed of the system and the magnitude of the control signal.

The LQ-controller has several good properties. It is applicable to multivariable and time-varying systems. Also, by changing the relative magnitude between the elements in the weighting matrices, it is easy to compromise between the speed of the recovery and the magnitudes of the control signals. Further, with the assumptions on the system (reachable) and the loss function (symmetric and positive definite), the resulting LQ-controller will always give a stable closed-loop system.

Theorem 11.3—Stability of the closed-loop system. Let the system of (11.5) be time-invariant and let the loss function of (11.12) be such that Q_1 and Q_2 are positive definite and that $Q_{12} = 0$. Assume that a positive-definite, steady-state solution, $\overline{S}$, to (11.18) exists. Then the steady-state, optimal-control strategy

$$u(k) = -Lx(k) = -(Q_2 + \Gamma^T \overline{S} \Gamma)^{-1} \Gamma^T \overline{S} \Phi x(k)$$

gives an asymptotically stable closed-loop system

$$x(k + 1) = (\Phi - \Gamma L)x(k)$$

Proof. Theorem 5.6 can be used to show that the closed-loop system is asymptotically stable. It is to be shown that the function

$$V(x(k)) = x^T(k) \overline{S} x(k)$$

is a Lyapunov function. V is positive definite and

$$
\begin{aligned}
\Delta V(x(k)) &= x^T(k + 1) \overline{S} x(k + 1) - x^T(k) \overline{S} x(k) \\
&= x^T(k)[\Phi - \Gamma L]^T \overline{S}[\Phi - \Gamma L]x(k) - x^T(k) \overline{S} x(k) \\
&= -x^T(k)[Q_1 + L^T Q_2 L]x(k)
\end{aligned}
$$

where (11.23) has been used. Because $Q_1 + L^T Q_2 L$ is positive definite, ΔV is negative definite. The closed-loop system is thus asymptotically stable.

The poles of the closed-loop system can be obtained in several ways. When the design is completed, the poles are obtained from

$$\det(\lambda I - \Phi + \Gamma L) = 0$$

It is possible to show that the poles are the n stable eigenvalues of the generalized eigenvalue problem

$$\det\left\{ \begin{bmatrix} I & 0 \\ Q_1 & \Phi^T \end{bmatrix} \lambda - \begin{bmatrix} \Phi & -\Gamma Q_2^{-1} \Gamma^T \\ 0 & I \end{bmatrix} \right\} = 0 \qquad (11.32)$$

Equation (11.32) is called the *Euler equation* of the LQ-problem.

Theorem 11.4 is given without proof for the single-input–single-output (SISO) case. A proof is given in Sec. 12.5.

Theorem 11.4—The closed-loop poles of an SISO system. Let the input and the output be scalar and assume that the steady-state, optimal feedback is used for a time-invariant system. Further assume that only the output and the control signal are penalized in the loss function; i.e., $Q_1 = C^T C$ and $Q_2 = \rho$. The poles of the closed-loop system are the n roots within the unit circle of the $2n$th-order equation

$$\rho + H(z^{-1})H(z) = 0 \qquad (11.33)$$

where

$$H(z) = C(zI - \Phi)^{-1}\Gamma$$

Example 11.2—LQ-control of the double integrator

To illustrate the dependence of the weighting matrices on the closed-loop poles, reconsider Example 11.1. Figure 11.2 shows the poles of the closed-loop system for different values of ρ. For $\rho = 0$ the root locus starts at $z = -1$ and $z = 0$. As ρ increases the roots move towards the poles of $H(z)$, $z = 1$. □

Theorem 11.3 shows that the LQ-controller gives a stable closed-loop system, i.e., all the poles of the closed-loop system are within the unit circle. It is also possible to get the poles inside a circle with a radius less than 1. This is done by introducing the transformation

$$\begin{cases} \Phi \to \Phi/r \\ \qquad\qquad r < 1 \\ \Gamma \to \Gamma/r \end{cases}$$

and then solving the linear quadratic problem for the system

$$x(k + 1) = \frac{1}{r}\Phi x(k) + \frac{1}{r}\Gamma u(k)$$

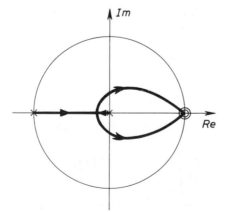

Figure 11.2 Closed-loop poles given by (11.33) when the double integrator is controlled with the optimal controller for ρ varying from 0 to ∞.

In the input-output case this implies that we make the substitution

$$z \rightarrow zr$$

This is further discussed in Section 12.6.

Gain Margin of the LQ-Controller

Theorem 11.3 shows that the closed-loop system is stable when the LQ-controller is used. It is also possible to determine the gain margin of the closed-loop system. Consider the system of (11.5) with $v(k) = e(k) = 0$. The pulse-transfer function of the open-loop system is

$$H(z) = C(zI - \Phi)^{-1}\Gamma$$

Assume that only inputs and outputs are penalized in the loss function of (11.12), i.e.,

$$Q_1 = C^T C$$

and that $Q_{12} = 0$. Let the system be controlled by the steady-state LQ state-feedback controller. The controller is then defined by the equations

$$
\begin{aligned}
S &= \Phi^T S \Phi + Q_1 - L^T R L \\
L &= R^{-1} \Gamma^T S \Phi \\
R &= \Gamma^T S \Gamma + Q_2
\end{aligned}
\tag{11.34}
$$

The algebraic Riccati equation, (11.34), can be written

$$Q_1 = (z^{-1}I - \Phi)^T S(zI - \Phi) + (z^{-1}I - \Phi)^T S\Phi + \Phi^T S(zI - \Phi) + L^T R L$$

The Riccati equation can now be used to rewrite an equation that corresponds to (11.33). This gives an expression for the closed-loop poles.

$$
\begin{aligned}
Q_2 + H^T(z^{-1})H(z) &= Q_2 + \Gamma^T(z^{-1}I - \Phi)^{-T}C^T C(zI - \Phi)^{-1}\Gamma \\
&= Q_2 + \Gamma^T \{ S + S\Phi(zI - \Phi)^{-1} \\
&\quad + (z^{-1}I - \Phi)^{-T}\Phi^T S \\
&\quad + (z^{-1}I - \Phi)^{-T}L^T R L(zI - \Phi)^{-1}\}\Gamma \\
&= R + RL(zI - \Phi)^{-1}\Gamma + \Gamma^T(z^{-1}I - \Phi)^{-T}L^T R \\
&\quad + \Gamma^T(z^{-1}I - \Phi)^{-T}L^T R L(zI - \Phi)^{-1}\Gamma \\
&= [I + L(z^{-1}I - \Phi)^{-1}\Gamma]^T R[I + L(zI - \Phi)^{-1}\Gamma] \\
&= (I + H_1(z^{-1}))^T R(I + H_1(z))
\end{aligned}
\tag{11.35}
$$

where

$$H_1(z) = L(zI - \Phi)^{-1}\Gamma$$

Equation (11.35) gives a spectral factorization of

$$Q_2 + H^T(z^{-1})H(z)$$

Consider the SISO case. Then

$$H(z) = C(zI - \Phi)^{-1}\Gamma = \frac{B(z)}{A(z)}$$

and the closed-loop system is defined by

$$H_2(z) = C[zI - (\Phi - \Gamma L)]^{-1}\Gamma = \frac{B(z)}{P(z)}$$

Further, the return difference of the system with the LQ-controller is

$$1 + L(zI - \Phi)^{-1}\Gamma = \frac{P(z)}{A(z)}$$

Hence

$$H_1(z) = \frac{P(z) - A(z)}{A(z)}$$

Now assume that the controller in (11.20) is replaced by

$$u(k) = -\beta L x(k) \tag{11.36}$$

where β is a positive scalar. The return difference when (11.36) is used is

$$1 + \beta H_1(z)$$

Thus the stability of the closed-loop sytem when (11.36) is used is determined from

$$A(z) + \beta(P(z) - A(z)) = 0 \tag{11.37}$$

The gain margin can now be determined from (11.37) by using root locus or by plotting the Nyquist curve for $(P - A)/A$. Because A and P are monic and $\deg A = \deg P$, it follows that $\deg(P - A) \leq n - 1$. This implies that the root locus of (11.37) with respect to β goes to infinity along at least one asymptote. Hence the discrete-time LQ controller has a finite gain margin, as opposed to the continuous-time LQ-controller, which has infinite gain margin.

In the scalar case, (11.35) can be written as

$$\rho A(z^{-1})A(z) + B(z^{-1})B(z) = rP(z^{-1})P(z) \tag{11.38}$$

where $r = \Gamma^T S \Gamma + \rho$. Using (11.35) gives the estimate for $|z| = 1$

$$\left| 1 + \frac{P(z) - A(z)}{A(z)} \right|^2 = |1 + H_1(z)|^2 \geq \rho/r$$

Thus

$$|1 + \beta H_1(z)| = \beta \left| \frac{1}{\beta} - 1 + 1 + H_1(z) \right| \geq \beta \left| \left| \frac{1}{\beta} - 1 \right| - |1 + H_1(z)| \right| > 0$$

The last equality holds if

$$\frac{1}{1 + \sqrt{\rho/r}} < \beta < \frac{1}{1 - \sqrt{\rho/r}} \tag{11.39}$$

Equation (11.39) gives an estimate of the values of β for which (11.37) is stable.

How to Find the Weighting Matrices

When using optimization theory, the loss function should ideally come from physical arguments. In such cases the LQG-control theory may be viewed as an approximation when the state equations are obtained from linearization of equations of motion and the loss function is obtained from a nonlinear loss function. Unfortunately, such formulations can be obtained only in a few cases. One example is Example 11.3.

Example 11.3—Ship steering

The linearized dynamics that describe the steering of ships can be described by the equation

$$\frac{d}{dt}\begin{bmatrix} v \\ r \\ \Psi \end{bmatrix} = \begin{bmatrix} a_{11} & a_{12} & 0 \\ a_{21} & a_{22} & 0 \\ 0 & 1 & 0 \end{bmatrix}\begin{bmatrix} v \\ r \\ \Psi \end{bmatrix} + \begin{bmatrix} b_1 \\ b_2 \\ 0 \end{bmatrix}\delta \tag{11.40}$$

where δ is rudder angle, Ψ is the heading angle, r is the turning rate, and v the sway velocity. The relative increase in the drag due to steering may be approximated by the expression

$$\frac{\Delta R}{R} = \frac{\alpha}{T}\int_0^T [vr + \rho\delta^2]\, dt \tag{11.41}$$

The first term represents the Coriolis force due to coupling of sway velocity and turning rate. The second term represents the drag induced by the rudder deflections. □

In many cases it is difficult to find natural quadratic loss functions. LQ-control theory has found considerable use even when this cannot be done. In such cases the control designer chooses a loss function. The feedback law is obtained directly by solving the Riccati equation. The closed-loop system obtained is then analyzed with respect to transient response, frequency response, robustness, and so on. The elements of the loss function are modified until the desired result is obtained. Such a procedure may seem like a strange use of optimization theory. The fact that other methods, such as direct search over the feedback gain or pole placement, are not used instead might be questioned. It has been found empirically that LQ-theory is quite easy to use in this way. The search will automatically guarantee stable closed-loop systems with reasonable margins. It is often fairly easy to see how the weighting matrices should be chosen to influence the properties of the closed-loop system. Variables z_i, which correspond to significant physical variables, are chosen first. The loss function is then chosen as a weighted sum of z_i. Large weights correspond to small responses. The responses of the closed-loop system to typical disturbances are then evaluated. A particular difficulty is to find the relative weights between state variables and control variables, which can be done by trial and error. Sometimes the specifications are given in terms of the maximum allowed deviations in the states and the control signals for a given disturbance. One rule of thumb to decide the weights in (11.6) is to choose the diagonal elements as the inverse value of the square of the allowed deviations. Another way is to consider only penalties on the state variables and constraints on the control deviations. If the constraints are quadratic, a method using a Lagrange multiplier gives a criterium such as (11.12).

11.3 Prediction and Filtering Theory

When using the LQ-controller, the full-state vector must be measurable. The problem of estimating the states of (11.5) from measurements of the output is discussed in this section. An estimator of the same structure as in Sec. 9.3 is postulated, but the gain vector is now determined differently. The problem is solved as a parametric optimization problem, where the variance of the estimation error is minimized.

Prediction, Filtering, and Smoothing

Different estimators for the states in (11.5) can be derived depending on the available measurements. Assume that the data

$$Y_k = \{y(i), u(i) \mid i \leq k\}$$

are known. Using Y_k we want to estimate $x(k + m)$. We have three cases:

Smoothing	$(m < 0)$.
Filtering	$(m = 0)$.
Prediction	$(m > 0)$.

Figure 11.3 illustrates the different cases. In this section the prediction and filtering problems are discussed. The resulting dynamic system is called a filter regardless of which of the problems is solved.

The Kalman Filter

In the one-step-ahead prediction problem, let the process be described by (11.5) with $h = 1$. For simplicity, it is first assumed that $R_{12} = 0$. Let the estimator have the form

$$\hat{x}(k + 1 \mid k) = \Phi\hat{x}(k \mid k - 1) + \Gamma u(k) + K(k)[y(k) - C\hat{x}(k \mid k - 1)] \quad (11.42)$$

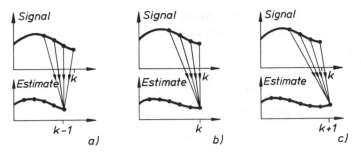

Figure 11.3 Smoothing, filtering, and prediction.

The reconstruction error $\tilde{x} = x - \hat{x}$ is governed by

$$\tilde{x}(k + 1) = \Phi\tilde{x}(k) + v(k) - K(k)[y(k) - C\hat{x}(k \mid k - 1)] \qquad (11.43)$$
$$= (\Phi - K(k)C)\tilde{x}(k) + v(k) - K(k)e(k)$$

In Sec. 9.3 K is used to give the system of (11.43) desired eigenvalues. The problem is approached differently here: The properties of the noise are taken into account and the criterion is to minimize the variance of the estimation error, which is denoted by $P(k)$.

$$P(k) = E(\tilde{x}(k) - E\tilde{x}(k))(\tilde{x}(k) - E\tilde{x}(k))^T$$

The mean value of $\tilde{x}$ is obtained from (11.43)

$$E\tilde{x}(k + 1) = (\Phi - K(k)C)E\tilde{x}(k)$$

Because $Ex(0) = m_0$, the mean value of the reconstruction error is zero for all times $k \geq 0$ independent of K if $E\hat{x}(0) = m_0$. Equation (11.43) now gives

$$P(k + 1) - E\tilde{x}(k + 1)\tilde{x}(k + 1)^T \qquad (11.44)$$
$$= (\Phi - K(k)C)P(k)(\Phi - K(k)C)^T + R_1 + K(k)R_2K(k)^T$$

since $\tilde{x}(k)$, $v(k)$, and $e(k)$ are independent. Further, $P(0) = R_0$. From (11.44) it follows that if $P(k)$ is positive semidefinite, then $P(k + 1)$ is also positive semidefinite. It is assumed that the criterion is to minimize the scalar $\alpha^T P(k + 1)\alpha$, where α is an arbitrary vector. It is also assumed that the optimal-gain vector, K, has been used up to time $k - 1$. From (11.44),

$$\alpha^T P(k + 1)\alpha = \alpha^T[\Phi P(k)\Phi^T + R_1 - K(k)CP(k)\Phi^T \qquad (11.45)$$
$$- \Phi P(k)C^T K^T(k) + K(k)(R_2 + CP(k)C^t)K^t(k)\}\alpha$$

The gain $K(k)$ can be determined from (11.45) by completing the squares.

$$\alpha^T P(k + 1)\alpha$$
$$= \alpha^T\{\Phi P(k)\Phi^T + R_1 - \Phi P(k)C^T(R_2 + CP(k)C^T)^{-1}CP(k)\Phi^T\}\alpha \qquad (11.46)$$
$$+ \alpha^T\{[K(k) - \Phi P(k)C^T(R_2 + CP(k)C^T)^{-1}][R_2 + CP(k)C^T]$$
$$\times [K(k) - \Phi P(k)C^T(R_2 + CP(k)C^T)^{-1}]^T\}\alpha$$

The criterion in (11.46) has two terms. The first part is independent of K and the second part is nonnegative because the matrix $R_2 + CPC^T$ is positive definite. The minimum is thus obtained if K is chosen such that the second part of (11.46) is zero. Then

$$K(k) = \Phi P(k)C^T(R_2 + CP(k)C^T)^{-1} \tag{11.47}$$

$$
\begin{aligned}
P(k + 1) &= \Phi P(k)\Phi^T + R_1 \\
&\quad - \Phi P(k)C^T(R_2 + CP(k)C^T)^{-1}CP(k)\Phi^T
\end{aligned}
\tag{11.48}
$$

Notice that the gain K is independent of α and that (11.44) also holds for the optimal $K(k)$. The reconstruction defined by (11.42), (11.47), and (11.48) is called the *Kalman filter*. This is summarized in the following theorem.

Theorem 11.5—The Kalman filter. Consider the process of (11.5). The reconstruction of the states using the model in (11.42) is optimal in the sense that the variance of the reconstruction error is minimized if the matrix $R_2 + CP(k)C^T$ is positive definite and if the gain matrix is chosen according to (11.47) and (11.48). The variance of the reconstructing error is given by (11.48).

Remark 1. The reconstruction problem has been solved as a parametric optimization problem by assuming the structure in (11.42) of the estimator. It is in fact true that the structure is optimal for Gaussian disturbances. The criterion can also be interpreted as a minimization of the estimation error of a linear combination of the states, $\alpha^T x(k)$.

Remark 2. Better than the traditional notation for the variance $P(k)$ is $P(k \mid k - 1)$. The latter notation indicates that measurements up to and including time $k - 1$ are used. The different terms in the variance equation of (11.48) can be interpreted in the following way: The term $\Phi P \Phi^T$ shows how the variance is changed due to the system dynamics, and R_1 represents the increase in the variance due to the noise v [compare with (6.20)]. The last term shows how the variance is decreased due to the information obtained through the measurements. Notice that $P(k)$ does not depend on the observations. Thus the gain can be precomputed in forward time and stored in the computer.

Remark 3. The Kalman filter can also be interpreted as the conditional mean of the state at time $k + 1$ given Y_k; i.e.,

$$\hat{x}(k + 1 \mid k) = E\{x(k + 1) \mid Y_k\}$$

$$P(k + 1) = E\{[x(k + 1) - \hat{x}(k + 1 \mid k)][x(k + 1) - \hat{x}(k + 1 \mid k)]^T \mid Y_k\}$$

Remark 4. If $R_{12} \neq 0$ then the gain and the variance equations of (11.47) and (11.48) become

$$K(k) = (\Phi P(k)C^T + R_{12})(CP(k)C^T + R_2)^{-1}$$

$$P(k + 1) = \Phi P(k)\Phi^T + R_1 - K(k)(CP(k)C^T + R_2)K^T(k)$$

The other equations remain unchanged.

Remark 5. The predictor in (11.42) has the property that the state at time k is reconstructed from $y(k - 1)$, $y(k - 2)$, ... It is also possible to derive the filter, which also uses $y(k)$, to estimate $x(k)$. The filter problem is solved by ($R_{12} \neq 0$).

$$\hat{x}(k|k) = \hat{x}(k|k - 1)$$
$$+ P(k|k - 1)C^T(CP(k|k - 1)C^T + R_2)^{-1}(y(k) - C\hat{x}(k|k - 1))$$

$$\hat{v}(k|k) = R_{12}(CP(k|k - 1)C^T + R_2)^{-1}(y(k) - C\hat{x}(k|k - 1)) \qquad (11.49)$$

$$\hat{x}(k + 1|k) = \Phi\hat{x}(k|k) + \Gamma u(k) + \hat{v}(k|k)$$
$$= \Phi\hat{x}(k|k - 1) + \Gamma u(k) + K(k)(y(k) - C\hat{x}(k|k - 1))$$

where

$$K(k) = (\Phi P(k|k - 1)C^T + R_{12})(CP(k|k - 1)C^T + R_2)^{-1}$$

$$P(k + 1|k) = \Phi P(k|k - 1)\Phi^T + R_1$$
$$- K(k)(CP(k|k - 1)C^T + R_2)K^T(k) \qquad (11.50)$$

$$P(k|k) = P(k|k - 1)$$
$$- P(k|k - 1)C^T(CP(k|k - 1)C^T + R_2)^{-1}CP(k|k - 1)$$

$$P(0|-1) = R_0$$

The notation $P(k|k - 1)$ is used here instead of $P(k)$ to specify the available data; $P(k|k)$ can be interpreted as the variance of the estimation error at time k given Y_k.

Example 11.4

Consider the scalar system

$$x(k + 1) = x(k)$$
$$y(k) = x(k) + e(k)$$

where e has standard deviation σ and $x(0)$ has mean value -2 and variance 0.5. The state is constant and has to be reconstructed from noisy measurements. The Kalman filter is given by

$$\hat{x}(k + 1 | k) = \hat{x}(k | k - 1) + K(k)(y(k) - \hat{x}(k | k - 1)) \qquad (11.51)$$

$$K(k) = \frac{P(k)}{\sigma^2 + P(k)}$$

$$P(k + 1) = \frac{\sigma^2 P(k)}{\sigma^2 + P(k)} \qquad (11.52)$$

The variance and the gain are decreasing with time. Figure 11.4 shows realizations of the estimation error when the Kalman filter is used and when (11.51) is used with constant gain. A large fixed gain gives a rapid initial decrease in the error, while the

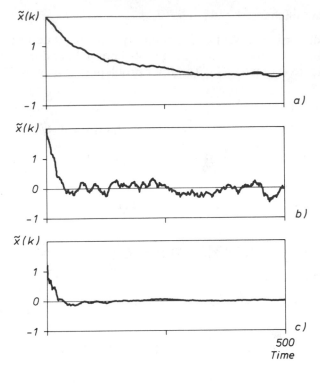

Figure 11.4 Estimation error for the system in Example 11.4 when $\sigma = 1$ and when using:

a. $K = 0.01$,

b. $K = 0.05$,

c. the optimal gain of (11.52).

steady-state variance is large. A small fixed gain gives a slow decrease in the error, but a better performance in steady state. ☐

Example 11.5

Consider the first-order system

$$y(k) + ay(k - 1) = e(k) + ce(k - 1) \qquad (11.53)$$

where e has standard deviation σ. Further assume that $|c| < 1$. A state-space representation of (11.53) is given by

$$x(k + 1) = -ax(k) + e(k)$$

$$y(k) = (c - a)x(k) + e(k)$$

It is easy to verify that the Kalman filter in steady state is characterized by $P = 0$ and $K = 1$. The one-step-ahead predictor of x is given by

$$\hat{x}(k + 1 \mid k) = -a\hat{x}(k \mid k - 1) + y(k) - (c - a)\hat{x}(k \mid k - 1)$$

$$= -c\hat{x}(k \mid k - 1) + y(k)$$

Further, in steady state the one-step-ahead prediction of the output is given by

$$\hat{y}(k + 1 \mid k) = (c - a)\hat{x}(k + 1 \mid k)$$

$$= \frac{c - a}{1 + cq^{-1}} y(k) \qquad ☐$$

Frequency-Domain Properties of Kalman Filters

Modeling is very important when design problems are solved using optimization techniques because the optimal regulator, or the optimal filter, is just a transformation of the model. It is thus useful to understand the properties of this transformation. In this section some insight into the design of Kalman filters is provided by analyzing the frequency-domain characteristics of a stationary Kalman filter.

Consider the problem of estimating the state of the system

$$x(k + 1) = \Phi_1 x(k) + v(k)$$

based on noisy observations

$$y(k) = C_1 x(k) + n(k)$$

where the noise n is given by

$$n(k) = C_2 z(k) + e(k)$$

$$z(k + 1) = \Phi_2 z(k) + w(k)$$

In these models, $\{v(k)\}$, $\{e(k)\}$, and $\{w(k)\}$ are sequences of uncorrelated random variables. The steady-state Kalman filter for one-step prediction of x is given by

$$\begin{bmatrix} \hat{x}(k + 1) \\ \hat{z}(k + 1) \end{bmatrix} = \begin{bmatrix} \Phi_1 & 0 \\ 0 & \Phi_2 \end{bmatrix} \begin{bmatrix} \hat{x}(k) \\ \hat{z}(k) \end{bmatrix} + \begin{bmatrix} K_1 \\ K_2 \end{bmatrix} [y(k) - C_1 \hat{x}(k) - C_2 \hat{z}(k)]$$

or

$$\begin{bmatrix} \hat{x}(k + 1) \\ \hat{z}(k + 1) \end{bmatrix} = \begin{bmatrix} \Phi_1 - K_1 C_1 & -K_1 C_2 \\ -K_2 C_1 & \Phi_2 - K_2 C_2 \end{bmatrix} \begin{bmatrix} \hat{x}(k) \\ \hat{z}(k) \end{bmatrix} + \begin{bmatrix} K_1 \\ K_2 \end{bmatrix} y(k) \quad (11.54)$$

The Kalman filter is thus characterized by the pulse-transfer function

$$H(z) = [I \quad 0] \begin{bmatrix} zI - \Phi_1 + K_1 C_1 & K_1 C_2 \\ K_2 C_1 & zI - \Phi_2 + K_2 C_2 \end{bmatrix}^{-1} \begin{bmatrix} K_1 \\ K_2 \end{bmatrix} \quad (11.55)$$

A frequency-response plot of the transfer function shows how the filter attenuates different frequencies. It is very useful to determine the frequency responses of the filter when designing Kalman filters. The properties of the frequency response will, in general, depend on the model in a complicated way. There are, however, some general properties that may be understood without detailed calculations.

Lemma 11.1   The transmission zeros of the pulse-transfer function, (11.55), of the stationary Kalman filter are given by

$$\det[zI - \Phi_2] = 0$$

Proof. A transmission zero is a complex number z such that an input signal of the form $z^k y_0$ gives zero output. For the system of (11.54) this implies that there exist y_o and $\hat{z}(k) = \hat{z}_0 z^k$ where $-\infty < k < \infty$ and $y_0 \neq 0$ such that

$$K_1 C_2 \hat{z}_0 - K_1 y_0 = 0$$

$$(zI - \Phi_2 + K_2 C_2)\hat{z}_0 - K_2 y_0 = 0$$

or

$$\begin{bmatrix} zI - \Phi_2 + K_2 C_2 & -K_2 \\ K_1 C_2 & -K_1 \end{bmatrix} \begin{bmatrix} \hat{z}_0 \\ y_0 \end{bmatrix} = \begin{bmatrix} zI - \Phi_2 & -K_2 \\ 0 & -K_1 \end{bmatrix} \begin{bmatrix} I & 0 \\ -C_2 & I \end{bmatrix} \begin{bmatrix} \hat{z}_0 \\ y_0 \end{bmatrix} = 0$$

There exist a nonzero solution to this equation only for those z that are eigenvalues of the matrix Φ_2.

Thus the Kalman filter will have zeros at the poles of the noise model. To obtain a Kalman filter that blocks certain frequencies (a notch filter) is just a matter of choosing a noise model with poles at those frequencies. The attenuation of certain frequencies by the Kalman filter is enhanced if the energy of the noise is increased at those frequencies in the noise model.

11.4 Linear Quadratic Gaussian Control

In the LQG-control problem, it is assumed that the system is governed by (11.5) and that the loss function is given by (11.12) with $Q_{12} = 0$. The admissible controls are assumed to be such that $u(k)$ is a function of Y_{k-1}. This means that there is a computational delay of one sampling period.

Theorem 11.2 and (11.29) still hold for the case of incomplete state information. Since (11.20) is not an admissible control strategy, the third term in (11.29) cannot be made equal to zero. The solution is given by the following theorem, which is given without proof.

Theorem 11.6—The separation theorem. Consider the system in (11.5). Let the admissible control strategies be such that $u(k)$ is a function of Y_{k-1}. Assume that (11.18) with initial condition $S(N) = Q_0$ has a solution $S(k)$ that is nonnegative definite and that $\Gamma^T S(k)\Gamma + Q_2$ is positive definite. Then there exists a unique admissible control strategy

$$u(k) = -L(k)\hat{x}(k \mid k - 1) \tag{11.56}$$

that minimizes the expected loss (11.12) for $Q_{12} = 0$. The minimum value of the loss function is given by

$$J = m_0^T S(0) m_0 + trS(0) R_0 + \sum_{k=0}^{N-1} trS(k + 1) R_1$$
$$+ \sum_{k=0}^{N-1} trP(k) L^T(k)[\Gamma^T S(k + 1)\Gamma + Q_2]L(k) \tag{11.57}$$

Remark 1. The difference in the minimal losses given by (11.30) and (11.57) is due to the estimation of the state variables.

Remark 2. It is possible to modify Theorem 11.6 to other admissible control strategies—for instance, the case when $u(k)$ is allowed to be a function of Y_{k-1} and $y(k)$. The stationary control law is given by

$$u(k) = -L\hat{x}(k \mid k) - L_v\hat{v}(k \mid k)$$

$$= -L\hat{x}(k \mid k - 1) - L_vK(y(k) - C\hat{x}(k \mid k - 1)) \qquad (11.58)$$

$$= -L_v(\Phi - KC)\hat{x}(k \mid k - 1) - L_vKy(k)$$

where $\hat{v}(k \mid k)$ is given by (11.49) and

$$L = (\Gamma^TS\Gamma + Q_2)^{-1}\Gamma^TS\Phi = L_v\Phi$$

$$S = \Phi^TS\Phi + Q_1 - \Phi^TS\Gamma(\Gamma^TS\Gamma + Q_2)^{-1}\Gamma^TS\Phi$$

$$L_v = (\Gamma^TS\Gamma + Q_2)^{-1}\Gamma^TS$$

One consequence of the separation theorem is that the synthesis problem can be split into two parts, which can be solved separately. First, the deterministic control problem is solved, giving $L(k)$. Second, the state is estimated using the Kalman filter. A block diagram of the system with the optimal control law is shown in Fig. 11.5.

Duality

The solutions to the LQ-control problem and the state estimation problem are very similar. It can be shown that the state estimation problem is equivalent to an LQ-problem. The equivalence is illustrated by the following table, which shows the substitutions required to convert the optimal control problem to a state estimation problem.

Optimal control problem	State estimation problem
k	$N - k$
Φ	Φ^T
Γ	C^T
Q_0	R_0
Q_1	R_1
Q_{12}	R_{12}
S	P
L	K^T

Properties of the Closed-Loop System

The closed-loop system with LQG-control is described by

$$x(k + 1) = \Phi x(k) + \Gamma u(k) + v(k)$$

$$y(k) = Cx(k) + e(k)$$

$$u(k) = -L\hat{x}(k \mid k - 1)$$

$$\hat{x}(k + 1 \mid k) = \Phi\hat{x}(k \mid k - 1) + \Gamma u(k) + K[y(k) - C\hat{x}(k \mid k - 1)]$$

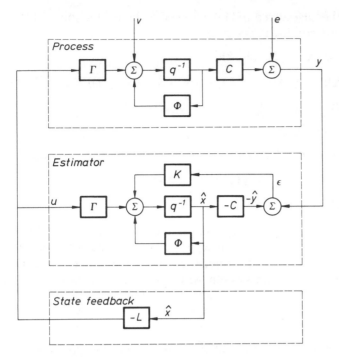

Figure 11.5 Separation theorem.

By introducing x and $\bar{x} = x - \hat{x}$, the equations can be written as

$$\begin{bmatrix} x(k+1) \\ \bar{x}(k+1) \end{bmatrix} = \begin{bmatrix} \Phi - \Gamma L & \Gamma L \\ 0 & \Phi - KC \end{bmatrix} \begin{bmatrix} x(k) \\ \bar{x}(k) \end{bmatrix} + \begin{bmatrix} I \\ I \end{bmatrix} v(k) + \begin{bmatrix} 0 \\ -K \end{bmatrix} e(k)$$

The dynamics of the closed-loop system are determined by $\Phi - \Gamma L$ and $\Phi - KC$—i.e., the dynamics of the corresponding deterministic LQ-control problem and the dynamics of the optimal filter (compare with Sec. 9.4).

For the case when $u(k)$ may be a function of Y_k we get from (11.58) that

$$u(k) = -Lx(k) + L_v(\Phi - KC)\bar{x}(k \mid k) - L_v K e(k)$$

and the closed-loop system becomes

$$\begin{bmatrix} x(k+1) \\ \bar{x}(k+1 \mid k) \end{bmatrix} = \begin{bmatrix} \Phi - \Gamma L & \Gamma L_v(\Phi - KC) \\ 0 & \Phi - KC \end{bmatrix} \begin{bmatrix} x(k) \\ \bar{x}(k \mid k-1) \end{bmatrix}$$

$$+ \begin{bmatrix} I \\ I \end{bmatrix} v(k) + \begin{bmatrix} -\Gamma L_v K \\ -K \end{bmatrix} e(k)$$

Notice that the closed-loop systems have the same poles in the two cases.

The Servo Problem

The servo problem is discussed in Sec. 9.5 for the state-feedback controller. For the LQG problem, the reference signal can be introduced in the same way as in

Fig. 9.8. The only difference is that the feedback matrix L is obtained by minimizing the quadratic loss function.

11.5 Practical Aspects

The previous sections show how the LQ- and the LQG-control problems can be solved. There are several practical problems when applying LQ-control. One occurs in choosing the design parameters—i.e., the weightings in the loss function—which is discussed in Sec. 11.2, and the sampling period. Another problem is the difficulty of obtaining good models for the process and the disturbances. Still another problem is making the numerical computations necessary to get the resulting controller. Some of these problems are discussed in this section.

Model Complexity

One criticism of LQ-control is that an accurate full-order model of the process must be available. Most physical processes are of high order. However, for control purposes it is often sufficient to use a low-order approximation. Ways to obtain mathematical models are discussed in Chapter 13.

One way to decrease the sensitivity to modeling errors is to decrease the desired bandwidth of the closed-loop system by changing the weightings in the loss function. Compare this with the robustness results in Sec. 5.2. Another way to decrease the sensitivity to modeling errors is to introduce artificial noise, which means that the noise covariances used in the design of the Kalman filter are larger than the true values.

Solution of the Riccati Equation

In many cases, only the steady-state optimal controller is implemented, which means that the steady-state values of the Riccati equations, (11.18) and (11.48), have to be determined. There are several ways to do this numerically. One way is to assume a constant S or P and solve the algebraic equations. A straightforward way to get the solution is to iterate the equations until a stationary condition is obtained. It is, however, important to make the computations so that the solution is guaranteed to be symmetric and positive definite. Special methods have been derived to solve the Riccati equation, such as square root and doubling algorithms. When using the square root method, the square root of S or P is calculated. This gives better numerical properties. Doubling algorithms or fast algorithms speed up the calculation of the stationary value by computing the solution at time $2k$ when the solution at time k is given. Many books and papers about different methods are available.

Choice of Sampling Period

The choice of the sampling period is influenced by how the specifications are given for the control problem. Two different cases are considered.

In the first case it is assumed that the specifications are given as a desired damping and response of the closed-loop system without using overly large control signals. It is then natural to determine the controller by iterating in the weightings of the sampled loss function of (11.12). To do this, a first choice of the sampling period has to be made based on the specifications. It is reasonable to choose the sampling period in relation to the dynamics of the closed-loop system, as discussed in Sec. 9.2. This means that it may be necessary to make one or two iterations in the sampling period. The closed-loop dynamics is a complicated function of the loss function.

In the second case it is assumed that the specifications are given in terms of the continuous-time loss function of (11.6). The continuous-time LQ-controller then minimizes the loss. For deterministic systems the loss increases quadratically with the sampling period h, for small h, which means that a limit exists when in practice it does not pay to decrease the sampling period further. For stochastic systems the loss function increases linearly with h, for small h. This implies that shorter sampling intervals are motivated for stochastic systems. It is possible to get an approximation of the increase in the loss due to an increase in the sampling period (see the references). When good interactive design programs are available, it is easy to check the loss and the performance for some sampling periods.

11.6 Conclusions

Optimal design based on state-space models are discussed in this chapter. The LQ-controllers and Kalman filters have many good properties. The main problem with LQ-control is translating the specifications on the system into a loss function. This is usually an iterative procedure, where it is necessary to have good interactive computer programs available.

11.7 Problems

11.1. Consider the first-order system

$$\dot{x} = -ax + bu$$

Assume that the loss function of (11.6) should be minimized with $Q_{1c} = 1$ and $Q_{2c} = \rho$. Determine the corresponding discrete-time loss function (11.12).

11.2. Consider the continuous-time double integrator in Example A.1. Assume that the loss function of (11.6) should be minimized with

$$Q_{1c} = \begin{bmatrix} 1 & 0 \\ 0 & 1 \end{bmatrix} \quad \text{and} \quad Q_{2c} = 1$$

Determine Q_1, Q_{12}, and Q_2 in the corresponding discrete-time loss function (11.12).

11.3. Given the system

$$x(k + 1) = ax(k) + bu(k)$$

with the loss function

$$J = \sum_{k=0}^{N} x^2(k)$$

Let the admissible control strategy be such that $u(k)$ is function of $x(k)$. Determine the strategy that minimizes the loss.

11.4. Consider the system in Problem 11.3. Determine the control strategy that minimizes the loss when the admissible control strategies are such that $u(k)$ is a function of $x(k-1)$.

11.5. The inventory model in Example A.5 is described by

$$x(k+1) = \begin{bmatrix} 1 & 1 \\ 0 & 0 \end{bmatrix} x(k) + \begin{bmatrix} 0 \\ 1 \end{bmatrix} u(k)$$

$$y(k) = \begin{bmatrix} 1 & 0 \end{bmatrix} x(k)$$

(a) Determine the steady-state LQ-controller when $Q_1 = C^T C$ and $Q_2 = \rho$.
(b) Determine the poles of the closed-loop system and investigate how they depend on the weight on the control signal, ρ.
(c) Simulate the system using the controller in (a). Assume that $x(0)^T = \begin{bmatrix} 1 & 1 \end{bmatrix}$ and consider the output and the control signal for different values of ρ.

11.6. Consider the two-tank system with the pulse-transfer operator given in Problem 3.10(c). Use (11.33) and plot the root locus that shows the closed-loop poles when the system is controlled by the steady-state LQ-controller for the loss function

$$J = \sum_{k=0}^{\infty} [y(k)^2 + \rho u(k)^2]$$

11.7. Show that a deadbeat control law, a control law such that the matrix $\Phi - \Gamma L$ has all its eigenvalues at the origin, can be obtained from the discrete-time optimization with $Q_2 = 0$, $Q_1 = 0$, and $Q_0 = I$.

11.8. Consider the ship-steering problem characterized by the model of (11.40) and the loss function in (11.41). Use the numbers $a_{11} = -0.454$, $a_{12} = -0.433$, $a_{21} = -4.005$, $a_{22} = -0.807$, $b_1 = 0.097$, $b_2 = -0.807$, $\alpha = 0.014$, and $\rho = 0.08$. Determine the optimal state-feedback when $h = 5$ s.

11.9. The ship-steering problem is sometimes approximated further by using the second-order model

$$\frac{d}{dt} \begin{bmatrix} \Psi \\ r \end{bmatrix} = \begin{bmatrix} 0 & 1 \\ 0 & -\alpha \end{bmatrix} \begin{bmatrix} \Psi \\ r \end{bmatrix} + \begin{bmatrix} 0 \\ k \end{bmatrix} \delta$$

and the following approximation of the loss function:

$$J = \lim_{T \to \infty} \frac{1}{T} \int_0^T [\Psi^2 + \rho \delta^2]\, dt$$

Determine the optimal feedback for a sampled regulator. Use the parameters $\alpha = 0.001$, $k = 0.0005$, and $\rho = 0.08$ and the sampling period $h = 5$ s.

11.10. Consider the LQ-controller determined in Problem 11.5 for the inventory model.
(a) Use (11.39) to get an estimate of the gain margin for the closed-loop system.

(b) Use (11.37) to determine the gain margin and compare with the estimate from (a).

11.11. A stochastic process is generated as

$$x(k + 1) = 0.5x(k) + v(k)$$

$$y(k) = x(k) + e(k)$$

Where v and e are uncorrelated white-noise processes with the covariances r_1 and r_2, respectively. Further, $x(0)$ is normally distributed with zero mean and standard deviation σ. Determine the Kalman filter for the system. What is the gain in steady state? Compute the pole of the steady-state filter and compare with the pole of the system.

11.12. The double integrator with process noise can be described by

$$x(k + 1) = \begin{bmatrix} 1 & 1 \\ 0 & 1 \end{bmatrix} x(k) + \begin{bmatrix} 0.5 \\ 1 \end{bmatrix} u(k) + \begin{bmatrix} 0 \\ 1 \end{bmatrix} v(k)$$

$$y(k) = [1 \quad 0]x(k)$$

where $v(k)$ is a sequence of independent, normal, zero-mean, random variables with unit variance. Assume that $x(0)$ is normal with mean $Ex(0) = [1 \quad 1]^T$ and unit covariance matrix.

(a) Determine the equations for the covariance matrix of the reconstruction error and the gain vector in the Kalman filter.

(b) Simulate the covariance and gain equations and determine the speed of convergence and the steady-state values.

11.13. Consider the double integrator in Problem 11.12, but let the output be

$$y(k) = [1 \quad 0]x(k) + v(k)$$

(a) Determine the equations for the covariance matrix of the reconstruction error and the gain vector in the Kalman filter.

(b) Simulate the covariance and gain equations and determine the speed of convergence and the steady-state values.

11.14. Given the system

$$x(k + 1) = \begin{bmatrix} 1 & 1 \\ 0 & 1 \end{bmatrix} x(k) + \begin{bmatrix} 0 \\ 1 \end{bmatrix} v(k) + \begin{bmatrix} 0.5 \\ 1 \end{bmatrix}$$

$$y(k) = [1 \quad 0]x(k)$$

where $v(k)$ is zero-mean white noise with standard deviation 0.1. Assume the $x(0)$ is known exactly. Determine the estimate of $x(k + 3)$, given $y(k)$ that minimizes the prediction error. Use that to determine the best estimate of $y(3)$ and its variance.

11.15. The signal $x(k)$ is defined as

$$x(k + 1) = ax(k) + v(k)$$

$$y(k) = x(k) + e(k)$$

where v and e are independent white-noise processes with zero mean. The variances are 1 and σ, respectively. The signal x is estimated using exponential smoothing as

$$\hat{x}(k \mid k) = \alpha \hat{x}(k - 1 \mid k - 1) + (1 - \alpha)y(k)$$

Determine an expression for how the variance of the estimation error depends on the parameters α and σ. Compare with the steady-state optimal Kalman filter.

11.16. Show that Theorem 11.5 can be generalized to the situation when the disturbances $e(k)$ and $v(k)$ have constant but *unknown* mean values. (Compare with Sec. 9.4.)

11.17. A constant variable x is measured through two different sensors. However, the measurements are noisy and have different accuracy. Assume that the system is described by

$$x(k + 1) = x(k)$$

$$y(k) = Cx(k) + e(k)$$

where $C^T = [1 \quad 1]$ and $e(k)$ is a zero-mean, white-noise vector with the covariance matrix

$$R_2 = \begin{bmatrix} 1 & 0 \\ 0 & 9 \end{bmatrix}$$

Estimate x as

$$\hat{x}(k) = a_1 y_1(k) + a_2 y_2(k)$$

Determine the constants a_1 and a_2 such that the mean value of the prediction error is zero and such that the variance of the prediction error is as low as possible. Compare the minimum variance with the cases when only one of the measurements is used. Compare the solution with the Kalman filter.

11.18. Prove that the filter estimate given by (11.49) and (11.50) is the optimal filter in the sense that the variance of the estimation error is minimized.

11.19. Consider the design of a Kalman filter for estimating the velocity in a motor drive based on angle measurements. The basic dynamics of the motor, which relate the angle to the current, are given by

$$G(s) = \frac{1}{s(s + 1)}$$

Assume that there are low-frequency disturbances (friction) which are modeled as

$$z_1(kh + h) - z_1(kh) + w_1(kh)$$

Also assume that it is desirable to filter out disturbances because of a mechanical resonance at the frequency ω. This signal is modeled as the signal obtained by driving a system with the transfer function

$$G(s) = \frac{\omega^2}{s^2 + 2\zeta\omega s + \omega^2}$$

with white noise. Determine the Bode diagrams for the Kalman filter for $\zeta = 0.05$, $\omega = 0.1$, and $\omega = 2$. Let the sampling period be 0.05 s. Also investigate the influence of different relative intensities of the low-frequency and the band-limited disturbance.

11.20. Consider the system

$$x(k + 1) = \begin{bmatrix} 1.45 & -0.45 \\ 1 & 0 \end{bmatrix} x(k) + \begin{bmatrix} 1 \\ 0 \end{bmatrix} u(k)$$

$$y(k) = [0.5 \quad 0.38]x(k)$$

Determine the stationary controller $u(k) = -Lx(k)$ that minimizes the loss function

$$J = \sum_{k=1}^{\infty} x^T(k)C^TCx(k)$$

11.21. A computer is used to control the velocity of a motor. Let the process be described by

$$x(k + 1) = 0.5x(k) + u(k)$$

$$y(k) = x(k) + e(k)$$

where x is the velocity, u is the input voltage, and y is the tachometer measurement of the velocity. The measurement noise is white noise with the variance σ^2. Assume that the initial speed is a stochastic variable with zero mean and unit variance. Construct a controller that minimizes the loss function

$$E(x(2)^2 + \sum_{k=0}^{1} \rho u^2(k))$$

The parameter ρ is used to control the amplitude of the control signal. It is further desired that the velocity be as small as possible after two sampling intervals.
(a) Determine the optimal controller when $\sigma = 0$ and the regulator parameters when $\rho = 1$, $\rho = 0.1$, and when $\rho \to 0$.
(b) Determine the optimal controller when the measurement noise has the variance $\sigma^2 = 1$.

11.22. Given the system

$$x(k + 1) = x(k) + v(k)$$

$$y_1(k) = x(k) + e_1(k)$$

$$y_2(k) = x(k) + e_2(k)$$

where $v \in N(0, 0.1)$, $e_1 \in N(0, \sigma_1)$, and $e_2 \in N(0, \sigma_2)$, and v, e_1, and e_2 are mutually uncorrelated.
(a) Determine the Kalman filter that gives $\hat{x}(k \mid k - 1)$ for the system.
(b) Compute the stationary variance when $\sigma_1 = 1$ and $\sigma_1 = 2$.
(c) Compute the stationary gain when $\sigma_1 = 1$ and $\sigma_2 = 2$.

11.8 References

LQG-control and optimal filters are the subjects of many textbooks, for instance,

KWAKERNAAK, H. and R. SIVAN (1972): *Linear Optimal Control Systems*. New York: Wiley-Interscience.

ANDERSON, B. D. O. and J. B. MOORE (1971): *Linear Optimal Control*. Englewood Cliffs, N.J.: Prentice Hall, Inc.

ÅSTRÖM, K. J. (1970): *Introduction to Stochastic Control Theory*. New York: Academic Press.

ANDERSON, B. D. O. and J. B. MOORE (1979): *Optimal Filtering*. Englewood Cliffs, N.J.: Prentice Hall, Inc.

Optimal Design Methods: State-Space Approach   Chap. 11

Kalman and Bucy made the main contributions to the development of the recursive optimal filters discussed in Sec. 11.3. See

KALMAN, R. E. (1960): "A New Approach to Linear Filtering and Prediction Problems," *Journal Basic Eng.*, 82, March, 34–45.

KALMAN, R. E. and R. S. BUCY (1961): "New Results in Linear Filtering and Prediction Theory," Trans. ASME, Ser. D, *Journal Basic Eng.*, 83, December, 95–107.

BUCY, R. S. (1959): "Optimum Finite Time Filters for a Special Nonstationary Class of Inputs," Internal memorandum BBD-600, Johns Hopkins University. Applied Physics Lab.

Numerical algorithms for solving the Riccati equation are discussed in

BIERMAN, G. (1977): *Factorization Methods for Dicrete Estimation.* New York: Academic Press.

PAPPAS, T., A. J. LAUB, and N. R. SANDELL, JR. (1980): "On the Numerical Solution of the Discrete Time Riccati Equation," *IEEE Trans. Autom. Contr.*, AC-25, August, 631–41.

VAN DOOREN, P. (1981): "A Generalized Eigenvalue Approach for Solving Riccati Equations," *SIAM Journal Sci. Stat. Comp.*, 2, 121–35.

Choice of the sampling interval for LQ-controllers is discussed in

ÅSTRÖM, K. J. (1963): "On the Choice of Sampling Rates in Optimal Linear Systems," Internal report IBM San Jose Research Laboratory.

MELZER, S. M. and B. C. KUO (1971): "Sampling Period Sensitivity of the Optimal Sampled Data Linear Regulator," *Automatica*, 7, 367–70.

LENNARTSON, B. (1987): "On the Choice of Controller and Sampling Period for Linear Stochastic Control," *Preprints 10th IFAC World Congress*, Munich, 9, 241–46.

The separation theorem appeared first in economic literature:

SIMON, H. A. (1956): "Dynamic Programming under Uncertainty with a Quadratic Criterion Function," *Econometrica*, 24, 74.

Discrete-time versions of the separation theorem can be found in

GUNKEL, III, T. L., G. F. FRANKLIN (1963): "A General Solution for Linear Sampled Data Control," *Trans. ASME Journal Basic Eng.* 85-D, 197–201.

Gain margin for discrete-time LQ-controllers is discussed in

WILLEMS, J. L. and H. VAN DE VOORDE (1978): "The Return Difference for Discrete-Time Optimal Feedback Systems," *Automatica*, 14, 511–13.

SAFONOV, M. G. (1980): *Stability and Robustness of Multivariable Feedback Systems.* Cambridge, Mass.: MIT Press.

Robustness of LQG controllers is discussed in

DOYLE, J. C., G. STEIN (1981): "Multivariable Feedback Design: Concepts for a Classical/Modern Synthesis," *IEEE Trans. Autom. Contr.*, AC-26, Feb., 4–16.

OPTIMAL DESIGN METHODS: INPUT-OUTPUT APPROACH

12

GOAL

To Show How Optimal Regulation Problems Can Be Formulated and Solved Using Input-Output Models. To Derive Formulas for Optimal Prediction, Minimum Variance Control, and LQG-Control.

12.1 Introduction

Optimal design methods based on input-output models are considered in this chapter. Design of regulators based on linear models and quadratic criteria is discussed. This is one class of problems that admits closed-form solutions. The problems are solved by other methods in Chapter 11. The input-output approach gives additional insight and different numerical algorithms are also obtained.

The problem formulation is given in Sec. 12.2. This includes discussion of models for dynamics, disturbances, and criteria, as well as specification of admissible controls. The model is given in terms of three polynomials. A very simple example is also solved using first principles. This example shows clearly that optimal control and optimal filtering problems are closely connected. The prediction problem is then solved in Sec. 12.3. The solution is easily obtained by polynomial division. A simple explicit formula for the transfer function of the optimal predictor is given. The minimum-variance control law is derived in Sec. 12.4. For systems with stable inverses, the control law is obtained in terms of the polynomials that characterize the optimal predictor. For systems with unstable inverses, the solution is obtained by solving a Diophantine equation in polynomials of the type discussed in Chapter 10. The minimum-variance control problem may thus be interpreted as a pole-placement problem. This gives insight into suitable choices of closed-loop poles and observer poles for the pole-placement problem.

The LQG-control problem is solved in Sec. 12.5. It is shown that the solution may be expressed in terms of spectral factorization and solution of a Diophantine equation. Practical aspects, such as selection of the sampling period, are given in Sec. 12.6.

12.2 Problem Formulation

It is assumed that the process to be controlled is linear and time-invariant and that it has one input u and one output y. The dynamics of the process are characterized by a combination of a time-delay and a rational-transfer function. It is also assumed that the disturbances may be described as filtered white noise. A steady-state regulation problem is considered. The criterion is based on the mean square deviations of the control signal and the output signal. In the formal problem statement given next, it is assumed that the model and the criterion are sampled (compare with Sec. 3.2 and 11.1).

Process Dynamics

Assume that the process dynamics are characterized by

$$x(k) = \frac{B_1(q)}{A_1(q)} u(k) \tag{12.1}$$

where $A_1(q)$ and $B_1(q)$ are polynominals in the forward-shift operator.

Disturbances

Assume that the influence of the environment on the process can be characterized by disturbances that are stochastic processes. Because the system is linear, the principle of superposition can be used to reduce all disturbances to an equivalent disturbance v at the system output. The output of the system is thus given by

$$y(k) = x(k) + v(k) \tag{12.2}$$

Further assume that the disturbance v may be represented as the output of a linear system driven by white noise—i.e.,

$$v(k) = \frac{C_1(q)}{A_2(q)} e(k) \tag{12.3}$$

where $C_1(q)$ and $A_2(q)$ are polynomials in the forward-shift operator and $\{e(k)\}$ is a sequence of independent or uncorrelated random variables with zero mean and standard deviation σ. The disturbance v may be a stationary random process. It may, however, also be drifting, because the polynomial $A_2(q)$ may be unstable. The model of the process and its environment can be reduced to a standard form.

Eliminate v and x among (12.1), (12.2), and (12.3) and introduce

$$A = A_1 A_2$$
$$B = B_1 A_2 \qquad (12.4)$$
$$C = C_1 A_1$$

The following model is then obtained.

$$A(q)y(k) = B(q)u(k) + C(q)e(k) \qquad (12.5)$$

This is the canonical model, which will be the basis of the control design. In the special case when there are no disturbances, the model is simply a rational pulse-transfer function (see Sec. 3.6). When there is no control signal, the model is a stochastic process with a rational spectral density or an ARMA process (see Sec. 6.5). The model (12.5) is a convenient canonical representation of a linear system perturbed by noise. When the polynomial $C(q)$ has all its zeros inside the unit disc, it is called an *innovations* representation, because the random variables $\{e(k)\}$ represent the innovations of the random process. Notice the symmetry between y and e. If e and u are known up to time k, then $y(k)$ can be computed, and if y and u are known up to time k, the innovation $e(k)$ can also be computed. Notice that the calculations of the residuals are governed by the dynamics of the polynomial $C(q)$. This polynomial can therefore be interpreted as the observer polynomial. Since (12.5) is an innovations model, the solutions to filtering problems become very simple.

Equation (12.5) can be normalized so that the leading coefficients of the polynomials $A(q)$ and $C(q)$ are unity. Such polynomials are called *monic*. The polynomial C may also be multiplied by an arbitrary power of q, as this does not change the correlation structure of $C(q)e(t)$. This may be used to normalize C so that deg C = deg A. The polynomials $A(q)$ and $B(q)$ may have zeros inside or outside the unit disc. It is assumed that all the zeros of the polynomial $C(q)$ are inside the unit disc. By spectral factorization (Theorem 6.3), the polynomial $C(q)$ may be changed so that all its zeros are inside the unit disc or on the unit circle. An example is used to show this important point.

Example 12.1—Modification of the polynomial C

Consider the polynomial

$$C(z) = z + 2$$

which has the zero $z = -2$ outside the unit disc. Consider the signal

$$n(k) = C(q)e(k)$$

where $\{e(k)\}$ is a sequence of uncorrelated random variables with zero mean and unit variance. The spectral density of n is given by

$$\phi(e^{i\omega h}) = \frac{1}{2\pi} C(e^{i\omega h})C(e^{-i\omega h})$$

Because

$$C(z)C(z^{-1}) = (z + 2)(z^{-1} + 2) = (1 + 2z^{-1})(1 + 2z)$$
$$= (2z + 1)(2z^{-1} + 1)$$
$$= 4(z + 0.5)(z^{-1} + 0.5)$$

the signal n may also be represented as

$$n(k) = C^*(q)e(k)$$

where

$$C^*(z) = 2z + 1$$

is the reciprocal of the polynomial $C(z)$ (see Sec. 3.3). □

If the calculations of (12.4) give a polynomial $C(q)$ that has zeros outside the unit disc, the polynomial C is factored as

$$C = C^+C^-$$

where C^- contains all factors with zeros outside the unit disc. The polynomial C is then replaced by C^+C^{-*}.

Criteria

In steady-state regulation it makes sense to express the criteria in terms of *steady-state variances* of the control variable and the process output. For regulation of systems with one output, the criterion may be to minimize the variance of the output. This is discussed in Sec. 7.6. Also compare with Fig. 7.7. This leads to the criterion

$$J_{mv} = Ey^2(k) \tag{12.6}$$

where it is assumed that the scales are chosen so that $y = 0$ corresponds to the desired set point. A control law that minimizes the criterion (12.6) is called *minimum-variance control*. The criterion may also be expressed as

$$J_\infty = \lim_{N\to\infty} E \frac{1}{N} \sum_{1}^{N} y^2(k)$$

Notice that this criterion is an approximation of the continuous-time loss function

$$J_c = \lim_{T\to\infty} \frac{1}{T} \int_{0}^{T} y^2(t)\, dt \tag{12.7}$$

A more-accurate approximation, which takes the behavior of the signals between the sampling instants into account, is given in Sec. 11.1. Some consequences of the approximation are discussed in Sec. 12.6.

The properties of the control signal under minimum-variance control depend critically on the sampling period. A short sampling period gives a large variance of the control signal and a long sampling period gives a small variance. In some cases it is desired to trade variances of control and output signals. This may be done by introducing the loss function

$$J_{lq} = E[y^2(k) + \rho u^2(k)] \tag{12.8}$$

The control law that minimizes this criterion is called the *linear quadratic control law*.

Admissible Controls

It is assumed that the control law is such that $u(k)$, i.e., the value of the control signal at time k, is a function of $y(k)$, $y(k - 1)$, ... and $u(k - 1)$, $u(k - 2)$, ... Thus the computational delay is negligible in comparison with the sampling period. It is very easy to modify the results to take delays in the computations into account.

There are two versions of the theory. A linear control law may be postulated. It is then sufficient to assume that the disturbances $e(i)$ and $e(j)$ are uncorrelated for $i \neq j$. If $e(i)$ and $e(j)$ are assumed to be independent, it can be shown that the optimal-control law is linear. The formula for the optimal-control law is the same in both cases.

An Example

The optimal-control problem defined by the model of (12.5) and the criterion of (12.6) is solved in a special case. The solution, which is easily obtained from first principles, gives good insight into the assumptions made. It also indicates how the general problem should be solved.

Consider the first-order system

$$y(k + 1) + ay(k) = bu(k) + e(k + 1) + ce(k) \tag{12.9}$$

where $|c| < 1$ and $\{e(k)\}$ is a sequence of independent random variables with unit variance.

Consider the situation at time k. The outputs $y(k)$, $y(k - 1)$, ... have been observed. The control $u(k)$ should be determined so that y is as close to zero as possible. It follows from (12.9) that $y(k + 1)$ may be changed arbitrarily by a proper choice of $u(k)$. Because $e(k + 1)$ is independent of $y(k)$ and of the terms of the right-hand side of (12.9), it follows that

$$\text{Var } y(k + 1) \geq \text{Var } e(k + 1) = 1 \tag{12.10}$$

The term $e(k)$ may be computed in terms of the known data $y(k)$, $y(k - 1)$, ... and $u(k - 1)$, $u(k - 2)$, ... When the variables $y(k)$ and $e(k)$ are known, the control law

$$u(k) = [ay(k) - ce(k)]/b \tag{12.11}$$

gives

$$y(k + 1) = e(k + 1) \tag{12.12}$$

which corresponds to the lower bound in (12.10). If the control law in (12.11) is used in each step, equation (12.12) holds for all k. The computation of $e(k)$ from the data available at time k is then trivial and the control law in (12.11) can be written as

$$u(k) = \frac{a - c}{b} y(k) \tag{12.13}$$

The optimal control is thus a simple proportional feedback with the gain $(a - c)/b$.

To analyze the properties of the closed-loop system under optimal control, eliminate u between (12.9) and (12.13). This gives

$$y(k + 1) + cy(k) = e(k + 1) + ce(k)$$

Notice that the closed-loop system has the characteristic polynomial

$$C(z) = z + c$$

This shows the importance of the assumption that the polynomial $C(z)$ is stable. This difference equation has the solution

$$y(k) = e(k) + (-c)^{k - k_0}[y(k_0) - e(k_0)]$$

Because c is less than one in magnitude, the last term goes to zero as $k - k_0$ increases towards infinity. Thus control law in (12.13) gives the minimum variance in steady state.

With this result, some observations are possible. The quantity $-ay(k) + bu(k) + ce(k)$ can be interpreted as the best estimate of $y(k + 1)$, given the data available at time k. The quantity $e(k + 1)$ is the prediction error. The control law in (12.13) implies that the control signal is chosen so that the predicted value is equal to the reference value, which is zero in this case. The control error is then equal to the prediction error. The solution to the minimum-variance control problem is thus closely related to the solution of a prediction problem. Therefore, the prediction problem is solved before the solution of the general minimum-variance control problem is attempted.

12.3 Optimal Prediction

Prediction theory can be stated in many different ways, which differ in the assumptions made on the process, the criterion, and the admissible predictors. One formulation is given in Sec. 11.3. In this section the following assumptions are made:

The process to be predicted is generated by filtered white Gaussian noise. The best predictor is the one that minimizes the mean-square prediction error.

An admissible m-step predictor for $y(k + m)$ is an arbitrary function of $y(k)$, $y(k - 1), \ldots$.

An intuitive derivation of a predictor is first given. The result is then formalized.

Heuristics

Consider the signal y generated by the model

$$y(k) = \frac{C(q)}{A(q)} e(k) = \frac{C^*(q^{-1})}{A^*(q^{-1})} e(k) \qquad (12.14)$$

where A^* and C^* are the reciprocals of A and C, and q^{-1} is the backward-shift operator. It is convenient to introduce this operator because the discussion is based on causality. It is assumed that A and C are of order n.

Consider the situation at time k. The variables $y(k), y(k-1), \ldots$ have been observed and it is desired to predict $y(k+m)$. A formal series expansion of C^*/A^* in q^{-1} gives

$$y(k+m) = \frac{C^*(q^{-1})}{A^*(q^{-1})} e(k+m)$$

$$= \{e(k+m) + f_1 e(k+m-1) + \cdots + f_{m-1} e(k+1)\} \qquad (12.15)$$

$$+ [f_m e(k) + f_{m+1} e(k-1) + \cdots]$$

The terms of the right-hand side are all independent because $\{e(k)\}$ is a sequence of independent random variables. It follows from the model of (12.14) that if the polynomial C is stable, then $e(i)$ can be computed exactly from $y(i), y(i-1),$ ... using

$$e(k) = \frac{A^*(q^{-1})}{C^*(q^{-1})} y(k)$$

The terms inside the brackets are thus known functions of the data available at time k. The terms in the braces are independent of the data at time k. Thus it follows that the optimal predictor is given by

$$\hat{y}(k+m \mid k) = f_m e(k) + f_{m+1} e(k-1) + f_{m+2} e(k-2) + \cdots$$

and that the prediction error is

$$\tilde{y}(k+m \mid k) = e(k+m) + f_1 e(k+m-1) + \cdots + f_{m-1} e(k+1)$$

To provide a formal proof it remains to show how the numbers f_i can be computed from A and C and how $e(k)$ can be expressed in terms of past data.

Main Result

The main result can be stated as follows.

Theorem 12.1—Optimal prediction. Let $\{y(k)\}$ be a random process generated by the model in (12.14), where all the zeros of the polynomial $C(z)$ are inside the unit disc and $\{e(k)\}$ is a sequence of independent random variables. The minimum-variance predictor over m steps is given by

$$\hat{y} = \hat{y}(k+m \mid k) = \frac{qG(q)}{C(q)} y(k) = \frac{G^*(q^{-1})}{C^*(q^{-1})} y(k) \qquad (12.16)$$

where the polynomials F and G are the quotient and the remainder when dividing $q^{m-1}C$ by A; i.e.,

$$q^{m-1}C(q) = A(q)F(q) + G(q) \tag{12.17}$$

The prediction error is a moving average

$$\tilde{y}(k + m \mid k) = y(k + m) - \hat{y}(k + m \mid k) = F(q)e(k + 1) \tag{12.18}$$

It has zero mean and the variance

$$E\tilde{y}(k + m \mid k)^2 = [1 + f_1^2 + \cdots + f_{m-1}^2]\sigma^2 \tag{12.19}$$

Proof. The polynomial F is monic of degree $m - 1$ and G is of degree less than n. Hence

$$F(q) = q^{m-1} + f_1 q^{m-2} + \cdots + f_{m-1}$$

and

$$G(q) = g_0 q^{n-1} + g_1 q^{n-2} + \cdots + g_{n-1}$$

Introduce

$$F^*(q^{-1}) = 1 + f_1 q^{-1} + \cdots + f_{m-1}q^{-m+1}$$

and

$$G^*(q^{-1}) = g_0 + g_1 q^{-1} + \cdots + g_{n-1}q^{-n+1}$$

It follows from (12.17) that

$$C^*(q^{-1}) = A^*(q^{-1})F^*(q^{-1}) + q^{-m}G^*(q^{-1}) \tag{12.20}$$

Equation (12.15) can then be written as

$$y(k + m) = \frac{C^*(q^{-1})}{A^*(q^{-1})} e(k + m) = F^*(q^{-1})e(k + m) + \frac{G^*(q^{-1})}{A^*(q^{-1})} e(k)$$

Using Equation (12.14) the signal e in the last term can be expressed in terms of the data available at time k. Hence,

$$y(k + m) = F^*(q^{-1})e(k + m) + \frac{G^*(q^{-1})}{C^*(q^{-1})} y(k) \tag{12.21}$$

The first term of the right-hand side is a linear function of $e(k + 1)$, $e(k + 2)$, $\ldots$, $e(k + m)$, which are all independent of the data $y(k)$, $y(k - 1)$, $y(k - 2)$, $\ldots$ available at time k. The last term is a linear function of the data. Let $\hat{y}$ be an arbitrary function of $y(k)$, $y(k - 1)$, $\ldots$. Then

$$E[y(k + m) - \hat{y}]^2 = E[F^*(q^{-1})e(k + m)]^2 + E\left[\frac{G^*(q^{-1})}{C^*(q^{-1})} y(k) - \hat{y}\right]^2$$

$$+ 2E[F^*(q^{-1})e(k + m)]\left[\frac{G^*(q^{-1})}{C^*(q^{-1})} y(k) - \hat{y}\right] \tag{12.22}$$

The last term is zero because $e(k + m)$, $e(k + m - 1)$, . . . , and $e(k + 1)$ have zero mean values and are independent of $y(k)$, $y(k - 1)$, ... The predictor that minimizes the mean-square prediction error is thus given by (12.16) and the prediction error by (12.18). The proof is completed by taking the mean value of the square of the prediction error (12.18). This gives (12.19). □

Remark 1. Notice that the best predictor is linear. The linearity does not depend critically on the minimum-variance criterion. If the probability density of $\{y(k)\}$ is symmetric, the predictor of (12.16) is optimal for all criteria of the form $Eg[(y(k + m) - \hat{y})^2]$ for symmetric g.

Remark 2. The assumption that $e(i)$ and $e(j)$ are independent for $i \neq j$ is essential for the last term in (12.22) to vanish. If the variables are uncorrelated, the term will still vanish if the predictor $\hat{y}$ is restricted to being linear.

Remark 3. It follows from (12.18) that

$$\tilde{y}(k + 1 \mid k) = y(k + 1) - \hat{y}(k + 1 \mid k) = e(k + 1)$$

The random variables $\{e(k)\}$ can thus be interpreted as the innovations of the process $\{y(k)\}$ (compare with Sec. 6.5).

Remark 4. Notice that the function

$$J(m) = \sigma^2[1 + f_1^2 + \cdots + f_{m-1}^2]$$

is the variance of the prediction error over the time interval mh. The function $J(m)$ approaches the variance of y as $m \to \infty$. A graph of the function J shows how well the process may be predicted over different horizons.

Remark 5. The predictor discussed in this section is equivalent to the steady-state predictor obtained using the Kalman filter in Sec. 11.3 (see Example 11.5).

Calculation of the Optimal Predictor

It follows from (12.17) that $F(q)$ is the quotient and $G(q)$ the remainder when dividing $q^{m-1}C(q)$ by $A(q)$. The polynomials F and G can thus be determined by polynomial division. An explicit formula for the coefficients of the polynomials can also be given. Equating the coefficients of equal powers of q in (12.17) gives the following equations:

$$c_1 = a_1 + f_1$$

$$c_2 = a_2 + a_1 f_1 + f_2$$

$$\vdots$$

$$c_{m-1} = a_{m-1} + a_{m-2}f_1 + \cdots + a_1 f_{m-2} + f_{m-1}$$

$$c_m = a_m + a_{m-1}f_1 + \cdots + a_1 f_{m-1} + g_0$$

$$c_{m+1} = a_{m+1} + a_m f_1 + \cdots + a_2 f_{m-1} + g_1 \qquad (12.23)$$

$$\vdots$$

$$c_n = a_n + a_{n-1}f_1 + \cdots + a_{n-m+1}f_{m-1} + g_{n-m}$$

$$0 = a_n f_1 + a_{n-1}f_2 + \cdots + a_{n-m+2}f_{m-1} + g_{n-m+1}$$

$$\vdots$$

$$0 = a_n f_{m-1} + g_{n-1}$$

These equations are easy to solve recursively. Compare the solution of the Diophantine equation in Chapter 10.

The Case When C Has Zeros on the Unit Circle

The predictor of (12.16) is a dynamic system with the characteristic polynomial $C(z)$. The assumption that C has all its zeros inside the unit disc thus guarantees that the predictor is stable in steady state. The initial conditions are irrelevant because their influence will decay exponentially. It follows from the spectral factorization that C may be chosen to have its zeros inside the unit disc or on the unit circle. Thus it remains to discuss the case when C has zeros on the unit circle.

Example 12.2—Zeros on the unit circle

Consider the process

$$y(k) = c(k) - c(k - 1) \tag{12.24}$$

In this case the polynomial $C(z) = z - 1$ has a zero on the unit circle. Applying the previous methods formally gives the one-step predictor

$$\hat{y}(k + 1 \mid k) = -e(k)$$

Attempting to calculate $e(k)$ from $y(k), y(k - 1), \ldots, y(k_0)$ as was done previously gives

$$e(k) = e(k_0 - 1) + \sum_{i=k_0}^{k} y(i) = e(k_0 - 1) + z(k) \tag{12.25}$$

The presence of the term $e(k_0 - 1)$, which does not go to zero as $k_0 \to -\infty$, shows the consequences of C being unstable. The Kalman filtering theory can, however, be used to determine the optimal predictor. The signal given by (12.24) can be written as

$$x(k + 1) = e(k)$$

$$y(k) = -x(k) + e(k)$$

where $R_1 = R_2 = R_{12} = \sigma^2$ with the notations used in Sec. 11.3. The Kalman filter is

$$\hat{x}(k + 1 \mid k) = K(k)[y(k) + \hat{x}(k \mid k - 1)]$$

$$P(k + 1) = \frac{\sigma^2 P(k)}{P(k) + \sigma^2}$$

$$K(k) = \frac{\sigma^2}{P(k) + \sigma^2}$$

with the initial conditions

$$\hat{x}(k_0 \mid k_0 - 1) = 0$$

$$P(k_0) = \sigma^2$$

The predictor for the output is

$$\hat{y}(k + 1 \mid k) = -\hat{x}(k + 1 \mid k) = -K(k)[y(k) - \hat{y}(k \mid k - 1)]$$

Simple calculations give

$$\hat{y}(k + 1 \mid k) = -\frac{1}{k - k_0 + 2} \sum_{n=0}^{k - k_0} (n + 1)y(k_0 + n)$$

The optimum predictor is thus a time-varying system. Notice that the influence of the initial condition $y(k_0)$ goes to zero at the rate $1/(k + 2 - k_0)$. This is much slower than in the case of stable polynomials C. ☐

It follows from the example that the optimal predictor is a *time-varying system* if the polynomial C has zeros on the unit circle. Such models should be avoided if time-invariant predictors are desired. Unfortunately, this fact is not always noticed, as Example 12.3 illustrates.

Example 12.3—How to model offsets

The model

$$A(q)y(k) = C(q)e(k) + b$$

where b is an unknown constant, represents a signal with an offset. The constant b can be eliminated by taking differences. Hence,

$$(q - 1)A(q)y(k) = (q - 1)C(q)e(k)$$

The common factor $q - 1$ can be eliminated by regarding $\Delta y(k) = (q - 1)y(k)$ as the output. The model

$$A(q)\Delta y(k) = (q - 1)C(q)e(k) = \tilde{C}(q)e(k)$$

is then obtained. In this model the polynomial $\tilde{C}$ apparently has a zero on the unit circle. This model is, however, *not* very desirable because the optimal predictor is a time-varying system. It is much better to model an offset as a Wiener process. This leads to a process model with $A(1) = 0$ that is unstable with a stationary predictor. ☐

Other reasons for avoiding models where the polynomial $C(z)$ has zeros close to the unit circle are given in Sec. 12.6.

12.4 Minimum-Variance Control

To determine the minimum-variance control law, the special case when the polynomial B in (12.5) is stable is discussed first. This means that the process dynamics have a stable inverse. With some abuse of language, this case is also called the minimum-phase case because the pulse-transfer function has all its zeros inside the unit disc. The solution to the control problem is very simple in this special case. The solution also gives insight into the properties of the control problem.

Systems with Stable Inverses

By introducing the backward-shift operator q^{-1}, the model in (12.5) can be written as

$$
\begin{aligned}
y(k) &= \frac{B(q)}{A(q)} u(k) + \frac{C(q)}{A(q)} e(k) \\
&= \frac{B^*(q^{-1})}{A^*(q^{-1})} q^{-d} u(k) + \frac{C^*(q^{-1})}{A^*(q^{-1})} e(k)
\end{aligned}
\tag{12.26}
$$

where

$$
d = \deg A - \deg B
$$

is the *pole excess* of the system (see Sec. 3.4). Further, $\deg A = \deg C = n$. The reciprocal polynomials are introduced to make the discussion based on causality arguments more transparent.

It follows from (12.26) that

$$
\begin{aligned}
y(k + d) &= \frac{C^*(q^{-1})}{A^*(q^{-1})} e(k + d) + \frac{B^*(q^{-1})}{A^*(q^{-1})} u(k) \\
&= F^*(q^{-1}) e(k + d) + \frac{G^*(q^{-1})}{A^*(q^{-1})} e(k) + \frac{B^*(q^{-1})}{A^*(q^{-1})} u(k)
\end{aligned}
\tag{12.27}
$$

where Equation (12.17) with $m = d$ has been used to obtain the last equality. The first term of the right-hand side is independent of the data available at time k and thus also of the second and third terms. The second term can be computed exactly in terms of data available at time k. To do this, the variable e is given by (12.26); i.e.,

$$
e(k) = \frac{A^*}{C^*} y(k) - q^{-d} \frac{B^*}{C^*} u(k)
$$

where the arguments have been dropped to simplify the writing. Using this expression for e, Equation (12.27) can be written as

$$
\begin{aligned}
y(k + d) &= F^* e(k + d) + \frac{G^*}{C^*} y(k) - q^{-d} \frac{B^* G^*}{A^* C^*} u(k) + \frac{B^*}{A^*} u(k) \\
&= F^* e(k + d) + \frac{G^*}{C^*} y(k) + \frac{B^* F^*}{C^*} u(k)
\end{aligned}
\tag{12.28}
$$

Now let $u(k)$ be an arbitrary function of $y(k)$, $y(k - 1)$, ... and $u(k - 1)$, $u(k - 2)$, ... Then

$$
Ey^2(k + d) = E[F^* e(k + d)]^2 + E\left[\frac{G^*}{C^*} y(k) + \frac{B^* F^*}{C^*} u(k) \right]^2
\tag{12.29}
$$

The mixed terms vanish because $e(k + d), \ldots, e(k + 1)$ are independent of $y(k), y(k - 1), \ldots$ and $u(k), u(k - 1), \ldots$. Since the last term in (12.29) is nonnegative, it follows that

$$Ey^2(k + d) \geq [1 + f_1^2 + \cdots + f_{d-1}^2]\sigma^2 \qquad (12.30)$$

where equality is obtained for

$$u(k) = -\frac{G^*(q^{-1})}{B^*(q^{-1})F^*(q^{-1})}\, y(k) = -\frac{G(q)}{B(q)F(q)}\, y(k) \qquad (12.31)$$

which is the desired minimum-variance control law. The result can be summarized as follows.

Theorem 12.2. Consider a process described by (12.5), where $\{e(k)\}$ is a sequence of independent random variables with zero mean values and standard deviations σ. Let the polynomials B and C have all their zeros inside the unit disc. The minimum-variance control law is then given by (12.31), where the polynomials F^* and G^* are given by (12.17) with $m = d$. This control law gives the output

$$\begin{aligned} y(k) &= F^*(q^{-1})e(k) \qquad\qquad\qquad\qquad\qquad (12.32)\\ &= e(k) + f_1 e(k - 1) + \cdots + f_{d-1}e(k - d + 1) \end{aligned}$$

in steady state. $\qquad\qquad\qquad\qquad\qquad\qquad\qquad\qquad\qquad$ $\square$

Remark 1. The theorem still holds when $e(i)$ and $e(j)$ are uncorrelated for $i \neq j$ if a linear control law is postulated.

Remark 2. The result is closely related to the solution of the prediction problem (Theorem 12.1). Identity (12.17) was used in both cases. The last two terms in (12.28) can be interpreted as the d-step prediction of the output. The minimum-variance strategy is thus obtained by predicting the output d steps ahead and choosing a control that makes the prediction equal to the desired output. The stochastic control problem can thus be separated into two problems, one sto-chastic-prediction problem and one deterministic-control problem. Theorem 12.2 is therefore called the *separation theorem*.

Remark 3. The error under minimum-variance control is a moving average of order $d - 1$. Thus the covariance function of the regulation error will vanish for arguments larger than $d - 1$. This fact can be used for diagnosis to determine if a minimum-variance strategy is used.

Remark 4. All process zeros are canceled when the control law of (12.31) is used. The consequences of this are discussed later.

It is very easy to calculate the minimum-variance control law for a given model (12.5), as illustrated by Example 12.4.

Example 12.4—Minimum-variance control

Consider a system given by (12.5), where

$$A(q) = q^3 - 1.7q^2 + 0.7q$$

$$B(q) = q + 0.5$$

$$C(q) = q^3 - 0.9q^2$$

The pole excess is $d = 2$. Division of $q^{d-1}C(q)$ by $A(q)$ gives the quotient

$$F(q) = q + 0.8$$

and the remainder

$$G(q) = 0.66q^2 - 0.56q$$

The minimum-variance control law is thus

$$u(k) = -\frac{q(0.66q - 0.56)}{(q + 0.5)(q + 0.8)} y(k) \qquad \square$$

Interpretation as Pole-Placement Design

The minimum-variance control law can be interpreted in terms of the pole-place-ment design discussed in Chapter 10. To see the relationships, the closed-loop system obtained when the control law of (12.31) is applied to the system of (12.5) are analyzed. Equations (12.5) and (12.31) can be written as

$$\begin{bmatrix} A(q) & -B(q) \\ G(q) & F(q)B(q) \end{bmatrix} \begin{bmatrix} y(k) \\ u(k) \end{bmatrix} = \begin{bmatrix} C(q) \\ 0 \end{bmatrix} e(k) \qquad (12.33)$$

The characteristic equation of the closed-loop system is obtained by setting the determinant of the matrix on the left-hand side of (12.33) equal to zero. Hence,

$$A(q)F(q)B(q) + G(q)B(q) - B(q)[A(q)F(q) + G(q)] \qquad (12.34)$$
$$= q^{d-1}B(q)C(q) = 0$$

where Equation (12.17), with $m = d$, is used to obtain the second equality. The closed-loop system thus has $2n - d$ poles at the zeros of B and C and an additional $d - 1$ poles at the origin.

The minimum-variance control strategy can be interpreted as a pole-place-ment design, where the poles are placed at the zeros given by (12.34). The sim-ilarities to pole placement are seen even more clearly if the control law of (12.31) is written as

$$u(k) = -\frac{G(q)}{B(q)F(q)} y(k) = -\frac{S(q)}{R(q)} y(k)$$

where $S = G$ and $R = FB$ [compare with Equation (10.4)]. Multiplication of (12.17) by B gives

$$q^{d-1}C(q)B(q) = A(q)F(q)B(q) + G(q)B(q)$$
$$= A(q)R(q) + B(q)S(q)$$

This equation is a special case of the Diophantine equation in (10.22) when B^+ = B and with $A_m = q^{d-1}$ and $A_o = C$.

Systems with Unstable Inverses

Remark 4 to Theorem 12.2, mentions that in the control law given by Theorem 12.2, all process zeros are canceled. If there are process zeros outside the unit disc, the closed-loop system will then have unstable modes that are unobservable from the output. The implications of this are discussed first. Other control laws that do not require all zeros of $B(z)$ to be inside the unit disc are then presented.

Solving Equation (12.33) for y and u gives

$$y(k) = \frac{F(q)}{q^{d-1}} e(k)$$

and

$$u(k) = -\frac{G(q)}{q^{d-1}B(q)} e(k)$$

The necessity of the assumption that B is stable is clearly seen from these equations. If the polynomial B is unstable, the system has unstable modes, which are excited by the disturbance. These unstable modes are coupled to the control signal and the control signal grows exponentially. However, the output signal remains bounded because the unstable modes are not coupled to the output. An example illustrates what happens.

Example 12.5—Cancellation of unstable process zero

Consider a system described by the polynomials

$$A(z) = (z - 1)(z - 0.7)$$

$$B(z) = 0.9z + 1$$

$$C(z) = z(z - 0.7)$$

The polynomial $B(z)$ has a zero $z = -\frac{10}{9}$, which is outside the unit disc. A simulation when using the minimum-variance controller is shown in Fig. 12.1. The presence of the unstable mode is clearly seen in the control signal, although it is not noticeable in the system output. This is called ringing. If the simulation is continued, the control signal will finally be so large that overflow or numerical errors occur. In a practical problem the signal will quickly be so large that the linear approximation is no longer valid. After a short time the unstable mode will then be noticeable in the output.

□

The minimum-variance control law is extended to the case when the polynomial B has zeros outside the unit disc in Theorem 12.3.

Theorem 12.3. Consider a system described by (12.5). Factor the polynomial $B(z)$ as

$$B(z) = B^+(z)B^-(z) \qquad (12.35)$$

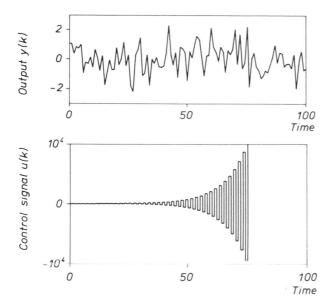

Figure 12.1 Simulation of the system in Example 12.5 with the control law given by Theorem 12.2 that cancels an unstable process zero.

All zeros of the polynomial $B^+(z)$ are inside the unit disc and all zeros of $B^-(z)$ are outside the unit disc or on the unit circle. Assume that all the zeros of polynomial $C(z)$ are inside the unit disc and that the polynomials $A(z)$ and $B^-(z)$ do not have any common factors. The minimum-variance control law is then given by

$$u(k) = -\frac{G(q)}{B^+(q)F(q)} y(k) \qquad (12.36)$$

where $F(q)$ and $G(q)$ are polynomials that satisfy the Diophantine equation

$$q^{d-1}C(q)B^{-*}(q) = A(q)F(q) + B^-(q)G(q) \qquad (12.37)$$

in which the polynomial $F(q)$ has the degree $d + \deg B^- - 1$ and $\deg G < \deg A = n$.

Proof. The proof is based on a clever trick introduced by Wiener in his original work on prediction. An alternative method is used in the proof of Theorem 12.4. Consider the operator

$$\frac{1}{q + a}$$

where $|a| > 1$. This operator is normally interpreted as a casual unstable (unbounded) operator. Because $|a| > 1$, and the shift operator has the norm $\|q\| = 1$, the series expansion

$$\frac{1}{q + a} = \frac{1}{a}\frac{1}{1 + q/a} = \frac{1}{a}\left[1 - \frac{q}{a} + \frac{q^2}{a^2} - \cdots\right]$$

converges. Thus the operator $(q + a)^{-1}$ can be interpreted as a noncausal stable operator; i.e.,

$$\frac{1}{q + a} y(k) = \frac{1}{a} \left[y(k) - \frac{1}{a} y(k + 1) + \frac{1}{a^2} y(k + 2) - \cdots \right]$$

With this interpretation, it follows that

$$(q + a) \left[\frac{1}{q + a} y(k) \right] = y(k)$$

The calculations required for the proof are conveniently done using the backward-shift operator. It follows from the process model of (12.5) that

$$y(k + d) = \frac{B^*(q^{-1})}{A^*(q^{-1})} u(k) + \frac{C^*(q^{-1})}{A^*(q^{-1})} e(k + d)$$

Introduce

$$w(k) = \frac{B^-(q^{-1})}{B^{-*}(q^{-1})} y(k)$$

where the operator $1/B^{-*}(q^{-1})$ is interpreted as a noncausal stable operator. The signals y and w have the same steady-state variance because B^- and B^{-*} are reciprocal polynomials and

$$\left| \frac{B^-(e^{i\omega})}{B^{-*}(e^{i\omega})} \right| = 1$$

An admissible control law that minimizes the variance of w also minimizes the variance of y. It follows that

$$w(k + d) = \frac{B^{+*}(q^{-1})B^-(q^{-1})}{A^*(q^{-1})} u(k) + \frac{C^*(q^{-1})B^-(q^{-1})}{A^*(q^{-1})B^{-*}(q^{-1})} e(k + d) \quad (12.38)$$

The assumption that $A(z)$ and $B^-(z)$ are relatively prime guarantees that (12.37) has a solution. Equation (12.37) implies that

$$C^*(q^{-1})B^-(q^{-1}) = A^*(q^{-1})F^*(q^{-1}) + q^{-d}B^{-*}(q^{-1})G^*(q^{-1})$$

Division by A^*B^{-*} gives

$$\frac{C^*(q^{-1})B^-(q^{-1})}{A^*(q^{-1})B^{-*}(q^{-1})} = \frac{F^*(q^{-1})}{B^{-*}(q^{-1})} + q^{-d} \frac{G^*(q^{-1})}{A^*(q^{-1})}$$

Using this equation (12.38) can be written as

$$w(k + d) = \frac{F^*(q^{-1})}{B^{-*}(q^{-1})} e(k + d) + \frac{B^{+*}(q^{-1})B^-(q^{-1})}{A^*(q^{-1})} u(k) + \frac{G^*(q^{-1})}{A^*(q^{-1})} e(k)$$

$$(12.39)$$

Since the operator $1/B^{-*}(q^{-1})$ is interpreted as a bounded, noncausal operator and since $\deg F^* = d + \deg B^- - 1$, it follows that

$$\frac{F^*(q^{-1})}{B^{-*}(q^{-1})} e(k + d) = \alpha_1 e(k + 1) + \alpha_2 e(k + 2) + \cdots$$

These terms are all independent of the last two terms in (12.39). Using the arguments given in detail in the proof of Theorem 12.2, we find that the optimal-control law is obtained by putting the sum of the last two terms in (12.39) equal to zero. This gives

$$u(k) = -\frac{G^*(q^{-1})}{B^{+*}(q^{-1})B^-(q^{-1})} e(k) \tag{12.40}$$

and

$$y(k) = \frac{B^{-*}(q^{-1})}{B^-(q^{-1})} w(k) = \frac{F^*(q^{-1})}{B^-(q^{-1})} e(k) = \frac{F(q)}{q^{d-1}B^{-*}(q)} e(k) \tag{12.41}$$

Elimination of $e(k)$ between (12.40) and (12.41) gives

$$u(k) = -\frac{G^*(q^{-1})}{B^{+*}(q^{-1})F^*(q^{-1})} y(k) \tag{12.42}$$

The numerator and the denominator have the same degree since $\deg G < n$, and the control law can then be rewritten as (12.36). $\square$

Remark 1. The stable process zeros are canceled by the optimal control-law.

Remark 2. It follows from the proofs of Theorems 12.2 and 12.3 that the variance of the output of a system such as (12.5) may have several local minima if the polynomial $B(z)$ has zeros outside the unit disc. There is one absolute minimum given by Theorem 12.2. However, this minimum will give control signals that are infinitely large. The local minimum given by Theorem 12.3 is the largest of the local minima. The control signal is bounded in this case.

Remark 3. The factorization of (12.35) is arbitrary because B^+ could be multiplied by a number and B^- could be divided by the same number. It is convenient to select the factors so that the polynomial $B^{-*}(q)$ is monic.

Example 12.6—Minimum-variance control with unstable process zero

Consider the system in Example 12.5 where $d = 1$ and

$$B^+(z) = 1$$

$$B^-(z) = B(z)$$

$$B^{-*}(z) = z + 0.9$$

Equation (12.37) becomes

$$z(z - 0.7)(z + 0.9) = (z - 1)(z - 0.7)(z + f_1) + (0.9z + 1)(g_0 z + g_1)$$

Let $z = 0.7$, $z = 1$, and $z = -\frac{10}{9}$. This gives

$$0.7g_0 + g_1 = 0$$

$$g_0 + g_1 = 0.3$$

$$f_1 = 1$$

The control law thus becomes

$$u(k) = -\frac{G(q)}{B^+(q)F(q)} y(k) = -\frac{q - 0.7}{q + 1} y(k)$$

The output is

$$y(k) = \frac{F(q)}{B^{-*}(q)} e(k + d - 1) = \frac{q + 1}{q + 0.9} e(k)$$

$$= e(k) + \frac{0.1}{q + 0.9} e(k)$$

The variance of the output is

$$Ey^2 = \left[1 + \frac{0.1^2}{1 - 0.9^2} \right] \sigma^2 = \frac{20}{19} \sigma^2$$

which is about 5% larger than using the controller in Example 12.5. A simulation of the control law is shown in Fig. 12.2. It is seen from the figure that the regulator performs well. Compare also with Fig. 12.1, which shows the effect of canceling the unstable zero. Also see Problem 12.33. ☐

A Pole-Placement Interpretation

Simple calculations show that the characteristic equation of the closed-loop system obtained from (12.5) and (12.36) is

$$z^{d-1}B^+(z)B^{-*}(z)*C(z) = 0 \tag{12.43}$$

Thus the control law of (12.36) can be interpreted as a pole-placement controller, which gives this characteristic equation.

Multiplication of (12.37) by B^+ gives the equation

$$A(z)R(z) + B(z)S(z) = z^{d-1}B^+(z)B^{-*}(z)C(z) \tag{12.44}$$

where $R(z) = B^+(z)F(z)$ and $S(z) = G(z)$. This equation is the same Diophantine equation that was used in the pole-placement design [compare with Equation (10.22)]. The closed-loop system has poles corresponding to the observer dynamics, to the stable process zeros, and to the reflections in the unit circle of the unstable process zeros. Notice that the transfer function $B(z)/A(z)$ may be interpreted as having $d = \deg A - \deg B$ zeros at infinity. The reflections of these zeros in the unit circle also appear as closed-loop poles.

Equation (12.44) shows that the closed-loop system is of order $2n - 1$ and that $d - 1$ of the poles are in the origin. A full Kalman filter observer and feedback from the observed states give a closed-loop system of order $2n$. The "missing"

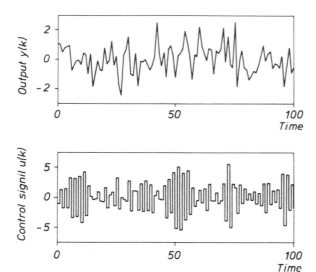

Figure 12.2 Simulation of the system in Example 12.6.

pole is due to a cancellation of a pole $z = 0$ in the controller. This is further discussed in Section 12.5.

12.5 LQG-Control

The optimal control problem for the system of (12.5) with the criterion of (12.8) is now solved. The minimum-variance control law discussed in Sec. 12.4 can be expressed in terms of a solution to a polynomial equation. The solution to the LQG-problem can be obtained in a similar way. Two polynomial equations are needed, however. These equations are discussed before the main result is given.

The name *Gaussian* in LQG is actually slightly misleading. The proofs show that the probability distribution is immaterial as long as the random variables $\{e(k)\}$ are independent.

Spectral Factorization

The LQ-problem is solved in Sec. 11.2 using the state-space approach, which led to a steady-state Riccati equation. It follows from the Riccati equation that

$$rP(z)P(z^{-1}) = \rho A(z)A(z^{-1}) + B(z)B(z^{-1}) \tag{12.45}$$

where the monic polynomial $P(z)$ is the characteristic polynomial of the closed-loop system. This result is derived in Sec. 11.2 [see Equation (11.38)]. The closed-loop characteristic polynomial can be obtained by solving a steady-state Riccati equation. An alternative is to find a polynomial $P(z)$ that satisfies (12.45) directly. A feedback that gives the desired closed-loop poles can then be determined by pole placement. The problem of finding a polynomial $P(z)$ that satisfies (12.45) is called *spectral factorization*.

First consider a polynomial of the form

$$F(z) = f_0 z^{2n} + f_1 z^{2n-1} + \cdots + f_{n-1} z^{n+1} + f_n z^n + f_{n-1} z^{n-1} + \cdots + f_1 z + f_0$$

Such a polynomial is self-reciprocal because

$$F^*(z) = z^{2n} F(z^{-1}) = F(z)$$

It then follows that if $z = a$ is a zero of $F(z)$, then $z = 1/a$ is also a zero. Moreover, if the coefficients f_i are real, then $z = \bar{a}$ and $z = 1/\bar{a}$ are also zeros. The following result can now be established.

Lemma 12.1. Let the real polynomials $A(z)$ and $B(z)$ be relatively prime with deg $A(z) >$ deg $B(z)$. Then there exists a unique polynomial $P(z)$ with deg $P(z) =$ deg $A(z) = n$ and all its zeros inside the unit disc or on the unit circle such that (12.45) holds. If $\rho > 0$, then $P(z)$ has no zeros on the unit circle.

Proof. A self-reciprocal polynomial is obtained if the right-hand side of (12.45) is multiplied by z^n. The zeros of the right-hand side are thus mirror images with respect to the unit circle. Because the coefficients are real, the zeros are also symmetric with respect to the real axis. The right-hand side of (12.45) cannot have zeros on the unit circle because if $z = e^{i\omega}$ is such a zero, then

$$\rho A(e^{i\omega})A(e^{-i\omega}) + B(e^{i\omega})B(e^{-i\omega}) = \rho \, | \, A(e^{i\omega}) \, |^2 + | \, B(e^{i\omega}) \, |^2 = 0$$

As $\rho > 0$, this implies that $z = \exp(i\omega)$ is a zero of both $A(z)$ and $B(z)$, which contradicts the assumption that $A(z)$ and $B(z)$ are relatively prime. The condition deg $P(z) = n$ ensures a unique $P(z)$. ☐

Remark 1. By introducing reciprocal polynomials, Equation (12.45) can be written as

$$rP(z)P^*(z) = \rho A(z)A^*(z) + z^d B(z)B^*(z) \tag{12.46}$$

Remark 2. Notice that condition deg $P = n$ is necessary to ensure that the polynomial P is unique. If $P(z)$ satisfies (12.45) so does $z^l P(z)$ where l is an arbitrary integer. To obtain a unique P we can either specify the degree of P or we can take the P of lowest degree that satisfies (12.45). The degree of P is normally equal to $n =$ deg A and we will therefore make P unique by choosing deg $P = n$ in all cases. Notice that it is possible to find a P of lower degree when $\rho = 0$ or when $A(0) = 0$.

Conceptually the spectral factorization problem can be solved by finding the zeros of the right-hand side of (12.45) and sorting them. There are also efficient recursive algorithms for solving the problem.

Heuristic Discussion

The LQG-problem will now be related to the pole-placement problem. We will first give the solution heuristically. A formal solution will be given later. First recall that the pole-placement problem required specifications of the closed-loop characteristic polynomial which were chosen as $A_r(z)A_o(z)$ when A_o was interpreted as the observer polynomial. In the LQG-problem the observer polynomial is simply $A_o(z) = C(z)$. Compare Theorem 12.1. The polynomial $A_r(z)$ is equal to the polynomial $P(z)$ obtained from the spectral factorization. The reason that we obtain the observer polynomial directly is because equation (12.5) is an innovations model. If (12.5) does not have this form, a spectral factorization is needed to bring it to the innovation form. Compare with Example 12.1. When the polynomials $A_o(z) = C(z)$ and $A_r(z) = P(z)$ are specified we can now expect that the optimal control law is given by

$$u(k) = -\frac{S(q)}{R(q)} y(k) \tag{12.47}$$

where $R(z)$ and $S(z)$ are solutions to the Diophantine equation

$$A(z)R(z) + B(z)S(z) = P(z)C(z) \tag{12.48}$$

The structure of the admissible control laws is determined by the polynomials $R(z)$ and $S(z)$. To describe a control law such that $u(k)$ is a function of $y(k)$, $y(k-1), \ldots$, and $u(k-1), u(k-2), \ldots$, i.e., no delay in the controller, the polynomials $R(z)$ and $S(z)$ should have the same degree. To describe a control law such that $u(k)$ is a function of $y(k-1), y(k-2), \ldots$, and $u(k-2), u(k-3)$, $\ldots$, i.e., one sampling period delay in the controller, the pole excess of $S(z)/R(z)$ should be one. The complexity of the control law is determined by the orders of the polynomials $R(z)$ and $S(z)$.

There are many polynomials $R(z)$ and $S(z)$ that satisfy (12.48). See Section 10.4. The control law that minimizes the loss function (12.8) is, however, unique. This means that the optimality condition leads to a unique solution to (12.48). Before showing this formally we will present the solution.

The case when there is a delay in the controller of one sampling interval, i.e., deg $S(z) = $ deg $R(z) - 1$, is the simplest case. Since deg $A > $ deg B, it follows from (12.48) that deg $R(z) = n$. Identification of coefficients of equal power of z then gives $2n$ equations to determine the $2n$ coefficients of the polynomials $R(z)$ and $S(z)$. It will be shown that this solution is in fact the optimal LQG-controller.

The problem is more complicated when there is no delay in the controller. Since deg $R = $ deg S, it follows from (12.48) that deg $R = n$. Identification of coefficients of equal powers of z in (12.48) gives $2n$ equations to determine the $2n + 1$ coefficients of the polynomials $R(z)$ and $S(z)$. It will be shown that the optimal LQG-controller corresponds to choosing the polynomial S such that $S(0) = 0$. The constant term of the polynomial S is thus equal to zero.

The conditions deg $R(z) = $ deg $S(z) = n$ and $S(0) = 0$ specify a unique solution to (12.48) if $A(0) \neq 0$. This can be seen as follows. If $S_0(z)$ and $R_0(z)$ is the unique solution such that deg $S_0(z) < n$, the general solution to (12.48) is

$$S(z) = S_0(z) - Q(z)A(z)$$

$$R(z) = R_0(z) + Q(z)B(z)$$

where $Q(z)$ is an arbitrary polynomial. Since deg $R_0(z) = n$, the solutions with deg $S(z) = n$ are obtained by choosing $Q(z)$ as a scalar. If $A(0) \neq 0$, the solution with $S(0) = 0$ is uniquely given by

$$S(z) = S_0(z) - \frac{S_0(0)}{A(0)} A(z)$$

$$R(z) = R_0(z) + \frac{S_0(z)}{A(0)} B(z)$$

A convenient way to find the solution to (12.48) where R and S have the same degrees is to rewrite (12.48) as

$$A^*(z)R^*(z) + z^d B^*(z)S^*(z) = P^*(z)C^*(z) \qquad (12.49)$$

where

$$d = \deg A(z) - \deg B(z)$$

Generically the polynomials $P^*(z)$ and $C^*(z)$ are of degree n and the polynomials $R^*(z)$ and $S^*(z)$ are of degree n and $n - 1$. Identification of coefficients of equal power of z then gives $2n$ equations to determine the coefficients of the polynomials. When $A(0) = 0$ the conditions $\deg S(z) = \deg R(z) = n$ and $S(0) = 0$ do not give a unique solution to (12.48). The solution to the optimization problem will, however, tell what auxiliary conditions are required.

To obtain a unique solution to (12.48) it must also be required that the polynomials A and B do not have any common factors. This will not necessarily be the case when the polynomials A and B are given by (12.4). The discussion of this case will be given later.

After this informal discussion we will now give a formal proof of the statements. For this purpose we will first prove a preliminary result.

Lemma 12.2. Let the polynomials $A(z)$, $C(z)$, $P(z)$, $R(z)$, and $S(z)$ all have degree n. Assume the polynomials $R(z)$, $S(z)$, and $P(z)$ satisfy (12.45) and (12.48) and that $S(0) = 0$; then there exists a polynomial $X(z)$ with $\deg X(z) < n$ such that

$$A^*(z)X(z) + rP(z)S^*(z) = B(z)C^*(z) \qquad (12.50)$$

and

$$P^*(z)X(z) + \rho A(z)S^*(z) = R^*(z)B(z) \qquad (12.51)$$

These equations can also be written as

$$A(z^{-1})X(z) + rP(z)S(z^{-1}) = B(z)C(z^{-1}) \qquad (12.50a)$$

$$P(z^{-1})X(z) + \rho A(z)S(z^{-1}) = R(z^{-1})B(z) \qquad (12.51a)$$

Proof. Consider

$$A^*(z)[R^*(z)B(z) - \rho A(z)S^*(z)]$$

$$= A^*(z)R^*(z)B(z) - S^*(z)[rP(z)P^*(z) - z^d B(z)B^*(z)]$$

$$= B(z)[A^*(z)R^*(z) + z^d B^*(z)S^*(z)] - rS^*(z)P(z)P^*(z) \qquad (12.52)$$

$$= B(z)P^*(z)C^*(z) - rS^*(z)P(z)P^*(z)$$

$$= P^*(z)[B(z)C^*(z) - rS^*(z)P(z)]$$

$$= A^*(z)P^*(z)X(z)$$

where the first equality follows from (12.46) and the third from (12.48).

First assume that $A(z)$ and $B(z)$ are relatively prime, as are $P(z)$ and $A(z)$. It then follows that $A^*(z)$ divides $B(z)C^*(z) - rS^*(z)P(z)$ and that $P^*(z)$ divides $R^*(z)B(z) - \rho A(z)S^*(z)$. There is then a polynomial $X(z)$ such that

$$B(z)C^*(z) - rS^*(z)P(z) = A^*(z)X(z)$$

and

$$R^*(z)B(z) - \rho A(z)S^*(z) = P^*(z)X(z)$$

which implies (12.50) and (12.51). Since deg $P(z) = n$ and $S(0) = 0$, it follows that the highest power of z in $P(z)S(z^{-1})$ is at most $n - 1$. Equation (12.50a) then implies that deg $X(z) < n$.

Now consider the case when $A(z)$ and $B(z)$ have common factors. Let $A_2(z)$ be the largest common divisor of $A(z)$ and $B(z)$. It follows from (12.45) that $A_2(z)$ also divides $P(z)$. Dividing (12.52) by $A_2(z)$ and repeating the argument above gives the result. Notice that in this case $A_2(z)$ also divides $X(z)$. ☐

Formal Proof

We will now give a formal solution to the LQG-problem from first principles.

Theorem 12.4. Consider the system in (12.5) with deg $A(z) = $ deg $C(z) = n$. Assume that all the zeros of polynomial $C(z)$ are inside the unit disc, that there are no factors common to all three of the polynomials $A(z)$, $B(z)$, and $C(z)$, and that a possible common factor of $A(z)$ and $B(z)$ has all its zeros inside the unit disc. Let the monic polynomial $P(z)$, which has all its zeros inside the unit disc, be the solution to (12.45) with deg $P(z) = n$. The admissible control law that minimizes the criterion of (12.8) is given by

$$u(k) = -\frac{S(q)}{R(q)} y(k) \qquad (12.53)$$

where polynomials $R(z)$ and $S(z)$ are of degree n such that the polynomial $S^*(z)$ is the unique solution to (12.50) with deg $X(z) < n$. The polynomial $R^*(z)$ satisfies (12.51) and polynomials $R(z)$ and $S(z)$ also satisfy (12.48).

With the control law of (12.53), the output becomes

$$y(k) = \frac{R(q)}{P(q)} e(k) \tag{12.54}$$

and the control signal is

$$u(k) = -\frac{S(q)}{P(q)} e(k) \tag{12.55}$$

The minimal value of the loss function is

$$\min E[y^2 + \rho u^2] = \frac{\sigma^2}{2\pi i} \oint \frac{R(z)R(z^{-1}) + \rho S(z)S(z^{-1})}{P(z)P(z^{-1})} \frac{dz}{z} \tag{12.56}$$

Proof. Introduce

$$u = v - \frac{S}{R} y \tag{12.57}$$

where v may be regarded as a transformed control variable. Equations (12.5) and (12.48) give

$$y = \frac{BRv + CRe}{AR + BS} = \frac{BRv + CRe}{PC} = \frac{BR}{PC} v + \frac{R}{P} e \tag{12.58}$$

It then follows from (12.58) that

$$u = v - \frac{SBv + SCe}{PC} = \frac{PC - BS}{PC} v - \frac{S}{P} e = \frac{AR}{PC} v - \frac{S}{P} e \tag{12.59}$$

The loss function of (12.8) can be written as

$$J = E[y^2 + \rho u^2] = E\left[\frac{BR}{PC} v + \frac{R}{P} e\right]^2 + \rho E\left[\frac{AR}{PC} v - \frac{S}{P} e\right]^2 \tag{12.60}$$

$$= J_1 + 2J_2 + J_3$$

where

$$J_1 = E\left[\left(\frac{BR}{PC} v\right)^2 + \rho \left(\frac{AR}{PC} v\right)^2\right]$$

$$J_2 = E\left[\left(\frac{BR}{PC} v\right)\left(\frac{R}{P} e\right) - \rho \left(\frac{AR}{PC} v\right)\left(\frac{S}{P} e\right)\right]$$

$$J_3 = E\left[\left(\frac{R}{P} e\right)^2 + \rho \left(\frac{S}{P} e\right)^2\right]$$

These terms are now evaluated separately. It follows from Remark 2 of Theorem 6.2 and (12.45) that

$$J_1 = \frac{1}{2\pi i} \oint \frac{(B(z)B(z^{-1}) + \rho A(z)A(z^{-1})) R(z)R(z^{-1})}{P(z)P(z^{-1})C(z)C(z^{-1})} V(z)V(z^{-1}) \frac{dz}{z}$$

$$= \frac{r}{2\pi i} \oint \frac{R(z)R(z^{-1})}{C(z)C(z^{-1})} V(z)V(z^{-1}) \frac{dz}{z} = rE\left(\frac{R(q)}{C(q)} v\right)^2$$

where $v(t) = V(q)e(t)$ and $V(q)$ is a rational function with zero pole excess.

$$J_2 = \frac{1}{2\pi i} \oint \frac{B(z)R(z)R(z^{-1}) - \rho A(z)R(z)S(z^{-1})}{P(z)C(z)P(z^{-1})} V(z)E(z^{-1}) \frac{dz}{z}$$

It follows from Lemma 12.2 that

$$B(z)R(z^{-1}) - \rho A(z)S(z^{-1}) = P(z^{-1})X(z)$$

Hence

$$J_2 = \frac{1}{2\pi i} \oint \frac{R(z)X(z)}{P(z)C(z)} V(z)E(z^{-1}) \frac{dz}{z}$$

$$= E\left[\left(\frac{R(q)X(q)}{P(q)C(q)} v(k) \right) e(k) \right]$$

It was assumed that $P(z)$ and $C(z)$ are stable and it follows from Lemma 12.2 that deg $X(z) < n$. This implies that

$$\deg R(z)X(z) < \deg P(z)C(z) = 2n$$

The quantity

$$\frac{R(q)X(q)}{P(q)C(q)} v(k)$$

is thus a function of $v(k-1)$, $v(k-2)$, ... Since all these terms are independent of $e(k)$, J_2 becomes zero. The loss function can thus be written as

$$J = rE\left(\frac{R(q)}{C(q)} v(k) \right)^2 + E\left(\frac{R(q)}{P(q)} e(k) \right)^2 + \rho E\left(\frac{S(q)}{P(q)} e(k) \right)^2 \quad (12.61)$$

where P and C are stable polynomials. It follows that the loss function achieves its minimum (12.56) for $v = 0$, which by (12.57) corresponds to the control law of (12.53). Equations (12.54) and (12.55) follow from (12.58) and (12.59), and Theorems 6.2 and 6.4 give the formula of (12.56). $\qquad\square$

Remark 1. The minimum-variance control law is a special case of Theorem 12.4 with $\rho = 0$. Since deg $X(z) < n$, it follows from (12.46) and (12.51) that deg $R^*(z) < n$ for $\rho = 0$. Since also deg $S^*(z) < n$, it follows that $R(0) = S(0) = 0$. The polynomials $R(z)$ and $S(z)$ thus have z as a common factor. Introducing $B(z) = B^+(z)B^-(z)$, where B^+ has all its zeros inside the unit disc and B^- all its zeros outside the unit disc, we get

$$P(z) = z^d B^+(z)B^{-*}(z)$$

The Diophantine equation (12.48) then becomes

$$A(z)R(z) + B(z)S(z) = z^d B^+(z)B^{-*}(z)C(z)$$

Cancelling the common factor z in $R(z)$ and $S(z)$ to give $\tilde{R}(z)$ and $\tilde{S}(z)$ we get

$$A(z)\tilde{R}(z) + B(z)\tilde{S}(z) = z^{d-1}B^+(z)B^{-*}(z)C(z)$$

which is identical to (12.37). Theorem 12.3 has thus been proven in a different way. The pole-zero cancellation at the origin of the control law explains that there are $d - 1$ instead of d closed-loop poles at the origin. Compare (12.17).

Remark 2. If the polynomial $A(z)$ has the form $A(z) = z^l A_1(z)$, where $l \le d = \deg A(z) - \deg B(z)$, it follows from (12.45) that $P(z) = z^l P_1(z)$. Equation (12.48) then imples that $S(z) = z^l S_1(z)$.

An Interpretation

Theorem 12.4 establishes the relation between LQG-control and pole-placement control because the polynomial $C(z)$ is the observer polynomial $A_o(z)$ and $P(z)$ is the polynomial $A_r(z)$. The LQG-controller may thus be considered as a pole-placement controller where the observer polynomial $A_o(z)$ is obtained from the noise characteristics and the polynomial $A_r(z)$ from the solution to an optimization problem. The solution to the optimization problem also tells what solution of the Diophantine equation we should choose.

A Computational Procedure

Theorem 12.4 gives a convenient way to compute the LQG-control law for SISO systems, which can be described as follows.

1. Rewrite the model of the process and the disturbance in the standard form (12.5), where $C(z)$ is a stable polynomial. It may be necessary to use a spectral factorization to obtain this form.
2. Use a spectral factorization to calculate $P(z)$. If the polynomials $A(z)$ and $B(z)$ have a stable common factor $A(z)$, the calculations of the control law can be simplified by first factoring $A(z)$ and $B(z)$ as $A(z) = A_1(z)A_2(z)$ and $B(z) = B_1(z)A_2(z)$. It follows from (12.45) that $A_2(z)$ also divides $P(z)$. This polynomial can thus be written as $P(z) = P_1(z)A_2(z)$, where $P_1(z)$ is given by

$$rP_1(z)P_1(z^{-1}) = \rho A_1(z)A_1(z^{-1}) + B_1(z)B_1(z^{-1}) \qquad (12.62)$$

The polynomial $P(z)$ is then equal to $P_1(z)A_2(z)$, which is stable, since $A_2(z)$ was assumed stable. Equation (12.49) can also be divided by $A_2(z)$ to give

$$P_1(z)C(z) = A_1(z)R(z) + B_1(z)S(z) \qquad (12.63)$$

where $\deg R(z) = \deg S(z) = \deg C(z) = n$ and $S(0) = 0$.
3a. If $A(0) \ne 0$, find the unique solution to the Diophantine equation (12.48) such that $\deg R(z) = \deg S(z) = n$ and $S(0) = 0$.
3b. If $A(0) = 0$ the solution is obtained from the Diophantine equation (12.48), and equations (12.50), and (12.51).

Theorem 12.4 is illustrated by an example.

Example 12.7

Consider a system characterized by

$$A(z) = z + a, \quad a \neq 0$$
$$B(z) = b$$
$$C(z) = z + c$$

To find the control law that minimizes the criterion of (12.8), the spectral factorization problem is first solved. Equation (12.45) can be written as

$$r(z + p_1)(z^{-1} + p_1) = \rho(z + a)(z^{-1} + a) + b^2$$

Equating coefficients of equal powers of z gives

$$rp_1 = \rho a$$
$$r(1 + p_1^2) = \rho(1 + a^2) + b^2$$

Elimination of p_1 gives

$$r^2 - r(\rho(1 + a^2) + b^2) + \rho^2 a^2 = 0 \tag{12.64}$$

This equation has the solution

$$r = \frac{1}{2} \left(\rho(1 + a^2) + b^2 + \sqrt{\rho^2(1 - a^2)^2 + 2\rho b^2(1 + a^2) + b^4} \right)$$

where the positive root is chosen to give $|p_1| < 1$. Furthermore

$$p_1 = \frac{\rho a}{r}$$

With $X(z) = x_0$ and $S^*(z) = s_0$, $S(z) = s_0 z$. Equation (12.50) becomes

$$(1 + az)x_0 + r(z + p_1)s_0 = b(1 + cz)$$

Putting $z = -p_1$ and $z = -1/a$ gives

$$x_0 = \frac{b(1 - cp_1)}{1 - ap_1}$$

$$s_0 = \frac{b(c - a)}{r(1 - ap_1)}$$

Solving (12.51) for $R^*(z)$ gives

$$R^*(z) = \frac{\rho c}{r} z + 1$$

which means that

$$R(z) = z + \frac{\rho c}{r}$$

Since $A(0) \neq 0$, the solution can also be found from the Diophantine equation (12.48). With deg $S = 1$ and $S(0) = 0$, equation (12.48) becomes

$$(z + a)(z + r_1) + bs_0 z = (z + p_1)(z + x_1)$$

Putting $z = -a$ we get

$$s_0 = - \frac{(p_1 - a)(c - a)}{ab}$$

It follows from (12.64) that

$$\rho a p_1^2 - \rho p_1 (1 + a^2) - p_1 b^2 + \rho a = 0$$

Hence

$$\rho(a p_1^2 - a^2 p_1 - p_1 + a) = p_1 b^2$$

or

$$\rho(a p_1 (p_1 - a) - (p_1 - a)) = p_1 b^2$$

Hence

$$p_1 - a = - \frac{p_1 b^2}{\rho(1 - a p_1)}$$

We thus get

$$s_0 = \frac{p_1 b^2 (c - a)}{\rho a b (1 - a p_1)} = \frac{b(c - a)}{r(1 - a p_1)}$$

Furthermore, equating the constant terms in (12.48) gives

$$r_1 = \frac{p_1 c}{a}$$

The control law thus becomes

$$u(k) = - \frac{S(q)}{R(q)} y(k) = - \frac{b(c - a)}{r(1 - p_1)} \frac{q}{q + p_1 c/a} y(k) \qquad \square$$

The calculations in Example 12.7 do not work when $a = 0$, because in this case the solution to the LQG-problem is not uniquely determined by the Diophantine equation (12.48) and it is necessary to use Lemma 12.2.

Example 12.8

Consider the case

$$A(z) = z$$
$$B(z) = b$$
$$C(z) = z + c$$

Solving the spectral factorization problem (12.45) with the constraint deg $P(z) = 1$ gives

$$P(z) = z, \qquad r = \rho + b^2$$

The Diophantine equation (12.48) with the constraint deg $S(z) = $ deg $R(z) = 1$ and $S(0) = 0$ gives

$$z(z + r_1) + b s_0 z = z(z + c)$$

This only gives one equation

$$r_1 + bs_0 = c$$

to determine two parameters r_1 and s_0. Equation (12.50a) gives

$$z^{-1}x_0 + rzs_0z^{-1} = b(z^{-1} + c)$$

Hence

$$rs_0 = bc$$

$$x_0 = b$$

or

$$s_0 = \frac{bc}{r} = \frac{bc}{\rho + b^2} \qquad\qquad \square$$

Uncontrollable and Unstable Modes

Models with the property that polynomials $A(z)$ and $B(z)$ have a common factor that is not a factor of $C(z)$ are important in practice. They appear when there are modes that are excited by disturbances and uncontrollable from the input. Compare Section 12.2. Because the modes are not controllable, they are not influenced by feedback.

Theorem 12.4 covers the case of stable common factors, but it does not work for unstable common factors. Unstable common factors are important in practice because they give one way of obtaining regulators with integral action.

To see what happens when there are unstable common factors, let A_2 denote the greatest common divisor of A and B and let A_2^- denote the factor of A_2 with zeros outside the unit disc or on the unit circle. Let the feedback be

$$u(k) = -\frac{S(q)}{R(q)}y(k)$$

where $R(z)$ and $S(z)$ are relatively prime. It follows from (12.5) that

$$y(k) = \frac{R(q)C(q)}{A(q)R(q) + B(q)S(q)}e(k) \qquad\qquad (12.65)$$

$$u(k) = -\frac{S(q)C(q)}{A(q)R(q) + B(q)S(q)}e(k) \qquad\qquad (12.66)$$

The unstable factor $A_2^-(z)$ divides the denominators of the right-hand sides of (12.65) and (12.66). Both y and u will be unbounded unless $R(z)$ or $S(z)$ are chosen in special ways. The signal y will be bounded if $R(z)$ is divisible by $A_2^-(z)$, and u will be bounded if $A_2^-(z)$ divides $S(z)$. Since $R(z)$ and $S(z)$ are relatively prime, it is not possible to make both y and u bounded. This is natural because infinitely large control actions are necessary to compensate for infinitely large disturbances.

To describe a problem of this type as a meaningful optimization problem, the criterion of (12.8) must be modified. One possibility is to introduce the variable

$$w(k) = q^{-m}A_2^-(q)u(k) \qquad\qquad (12.67)$$

where $m = \deg A_2^-(z)$, and to introduce the criterion

$$J'_{lq} = E[y^2(k) + \rho w^2(k)] \tag{12.68}$$

Example 12.9—Integral action

Let the system be described by

$$y(k) = \frac{B_1(q)}{A_1(q)} u(k) + \frac{C_1(q)}{q-1} e(k)$$

which is a special case of equations (12.1)–(12.4) with a drifting disturbance. Hence

$$A(q) = (q-1)A_1(q)$$

$$B(q) = (q-1)B_1(q)$$

$$C(q) = A_1(q)C_1(q)$$

Unbounded control signals are necessary to compensate for the unbounded disturbance. This implies that the modified loss function (12.68) becomes

$$J'_{lq} = E[y^2(k) + \rho(\Delta u(k))^2]$$

where

$$\Delta u(k) = u(k) - u(k-1)$$

This means that the difference and not the absolute value of the control signal is penalized. The solution to the LQG-problem gives a controller with integral action. □

The following result can then be established.

Theorem 12.5   Consider the system described by (12.5) where $A(z)$ and $C(z)$ are monic polynomials of degree n. Assume that all zeros of $C(z)$ are inside the unit disc and that there is no nontrivial polynomial that divides $A(z)$, $B(z)$, and $C(z)$. Let $A_2(z)$ be the greatest common divisor of $A(z)$ and $B(z)$, let $A_2^+(z)$ of degree l be the factor of $A_2(z)$ with all its zeros inside the unit disc, and let $A_2^-(z)$ of degree m be the factor of $A(z)$ which has zeros on the unit circle or outside the unit disc. The admissible control law that minimizes (12.68) is given by

$$u(k) = - \frac{S(q)}{R(q)} y(k) \tag{12.69}$$

where $R(z)$ and $S(z)$ are of degree $n + m$

$$R(z) = A_2^-(z)\tilde{R}(z) \tag{12.70}$$

$$S(z) = z^m \tilde{S}(z)$$

and $\tilde{R}(z)$ and $\tilde{S}(z)$ satisfies

$$A_1(z)A_2^-(z)\tilde{R}(z) + z^m B_1(z)\tilde{S}(z) = P_1(z)C(z) \tag{12.71}$$

with deg $\tilde{R}(z) = $ deg $\tilde{S}(z) = n$ and $\tilde{S}(0) = 0$. Furthermore

$$A(z) = A_1(z)A_2(z)$$

$$B(z) = B_1(z)A_2(z) \tag{12.72}$$

$$\tilde{B}(z) = B_1(z)A_2^+(z)$$

and $P_1(z)$ is the solution of the spectral factorization problem

$$rP_1(z)P_1(z^{-1}) = \rho A_1(z)A_2^-(z)A_1(z^{-1})A_2^-(z^{-1}) + B_1(z)B_1(z^{-1}) \tag{12.73}$$

with deg $P_1(z) = $ deg $A_1(z) + $ deg $A_2^-(z)$. $\qquad\square$

Proof: Introducing the signal (12.67), the model (12.5) can be written as

$$A(q)y(k) = \tilde{B}(q)q^m w(k) + C(q)e(k)$$

The polynomials $A(q)$ and $\tilde{B}(q)$ have the common factor $A_2^+(z)$, which has all its zeros inside the unit disc, but no other common factors with zeros outside the unit disc or on the unit circle. It then follows from Theorem 12.4 that the optimal control law

$$w(k) = -\frac{\tilde{S}(q)}{\tilde{R}(q)} y(k) \tag{12.74}$$

is obtained from (12.48). Since $A(z)$ and $\tilde{B}(z)$ have the stable common factor $A_2^!(z)$, the polynomial $P(z)$ has the form

$$P(z) = A_2^+(z)P_1(z)$$

where $P_1(z)$ is the solution to the spectral factorization problem (12.73). The polynomials $\tilde{R}(z)$ and $\tilde{S}(z)$ satisfy the Diophantine equation

$$A(z)\tilde{R}(z) + z^m\tilde{B}(z)\tilde{S}(z) = A_2^+(z)P_1(z)C(z)$$

with deg $\tilde{R}(z) = $ deg $\tilde{S}(z) = n$. Since A_2^+ divides $A(z)$ and $\tilde{B}(z)$ we get (12.71). Using (12.67) to express the control law in terms of the control variable u gives the result. $\qquad\square$

Remark 1. Notice that using (12.70), Equation (12.71) can be written as

$$A(z)R(z) + B(z)S(z) = A_2(z)P_1(z)C(z)$$

The LQG-solution can thus be interpreted as a pole-placement controller, where the poles are positioned at the zeros of A_2, P_1, and C. The controller also has the property that A_2^- divides R. This is an example of the internal model principle.

Remark 2. A result analogous to Lemma 12.2 also holds.

Comparisons with the State-Space Approach

The problems discussed in this chapter can also be solved by the state-space methods of Chapter 11. To compare the different approaches, a state-space representation of the model of (12.5) is first given. For this purpose it is assumed that the model is normalized, so that deg $C(z)$ = deg $A(z)$. The model of (12.5) can then be represented as

$$x(k + 1) = \Phi x(k) + \Gamma u(k) + Ke(k)$$
$$y(k) = Cx(k) + e(k) \tag{12.75}$$

where

$$\Phi = \begin{bmatrix} -a_1 & 1 & 0 & \cdots & 0 \\ -a_2 & 0 & 1 & \cdots & 0 \\ \vdots & & & & \\ -a_{n-1} & 0 & 0 & \cdots & 1 \\ -a_n & 0 & 0 & \cdots & 0 \end{bmatrix}, \Gamma = \begin{bmatrix} b_1 \\ b_2 \\ \vdots \\ b_n \end{bmatrix}, K = \begin{bmatrix} c_1 - a_1 \\ c_2 - a_2 \\ \vdots \\ c_n - a_n \end{bmatrix} \tag{12.76}$$

$$C = \begin{bmatrix} 1 & 0 & \cdots & 0 \end{bmatrix}$$

Since this is an innovation representation, the steady-state Kalman filter is obtained by inspection:

$$\hat{x}(k + 1 \mid k) = \Phi \hat{x}(k \mid k - 1) + \Gamma u(k) + K[y(k) - C\hat{x}(k \mid k - 1)] \tag{12.77}$$

The Kalman filter has the characteristic polynomial

$$\det[zI - (\Phi - KC)] = C(z) \tag{12.78}$$

The control laws derived in Chapter 11 assume a computational delay of one sampling period. The optimal control law is then

$$u(k) = -L\hat{x}(k \mid k - 1)$$

The transfer function of the optimal regulator is

$$H_r(z) = L(zI - \Phi + KC + \Gamma L)^{-1}K = \frac{S(z)}{R(z)}$$

where $S(z)$ and $R(z)$ form the solution to (12.49) with deg $S(z) < n$ and deg $R(z)$ = n. It follows from this discussion and Section 11.4 that

$$R(z) = \det(zI - \Phi + KC + \Gamma L) \tag{12.79}$$

and

$$P(z) = \det(zI - \Phi + \Gamma L) \tag{12.80}$$

Notice that the steady-state covariance matrix of the estimate is zero. This can be understood intuitively because $x(k + 1)$ can be computed exactly from $y(k)$, $y(k - 1)$, ... This also implies that

$$\hat{x}(k \mid k) = \hat{x}(k \mid k - 1)$$

because $y(k)$ gives no additional information about $x(k)$.

It is more complicated to derive the control law when the admissible control is such that $u(k)$ is a function of $y(k)$, $y(k - 1)$, ... In this case we use the results from state-space theory (Remark 2 of Theorem 11.6) that the control law is

$$u(k) = -L\hat{x}(k \mid k) - L_v\hat{v}(k \mid k)$$

$$= -L\hat{x}(k \mid k - 1) - L_vK(y(k) - C\hat{x}(k \mid k - 1)) \quad (12.81)$$

$$= -L_v(\Phi - KC)\hat{x}(k \mid k - 1) - L_vKy(k)$$

where $\hat{x}(k \mid k - 1)$ is given by (12.77). Recall that $L = L_v\Phi$. Eliminating $\hat{x}$ between (12.77) and (12.81), we find that the controller can be described by the transfer function

$$H_r(z) = L_v(\Phi - KC)(zI - (I - \Gamma L_v)(\Phi - KC))^{-1}(I - \Gamma L_v)K + L_vK$$

$$= zL_v(zI - (\Phi - KC)(I - \Gamma L_v))^{-1} K \quad (12.82)$$

where the second equality is obtained after straightforward but tedious calculations.

Command Signals

The discussion in this chapter has so far been limited to the regulator problem. To introduce command signals, refer to the discussion in Chapter 10. The key issue is to introduce the command signals in such a way that they do not generate unneccessary reconstruction errors. This is achieved by the control law

$$R(q)u(k) = t_0A_o(q)u_c(k) - S(q)y(k)$$

where A_o is the observer polynomial and t_0 a constant. For the optimal Kalman filter $A_o = C$, where C is given by (12.78). It then follows from (12.5) that the output of the system is given by

$$y(k) = t_0 \frac{B(q)}{P(q)} u_c(k) + \frac{R(q)}{P(q)} e(k)$$

where $\deg R = n$.

The pulse-transfer function from the command signal is $B(z)/P(z)$. This response may be shaped further by cascading with a precompensator that has an arbitrary, stable transfer function $H_p(z)$. The control law becomes

$$u(k) = \frac{A_o(q)}{R(q)} H_p(q)u_c(k) - \frac{S(q)}{R(q)} y(k) \quad (12.83)$$

which gives

$$y(k) = \frac{B(q)}{P(q)} H_p(q)u_c(k) + \frac{R(q)}{P(q)} e(k) \quad (12.84)$$

Because the polynomial P is stable, this may be canceled by the precompensator. It thus follows that the response for disturbances and command signals may be shaped differently.

The feedback S/R is first designed to ensure a good response to disturbances. The precompensator H_p is then chosen to obtain the desired response to command signals.

12.6 Practical Aspects

Much of the arbitrariness of design seems to disappear when design problems are formulated as optimization problems. The model and the criteria are stated, and the control law is obtained simply as the solution to an optimization problem. This simplicity is deceptive because the arbitrariness is instead transferred to the modeling and the formulation of criteria. A successful application of optimization theory requires insight into how the properties of the model and the criteria are reflected in the control law. Typical questions are: What should the model look like in order to get a regulator with integral action? What problem statements give regulators with a PID-structure? Some of these issues are discussed in this section, which also gives insight into the properties of the optimal-control laws. It turns out that some results can be formulated as design rules. The polynomial approach, which operates directly with the transfer functions, is well suited to do this.

Other aspects of practical relevance, such as sensitivity and robustness, are also discussed. A brief treatment of the intersample ripple of the loss function is given, together with some aspects of the choice of the sampling period.

Properties of the Optimal Regulator

Some properties of the model influence the optimal-control laws. The basic model used is given by (12.5)—i.e.,

$$A(q)y(k) = B(q)u(k) + C(q)e(k) \qquad (12.85)$$

The ratio B/A represents the pulse-transfer function of the process, and the ratio C/A represents the pulse-transfer function that generates the disturbance of the process output. The polynomials A, B, and C may have common factors that reflect the way the control signal and the disturbance are coupled to the system. There are, however, no factors common to all three polynomials. Compare this with the discussion in Sec. 12.2, where the model is derived. The presence of common factors which will directly influence the properties of the regulators will now be investigated.

The internal-model principle.   Factors that are common to polynomials A and B correspond to disturbance modes that are not controllable from u. Such modes will appear as factors of P. Let

$$A_2 = gcd(A, B) \qquad (12.86)$$

be the greatest common divisor of polynomials A and B. If A_2 is stable, it follows from Theorem 12.4 that A_2 also divides P. If A_2 has a factor A_2^- with all its zeros outside the unit disc, the corresponding result follows from Theorem 12.5. In this case it also follows from Theorem 12.5 that A_2^- divides R. This observation is called *the internal-model principle*; it says that to regulate a system with unstable disturbances, the disturbance dynamics must also appear in the dynamics of the regulator. A few examples illustrate this idea.

Example 12.10—Integral action

A regulator has integral action if $z - 1$ divides $R(z)$. It follows from Theorem 12.5, and the internal-model principle, that this will occur if $z - 1$ divides both A and B, which means that the model is of the form

$$A_1(q)(q - 1)y(k) = B_1(q)(q - 1)u(k) + C(q)e(k) \qquad (12.87)$$

This means that there is a drifting disturbance. □

Example 12.11—Elimination of a sinusoidal disturbance

A narrow-band sinusoidal disturbance with frequency centered at ω may be represented as white noise driving a system with the denominator

$$D(q) = q^2 - 2q \cos \omega h + 1$$

If the poles of the system dynamics do not correspond to D, the model becomes

$$A_1(q)D(q)y(k) = B_1(q)D(q)u(k) + C(q)e(k) \qquad (12.88)$$

The optimal regulator is then such that $D(z)$ divides $R(z)$. □

Cancellation of process poles. A common factor of A and C corresponds to controllable modes that are not excited by the disturbances. Let A_2 be the greatest common divisor of A and C. The polynomial A_2 is stable because C is stable, and it does not divide B because there is no factor which divides all of A, B, and C. It follows from (12.49) that A_2 also divides the polynomial S, which is the numerator of the regulator transfer function. Thus *stable process poles that are not excited by the disturbances may be canceled*.

Cancellation of process zeros. Common factors of B and C correspond to process zeros that block transmission both for the control signal u and for the disturbance e. Let B_2 be the greatest common divisor of B and C. The polynomial B_2 is stable and it does not divide A. It then follows from (12.49) that B_2 divides R. This means that the zeros corresponding to $B_2 = 0$ are canceled by the regulator. Therefore, *process zeros that are also transmission zeros for the disturbance C are canceled by the regulator*.

For the minimum-variance control, it follows from (12.46) with $\rho = 0$ that

$$P = q^d B^+ B^{-*}$$

and from (12.49) that B^+ divides R. All stable zeros are thus canceled by the minimum-variance control law.

An analysis of the properties of the optimal-control law thus gives partial answers to the classic cancellation problem.

Sensitivity and Robustness

It is important that a control system be insensitive or robust with respect to measurement errors, plant disturbances, and modeling errors. This may be analyzed as in Sec. 10.6 for the pole-placement problem. The robustness properties are conveniently expressed in terms of the loop gain:

$$H_{lg} = \frac{BS}{AR} \tag{12.89}$$

or the return difference

$$H_{rd} = 1 + \frac{BS}{AR} = \frac{AR + BS}{AR} = \frac{PC}{AR} \tag{12.90}$$

The loop gain $H_{lg}(\exp i\omega h)$ is normally high for low frequencies and small for high frequencies. The crossover frequency ω_c is the lowest frequency, where

$$| H_{lg}(e^{i\omega_c h}) | = 1$$

The closed-loop system is insensitive to plant disturbances at those frequencies where the loop gain is high. To have low sensitivity to poor modeling of the high-frequency dynamics of the plant, it is desirable that the loop gain decrease rapidly above the crossover frequency. It is possible to make sure that the loop gain is high for certain frequencies by choosing models with special structure, as was done in Examples 12.10 and 12.11. Plots similar to those in Fig. 10.5 are also useful in evaluating the sensitivity. In a properly designed sample-data system, there will be antialiasing filters, which eliminate signal transmission above the Nyquist frequency. The selection of a proper sampling rate is one way to make sure that the loop gain is low over a given frequency. This also means that high-frequency modeling errors have little influence. Notice, however, that plots of the loop gain and the return difference will not give the complete picture because there may be pole-zero cancellations that do not show up in these plots.

An analysis of the characteristic equations is useful in such a case. To perform such an analysis, assume that the system is governed by

$$A^0(q)y(k) = B^0(q)u(k) + C^0(q)e(k) \tag{12.91}$$

but that a regulator is designed based on a different model, as in (12.85). The regulators given by Theorems 12.4 and 12.5 give a closed-loop system with the characteristic polynomial

$$
\begin{aligned}
A^0R + B^0S &= A^0R - AR + B^0S - BS + AR + BS \\
&= PC + (A^0 - A)R + (B^0 - B)S
\end{aligned}
\tag{12.92}
$$

When the model of (12.85) is equal to the system of (12.91) the characteristic polynomial is $PC = P_1 A_2 C$, as expected. By continuity it also follows that small changes in the system give small changes in the closed-loop poles. The system is sensitive to changes in the parameters if polynomial P_1 or C have zeros close to the unit circle.

To guarantee systems with a low sensitivity, it is necessary to impose further constraints. Recall that both C and P_1 were obtained as solutions to a spectral factorization problem.

Closed-Loop Systems with Guaranteed Exponential Stability

The control laws given by Theorems 12.2, 12.3, 12.4, and 12.5 give closed-loop systems with poles inside the unit disc. It is sometimes desirable to have control laws such that the closed-loop system has its poles inside a circle with radius r. It is straightforward to formulate optimization problems that give such control laws.

Introduce the criterion

$$J = E\left[\left(\frac{1}{r}\right)^{2k} (y^2(k) + \rho u^2(k))\right] \tag{12.93}$$

If a control law that minimizes this criterion can be found, the variables $y(k)$ and $u(k)$ must converge to zero at least as fast as r^k when k increases. To obtain such a result, it must be assumed that the model of (12.5) is such that the covariance of $e(k)$ also goes to zero as r^k.

Introduce the scaled variables η, μ, and ϵ defined by

$$y(k) = r^k \eta(k)$$
$$u(k) = r^k \mu(k) \tag{12.94}$$
$$e(k) = r^k \epsilon(k)$$

Because

$$q^l y(k) = q^l [r^k \eta(k)] = r^{k+l} \eta(k+l) = r^k (rq)^l \eta(k)$$

It follows that

$$A(q)y(k) = A(q)[r^k \eta(k)] = r^k A(rq)\eta(k)$$

Introducing the transformed polynomials

$$\tilde{A}(z) = A(rz)$$
$$\tilde{B}(z) = B(rz)$$
$$\tilde{C}(z) = C(rz)$$

the model of (12.5) can be written as

$$\tilde{A}(q)\eta(k) = \tilde{B}(q)\eta(k) + \tilde{C}(q)\epsilon(k) \tag{12.95}$$

and the criterion of (12.93) becomes

$$J = E[\eta^2(k) + \rho\mu^2(k)] \tag{12.96}$$

The control law that minimizes (12.96) for the system of (12.95) is then given by Theorem 12.4. This control law gives a closed-loop system in which all the zeros of the characteristic equation

$$\tilde{P}(z)\tilde{C}(z) = 0$$

are inside the unit disc. Going back to the original variables results in the characteristic equation

$$P(z)C(z) = \tilde{P}\left(\frac{z}{r}\right)\tilde{C}\left(\frac{z}{r}\right) = 0$$

All the zeros of this equation are inside the circle $|z| = r$.

A simple procedure for obtaining feedback laws that give closed-loop systems with all poles inside the circle $|z| = r$ has thus been devised.

Disturbance Reduction

The return difference is

$$H_{rd}(z) = 1 + H_{lg}(z) = 1 + \frac{BS}{AR} = \frac{AR + BS}{AR}$$

The inverse of the return difference is a measure of how effectively the closed-loop system eliminates disturbances.

Consider the model of (12.85). Without control the output is

$$y_{ol} = \frac{C}{A}e$$

With the LQ-control law, the output becomes

$$y_{lqg} = \frac{R}{P}e$$

Elimination of e between these equations gives

$$y_{lqg} = \frac{AR}{PC}y_{ol} = \frac{1}{\dfrac{PC}{AR}}y_{ol} = \frac{1}{1 + \dfrac{BS}{AR}}y_{ol} = \frac{1}{H_{rd}}y_{ol} \qquad (12.97)$$

The pulse-transfer function

$$H(z) = \frac{A(z)R(z)}{P(z)C(z)} = \frac{1}{H_{rd}(z)} \qquad (12.98)$$

which is the inverse of the return difference, thus tells how much disturbances of different frequencies are attenuated.

Selection of the Sampling Period

There is a substantial difference between the minimum-variance control law discussed in Sec. 12.4 and the LQG-control law discussed in Sec. 12.5 in terms of the influence of the sampling period. The choice of sampling period is critical for the minimum-variance control. A short sampling period gives a high-bandwidth system, which settles quickly. The control actions will also be large when the sampling period is short. In this respect, the minimum-variance control law is similar to the deadbeat control law discussed in Sec. 9.2. The sampling period is less critical for LQG-control. It follows from the analysis of Sec. 11.5 that the control law approaches continuous-time control as the sampling period h goes to zero. For small sampling periods, the sampled loss is linear in h. The following discussion therefore concentrates on the minimum-variance control law.

Intersample Variation of the Output Variance

The minimum-variance control law minimizes the variance of the output *at the sampling instants*. However, the main objective may be to minimize the continuous-time loss function of (12.7). This may be achieved by first sampling the continuous-time loss function and to minimize the corresponding discrete-time loss function. This results in a more complicated design procedure. The minimum-variance control laws are in many cases a sufficiently good approximation. It is useful to investigate the intersample variation of the loss function. This analysis is similar to the analysis of intersample ripple for deterministic systems of Sec. 10.5. An example is used to illustrate the idea.

Example 12.12—Intersample variation of the loss function

Consider the continuous-time system

$$dx = u\, dt + dv \tag{12.99}$$

where $\{v(t)\}$ is a Wiener process with incremental covariance $\sigma_v^2\, dt$. Assume that the output is observed without antialiasing filters at times $t_k = k \cdot h$, where h is the sampling period. Hence,

$$y(t_k) = x(k) + \epsilon(t_k)$$

where $\{\epsilon(t_k)\}$ is a sequence of independent random variables with zero mean and covariance σ_ϵ^2. Sampling of the system gives

$$x(kh + h) = x(kh) + hu(kh) + v(kh + h) - v(kh)$$

$$y(kh) = x(kh) + \epsilon(kh)$$

Hence,

$$y(kh + h) = y(kh) + hu(kh) + \epsilon(kh + h) - \epsilon(kh) + v(kh + h) - v(kh)$$

The disturbance on the right-hand side may be represented as

$$w(kh + h) = e(kh + h) + ce(kh)$$

where $\{e(kh)\}$ is a sequence of independent, zero-mean, random variables with standard derivation σ.

Simple calculations give

$$c = -1 - \frac{h\sigma_v^2}{2\sigma_\epsilon^2} + \sqrt{\frac{h\sigma_v^2}{\sigma_\epsilon^2} + \frac{h^2\sigma_v^4}{4\sigma_\epsilon^4}} \tag{12.100}$$

$$\sigma^2 = -\frac{\sigma_\epsilon^2}{c} \tag{12.101}$$

The minimum-variance control law for the system is

$$u(kh) = -\frac{1+c}{h} y(kh)$$

The standard deviation of the output under minimum variance control is

$$Ey^2(t) = \sigma^2 \qquad t = h, 2h, \ldots$$

The standard deviation of the state variable x is

$$Ex^2(t) = \sigma^2 - \sigma_\epsilon^2 \qquad t = h, 2h, \ldots$$

Equation (12.99) is integrated to determine the variance of the state variable between the sampling instants. This gives

$$x(kh + s) = x(kh) + su(kh) + v(kh + s) - v(kh)$$
$$= (1 - \alpha s)x(kh) - \alpha s\epsilon(kh) + v(kh + s) - v(kh)$$

where

$$\alpha = (1 + c)/h$$

Introduce

$$P(s) = Ex^2(kh + s)$$

It then follows that

$$P(s) = (1 - \alpha s)^2 (\sigma^2 - \sigma_\epsilon^2) + (\alpha s)^2 \sigma_\epsilon^2 + s\sigma_v^2$$

The function $P(s)$ is shown in Fig. 12.3. Notice that

$$\max_s [P(0) - P(s)] = h^2 \sigma_v^2/2$$

The variation in P over a sampling interval thus decreases with decreasing h. □

The analysis is similar in the general case. The only difference is that Theorem 6.5 must be used to compute the state covariance. In the example the variance

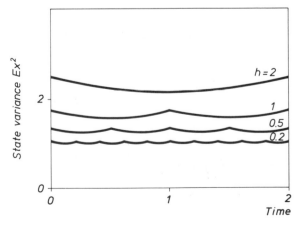

Figure 12.3 Variations of the output variance in Example 12.12 with time for regulators having different sampling periods.

Optimal Design Methods: Input-Output Approach   Chap. 12

is largest at the sampling instants. This is not always the case. Also notice that the correct way of dealing with intersample ripple is to sample the continuous-time system and the continuous-time loss function as was discussed in Section 11.1.

Computational Aspects

The LQ-control law can be determined by a combination of spectral factorization and solution of linear Diophantine equations. Recall, however, the fundamental difficulty that arises from poor numerical conditioning of polynomial equations (see Sec. 15.6).

12.7 Conclusions

In this chapter optimal control problems are solved for systems described by input-output models. The results given are limited to single-input–single-output systems. A canonical model for the system, Equation (12.5), is derived first. This model is characterized by three polynomials, A, B, and C. The underlying continuous-time model may be described as a combination of a time delay and a system with rational transfer functions. The disturbances are characterized as filtered white noise. There are many physical systems that can be described by such models.

Optimal-control problems characterized by quadratic loss functions are solved for the system. A special case where the loss function simply is the variance of the output is considered first. The general problem, in which there is also a penalty on the control variable, is then treated. Both these problems are closely related to the prediction problem for a random process with rational spectral density. This problem is also solved. Practical aspects, such as selection of the sampling period, are also discussed.

The solutions to the optimal-control problems give design tools. The solutions also give insight into the character of the optimal solutions. In particular, they tell that the optimal regulator always cancels stable process zeros that are also zeros for the process disturbances. Stable process poles are cancelled only if they are not excited by disturbances. The results also give insight into the relationships between the different design methods. For instance, the LQG solutions can be interpreted as pole-placement regulators, where the process poles and the observer poles are chosen in special ways.

Calculation of the optimal solution is expressed in terms of two polynomial operations, spectral factorization and solution of Diophantine equations.

12.8 Problems

12.1. Consider the process

$$y(k) = 2 \frac{q^2 - 1.4q + 0.5}{q^2 - 1.2q + 0.4} e(k)$$

where $e(k)$ is white noise with zero mean and unit variance. Determine the optimal m-step-ahead predictor and the variance of the prediction error when $m = 1, 2,$ and 3.

12.2. Determine the m-step-ahead predictor for the process

$$y(k) + ay(k - 1) = e(k) + ce(k - 1)$$

Determine also the variance of the prediction error as a function of m.

12.3. A stochastic process is described by

$$y(k) - 0.9y(k - 1) = e(k) + 5e(k - 1)$$

(a) Determine an equivalent description such that the zero of a corresponding polynomial C is inside the unit circle.
(b) Determine the two-step-ahead predictor for the process and the variance of the prediction error.

12.4. Assume that the demand for a product in an inventory, $z(k)$, can be described as

$$z(k) = 300 + 10k + y(k)$$

where the time unit is months and $y(k)$ is described by the process

$$y(k) - 0.7y(k - 1) - 0.1y(k - 2) = 5e(k)$$

where $e(k)$ is white noise with zero mean and unit variance. Make a prediction and determine the expected standard deviation of the prediction error for August through November when the following data are available:

Month	k	$z(k)$
January	1	320
February	2	320
March	3	325
April	4	330
May	5	350
June	6	370
July	7	375

12.5. Consider the process

$$y(k) - y(k - 1) + 0.5y(k - 2)$$
$$= u(k - 2) + 0.5u(k - 3) + 0.5[e(k) + 0.8e(k - 1) + 0.25e(k - 2)]$$

Determine the minimum-variance controller and the minimum achievable variance.

12.6. Determine the minimum-variance controller for the system

$$y(k) - 0.5y(k - 1) = u(k - 2) + e(k) - 0.7e(k - 1)$$

where $e(k)$ is white noise with mean 2 and unit variance.

12.7. Consider the process

$$y(k) + ay(k - 1) = u(k - 2) + e(k) + ce(k - 1)$$

(a) Determine the minimum-variance controller.
(b) Discuss the special case $a = 0$.

12.8. Given the system

$$y(k) - 1.7y(k - 1) + 0.7y(k - 2)$$
$$= u(k - d) + 0.5u(k - d - 1) + e(k) + 1.5e(k - 1) + 0.9e(k - 2)$$

(a) Determine the minimum-variance controller and the variance of the output for $d = 1$ and 2.
(b) Simulate the open-loop system and the system controlled with the minimum-variance controller. Compare the output and the control signal for the different cases.

12.9. Consider the process in Fig. 12.4. The disturbance z has the spectral density

$$\phi_z(\omega) = \frac{1}{2\pi} \frac{1}{1.36 + 1.2 \cos \omega}$$

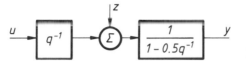

Figure 12.4

(a) Determine a pulse-transfer function $H(z)$, which gives an output with spectral density ϕ_z when driven by zero-mean white noise with unit variance.
(b) What is the steady state variance of y when

$$u(k) = -Ky(k)$$

for $K = 1$?
(c) What is the minimum achievable variance for a proportional controller and how large is the corresponding value of K?
(d) How large is the variance of y when a minimum-variance controller is used?

12.10. Given the system

$$y(k) - 0.25y(k - 1) + 0.5y(k - 2) - u(k - 1) + e(k) + 0.5e(k - 1)$$

where $e(k)$ is white noise with unit variance. Assume that the process is controlled with the proportional controller

$$u(k) = -Ky(k)$$

(a) Show that the variance of the output is

$$\frac{2.125 - K}{0.5(1.75 - K)(1.25 + K)}$$

and that the minimum-variance control is obtained for $K = 1$, which gives the variance $\frac{4}{3}$.
(b) The expression above is zero for $K = 2.125$. Explain the paradox.

12.11. Given the process

$$y(k) - 1.5y(k - 1) + 0.7y(k - 2) = u(k - 2) - 0.5u(k - 3) + v(k)$$

(a) Assume that $v(k) = 0$ and compute the deadbeat controller for the system.

(b) Assume that

$$v(k) = e(k) - 0.2e(k - 1)$$

where $e(k)$ is white noise. Compute the minimum-variance control law.

(c) What is the steady-state variance of y when the deadbeat and the minimum-variance controllers are used on the system when v is as in (b)?

(d) Simulate the system using the different controllers. Study the output and the accumulated loss, i.e., the sum of the square of the output.

12.12. Consider the dynamic system

$$y(k) = \frac{B(q)}{A(q)} u(k) + \lambda \frac{C(q)}{D(q)} e(k)$$

where $e(k)$ is white noise and B is stable. The polynomials A, C, and D are assumed to be monic. Determine the minimum-variance controller for the system.

12.13. Use the result from Problem 12.12 to determine the minimum-variance controller for the system

$$y(k) = \frac{bq^{-1}}{1 + aq^{-1}} u(k) + (1 + cq^{-1})e(k)$$

12.14. Consider the process in Problem 12.13. Assume that the sampling period is doubled; i.e., the control signal can be changed only at every second time unit. Determine the minimum-variance controller and compare with the case when the control period is one time unit.

12.15. Consider the system in Fig. 12.5, where e is white noise with zero mean and unit variance. Further,

$$A(q) = q - 0.7 \qquad B(q) = q$$

$$C(q) = 1 - 0.5q \qquad \alpha = -0.8$$

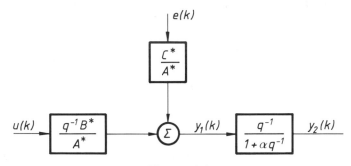

Figure 12.5

(a) Determine a controller that minimizes the variance of y_1.

(b) Determine the variances of y_1 and y_2 when the controller in (a) is used.

(c) Determine a controller that minimizes the variance of y_2 if only y_2 is measurable, and compute the variances of y_1 and y_2.

(d) Determine a controller that minimizes the variance of y_2 if both y_1 and y_2 are measurable.

(e) What are the variances of y_1 and y_2 when the controller in (d) is used?

12.16. Given the process

$$A(q)y(k) = B(q)u(k) + C(q)e(k) + D(q)v(k)$$

where $v(k)$ is a known disturbance. Determine the minimum variance controller for the process when deg B = deg D.

12.17. Determine the LQG-controller given by Theorem 12.4 for the process

$$(1 - 0.9q^{-1})y(k) = u(k) + (1 - 0.5q^{-1})e(k)$$

when $\rho = 1$. Calculate the variance of the output and the input for different values of ρ.

12.18. Consider a system with stable inverse. Derive the minimum-variance controller, where the control signal $u(k)$ is allowed to be a function of $y(k - 1)$, $y(k - 2)$, ... $u(k - 1)$, ... Derive the characteristic equation of the closed-loop system.

12.19. Show that the pulse-transfer function from e to y for (12.5) and (12.53) is given by (12.54). Use (12.45) to derive the minimum-variance controller for a system where

$$A(q) = q^2 \quad 1.5q + 0.7$$

$$B(q) = q + 0.5$$

$$C(q) = q^2 - q + 0.24$$

Compare with the controller obtained through the identity in (12.17).

12.20. Determine for which systems a digital PID-controller has the same structure as the optimal minimum-variance controller.

12.21. Consider a system described by

$$y(k) = \frac{1}{q - a} [bu(k) + \epsilon(k)] + \frac{1}{q - 1} w(k)$$

where ϵ and w are white-noise processes with zero mean and standard deviations σ_ϵ and σ_w, respectively.

(a) Reduce the system to standard form and determine the minimum-variance controller.

(b) Interpret the controller in (a) as a PI-controller and determine how the gain and the reset time depend on the ratio $\sigma_w^2/\sigma_\epsilon^2$.

12.22. Consider the minimum-variance control law of (12.36) for a system with an unstable inverse. The output of the closed-loop system is given by

$$y(k) = \frac{F(q)}{q^{d-1}B^{-*}(q)} e(k)$$

Show that the function F/B^{-*} has the series expansion

$$\frac{F(q)}{B^{-*}(q)} = q^{d-1} + f_1 q^{d-2} + \cdots + f_{d-1} + \frac{F_2(q)}{B^{-*}(q)}$$

where deg $F_2(q) <$ deg B^{-*} and

$$F_1(q) = q^{d-1} + f_1 q^{d-2} + \cdots + f_{d-1}$$

is the quotient of $q^{d-1}C(q)$ and $A(q)$. Give a convenient way of computing F_2. Use the results of the problem to determine the increase of the minimum variance due to unstable system zeros.

12.23. Determine the intersample ripple of the loss function when the process

$$dx_1 = x_2 \, dt$$

$$dx_2 = u \, dt + dv$$

$$y(t_k) = x_1(t_k) + \epsilon(t_k)$$

is controlled by the minimum-variance regulator. The process $\{v(t)\}$ is a Wiener process with incremental covariance $\sigma_v^2 \, dt$, and $\{\epsilon(t_k)\}$ is white measurement noise with zero mean and variance σ_ϵ^2.

12.24. Consider the process in Example 12.12. Determine the control law with sampling period h that minimizes

$$\lim_{T \to \infty} E \, \frac{1}{T} \int_0^T y^2(s) \, ds$$

and compare it with the minimum-variance control.

12.25. Consider a process subject to a disturbance that is characterized as a Wiener process with incremental covariance dt. Determine the prediction error of the minimum variance in each case. Use different prediction horizons and sampling periods.
(a) The process has an unstable zero $z = b > 1$.
(b) The process has an unstable pole $z = a > 1$.

12.26. Consider the system in Problem 12.23 with an extra time delay of 1 s. Determine the minimum variance as a function of the sampling period.

12.27. Consider the system in Problem 12.23. Determine the output variance as a function of the input covariance for different sampling periods.

12.28. Consider the system

$$y(k) = \frac{1}{q - 0.999} u(k) + \frac{q}{q - 0.7} e(k)$$

Determine the minimum-variance control law for the system. Compare it with a proportional feedback that gives a corresponding response rate. Discuss the relative merits of the control laws by calculating their loop gains and return differences. Explain why the minimum-variance control is inferior. (*Hint:* A bad optimization problem gives a bad optimal regulator.)

12.29. Given the system

$$y(k) = 1.4y(k - 1) - 0.65y(k - 2) + u(k - 1) - 0.2u(k - 2) + e(k) + 0.4e(k - 1)$$

where $e \in N(0, 2)$
(a) Determine the minimum-variance controller.
(b) Determine the deadbeat controller.
(c) Compute the variance of y when the controllers in (a) and (b) respectively are used.

12.30. Consider the system

$$y(k) + ay(k - 1) = u(k - 1) + e(k) + ce(k - 1)$$

where $e \in N(0, 1)$. We want to determine the minimum-variance controller for the process but the value of c is unknown.
(a) Assume in the design that $c = 0$ and determine the minimum-variance controller

for the system

$$y(k) + ay(k - 1) = u(k - 1) + e(k)$$

How large will the output variance be if this controller is used on the true system?

(b) Assume instead that $c = \hat{c}$ and redo the calculations in (a).

12.31. Consider the stochastic process

$$y(k + 2) - 1.1y(k + 1) + 0.3y(k) = e(k + 2) - 1.25e(k + 1)$$

where $e \in N(0, 1)$.

(a) Determine the two-step-ahead predictor for $y(k)$.

(b) Calculate the variance of the prediction error.

12.32. Given the system

$$A(q)y(k) = B(q)u(k) + C(q)e(k)$$

where

$$A(q) - q^3 - 1.7q^2 + 0.8q - 0.1$$

$$B(q) = 2(q - 0.9)$$

$$C(q) = q^2(q - 0.1)$$

and $e(k) \in N(0, 1)$.

(a) Determine the minimum-variance controller for the system.

(b) Determine the variance of the output when controlling the system with the controller in (a).

(c) Redo the calculations in (a) and (b) when

$$B(q) = 2(0.9q - 1)$$

12.33. Consider the process in Example 12.6. Compute the output variance when the controller does not cancel the zero, i.e., when the controller is obtained from the identity

$$zC = AR + BS$$

Compare the variances.

12.34. Consider the process in Example 12.6. Compute the controller that minimizes the loss function (12.7).

12.35. Show that a system with the input-output description

$$A(q)y(k) = B(q)u(k) + C(q)e(k)$$

where

$$A(q) = q^n + a_1q^{n-1} + \cdots + a_n$$

$$B(q) = b_1q^{n-1} + \cdots + b_n$$

$$C(q) = q^n + c_1q^{n-1} + \cdots + c_n$$

has the following state-space description

$$x(k + 1) = \Phi x(k) + \Gamma u(k) + Ke(k + 1)$$

$$y(k) = Cx(k)$$

where the state vector has dimension $n + 1$ and

$$
\Phi = \begin{bmatrix} -a_1 & 1 & 0 & \dots & 0 \\ -a_2 & 0 & 1 & \dots & 0 \\ \vdots & & & & \\ -a_n & 0 & 0 & \dots & 1 \\ 0 & 0 & 0 & \dots & 0 \end{bmatrix}, \Gamma = \begin{bmatrix} b_1 \\ b_2 \\ \vdots \\ b_n \\ 0 \end{bmatrix}, K = \begin{bmatrix} 1 \\ c_1 \\ \vdots \\ c_{n-1} \\ c_n \end{bmatrix}
$$

$$
C = [1 \quad 0 \quad 0 \quad \dots \quad 0]
$$

12.36. Consider the system in Problem 12.35. Assume that the polynomial $C(z)$ has all its zeros inside the unit disc. Show that the Kalman filter for the system can be written as

$$
\hat{x}(k + 1 \mid k) = \Phi \hat{x}(k \mid k) + \Gamma u(k)
$$

$$
\hat{x}(k + 1 \mid k + 1) = \hat{x}(k + 1 \mid k) + K[y(k + 1) - C\hat{x}(k + 1 \mid k)]
$$

and that the characteristic polynomial of the filter is $zC(z)$.

12.37. Consider the system in Problem 12.35. Assume that minimization of a quadratic loss function gives the feedback law

$$
u(k) = -L\hat{x}(k \mid k)
$$

Show that the controller has the pulse-transfer function

$$
H_r(z) = zL[zI - (I - KC)(\Phi - \Gamma L)]^{-1}\Gamma
$$

Show that the results are the same as those given by equation (12.82).

12.38. Consider the system in Problem 12.35. Assume that $b_1 \neq 0$. Determine the minimum-variance strategy using the state-space representations in (12.76) and in Problem 12.37. Compare the results. (*Hint:* The minimum-variance control corresponds to $L = [-a_1 \quad 1 \quad 0 \quad \dots \quad 0]$.)

12.39. Derive the expression for the transfer function $H_r(z)$ in equation (12.82) using the matrix inversion Lemma 13.1 in Section 13.5.

12.40. Show that the transfer function $H_r(z)$ in equation (12.82) can be written as

$$
H_r(z) = \frac{S(z)}{R(z)} = \alpha + (L - \alpha C)(zI - \Phi + \Gamma L + KC - \alpha \Gamma C)^{-1}(K - \Gamma \alpha)
$$

where $\alpha = L_v K$. Show that this expression is equivalent to

$$
\frac{S(z)}{R(z)} = \frac{S_0(z) + \alpha A(z)}{R_0(z) - \alpha B(z)}
$$

where $S_0(z)$ and $R_0(z)$ is the solution to the Diophantine equation

$$
A(z)R(z) + B(z)S(z) = P(z)C(z)
$$

with deg $R(z) = n$ and deg $S(z) < n$.

12.9 References

The treatment of the linear quadratic case is in the spirit of Wiener's work; see

WIENER, N. (1949): *Extrapolation, Interpolation & Smoothing of Stationary Time Series.* Cambridge, Mass.: MIT Press.

NEWTON, G. C., L. A. GOULD, and J. F. KAISER (1957): *Analytical Design of Linear Feedback Controls*. New York: John Wiley.

YOULA, D. C., J. J. BONGIORNO, JR., and H. A. JABR (1976): "Modern Wiener-Hopf Design of Optimal Controllers. Part I: The Single-Input–Single-Output Case; Part II: The Multivariable Case," *IEEE Trans. Autom. Control,* AC-21, 3–13 and 319–38.

A thorough discussion of prediction and minimum-variance control is found in

ÅSTRÖM, K. J. (1970): *Introduction to Stochastic Control Theory*. New York: Academic Press.

which is based on

ÅSTRÖM, K. J. (1965): Notes on the Regulation Problem, Report CT211, IBM Nordic Laboratory.

ÅSTRÖM, K. J. (1967): "Computer Control of a Paper Machine: An Application of Linear Stochastic Control Theory," *IBM Journal Res. Develop.,* II, 389–405.

A similar approach to the stochastic-control problem is found in

BOX, G. E. P. and G. M. JENKINS (1970): *Time Series Analysis, Forecasting, and Control,* San Francisco: Holden-Day.

The theorem for minimum-variance control of systems with unstable inverses was first published in

PETERKA, V. (1972): "On Steady-State Minimum Variance Control Strategy," *Kybernetika,* 8, 219–32.

An algebraic approach to the multivariable LQ- and minimum-variance control problems is given in

KUČERA, V. (1979): *Discrete Linear Control*. Prague: Academia.

Choice of sampling interval for stochastic control is discussed in the books mentioned above, and also in

MACGREGOR, J. F. (1976): "Optimal Choice of the Sampling Interval for Discrete Process Control," *Technometrics,* 18, no. 2, 151–60.

The intersample variation of the variance is discussed in

DE SOUZA, C. E. and G. C. GOODWIN (1984): "Intersample Variances in Discrete Minimum Variance Control," *IEEE Trans. Autom. Control,* AC-29, 759–61.

LENNARTSON, B. and T. SÖDERSTRÖM (1986): "An Investigation of the Intersample Variance for Linear Stochastic Control," *Preprints 25th IEEE Conf. on Decision and Control,* 1770–75.

13 IDENTIFICATION

GOAL

To Discuss How Process Models Can Be Obtained from Experimental Data. To Present the Least-Squares Method.

13.1 Introduction

The notion of a mathematical model is fundamental to science and engineering. A model is a very useful and compact way to summarize the knowledge about a process. A model is also a very effective tool for education and communication. The design methods in the previous chapters assume that models for the process and the disturbances are given. The process models can sometimes be obtained from first principles of physics. It is more difficult to get the models of the disturbances, which are equally important. These models often have to be obtained from experiments. The types of models that are needed for the design methods presented here are either state-space models (internal models) or input-output models (external models). The models for the disturbances are for the internal models given as dynamic systems driven by white noise. For external models the disturbances are given in terms of spectral densities and covariance functions. Models for disturbances can, however, only rarely be determined from first principles. Experiments are thus often the only way to get models for the disturbances.

A process cannot be characterized by *one* mathematical model. A process should be represented by a *hierarchy* of models ranging from detailed and complex simulation models to very simple models, which are easy to manipulate analytically. The simple models are used for exploratory purposes and to obtain the gross features of the system behavior. The complicated models are used for a detailed check of the performance of the control system. The complicated models take a long time to develop. Between the two extremes, there may be many different types of models. The trademark of good engineering is to choose the right model for each specific purpose.

Example 13.1

> To describe a drum boiler power unit, several different models may be needed. For production planning and frequency control, it may be sufficient to characterize the unit using two or three states describing the energy storage in the drum and the superheaters. To construct security systems and control systems, it may be necessary to have a model with 20 to 50 states. Finally, to model temperature and stresses in the turbine unit, several hundred states must be used. □

In principle, there are two different ways in which models can be obtained: from prior knowledge—e.g., in terms of physical laws—or by experimentation on a process. When attempting to obtain a specific model, it is often beneficial to combine both approaches.

Mathematical model-building based on physical laws is discussed briefly in Sec. 13.2. In most cases it is not possible to make a complete model only from physical knowledge. Some parameters must be determined from experiments. This approach is called *system identification* and is discussed in Sec. 13.3. There are many methods for analyzing data obtained from experiments. One basic approach is *the principle of least squares* (LS), discussed in Sec. 13.4. Recursive ways to make the computations are given in Sec. 13.5. Two examples are given in Sec. 13.6.

13.2 Mathematical Model-Building

There are no general methods that always can be used to get a complete model. Each process or problem has its own characteristics. Some general guidelines can be given, but under no circumstances can they replace experience. Model-building using physical laws require knowledge and insight about the process.

The main problem when making a mathematical model is to find the states of the system. The state variables essentially describe storage of energy and mass in the system. Typical variables that are chosen as states are positions and velocities (mechanical systems); voltages and currents (electrical systems); levels and flows (hydraulic systems); and temperatures, pressures, and densities (thermal systems). The relationship between the states is determined using balance equations for force, moment, mass, energy, and constitutive equations.

The advantage of model-building from physics is that it gives insight; also, the different parameters and variables have physical interpretations. The drawback is that it may be difficult and time-consuming to build the model from first principles. Mathematical model-building often has to be combined with experiments. The references give a more detailed treatment of mathematical model-building.

13.3 System Identification

System identification is the experimental approach to process-modeling. System identification includes the following:

Experimental planning.
Selection of model structure.
Parameter estimation.
Validation.

In practice, the procedure of system identification is iterative. When investigating a process where the a priori knowledge is poor, it is reasonable to start with transient or frequency-response analysis to get crude estimates of the dynamics and the disturbances. The results can then be used to plan further experiments. The data obtained are then used to estimate the unknown parameters in the model. Based on the results, the model structure can be improved and new experiments. may be necessary.

Experimental Planning

It is often difficult and costly to experiment with industrial processes. Therefore, it is desirable to have identification methods that do not require special input signals. Many "classic" methods depend strongly on having the input be of a precise form, e.g., sinusoid or impulses. Other techniques can handle virtually any type of input signal, at the expense of increased computations. One requirement of the input signal is that it should excite all the modes of the process sufficiently. A good identification method should thus be insensitive to the characteristics of the input signal.

It is sometimes possible to base system identification on data obtained under closed-loop control of the process. This is useful from the point of view of applications. For instance, adaptive controllers are based mostly on closed-loop identification. The main difficulty with data obtained from a process under feedback is that it may be impossible to determine all the parameters in the desired model; i.e., the system is not identifiable, even if the parameters can be determined from an open-loop experiment. Identifiability can be recovered if the feedback is sufficiently complex. It helps to make the feedback nonlinear and time-varying and to change the set points.

Model Structures

The model structures are derived from prior knowledge of the process and the disturbances. In some cases the only a priori knowledge is that the process can be described as a linear system in a particular operating range. It is then natural to use general representations of linear systems. Such representations are called *black-box models.* A typical example is the difference-equation model

$$A(q)y(k) = B(q)u(k) + C(q)e(k) \tag{13.1}$$

where u is the input, y is the output, and e is a white-noise disturbance. The parameters, as well as the order of the models, are considered as the unknown parameters.

Sometime it is possible to apply physical laws to derive models of the process that contain only a few unknown parameters. The model may then be of the form

$$\begin{cases} \dot{x} = f(x, u, v, \theta) \\ y = g(x, u, e, \theta) \end{cases} \tag{13.2}$$

where θ is a vector of unknown parameters, x is the state of the system, and v and e are disturbances.

Criteria

When formulating an identification problem, a criterion is introduced to give a measure of how well a model fits the experimental data. The criteria can be postulated. By making statistical assumptions, it is also possible to derive criteria from probabilistic arguments. The criteria for discrete-time systems are often expressed as

$$J(\theta) = \sum_{k=1}^{N} g(\epsilon(k))$$

where ϵ is the input error, the output error, or a generalized error. The prediction error is a typical example of a generalized error. The function g is frequently chosen to be quadratic, but it is possible for it to be of many other forms.

The first formulation, solution, and application of an identification problem were given by Gauss in his famous determination of the orbit of the asteroid Ceres. Gauss formulated the identification problem as an optimization problem and introduced the principle of least squares, a method based on the minimization of the sum of the squares of the error. Since then, the least-squares criterion has been used extensively.

The least-squares method is very simple and easy to understand. Under some circumstances it gives estimates with the wrong mean values (bias). However, this can be overcome by using various extensions. The least-squares method is restricted to model structures that are linear in the unknown parameters.

When the disturbances of a process are described as stochastic processes, the identification problem can be formulated as a statistical parameter-estimation problem. It is then possible to use the maximum-likelihood method, for example; this method has many attractive statistical properties. It can be interpreted as a least-squares criterion if the quantity to be minimized is taken as the sum of squares of the prediction error. The maximum-likelihood method is a very general technique that can be applied to a wide variety of model structures.

Parameter Estimation Methods

Solving the parameter estimation problem requires the following:

- Input-output data from the process.
- A class of models.
- A criterion.

Parameter estimation can then be formulated as an optimization problem, where the best model is the one that best fits the data according to the given criterion.

The result of the estimation problem depends, of course, on how the problem is formulated. For instance, the obtained model depends on the amplitude and frequency content of the input signal. There are many possibilities for combining experimental conditions, model classes, and criteria. There are also many different ways to organize the computations. Consequently, there is a large number of different identification methods available. One broad distinction is between *on-line* methods and *off-line* methods. The on-line methods give estimates recursively as the measurements are obtained and are the only alternative if the identification is going to be used in an adaptive controller or if the process is time-varying. In many cases the off-line methods give estimates with higher precision and are more reliable, for instance in terms of convergence.

The large number of methods is confusing for an industrial engineer who is primarily interested in having a tool to obtain a model. Several attempts to compare different identification methods have been made. The comparisons are largely inconclusive in the sense that there is no method that is universally best. Fortunately, it appears that the choice of method is not crucial. Therefore, it can be recommended that a prospective user learn the classic methods (frequency and transient-response analysis and correlation and spectral analysis), the least-squares method with extensions, and the maximum-likelihood method.

Model Validation

When a model has been obtained from experimental data, it is necessary to check the model in order to reveal its inadequacies. For model validation, it is useful to determine such factors as step responses, impulse responses, poles and zeros, model errors, and prediction errors. Since the purpose of the model validation is to scrutinize the model with respect to inadequacies, it is useful to look for quantities that are sensitive to model changes.

13.4 The Principle of Least Squares

According to Gauss the principle of least squares is that the unknown parameters of a model should be chosen in such a way that

> the sum of the squares of the differences between the actually observed and computed values multiplied by numbers that measure the degree of precision is a minimum.

To be able to give an analytic solution, the computed values must be linear functions of the unknown parameters. In the framework of the general formulation of the identification problem given in the previous sections, the class of models is such that the model output is linear in the parameters and the criterion is a quadratic function. The purpose of this section is to formulate the least-squares problem and to give its solution.

The General Problem

In the general least-squares problem, it is assumed that "the computed variable," $\hat{y}$, in Gauss' terminology is given by the model

$$\hat{y} = \theta_1 \varphi_1(x) + \theta_2 \varphi_2(x) + \cdots + \theta_n \varphi_n(x) \tag{13.3}$$

where $\varphi_1, \varphi_2, \ldots, \varphi_n$ are known functions and $\theta_1, \theta_2, \ldots, \theta_n$ are unknown parameters. Pairs of observations $\{(x_i, y_i), i = 1, 2, \ldots, N\}$ are obtained from an experiment. The problem is to determine the parameters in such a way that the variables $\hat{y}$, computed from the model of (13.3) and the experimental values x_i agree as closely as possible with the measured variables y_i. Assuming that all measurements have the same precision, the principle of least squares says that the parameters should be selected in such a way that the loss function

$$J(\theta) = \tfrac{1}{2} \sum_{i=1}^{N} \epsilon_i^2$$

is minimal where

$$\epsilon_i = y_i - \hat{y}_i = y_i - \theta_1 \varphi_1(x_i) - \cdots - \theta_n \varphi_n(x_i) \qquad i = 1, 2, \ldots, N$$

To simplify the calculations, the following vector notations are introduced:

$$\varphi = [\varphi_1 \quad \varphi_2 \quad \cdots \quad \varphi_n]^T$$

$$\theta = [\theta_1 \quad \theta_2 \quad \cdots \quad \theta_n]^T$$

$$y = [y_1 \quad y_2 \quad \cdots \quad y_N]^T$$

$$\epsilon = [\epsilon_1 \quad \epsilon_2 \quad \cdots \quad \epsilon_N]^T$$

$$\Phi = \begin{bmatrix} \varphi^T(x_1) \\ \vdots \\ \varphi^T(x_N) \end{bmatrix}$$

The least-squares problem can now be formulated in a compact form. The loss function J can be written as

$$J(\theta) = \tfrac{1}{2}\epsilon^T \epsilon = \tfrac{1}{2} \| \epsilon \|^2 \tag{13.4}$$

where

$$\epsilon = y - \hat{y} \tag{13.5}$$

and
$$\hat{y} = \Phi\theta$$

Determine the parameter θ in such a way that $\| \epsilon \|^2$ is minimal. The solution to the least-squares problem is given by Theorem 13.1.

Theorem 13.1 The function of (13.4) is minimal for parameters $\hat{\theta}$ such that
$$\Phi^T\Phi\hat{\theta} = \Phi^Ty \qquad (13.6)$$

If the matrix $\Phi^T\Phi$ is nonsingular, the minimum is unique and given by
$$\hat{\theta} = (\Phi^T\Phi)^{-1}\Phi^Ty = \Phi^\dagger y \qquad (13.7)$$

Proof. The loss function of (13.4) can be written as
$$2J(\theta) = \epsilon^T\epsilon = [y - \Phi\theta]^T[y - \Phi\theta]$$
$$= y^Ty - y^T\Phi\theta - \theta^T\Phi^Ty + \theta^T\Phi^T\Phi\theta.$$

Because the matrix $\Phi^T\Phi$ is always nonnegative definite, the function J has a minimum. Using (11.15) the minimum is obtained for
$$\theta = \hat{\theta} = (\Phi^T\Phi)^{-1}\Phi^Ty$$

and the theorem is proved. $\square$

Remark 1. Equation (13.6) is called the *normal equation*.

Remark 2. The matrix $\Phi^\dagger = (\Phi^T\Phi)^{-1}\Phi^T$ is called the *pseudoinverse* of Φ if the matrix $\Phi^T\Phi$ is nonsingular.

System Identification

The least-squares method can be used to identify parameters in dynamic systems. Let the system be described by (13.1) with $C(q) = q^h$. Further, assume that A and B are of order n and $n - 1$, respectively. Assume that a sequence of inputs $\{u(1), u(2), \ldots, u(N)\}$ has been applied to the system and that the corresponding sequence of outputs $\{y(1), y(2), \ldots, y(N)\}$ has been observed. The unknown parameters are then
$$\theta = [a_1 \quad \ldots \quad a_n \quad b_1 \quad \ldots \quad b_n]^T \qquad (13.8)$$

Further
$$\varphi^T(k + 1) = [-y(k) \quad \ldots \quad -y(k - n + 1) \quad u(k) \quad \ldots \quad u(k - n + 1)] \qquad (13.9)$$

and
$$\Phi = \begin{bmatrix} \varphi^T(n + 1) \\ \vdots \\ \varphi^T(N) \end{bmatrix}$$

The least-squares estimate is then given by (13.7) if $\Phi^T\Phi$ is nonsingular. For instance, this is the case if the input signal is, loosely speaking, sufficiently rich.

Example 13.2

Determine the least-squares estimate of the parameters a and b in the model

$$\hat{y}(k) = -ay(k-1) + bu(k-1)$$

in such a way that the criterion

$$J(a, b) = \tfrac{1}{2}\sum_{i=2}^{N} \epsilon(k)^2$$

is minimal, where

$$\epsilon(k) = y(k) - \hat{y}(k) = y(k) + ay(k-1) - bu(k-1)$$
$$= y(k) - \varphi^T(k)\theta$$

A comparison with the general case gives

$$y = \begin{bmatrix} y(2) \\ y(3) \\ \vdots \\ y(N) \end{bmatrix}, \quad \Phi = \begin{bmatrix} -y(1) & u(1) \\ -y(2) & u(2) \\ \vdots & \vdots \\ -y(N-1) & u(N-1) \end{bmatrix}, \quad \epsilon = \begin{bmatrix} \epsilon(2) \\ \epsilon(3) \\ \vdots \\ \epsilon(N) \end{bmatrix}$$

and

$$\theta = [a \quad b]^T$$

Hence

$$\Phi^T\Phi = \begin{bmatrix} \sum\limits_{k=1}^{N-1} y(k)^2 & -\sum\limits_{k=1}^{n-1} y(k)u(k) \\ -\sum\limits_{k=1}^{N-1} y(k)u(k) & \sum\limits_{k=1}^{N-1} u(k)^2 \end{bmatrix}$$

$$\Phi^T y = \begin{bmatrix} -\sum\limits_{k=1}^{N-1} y(k+1)y(k) \\ \sum\limits_{k=1}^{N-1} y(k+1)u(k) \end{bmatrix}$$

Provided the matrix $\Phi^T\Phi$ is nonsingular, the least-squares estimate of the parameters a and b is now easily obtained. The matrix $\Phi^T\Phi$ will be nonsingular if conditions (*sufficient richness* or *persistent excitation*) are imposed on the input signal. □

Statistical Interpretation

To analyze the properties of the least-squares estimator, it is necessary to make some assumptions. Let the data be generated from the process

$$y = \Phi\theta_0 + \epsilon \tag{13.10}$$

where θ_0 is the vector of "true" parameters and ϵ is a vector of noise with zero mean value. The following theorem is given without proof.

Theorem 13.2. Consider the estimate (13.7) and assume that the data is generated from (13.10) where ϵ_i is white noise with variance σ^2. Then, if n is the number of parameters of $\hat{\theta}$ and θ_0 and N is the number of data, the following conditions hold.

1. $E\hat{\theta} = \theta_0$.
2. $\text{Var } \hat{\theta} = \sigma^2(\Phi^T\Phi)^{-1}$.
3. $s^2 = 2J(\theta)/(N - n)$ is an unbiased estimate of σ^2. □

Theorem 13.2 implies that the parameters in (13.1) can be estimated without bias if $C(q) = q^n$. If $C(q) \neq q^n$, then the estimates will be biased. This is due to the correlation between the noise $C^*(q^{-1})e(k)$ and the data in $\varphi(k)$.

Extensions of the Least-Squares Method

The least-squares method gives unbiased results of the parameters in (13.1) only if $C(q) = q^n$. However, the maximum likelihood method can be used for the general case. It can be shown that maximizing the likelihood function is equivalent to minimizing the loss function of (13.4), where the residuals, ϵ, are related to the inputs and outputs by

$$C(q)\epsilon(k) = A(q)y(k) - B(q)u(k)$$

The residuals can be interpreted as the one-step-ahead prediction error. However, the loss function is not linear in the parameters and it has to be minimized numerically. This can be done using a Newton-Raphson gradient routine, which involves computation of the gradient of J with respect to the parameters, as well as the matrix of second partial derivatives. The maximum-likelihood method is thus an off-line method. It is possible to make approximations of the maximum-likelihood method that allow on-line computations of the parameters of the model in (13.1). Some common methods are Extended Least Squares (ELS), Generalized Least Squares (GLS), and Recursive Maximum Likelihood (RML).

13.5 Recursive Computations

In many cases the observations are obtained sequentially. It may then be desirable to compute the least-squares estimate for different values of N. If the least-squares problem has been solved for N observations, it seems to be a waste of computational resources to start from scratch when a new observation is obtained. Hence, it is desirable to arrange the computations in such a way that the results obtained for N observations can be used in order to get the estimates for $N + 1$ observations. An analogous problem occurs when the number of parameters is not known in advance. The LS estimate may then be needed for a different number of parameters. The possibility of calculating the least-squares estimate recursively is pursued in this section.

Recursion in the Number of Observations

Recursive equations can be derived for the case when the observations are obtained sequentially. The procedure is often referred to as *recursive identification*. The solution in (13.7) to the LS problem can be rewritten to give recursive equations. Let $\hat{\theta}(N)$ denote the least-squares estimate based on N measurements. To derive the equations, N is introduced as a formal parameter in the functions, i.e.,

$$\Phi(N) = \begin{bmatrix} \varphi^T(x_1) \\ \vdots \\ \varphi^T(x_N) \end{bmatrix}, \qquad y(N) = \begin{bmatrix} y_1 \\ \vdots \\ y_N \end{bmatrix}$$

It is assumed that the matrix $\Phi^T\Phi$ is regular for all N. The least-squares estimate $\hat{\theta}(N)$ is then given by Equation (13.7):

$$\hat{\theta}(N) = [\Phi^T(N)\Phi(N)]^{-1}\Phi^T(N)y(N)$$

When an additional measurement is obtained, a row is added to the matrix Φ and an element is added to the vector y. Hence

$$\Phi(N + 1) = \begin{bmatrix} \Phi(N) \\ \varphi^T(N + 1) \end{bmatrix}, \qquad y(N + 1) = \begin{bmatrix} y(N) \\ y_{N+1} \end{bmatrix} \qquad (13.11)$$

The estimate $\hat{\theta}(N + 1)$ given by (13.7) can then be written as

$$\hat{\theta}(N + 1) = [\Phi^T(N + 1)\Phi(N + 1)]^{-1}\Phi^T(N + 1)y(N + 1)$$

$$= |\Phi^T(N)\Phi(N) + \varphi(N + 1)\varphi^T(N + 1)|^{-1} \qquad (13.12)$$

$$\times [\Phi^T(N)y(N) + \varphi(N + 1)y_{N+1}]$$

The solution is given by the following theorem.

Theorem 13.3—Recursive least-squares estimation. Assume that the matrix $\Phi^T(N)\Phi(N)$ is positive definite. The least-squares estimate $\hat{\theta}$ then satisfies the recursive equation

$$\hat{\theta}(N + 1) = \hat{\theta}(N) + K(N)[y_{N+1} - \varphi^T(N + 1)\hat{\theta}(N)] \qquad (13.13)$$

$$K(N) = P(N + 1)\varphi(N + 1)$$

$$= P(N)\varphi(N + 1)[1 + \varphi^T(N + 1)P(N)\varphi(N + 1)]^{-1} \qquad (13.14)$$

$$P(N + 1) = [I - K(N)\varphi^T(N + 1)]P(N) \qquad (13.15)$$

The following lemma is useful to prove the theorem.

Lemma 13.1—Matrix inversion lemma. Let A, C, and $C^{-1} + DA^{-1}B$ be nonsingular square matrices; then

$$[A + BCD]^{-1} = A^{-1} - A^{-1}B[C^{-1} + DA^{-1}B]^{-1}DA^{-1}$$

Proof. By direct substitution,

$$[A + BCD]\{A^{-1} - A^{-1}B[C^{-1} + DA^{-1}B]^{-1}DA^{-1}\}$$

$$= I + BCDA^{-1} - B[C^{-1} + DA^{-1}B]^{-1}DA^{-1}$$

$$\quad - BCDA^{-1}B\,[C^{-1} + DA^{-1}B]^{-1}DA^{-1}$$

$$= I + BCDA^{-1} - BC[C^{-1} - DA^{-1}B][C^{-1} + DA^{-1}B]^{-1}DA^{-1}$$

$$= I + BCDA^{-1} - BCDA^{-1}$$

$$= I \qquad\qquad\qquad\qquad \square$$

Proof of Theorem 13.3. To simplify the notation in the manipulations that follow, the argument N of $\Phi(N)$ and $y(N)$ and the argument $N + 1$ of $\varphi^T(N + 1)$ will be suppressed. Equation (13.12) can then be written as

$$\hat{\theta}(N + 1) = [\Phi^T\Phi + \varphi\varphi^T]^{-1}[\Phi^T y + \varphi y_{N+1}]$$

$$= (\Phi^T\Phi)^{-1}\Phi^T y + [(\Phi^T\Phi + \varphi\varphi^T)^{-1} \qquad (13.16)$$

$$\quad - (\Phi^T\Phi)^{-1}]\Phi^T y + (\Phi^T\Phi + \varphi\varphi^T)^{-1}\varphi y_{N+1}$$

Observe that

$$\hat{\theta}(N) = (\Phi^T\Phi)^{-1}\Phi^T y$$

and

$$[(\Phi^T\Phi + \varphi\varphi^T)^{-1} - (\Phi^T\Phi)^{-1}]\Phi^T y$$

$$= (\Phi^T\Phi + \varphi\varphi^T)^{-1}(\Phi^T\Phi - \Phi^T\Phi - \varphi\varphi^T)(\Phi^T\Phi)^{-1}\Phi^T y$$

$$= -(\Phi^T\Phi + \varphi\varphi^T)^{-1}\varphi\varphi^T(\Phi^T\Phi)^{-1}\Phi^T y$$

$$= -(\Phi^T\Phi + \varphi\varphi^T)^{-1}\varphi\varphi^T\theta$$

Equation (13.16) can be written as

$$\hat{\theta}(N + 1) = \hat{\theta}(N) + K(N)[y_{N+1} - \varphi^T(N + 1)\hat{\theta}(N)]$$

where

$$K(N) = [\Phi^T(N)\Phi(N) + \varphi(N + 1)\varphi^T(N + 1)]^{-1}\varphi(N + 1)$$

$$= [\Phi^T(N + 1)\Phi(N + 1)]^{-1}\varphi(N + 1)$$

In order to obtain a recursive equation for the weighting factor $K(N)$, it is convenient to introduce the quantity P defined by

$$P(N) = [\Phi^T(N)\Phi(N)]^{-1}$$

P is proportional to the variance of the estimates (compare with Theorem 13.2).

Applying Lemma 13.1 to the matrix $P(N + 1)$ gives

$$P(N + 1) = [\Phi^T(N + 1)\Phi(N + 1)]^{-1} = [\Phi^T\Phi + \varphi\varphi^T]^{-1}$$
$$= (\Phi^T\Phi)^{-1} - (\Phi^T\Phi)^{-1}\varphi[I + \varphi^T(\Phi^T\Phi)^{-1}\varphi]^{-1}\varphi^T(\Phi^T\Phi)^{-1}$$

Hence

$$P(N + 1) = P(N) - P(N)\varphi(N + 1)$$
$$\times [I + \varphi^T(N + 1)P(N)\varphi(N + 1)]^{-1}\varphi^T(N + 1)P(N)$$

Simple calculations now give

$$K(N) = P(N + 1)\varphi(N + 1)$$
$$= P(N)\varphi(N + 1)[I + \varphi^T(N + 1)P(N)\varphi(N + 1)]^{-1}$$

Notice that a matrix inversion is necessary to compute P. However, the matrix to be inverted is of the same dimension as the number of measurements; i.e., for a single-output system, it is a scalar. □

Remark 1. Equation (13.13) has a strong intuitive appeal. The estimate $\hat{\theta}(N + 1)$ is obtained by adding a correction to the previous estimate $\hat{\theta}(N)$. The correction is proportional to $y_{N+1} - \varphi^T(N + 1)\hat{\theta}(N)$, where the last term can be interpreted as the value of y at time $N + 1$ predicted by the model (13.3). The correction term is thus proportional to the difference between the measured value of y_{N+1} and the prediction of y_{N+1} based on the previous estimates of the parameters. The components of the vector $K(N)$ are weighting factors that tell how the correction and the previous estimate should be combined. Notice that the component $K_i(N)$ is proportional to $\varphi_i^T(N + 1)$.

Remark 2. The least-squares estimate can be interpreted as a Kalman filter for the process

$$\theta(k + 1) = \theta(k)$$
$$y(k) = \varphi^T(k)\theta(k) + e(k)$$

See Section 11.3.

Notice that the matrix $P(N)$ is defined only when the matrix $\Phi^T(N)\Phi(N)$ is nonsingular. Because

$$\Phi^T(N)\Phi(N) = \sum_{k=1}^{N} \varphi(k)\varphi^T(k)$$

it follows that $\Phi^T\Phi$ is always singular if N is sufficiently small. In order to obtain an initial condition for P, it is necessary to choose an $N = N_0$ such that $\Phi^T(N_0)\Phi(N_0)$ is nonsingular and determine

$$P(N_0) = [\Phi^T(N_0)\Phi(N_0)]^{-1}$$
$$\hat{\theta}(N_0) = P(N_0)\Phi^T(N_0)y(N_0)$$

The recursive equations can then be used from $N \geq N_0$. It is, however, often convenient to use the recursive equations in all steps. If the recursive equations are begun with the initial condition

$$P(0) = P_0$$

where P_0 is positive definite, then

$$P(N) = [P_0^{-1} + \Phi^T(N)\Phi(N)]^{-1}$$

This can be made arbitrarily close to $[\Phi^T(N)\Phi(N)]^{-1}$ by choosing P_0 sufficiently large.

Using the statistical interpretation of the least-squares method shows that this way of starting the recursion corresponds to the situation when the parameters have a prior covariance proportional to P_0.

Time-Varying Systems

Using the loss function of (13.4), all data points are given the same weight. If the parameters are time-varying, it is necessary to eliminate the influence of old data. This can be done by using a loss function with exponential weighting, i.e.,

$$J(\theta) = \sum_{k=1}^{N} \lambda^{N-k}[y(k) - \varphi^T(k)\theta]^2 \tag{13.17}$$

The "forgetting factor," λ, is less than one and is a measure of how fast old data are forgotten. The least-squares estimate when using the loss function of (13.17) is given by

$$\hat{\theta}(k+1) = \hat{\theta}(k) + K(k)[y_{k+1} - \varphi^T(k+1)\hat{\theta}(k)]$$

$$K(k) = P(k)\varphi(k+1)[\lambda + \varphi^T(k+1)P(k)\varphi(k+1)]^{-1} \tag{13.18}$$

$$P(k+1) = [I - K(k)\varphi^T(k+1)]P(k)/\lambda$$

It is also possible to model the time-varying parameters by a Markov process,

$$\theta(k+1) = \Phi\theta(k) + v(k)$$

and then use a Kalman filter to estimate θ. See Remark 2 of Theorem 13.3.

Recursion in the Number of Parameters

When extra parameters are introduced, the vector $\hat{\theta}$ will have more components and there will be additional rows in the matrix Φ. The calculations can be arranged so that it is possible to make a recursion in the number of parameters in the model. The recursion involves an inversion of a matrix of the same dimension as the number of added parameters.

U-D Covariance Factorization

Equation (13.18) is one way to mechanize the recursive update of the estimates and the covariance matrix. These equations are not well-conditioned from a numerical point of view, however. A better way of doing the calculation is to update the square root of P instead of updating P. Another way to do the calculations is to use the U-D algorithm by Bierman and Thorton. This method is based on a factorization of P as

$$P = UDU^T$$

where D is diagonal and U is an upper-triangular matrix. This method is a square-root type as $UD^{1/2}$ is the square root of P. The U-D factorization method does not include square-root calculations and is therefore well suited for small computers and real-time applications. Details about the algorithm can be found in the references.

A Pascal program for least-squares estimation based on U-D factorization is given in Listing 13.1. The program gives estimates of the parameters of the process

$$y(k) + a_1 y(k-1) + \cdots + a_{na} y(k - na)$$
$$= b_1 u(k-1) + \cdots + b_{nb} u(k - nb) + e(k) \quad (13.19)$$

LISTING 13.1 Pascal program for least-squares estimation of the parameters of the process of (13.19) using U-D factorization

```
const  npar=10;{maximum number of estimated parameters}
       noff=45;{noff=npar*(npr-1)/2}
type vec1=array[1 .. npar] of real;
     vec2=array[1 .. noff] of real;
     estpartyp = rooord
       n, na:integer;
       theta:vec1;
       fi:vec1;
       diag:vec1;
       offdiag:vec2;
       end;
var  y,u,lambda:real;
     eststate:estpartyp;


Procedure LS(u,y,lambda:real;var eststate:estpartyp);
{Computes the least-squares estimate using the U-D method
after Bierman and Thornton}


var  kf,ku,i,j:integer;
     perr,fj,vj,alphaj,ajlast,pj,w:real;
     k:vec1;


begin
   with eststate do {Calculate prediction error}
```

LISTING 13.1 (Continued)

```
                     begin
                       perr: = y;
                       for i: = 1 to n do perr: = perr - theta[i]*fi[i];
                       {Calculate gain and covariance using U-D method}
                       fj: = fi[1];
                       vj: = diag[1]*fj;
                       k[1]: = vj;
                       alphaj: = 1.0 + vj*fj;
                       diag[1]: = diag[1]/alphaj/lambda;
                       if n>1 then
                       begin
                         kf: = 0;
                         ku: = 0;
                         for j: = 2 to n do
                         begin
                           fj:fi[j];
                           for i: = 1 to j ┬ 1 do
                           begin {f = fi*U}
                             kf: = kf + 1;
                             fj: = fj + fi[i]*offdiag[kf]
                           end; {i}
                           vj: = fj*diag[j]; {v = D*f}
                           k[j]: = vj;
                           ajlast: = alphaj;
                           alphaj: = ajlast + vj*fj;
                           diag[j]: = diag[j]*ajlast/alphaj/lambda;
                           pj: = - fj/ajlast;
                           for i: = 1 to j - 1 do
                           begin
                             {kj + 1: = kj + vj*uj}
                             {uj: = uj + pj*kj}
                             ku: = ku + 1;
                             w: = offdiag[ku] + k[i]*pj;
                             k[i]: = k[i] + offdiag[ku]*vj;
                             offdiag[ku]: = w
                           end; {i}
                         end; {j}
                       end; {if n>1 then}
                       {Update parameter estimates}
                       for i: = 1 to n do theta[i]: = theta[i] + perr*k[i]/alphaj;
                       {Updating of fi}
                       for i: = 1 to n - 1 do fi[n + 1 - i]: = fi[n - i];
                       fi[1]: = - y;
                       fi[na + 1]: = u
                     end {with eststate do}
                   end; {LS}
```

Identification   Chap. 13

The notations used in the program are

Variable	Notation in the program
$u(k)$	u
$y(k)$	y
na	na
$na + nb$	n
$n(n - 1)/2$	noff
$\hat{\theta}(k)$ compare (13.8)	theta
$\varphi^T(k)$ compare (13.9)	fi
λ	lambda

13.6 Examples

Two examples show the use of identification methods.

Example 13.3

Let the system be described by the model

$$y(k) - 1.5y(k - 1) + 0.7y(k - 2)$$
$$= u(k - 1) + 0.5u(k - 2) + e(k) - e(k - 1) + 0.2e(k - 2) \quad (13.20)$$

where e has zero mean and standard deviation 0.5. This is a "standard" system that has been used often in the literature to test different identification methods. In (13.20), $C(q) \neq q^n$, which implies that the least-squares method will give biased estimates. However, the input-output relation of the process can be approximated by using the least-squares method for a higher-order model. Figure 13.1 shows a simulation of the system. The input is a Pseudo Random Binary Signal (PRBS) sequence with amplitude ± 1. The data have been used to identify models of different orders using the least-squares and maximum-likelihood methods. Figure 13.2 shows the step responses of the true system in (13.20) and of the estimated models when using the least-squares method with model orders $n = 1, 2,$ and 4 and the maximum-

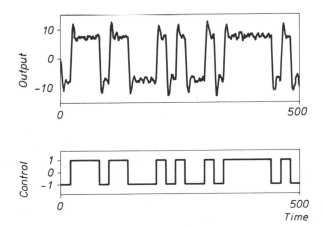

Figure 13.1 Input and output when the system of (13.20) is simulated. The input is a PRBS sequence.

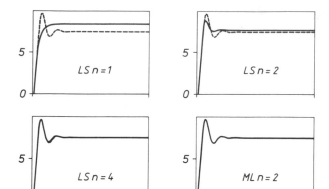

Figure 13.2 Step responses of the deterministic part of the system of (13.20) and of the estimated models obtained when using the least-squares method (LS) with n = 1, 2, and 4 and the maximum-likelihood method (ML) with n = 2.

llkelihood method when the model order is 2. The least-squares method gives a poor description for a second-order model, while a good model is obtained when the model order is increased to four. The maximum-likelihood method gives very good estimates of the dynamics and the noise characteristics for a second-order model. The estimated parameters for second-order models when using the least-squares method and the maximum-likelihood method are shown in Table 13.1.

□

TABLE 13.1 Estimated parameters and standard deviations for second-order models of the process in (13.20) when using the least-squares (LS) and the maximum-likelihood (ML) methods

Parameter	True value	LS $n = 2$	ML $n = 2$
a_1	−1.5	−1.285 ± 0.027	−1.497 ± 0.009
a_2	0.7	0.540 ± 0.021	0.699 ± 0.006
b_1	1	1.056 ± 0.091	1.019 ± 0.051
b_2	0.5	0.913 ± 0.121	0.497 ± 0.075
c_1	−1	—	−0.964 ± 0.045
c_2	0.2	—	0.174 ± 0.044

Example 13.4—Recursive estimation

Consider the process

$$y(k) + ay(k - 1) = bu(k - 1) + e(k)$$

with $a = -0.8$ and $b = 1$. The variance of the noise e is 1. The input signal is assumed to be a PRBS signal with amplitude ±1. The input and the output data are shown in Fig. 13.3. The recursive equations (13.13)–(13.15) have been used to estimate a and b. The estimates are shown in Fig. 13.4 when the initial values of the parameters are zero and when $P(0)$ is 10 times the unit matrix. The estimates are after a few observations close to the true values.

□

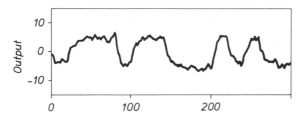

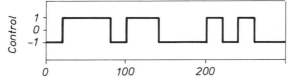

Figure 13.3 Input-output data for the process in Example 13.4.

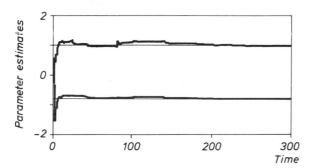

Figure 13.4 Recursive parameter estimates when (13.13)–(13.15) are used on the data shown in Fig. 13.3.

13.7 Summary

This chapter gives a short review of the identification problem. The presentation is concentrated on the least-squares method, because it is the basis for many other methods. On many occasions it is important to make the estimation in real time, and it is shown how the least-squares estimate can be obtained recursively. This is used in connection with adaptive controllers in Chapter 14.

13.8 Problems

13.1 The following experiment has been made to determine the normal acceleration, g. A steel ball has been dropped without initial velocity from a high TV antenna. The position of the ball, 1, has been determined at different times, giving the following measurements:

Time, s	Length of fall, in meters
1	8.49
2	20.05
3	50.65
4	72.19
5	129.85
6	171.56

The times of the measurements are exact, but there is an error in the measurement of the position. Determine the normal acceleration using the method of least squares from the model

$$l = \frac{gt^2}{2} + e$$

13.2. Derive recursive equations for increasing the number of parameters for the method of least squares. (*Hint:* Use the same idea as when making the observations recursively.)

13.3. Consider the process

$$y(k) + ay(k - 1) = bu(k - 1) + e(k) + ce(k - 1)$$

where u and e are independent white-noise processes with zero mean and unit variance. Assume that the method of least squares is used to estimate a and b, as in Example 13.2. Determine the expected values of $\hat{a}$ and $\hat{b}$ as a function of a, b, and c.

13.4. The parameters b_1 and b_2 in the system

$$y(k) = b_1u(k - 1) + b_2u(k - 2) + e(k)$$

are determined using the method of least squares. Let the input be a step at time $k = 0$. Can the parameters b_1 and b_2 be determined with arbitrary accuracy when the number of observations increases? Will there be any changes if it is known that $b_2 = 0$?

13.5. The English mathematician Richardson has proposed the following simple model for an arms race between two countries:

$$x(k + 1) = ax(k) + by(k) + f$$

$$y(k + 1) = cx(k) + dy(k) + g$$

where $x(k)$ and $y(k)$ are the expenditures on arms of the two nations and a, b, c, d, f, and g are constants. The following data have been completed by SIPRI. (World Armaments and Disarmaments, SIPRI Yearbook 1982).

Millions of U.S. dollars at 1979 prices and 1979 exchange rates

	Iran	Iraq	NATO	WTO
1972	2891	909	216478	112893
1973	3982	1123	211146	115020
1974	8801	2210	212267	117169
1975	11230	2247	210525	119612
1976	12178	2204	205717	121461
1977	9867	2303	212009	123561
1978	9165	2179	215988	125498
1979	5080	2675	218561	127185
1980	4040		225411	129000
1981			233957	131595

Determine the parameters of the model by least squares and investigate the stability of the model.

13.6. Consider Richardson's arms-race model in Problem 13.5.

(a) Determine the estimates of the parameters based on three consecutive years. Look at the validity of the estimates.

(b) Determine a recursive estimate of the parameters. Start at 1975 with initial values determined for 1972–74.

13.7. Consider the system

$$y(k) = -ay(k - 1) + bu(k - 1) + e(k)$$

where e is zero-mean white noise. An experiment is done on the system to estimate a and b. The following data were calculated:

$$\Sigma y^2(k) = 30 \qquad \Sigma u^2(k) = 50$$
$$\Sigma y(k + 1)y(k) = 1 \qquad \Sigma y(k)u(k) = 20$$
$$\Sigma y(k + 1)u(k) = 36$$

All sums are from $k = 1$ to $k = 999$. Determine the least-squares estimate of a and b.

13.9 References

There are many books and papers dealing with identification methods. Some basic references in book form are

JENKINS, G. M. and D. G. WATTS (1968): *Spectral Analysis and Its Applications*. San Francisco: Holden-Day.

EYKHOFF, P. (1974): *System Identification: Parameter and State Estimation*. London: John Wiley.

Box, G. E. P. and G. M. JENKINS (1976): *Time Series Analysis and Control* (rev. ed.). San Francisco: Holden-Day.

GOODWIN, G. C. and R. L. PAYNE (1977): *Dynamic System Identification: Experiment Design and Data Analysis*. New York: Academic Press.

LJUNG, L. and T. SÖDERSTRÖM (1983): *Theory and Practice of Recursive Identification*. Cambridge, Mass.: MIT Press.

NORTON, J. P. (1986): *An Introduction to Identification*. London: Academic Press.

LJUNG, L. (1987): *System Identification: Theory for the User*. Englewood Cliffs, N.J.: Prentice Hall, Inc.

SÖDERSTRÖM, T. and P. STOICA (1989): *System Identification*. Hemel Hempstead U.K.: Prentice Hall International.

A survey of system identification is given in

ÅSTRÖM, K. J. and P. EYKHOFF (1971): "System Identification: A Survey." *Automatica*, 7, 123–62.

Good sources for further references are

BIERMAN, G. J. (1977): *Factorization Methods for Discrete Sequential Estimation*. New York: Academic Press.

EYKHOFF, P., ed., (1981): *Trends and Progress in System Identification*. Oxford: Pergamon Press.

Isermann, R., ed., (1981): "System Identification," Tutorials presented at the 5th IFAC Symposium on Identification and System Parameter Estimation, Darmstadt, September 1979, Oxford: Pergamon Press.

Lawson, C. L. and R. J. Hanson (1974): *Solving Least Squares Problems*. Englewood Cliffs, N.J.: Prentice Hall, Inc.

Special issue on Identification and System Parameter Estimation, *Automatica,* 17, no. 1 (January 1981).

For someone interested in historical notes, see

Gauss, K. F. (1809): *Theoria Motus Corporum Coelestium* (in Latin). Eng. trans. (1963): *Theory of Motion of the Heavenly Bodies*. New York: Dover.

Further historical remarks are found in

Sorensen, H. W. (1970): "Least-squares Estimation: From Gauss to Kalman," *IEEE Spectrum,* 7, no. 7, 63–68.

ADAPTIVE CONTROL

14

GOAL

To Combine Control Design and Recursive Estimation Methods to Obtain Self-Tuning Regulators. To Outline the Principles of Adaptive Control.

14.1 Introduction

To apply the techniques discussed in the previous chapters to a practical problem is a substantial effort. It is necessary to carry out the steps of modeling, identification, control design, and sensitivity analysis. It may even be necessary to repeat the steps a few times until a satisfactory result is obtained. It is of considerable interest to explore the possibility of simplifying the procedure by admitting more complex regulators. One possibility is to try to automate the whole procedure. This could be done by providing the regulator with algorithms for parameter estimation and control design. Such an approach leads to the so-called Self-Tuning Regulator (STR), which has the potential of tuning itself. This is discussed in Sec. 14.2. Such regulators will be more complex than constant gain regulators. However, they can conveniently be implemented using microprocessors.

The closed-loop systems obtained with a self-tuning regulator are nonlinear, time-varying systems. Analysis of such systems requires approaches quite different from those developed in this book. Some key problems are outlined in Sec. 14.3. The properties are also illustrated by simple examples.

Small modifications of a self-tuning regulator give an adaptive controller, which can handle systems with large parameter variations. A review of some different approaches to adaptive control is given in Sec. 14.4. This includes gain scheduling, model-reference adaptive control, and dual control.

There is now good experience in the industrial use of adaptive techniques. Some remarks on uses and abuses are given in Sec. 14.5.

14.2 Self-Tuning Control

One way of automating modeling and design is the following: Determine a suitable model structure. Estimate the parameters of the model recursively using the methods discussed in Chapter 13. Use the estimates to calculate the control law by a suitable design method, e.g., one of the techniques discussed in Chapter 9, 10, 11, or 12. A block diagram of such a system is shown in Fig. 14.1. The regulator obtained is called a *self-tuning regulator* because it has facilities for tuning its own parameters. The regulator can be thought of as being composed of two loops. The inner loop consists of the process and an ordinary linear-feedback regulator. The parameters of the regulator are adjusted by the outer loop, which is composed of a recursive parameter estimator and a design calculation. The design calculation box in Fig. 14.1 represents an on-line solution to a design problem for a system with known parameters.

The self-tuning regulator is very flexible with respect to the design method. Virtually any design technique can be accommodated. Self-tuners based on phase and amplitude margins, pole-placement, minimum-variance control, and LQG-control have been considered. Many different parameter-estimation schemes may be used—e.g., stochastic approximation, least squares, extended and generalized least squares, instrumental variables, extended Kalman filtering, and maximum-likelihood.

The self-tuning algorithms can be divided into two major classes: direct and indirect algorithms. In an indirect algorithm, there is an estimation of an indirect process model. The controller parameters are obtained through the design procedure. It is sometimes possible to reparameterize the process so that it can be expressed in terms of the regulator parameters. This gives a significant simplification of the algorithm because the design calculations are eliminated. Such a

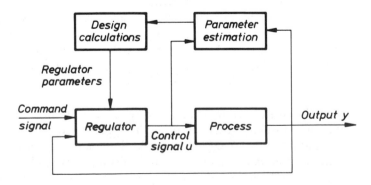

Figure 14.1 Block diagram of a self-tuning regulator (STR).

self-tuner is called a direct self-tuning regulator as it is based on direct estimation of the controller parameters.

The Design Problem

Consider a pole-placement design problem of the type discussed in Sec. 10.5. Let the process be described by the model

$$A(q)y(k) = B(q)u(k) \tag{14.1}$$

where u is the control signal, y is the measured signal, and disturbances are neglected. Further, $d = \deg A - \deg B$. Assume that it is desired to find a regulator such that the transfer function from command signal to output signal is given by

$$H_m = \frac{B_m}{A_m} \tag{14.2}$$

The observer polynomial is assumed to be A_o. The solution to the design problem is given in Sec. 10.5. The regulator is given by

$$Ru = Tu_c - Sy \tag{14.3}$$

where u_c is the command signal. The polynomials R, S, and T are obtained by solving the Diophantine equation

$$AR_1 + B^-S = A_mA_o \tag{14.4}$$

with respect to R_1 and S. The desired feedback is given by (14.3) with

$$R = R_1B^+ \quad \text{and} \quad T = A_oB'_m \tag{14.5}$$

where $B_m = B^-B'_m$ and $B = B^-B^+$.

There are many different ways to estimate the parameters of the model of (14.1) recursively. The least-squares method discussed in Sec. 13.4 is one possibility.

An Indirect Self-Tuner

A simple self-tuner may be described by the following algorithm.

Algorithm 14.1

Step 1. Estimate the coefficients of the polynomials A and B in (14.1) by recursive least squares.

Step 2. Substitute A and B by their estimates obtained in step 1 and solve Equation (14.4) for R_1 and S. Calculate R and T from (14.5).

Step 3. Calculate the control signal from (14.3).

Steps 1, 2, and 3 are repeated at each sampling period.

Some precautions must be taken when using this algorithm. To obtain good estimates, it is necessary that the input signal to the process be sufficiently rich in frequencies. This will not always be the case, because the input is generated by feedback. Solution of the Diophantine equation of (14.4) requires that polynomials A and B not have common factors.

A Direct Self-Tuner

It is possible to eliminate step 2 in Algorithm 14.1 by reparameterizing the model (14.1). It follows from (14.4) that

$$A_oA_my = AR_1y + B^-Sy = BR_1u + B^-Sy = B^-[Ru + Sy] \quad (14.6)$$

where the second equality follows from (14.1) and the third from (14.5). Equation (14.6) can be interpreted as a process model, which is parameterized in B^-, R, and S. An estimation of the parameters of the model in (14.6) gives the regulator parameters directly. Also, the model of (14.6) is linear in the parameters only if $B^- = 1$. The direct algorithm can be expressed as follows.

Algorithm 14.2

Step 1. Estimate the coefficients of the polynomials R, S, and B^- in (14.6) recursively.

Step 2. Calculate the control signal from (14.3), where R and S are replaced by the estimates obtained in step 1. T is chosen as in (14.5).

Steps 1 and 2 are repeated each sampling period.

In this case it is also necessary to take some precautions, because the control law of (14.3) is not causal if the leading coefficient of the estimate of polynomial R is zero. Minor modifications are required to avoid this difficulty.

Algorithm 14.2 becomes particularly simple if $B^- = 1$, because the estimation problem can then be solved by recursive least squares. Most simple self-tuners are based on this assumption. It follows from (14.5) that all process zeros then are canceled in the design. Thus algorithms based on this assumption will not work for processes with unstable inverses, even in the ideal case of known parameters (compare with Sec. 10.5).

14.3 Analysis

The closed-loop systems obtained with self-tuning control are nonlinear, which makes it difficult to understand how the system works. A reasonably complete treatment is far outside the scope of this book. There are some properties of this interesting and important class of computer-control systems worth discussing, however.

A model of the process to be controlled is necessary for the discussion. It is assumed that the process is governed by the model

$$Ay = Bu + Ce \tag{14.7}$$

where e is white noise.

Key Issues

The essential problems are to find out how the closed-loop systems behave under self-tuning control. This includes analysis of stability, convergence, and performance. Another important objective is to find out if a control system such as the self-tuner is reasonable or if there are better schemes.

Indirect Algorithms

An indirect self-tuner, such as the one given by Algorithm 14.1, converges if the parameter estimates converge. This requires that the model structure used in the estimation be correct and that the input signal be sufficiently rich in frequencies. Because the least-squares method is used, it is necessary that there be no correlation in the disturbances, i.e., $C = 1$ in the model of (14.7). Because the control signal is generated by feedback, there is no guarantee that it is sufficiently rich in frequencies. It may be necessary to introduce perturbation signals to ensure this.

Analysis of a First-Order System

To illustrate some principles of analysis of self-tuning regulators, a first-order example is considered.

Assume that the dynamics of the process and its environment can be described by the simple first-order system

$$y(k) + ay(k - 1) = bu(k - 1) + e(k) + ce(k - 1) \tag{14.8}$$

where u is the control variable, y is the output, and $\{e(k)\}$ is a sequence of independent Gaussian random variables. Furthermore, assume that the criterion is to minimize the quadratic loss function

$$J = \lim_{N \to \infty} E \frac{1}{N} \sum_{k=1}^{N} y^2(k) \tag{14.9}$$

The admissible controls are assumed to be such that $u(k)$ is a function of all past outputs $y(k), y(k - 1), \ldots$ If the parameters are known, it follows from Theorem 12.2 that the optimal control is the proportional feedback

$$u(k) = \frac{a - c}{b} y(k) \tag{14.10}$$

Assume that $b = 1$ and consider a self-tuner that is based on least-squares estimation of the parameter θ in the model

$$y(k) + \theta y(k - 1) = u(k - 1) + e(k) \qquad (14.11)$$

The least-squares estimate $\hat{\theta}$ based on data available up to time k—i.e., $y(k)$, $y(k - 1), \ldots, y(1), u(k - 1), u(k - 2), \ldots, u(1)$—is given by

$$\hat{\theta}(k) = \frac{-\left[\sum_{i=1}^{k-1} [y(i + 1) - u(i)]y(i) \right]}{\sum_{i=1}^{k-1} y^2(i)} \qquad (14.12)$$

The minimum-variance control law for (14.11) is

$$u(k) = \theta(k)y(k)$$

if θ is known. When θ is not known, it is replaced by its estimate. The control law then becomes

$$u(k) = \hat{\theta}(k)y(k) \qquad (14.13)$$

The control algorithm given by (14.12) and (14.13) can be expected to work nicely for the system (14.8) if $c = 0$ and $b = 1$. In this case the least-squares estimate $\hat{\theta}$ will converge to a as $k \to \infty$, and the control law (14.13) will converge to

$$u(k) = ay(k)$$

which is the desired control law. Compare (14.10).

It is a remarkable property of the feedback law described by (14.12) and (14.13) that it will also converge to the optimal law (14.10) when $c \neq 0$. This is illustrated in Fig. 14.2, which shows the parameter estimate θ for $a = -0.9$, $b = 3$, and $c = -0.3$. Notice that the estimate θ appears to converge to $\hat{\theta} = -0.2$ and not to the value of a, which is -0.9. The value -0.2 corresponds to the minimum-variance control. To compare the self-tuning regulator with the optimal

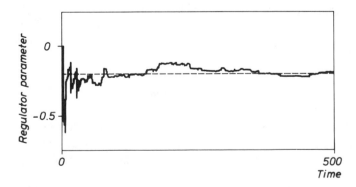

Figure 14.2 Parameter estimate $\hat{\theta}$ obtained in a simulation of the control law of (14.12) and (14.13) applied to the system in (14.8) with $a = -0.9$, $b = 3$, and $c = -0.3$.

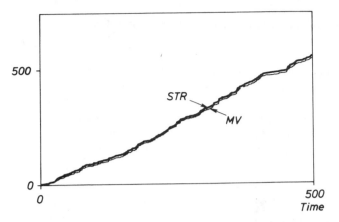

Figure 14.3 Accumulated loss functions for the self-tuning regulator of (14.12) and (14.13) and the optimal regulator based on known parameters.

regulator for known parameters, the accumulated loss function

$$J(k) = \sum_{i=1}^{k} y^2(i)$$

has been calculated for the self-tuning regulator and the optimal regulator

$$u(k) = -0.2y(k)$$

The results are shown in Fig. 14.3. It is clear from this figure that there is not a large difference between the performance of the two regulators.

The example shows that the simple self-tuning regulator of (14.12) and (14.13) will perform very well. After a short transient period, it will give almost the same performance as a regulator based on exact knowledge of the system parameters. The parameter θ appears to converge to a value that corresponds to a minimum-variance regulator. These empirical observations are investigated next. Assume first that the system is governed by

$$y(k + 1) + ay(k) = u(k) + n(k) \tag{14.14}$$

where n is a disturbance. If the disturbance is bounded in the sense

$$\lim_{k \to \infty} \frac{1}{k} \sum_{i=1}^{k} n^2(i) < \infty \tag{14.15}$$

then the mean-square output of the closed-loop system

$$\frac{1}{t} \sum_{k=1}^{t} y^2(k) \tag{14.16}$$

is also bounded.

This statement is shown by contradiction along the following lines. If y is unbounded, then the influence of $n(k)$ in (14.14) can be disregarded. This then implies that $\hat{\theta}(k)$ will converge to a. Using the control law

$$u(k) = ay(k)$$

will, however, give a bounded output because n is bounded; this gives a contradiction.

The self-tuning regulator of (14.12), and (14.13) will always stabilize the system in (14.14) in the mean-square sense. It can be shown in this case that the parameter estimate $\hat{\theta}(k)$ also converges. If $\theta(k)$ converges as $k \to \infty$, it is easy to find the convergence point. The normal equations can be written

$$\sum_{i=1}^{k} y(i + 1)y(i) = \hat{\theta}(k + 1) \sum_{i=1}^{k} y^2(i) + \sum_{i=1}^{k} y(i)u(i)$$

Equation (14.13) now gives

$$\frac{1}{k} \sum_{i=1}^{k} y(i + 1)y(i) = \frac{1}{k} \sum_{i=1}^{k} [\hat{\theta}(k + 1) - \hat{\theta}(i)]y^2(i)$$

The right member converges to zero as $k \to \infty$ because $\hat{\theta}(k)$ converges and $y(k)$ is mean-square bounded. Thus if the parameter estimates converge, then

$$\lim_{k \to \infty} \frac{1}{k} \sum_{i=1}^{k} y(i + 1)y(i) = 0 \tag{14.17}$$

The self-tuning regulator attempts to make the correlation of the closed-system output, $r_y(\tau)$, zero for $\tau = 1$. Assuming that the process to be controlled is given by (14.8), it now follows that (14.16) is bounded and (14.17) holds for only one value of $\hat{\theta}$, namely,

$$\hat{\theta} = a - c$$

It has thus been established that the regulator of (14.12), and (14.13) is self-tuning for the system in (14.8) and the minimum-variance criterion. The analysis can be extended to the case when $b \neq 1$. The results can be extended to control of an nth-order process of the form (14.7). Additional conditions are then required both for stability and convergence. There are also cases in which the parameter estimates do not converge.

General Results

The analysis of the simple example may be generalized. The asymptotic properties are summarized in two theorems.

Theorem 14.1. Consider Algorithm 14.2 and assume that $B^- = 1$. Assume that the parameter estimates s_i, $i = 0, \ldots, n_s$, and r_i, $i = 1, \ldots, n_r$, converge and that the closed-loop system is ergodic (in the second moments); then the closed-loop system has the properties

$$Ey(k + \tau)y(k) = r_y(\tau) = 0, \qquad \tau = d, \ldots, d + n_s$$

$$Ey(k + \tau)u(k) = r_{yu}(\tau) = 0, \qquad \tau = d, \ldots, d + n_r \qquad \square$$

Theorem 14.2. Let the system to be controlled be governed by (14.7), where $d = \deg A - \deg B$ is known. Assume that Algorithm 14.2 is used with $n_s \geq n - 1$ and $n_r \geq n + d - 1$. If the parameter estimates converge, then the regulator of (14.3) with $T = 0$ will converge to the minimum-variance regulator for the process (14.7). $\qquad\qquad\qquad\qquad\qquad\qquad\qquad\qquad\qquad\qquad\qquad$ □

Theorem 14.1 implies that the stationary points of the algorithm are characterized by the condition that some values of $r_y(\tau)$ and $r_{yu}(\tau)$ are equal to zero. The theorem holds independently of what the process looks like. Theorem 14.2 implies that if the process is governed by (14.7), if d is known, and if there are sufficiently many parameters in the regulator, then the algorithm will converge to the minimum-variance controller if it converges at all.

The stability and convergence properties of self-tuning regulators can also be analyzed. It can be shown that a self-tuner based on least squares estimation and minimum-variance control will converge to the minimum-variance regulator for (14.7) provided that

$$\frac{1}{C(z)} - \frac{1}{2}$$

is strictly positive real. The algorithm will be locally stable provided that

$$C(z_i) > 0$$

for all z_i such that $B(z_i) = 0$.

14.4 Other Approaches to Adaptive Control

The self-tuner discussed in Sec. 14.2 is motivated by the desire to obtain automatic tuning of a control loop. The name *self-tuning* was in fact introduced to emphasize this. It seems reasonable, however, that the regulator in Fig. 14.1 could also be made to control a process with varying parameters—i.e., an adaptive regulator could be achieved. To do this it is necessary to change the algorithm so that the parameter estimator can track varying process parameters. One way to do this is to introduce discounting of old data, as discussed in Sec. 13.5. There are also other schemes for adaptive control that are closely related to the self-tuner. Three of these will be described in this section. The starting point is an ordinary feedback-control loop with a process and a regulator with adjustable parameters. The schemes represent different ways to alter the regulator parameters in response to changes in process and disturbance dynamics. The schemes differ only in the way the parameters of the regulator are adjusted.

Gain Scheduling

It is sometimes possible to find auxiliary process variables that correlate well with the changes in process dynamics. It is then possible to eliminate the influences of parameter variations by changing the parameters of the regulator as functions

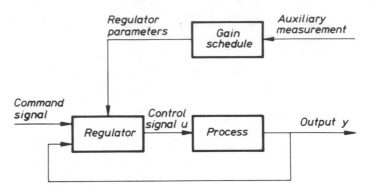

Figure 14.4 Block diagram of a system where parameter variations are eliminated by gain scheduling.

of the auxiliary variables (see Fig. 14.4). This approach is called *gain scheduling* because the system was originally used to accommodate changes only in process gain.

One drawback of gain scheduling is that it is an open-loop compensation. There is no feedback that compensates for an incorrect schedule. Gain scheduling can thus be viewed as an extension of feedforward compensation. Another drawback is that the design is time-consuming. The regulator parameters must be determined for many operating conditions. The performance must be checked by extensive simulations.

Gain scheduling has the advantage that the parameters can be changed quickly in response to process changes. The limiting factors depend on how quickly the auxiliary measurements respond to process changes.

There is a controversy in nomenclature concerning whether gain scheduling should be considered as an adaptive scheme or not, because the parameters are changed in open loop. Irrespective of this discussion, gain scheduling is a very useful technique for reducing the effects of parameter variations.

Model-Reference Adaptive Systems (MRAS)

Another way to adjust the parameters of the regulator is illustrated in Fig. 14.5. This scheme was originally developed for the servo problem. The specifications are given in terms of a reference model, which tells how the process output ideally should respond to the command signal. Notice that the reference model is part of the control system. The regulator can be thought of as consisting of two loops. The inner loop is an ordinary control loop composed of the process and the regulator. The parameters of the regulator are adjusted by the outer loop in such a way that the error e between the model output y_m and the process output y becomes small. The outer loop thus also looks like a regulator loop. The key problem is to determine the adjustment mechanism so that a stable system, which brings the error to zero, is obtained. This problem is nontrivial. It is easy to show that it cannot be solved with a simple linear feedback from the error to the controller parameters.

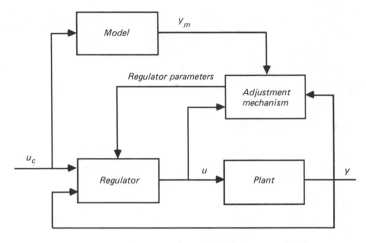

Figure 14.5 Block diagram of model-reference adaptive system.

The following parameter adjustment mechanism, called the *MIT rule*, was used in the original MRAS:

$$\frac{d\theta}{dt} = -\alpha e \ \text{grad}_\theta \ e \qquad (14.18)$$

where e is the model error and the components of the vector θ are the adjustable parameters. The number α is a parameter that determines the adaptation rate. Equation (14.18) represents an adjustment mechanism, which is composed of three parts: a linear filter for computing the sensitivity derivatives from process inputs and outputs, a multiplier, and an integrator. This configuration is typical for many adaptive systems.

The MIT rule will perform well if the parameter α is small. The allowable size depends on the magnitude of the reference signal. Consequently, it is not possible to give fixed limits that guarantee stability. Thus the MIT rule can give an unstable closed-loop system. Modified adjustment rules can be obtained using stability theory. These rules are similar to the MIT rule. The sensitivity derivatives in (14.18) will be replaced by other functions.

Relationships between MRAS and STR

It is clear from Figs. 14.1 and 14.5 that the MRAS and the STR are closely related. Both systems have two feedback loops. The inner loop is an ordinary feedback loop with a process and a regulator. The regulator has adjustable parameters which are set by the outer loop. The adjustment is based on feedback from the process inputs and outputs. However, the methods for design of the inner loop and the techniques used to adjust the parameters in the outer loop may be different. When the underlying design methods are the same, the direct self-tuner and the MRAS are identical.

The regulator shown in Fig. 14.1 can also be derived from the MRAS approach if the parameter estimation is done by updating a reference model. The scheme is then called an *indirect MRAS* because the regulator parameters are updated indirectly via the design calculation. The *direct MRAS*, where the regulator parameters are updated directly, is closely related to the direct STR.

Stochastic Control Theory

Regulator structures such as MRAS and STR are based on heuristic arguments. It would be appealing to obtain the regulators from a unified theoretical framework. This can be done using nonlinear stochastic-control theory. The system and its environment are then described by a stochastic model. The criterion is formulated to minimize the expected value of a loss function, which is a scalar function of states and controls.

The problem of finding a control that minimizes the expected loss function is difficult. Conditions for existence of optimal control are not known. Under the assumption that a solution exists, a functional equation for the optimal loss function can be derived using dynamic programming. This equation, which is called the *Bellman equation*, can be solved numerically only in very simple cases. The structure of the optimal regulator obtained is shown in Fig. 14.6. The controller can be considered to be composed of two parts: an estimator and a feedback regulator. The estimator generates the conditional probability distribution of the state from the measurements. This distribution is called the *hyperstate* of the problem. The feedback regulator is a nonlinear function that maps the hyperstate into the space of control variables.

The structural simplicity of the solution is obtained at the price of introducing the hyperstate, which is a quantity of very high dimension. Observe that the self-tuner in Fig. 14.1 can be regarded as an approximation of the regulator in Fig. 14.6, where the hyperstate is approximated by the process state and the parameter estimate.

The optimal-control law has an interesting property. The control will not only try to drive the output to its desired value; when the parameters are uncertain,

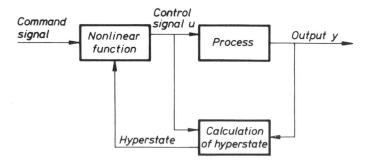

Figure 14.6 Block diagram of an adaptive regulator obtained from stochastic-control theory.

the regulator also introduces perturbations, which will improve the estimates and the future controls. The optimal control gives the correct balance between maintaining small control signals and estimation errors. This property is called *dual control*.

Example 14.1

A simple example is used for illustration. Consider a system described by

$$y(k + 1) = y(k) + bu(k) + e(k)$$

where u is the control, y is the output, e is white noise, and b is a constant parameter or a Markov process. Let the criterion be to minimize the mean-square deviation of the output y.

If the parameter b has a prior Gaussian distribution, it can be shown that the conditional distribution of b, given inputs and outputs up to time k, is Gaussian with mean $\hat{b}(k)$ and standard deviation $\sigma(k)$. The equations for updating $\hat{b}$ and σ are the same as the ordinary Kalman filtering equations.

The optimal-control law can be computed numerically using dynamic programming. An approximation of the dual-control law is given by

$$u(k) = -y(k) \frac{\hat{b}(k) + 0.56\sigma(k)}{\hat{b}^2(k) + 0.08\hat{b}(k)\sigma(k) + 2.2\sigma^2(k)} - \frac{1.9\sigma^3(k)}{\hat{b}^4(k) + 1.7\sigma^4(k)} \quad (14.19)$$

for $\hat{b} > 0$ and $y \geq 0$. The control law is asymmetric in $\hat{b}$ and y. This control law has dual-control action because of the last term.

Some approximations of the optimal control law can be done. The *certainty-equivalence controller*

$$u(k) = -y(k)/\hat{b}(k) \quad (14.20)$$

is obtained simply by solving the control problem in the case of known parameters and substituting the known parameters with their estimates. The self-tuning regulator can be interpreted as a certainty-equivalence controller.

The control law

$$u(k) = -\frac{1}{\hat{b}(k)} \cdot \frac{\hat{b}^2(k)}{\hat{b}^2(k) + \sigma^2(k)} y(k) \quad (14.21)$$

is another approximation, called a *cautious controller*, because it hedges and uses lower gain when the estimates are uncertain. The controllers of (14.19) and (14.21) are the same as (14.20) when there are no uncertainties about b.

Notice that the certainty-equivalence controller and the cautious controller do not have the dual property. ☐

14.5 Uses and Abuses of Adaptive Techniques

Self-tuning and adaptive techniques are still in their infancy. Such methods could not be used at reasonable costs before the advent of microprocessors. Because the techniques offer very interesting possibilities for obtaining new types of regulators and new functions in control systems, possible uses of the techniques are of interest.

Auto-Tuning

It is possible to tune regulators with three to four parameters by hand if there is not too much interaction between adjustments of different parameters. For more complex regulators, however, it is necessary to have suitable tuning tools. Traditionally, tuning of more-complex regulators has followed the route of modeling or identification and regulator design. This is often a time-consuming and costly procedure, which can be applied only to important loops or to systems that are made in large quantities.

Both the MRAS and the STR become constant-gain feedback controls when the estimated parameters are constant. Hence, the adaptive loop can be used as a *tuner* for a control loop. In such applications the adaptation loop is simply switched on; perturbation signals may be added. The adaptive regulator is run until the performance is satisfactory. In such cases the identification loop is also run with decreasing gain. The adaptation loop is then disconnected and the system is left running with fixed regulator parameters. Auto-tuning can be considered as a convenient way to incorporate automatic modeling and design into a regulator. It widens the class of problems where systematic design methods can be used cost effectively. Auto-tuning is particularly useful for design methods, such as feedforward, which depend critically on good models. Automatic tuning can be applied to simple PID-controllers, as well as to more complicated regulators.

It is very convenient to introduce auto-tuning in a DDC package. One tuning algorithm can then serve many loops. Because a good tuning algorithm requires only a few thousand bytes of memory in a control computer, substantial benefits are obtained at marginal costs.

Auto-tuning can also be included in single-loop regulators. For example, it is possible to design regulators where the mode switch has three positions: manual, automatic, and tuning. However, the tuning algorithm represents a major part of the software of a single-loop regulator. Memory requirements will typically be more than doubled when auto-tuning is introduced.

A single-loop PID-controller with auto-tuning is shown in Fig. 1.3. When the operator pushes the 'tune' button, the PID-controller is replaced by a relay. The relay induces a limit cycle oscillation in the control loop. The period and amplitude of the oscillation are estimated. This gives one point on the Nyquist curve. A modified Ziegler-Nichols rule is then used to tune the parameters in the PID-controller.

Auto-tuning bypasses many of the remaining theoretical difficulties currently associated with adaptive control. The only theoretical difficulty arising is identification under closed-loop conditions. This may be bypassed by introducing perturbation signals. Because the tuning can be done under human supervision, it is often possible to do it safely.

Performance-Related Tuning Knobs

The operator interface is important because adaptive regulators may also have parameters, which must be chosen. When applying adaptive techniques, it is often desirable to have the absolute black box without any external adjustments. Such regulators may be designed for specific applications, where the purpose of control can be stated a priori; in many cases, however, it is not possible to specify the purpose of control a priori. It is at least necessary to tell the regulator what it is expected to do. This can be done by introducing *performance-related* dials that give the desired properties of the closed-loop system. New types of regulators can be designed using this concept. For example, it is possible to have a regulator with one dial, which is labeled with the desired closed-loop bandwidth. Another possibility is to have a regulator with a dial, which is labeled with the weighting between state deviation and control action in a LQG-problem. A third possibility is to have a dial labeled with the phase margin or amplitude margin. The characteristics of a regulator with performance-related knobs are illustrated by an example.

Example 14.2—The bandwidth self-tuner

The bandwidth self-tuner is an adaptive regulator that has one adjustable parameter on the front panel, which is labeled with the *desired closed-loop bandwith*. The particular implementation is in the form of a pole-placement self-tuner.

The response of the servo to a square-wave command signal is shown in Fig. 14.7. From the beginning the regulator does not have any information about the

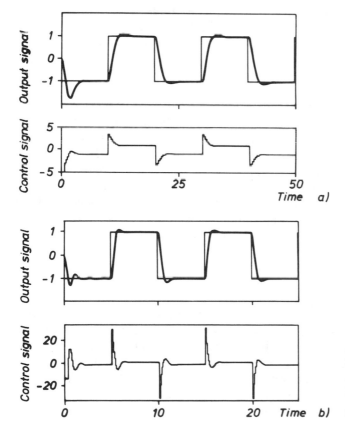

Figure 14.7 Simulation of the bandwidth self-tuner. The process has the transfer function $1/(s + 1)^2$. The requested bandwidth is (a) 1.5 rad/s and (b) 4.5 rad/s.

system. It is clear from the figure that the servo performs very well after the first command step. Figure 14.7 shows that the control signal gives a moderate "kick" after a command step when the requested bandwidth is low (1.5 rad/s). The control signal then decreases gradually to a steady-state level. When a larger bandwidth (4.5 rad/s) is demanded, as in Fig. 14.7(b), the value of the control signal immediately after the step is more than 30 times larger than its steady-state value. A low bandwidth gives a slow response, small control signals, and low sensitivity to measurement noise. A high bandwidth gives a fast response. However, the control actions will be large and the system will be sensitive to noise. It is easy for an operator to determine the bandwidth suitable for a particular application by experimentation. Notice that apart from the specifications in terms of the bandwidth, all necessary adjustments— for instance selection of the sampling interval—are handled automatically by the self-tuner. □

Abuses of Adaptive Control

An adaptive regulator, being inherently nonlinear, is more complicated than a fixed-gain regulator. Before attempting to use adaptive control, it is important to make sure that the control problem cannot be solved by constant-gain feedback. In the vast literature on adaptive control, there are many cases in which a constant-gain feedback can do as well as an adaptive regulator. Notice that it is not possible to judge the need for adaptive control from the variations of the open-loop dynamics over the operating range. Many cases are known where a constant-gain feedback can cope well with considerable variations in system dynamics.

Supervisory Loops

The adaptive systems in Figs. 14.1 and 14.5 can be regarded as hierarchical systems with two levels. The lower level is the ordinary-feedback loop with the process and the regulator. The adaptation loop, which adjusts the parameters, represents the higher level. The adaptation loop in typical STR or MRAS requires parameters such as model orders, forgetting factors, and sampling periods. A third layer can be added to the hierarchy to set these parameters. For instance, a suitable forgetting factor can be determined by monitoring the excitation of the process. By storing process inputs and outputs, it is also possible to estimate models with different sampling periods, different orders, and different structures.

14.6 Conclusions

The adaptive technique is slowly emerging after many years of research and experimentation. Important theoretical results on stability and structure have been established. Much theoretical work still remains to be done. The advent of microprocessors has been a strong driving force for the applications. Laboratory experiments and industrial feasibility studies have contributed to a better understanding of the practical aspects of adaptive control. There are also a number of adaptive regulators appearing on the market.

14.7 Problems

14.1. Consider the process

$$y(k) + ay(k - 1) = bu(k - 1) + e(k) + ce(k - 1)$$

and use the least-squares method to estimate the parameter α in the model

$$y(k + 1) + \alpha y(k) = \beta_0 u(k) + \epsilon(k + 1)$$

where β_0 is assumed to be known. Let the control signal be generated as

$$u(k) = \alpha/\beta_0 y(k)$$

(a) Show that a possible convergence point for the least-squares estimate is characterized by

$$r_y(1) = 0$$

$$r_{yu}(1) = 0$$

(b) Show that the minimum-variance controller given by

$$\alpha = \frac{a - c}{b} \beta_0$$

is a possible convergence point.

14.2. Simulate the system in Problem 14.1 where α is estimated recursively. Investigate the influence of β_0. For instance, can b and β_0 have different signs? (*Hint:* It is useful to constrain the control signal, for instance, to 3–5 times the standard deviation of the control signal, when the optimal minimum variance controller is used.)

14.3. Derive conditions for

$$y(k) + a_1 y(k - 1) + \cdots + a_n y(k - n) = n(k)$$

to be mean-square bounded when n is mean-square bounded, i.e.,

$$\lim_{k \to \infty} \frac{1}{k} \sum_{j=1}^{k} n(j)^2 < \infty$$

14.4. Consider the process

$$y(k + 1) - y(k) = u(k - 1) + e(k + 1) - 0.9e(k)$$

(a) Determine the minimum-variance control law for the system.
(b) Determine a self-tuner for the model

$$y(k + 1) = s_0 y(k) + s_1 y(k - 1) + r_0[u(k) + r_1 u(k - 1)] + \epsilon(k)$$

based on minimum-variance control and least-squares estimation. Show that the regulator is compatible with the minimum-variance regulator for the system in (a) and compare the performance of the self-tuner with the minimum-variance regulator based on known parameters.
(c) Determine the possible equilibrium points for the self-tuner in (b) using Theorems 14.1 and 14.2.

14.5. Simulate a self-tuner based on estimation of the parameters of the model

$$y(k) + a_1 y(k - 1) + a_2 y(k - 2) = b_1 u(k - 1) + b_2 u(k - 2)$$

and the pole-placement design given in Chapter 10. Verify the results of Fig. 14.7.

14.6. Prove that (14.14) will have a bounded output if the disturbance is bounded when the simple self-tuner (14.12) and (14.13) is used.

14.8 References

Surveys of the theory and applications of self-tuning regulators can be found in

ÅSTRÖM, K. J. (1980): "Self-tuning Regulators: Design Principles and Applications," in *Applications of Adaptive Control,* eds. Narendra and Monopoli. New York: Academic Press.

ISERMANN, R. (1982): "Parameter Adaptive Control Algorithms-A Tutorial," *Automatica,* 18, 513–528.

ÅSTRÖM, K. J. (1983): "Theory and Applications of Adaptive Control," *Automatica,* 19 471–86.

GOODWIN, G. C. and K. S. SIN (1984): *Adaptive Filtering Prediction and Control.* Englewood Cliffs N.J.: Prentice Hall, Inc.

ÅSTRÖM, K. J. and B. WITTENMARK (1989): *Adaptive Control.* Reading, Mass.: Addison-Wesley.

The self-tuning regulator was originally proposed in

KALMAN, R. E. (1958): "Design of a Self-optimizing Control System," *Trans. ASME,* 80, 468–78.

Self-tuning regulators have received much attention since then. Some basic references are

PETERKA, V. (1970): Adaptive Digital Regulation of Noisy Systems. Preprint 2nd IFAC Symposium on Identification and Process Parameter Estimation, June 1970, Prague, Czechoslovakia.

ÅSTRÖM, K. J. and B. WITTENMARK (1973): "On Self-tuning Regulators," *Automatica,* 9, 185–99.

CLARKE, D. W. and P. J. GAWTHROP (1975): "A Self-tuning Controller," *Proc. IEE,* 122, 929–34.

WELLSTEAD, P. E., J. M. EDMUNDS, D. PRAGER, and P. ZANKER (1979): "Self-tuning Pole/ Zero Assignment Regulators," *Int. Journal Control,* 30, 1–26.

ÅSTRÖM, K. J. and B. WITTENMARK (1980): "Self-tuning Controllers Based on Pole-Zero Placement," *Proc. IEE,* 127, 120–30.

ÅSTRÖM, K. J. and B. WITTENMARK (1985): "The Self-Tuner Revisited," *Preprints IFAC 7th Symposium on Identification and System Parameter Estimation,* York, U.K.

Adaptive control in general is treated in

BELLMAN, R. (1961): *Adaptive Control: A Guided Tour.* Princeton, N.J.: Princeton University Press.

SARIDIS, G. N. (1977): *Self-organizing Control of Stochastic Systems.* New York: Marcel Dekker.

WITTENMARK, B. (1975): "Stochastic Adaptive Control Methods: A Survey," *Int. Journal Control,* 21, 705–30.

Model-reference adaptive control is discussed in

WHITAKER, H. P., J. YAMROM, and A. KEZER (1958): Design of Model-reference Adaptive Control Systems for Aircraft. Report R-164, Instrumentation Laboratory, MIT, Cambridge, Mass.

PARKS, P. C. (1966): "Lyapunov Redesign of Model Reference Adaptive Control Systems," *IEEE Trans. Autom. Control,* AC-11, 362–67.

MONOPOLI, R. V. (1974): "Model Reference Adaptive Control with an Augmented Error Signal," *IEEE Trans. Autom. Control,* AC-19, 474–84.

LANDAU, Y. D. (1979): *Adaptive Control: The Model Reference Approach.* New York: Marcel Dekker.

Analysis of stability and convergence properties of self-tuning regulators is found in

LJUNG, L. (1977): "On Positive Real Transfer Functions and the Convergence of Some Recursive Schemes," *IEEE Trans. Autom. Control,* AC-22, 539–51.

EGARDT, B. (1979): *Stability of Adaptive Controllers.* Berlin: Springer-Verlag.

MORSE, A. S. (1980): "Global Stability of Parameter Adaptive Systems," *IEEE Trans. Autom. Control,* AC-25, 433–39.

NARENDRA, K. S., Y. H. LIN, and L. S. VALAVANI (1980): "Stable Adaptive Controller Design. Part II: Proof of Stability," *IEEE Trans. Autom. Control,* AC-25, 440–48.

GOODWIN, G. C., P. J. RAMADGE, and P. E. CAINES (1980): "Discrete Time Multivariable Adaptive Control," *IEEE Trans. Autom. Control,* AC-25, 449–56.

GOODWIN G. C., P. J. RAMADGE, and P. E. CAINES (1981): "Discrete Time Stochastic Adaptive Control," *SIAM Journal on Control and Optimization,* 19, 829–53.

Dual control was introduced by Feldbaum in

FELDBAUM, A. A. (1960): "Dual Control Theory I–IV," *Automation & Remote Control,* 21, 874–80, 1033–39, 22, 1–12, 109–21.

Further discussions of the dual-control problem are found in

ÅSTRÖM, K. J. and B. WITTENMARK (1971): "Problems of Identification and Control," *J. Math. Analysis and Appl.,* 34, 90–113.

BAR-SHALOM, Y. and E. TSE (1974): "Dual Effect, Certainty Equivalence and Separation in Stochastic Control," *IEEE Trans. on Autom. Control,* AC-19, 494–500.

ÅSTRÖM, K. J. and A. HELMERSSON (1982): "Dual Control of a Low Order System," *Proc. CNRS Colloque National,* Belle Île, France, September.

WITTENMARK, B. and C. ELEVITCH (1985): "An Adaptive Control Algorithm with Dual Features," *Preprints IFAC 7th Symposium on Identification and System Parameter Estimation,* York, U.K.

Applications of adaptive control are discussed in

HARRIS, C. J. and S. A. BILLINGS, eds., (1981): *Self-Tuning and Adaptive Control: Theory and Applications.* London: Peter Peregrinus.

NARENDRA, K. S. and R. V. MONOPOLI, eds., (1980): *Applications of Adaptive Control.* New York: Academic Press.

UNBEHAUEN, H., ed., (1980): *Methods and Applications in Adaptive Control.* Berlin: Springer-Verlag.

15

IMPLEMENTATION OF DIGITAL CONTROLLERS

GOAL

To Find Effects of Quantization in A-D and D-A Converters and Computation with Finite Wordlength. To Explore Prefiltering and Computational Delay. To Discuss Effects of Saturating Actuators. To Discuss Coding of Algorithms.

15.1 Introduction

Design of algorithms for computer control is discussed in the previous chapters. The problem of implementing a control algorithm on a digital computer is discussed in this chapter.

The control algorithms obtained in the previous chapters are discrete-time, dynamic systems. The key problem is to implement a discrete-time, dynamic system using a digital computer. An overview of this problem is given in Sec. 15.2, which shows that it is straightforward to obtain a computer code from the discrete-time algorithm. There are, however, several issues that must be considered. It is necessary to take the interfaces to the sensors, the actuators, and the human operators into account. It is also necessary to consider the numerical precision required.

The sensor interface is discussed in Sec. 15.3. This covers prefiltering and computational delays and shows that the computational delay depends critically on the organization of the algorithm. Different ways to shorten the computational delay by reorganizing the code are discussed. Excellent methods of filtering the signals effectively by introducing nonlinearities, which may reduce the influence of unreliable sensors, are shown. This is one of the major advantages of computer control. Most theory in this book deals with linear theory. There are, however, a few nonlinearities that must be taken into account. Actuator saturation is such a nonlinearity. Different ways of handling this are discussed in Sec. 15.4. This leads to extensions of the methods for antireset windup used in classical process control.

The operator interface is another important factor; it is discussed in Sec. 15.5. This includes treatment of operational modes and different ways to avoid switching transients. The information that should be displayed and different ways of influencing the control loop are also discussed. Digital computers offer many interesting possibilities; so far they have been used only to a very modest degree. There are many opportunities for innovations in this field.

It is, of course, important to have sound numerics in the control algorithm, which is the topic of Sec. 15.6. Effects of a finite word length are also discussed. Realization of digital controllers is treated in Sec. 15.7. Programming of control algorithms is discussed in Sec. 15.8. For more ambitious systems in which parameters and control algorithms are changed on-line, it is necessary to have an understanding of the issues of concurrent programming. A brief introduction is given in Sec. 15.9.

15.2 An Overview

This section gives an overview of implementation of digital control laws. Different representations of the control laws obtained from the design methods in Chapters 8–12 are first given; the algorithms are then implemented. A list of some important problems is given. These problems are discussed in greater detail in the following sections.

Different Representations of the Regulator

The design methods of the previous chapters give control laws in the form of a discrete time, dynamic system. Different representations are obtained, depending on the approaches used.

The design methods based on pole placement by state-feedback (Chapter 9) and LQG-control (Chapter 11) give a regulator of the form

$$\hat{x}(k \mid k) = \hat{x}(k \mid k - 1) + K[y(k) - \hat{y}(k \mid k - 1)]$$

$$u(k) = L[x_m(k) - \hat{x}(k \mid k)] + Du_c(k)$$

$$\hat{x}(k + 1 \mid k) = \Phi\hat{x}(k \mid k) + \Gamma u(k) \qquad (15.1)$$

$$x_m(k + 1) = f(x_m(k), u_c(k))$$

$$\hat{y}(k + 1 \mid k) = C\hat{x}(k + 1 \mid k)$$

In this representation the state of the regulator is $\hat{x}$ and x_m, where $\hat{x}$ is an estimate of the process state and x_m is the state of the model that generates the desired response to command signals u_c. The form in (15.1) is called a *state representation with an explicit observer* because of the physical interpretation of the regulator state. It is easy to include a nonlinear model for the desired state in this representation.

If the function f in (15.1) is linear, the regulator of (15.1) is a linear system with the inputs y and u_c and the output u. Such a regulator may always be represented as

$$u(k) = Cx(k) + Dy(k) + D_c u_c(k) \qquad (15.2)$$
$$x(k + 1) = Fx(k) + Gy(k) + G_c u_c(k)$$

(see Problem 9.7). Equation (15.2) is a *general-state representation* of a discrete-time, dynamic system. This form is more compact than (15.1). The state does not necessarily have a simple physical interpretation.

The design methods for single-input–single-output systems discussed in Chapters 10 and 12, which are based on external models, give a regulator in the form of a general *input-output representation*

$$R(q)u(k) = T(q)u_c(k) - S(q)y(k) \qquad (15.3)$$

where $R(q)$, $S(q)$, and $T(q)$ are polynomials in the forward-shift operator q. There are simple transformations between the different representation (compare with Chapter 3).

Implementing a Discrete-Time System

The implementation of a discrete-time system described by (15.1), (15.2), or (15.3) using a digital computer is straightforward. The details depend on the hardware and software available. To show the principles, it is assumed that the system described by (15.2) should be implemented using a digital computer with A-D and D-A converters and a real-time clock. A graphical representation of the program is shown in Fig. 15.1. The execution of the program is controlled by the clock. The horizontal bar indicates that execution is halted until an interrupt comes from the clock. The clock is set so that an interrupt is obtained at each sampling instant. The code in the box is executed after each interrupt. The body of the code required to implement the algorithm is given in Listing 15.1.

Analog-to-digital conversion is commanded in the first line. The appropriate

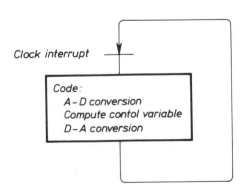

Clock interrupt

Code:
  A–D conversion
  Compute contol variable
  D–A conversion

Figure 15.1 Graphical representations of a program used to implement a discrete-time system.

Implementation of Digital Controllers   Chap. 15

LISTING 15.1 Computer code skeleton for the control law of (15.2). Line numbers are introduced only for purposes of referencing.

```
        Procedure Regulate
        begin
1          Adin y  uc
2          u := C*x + D*y + Dc*uc
3          x := F*x + G*y + Gc*uc
4          Daout u
        end
```

values are stored in the arrays y and uc. The control signal u is computed in the second line using matrix vector multiplication and vector addition. The state vector x is updated in the third line, and the digital-to-analog conversion is performed on the fourth line. To obtain a complete code, it is also necessary to have type declarations for the vectors u, uc, x, and y and the matrices F, G, Gc, C, D, and Dc. It is also necessary to assign values to the matrices and the initial value for the state x. When using computer languages that do not have matrix operations, it is necessary to write appropriate procedures for generating matrix operations using operations on scalars. Notice that the second and third line of the code correspond exactly to the algorithm in (15.2).

To obtain a good control system, it is also necessary to consider the following issues:

Numerics.

Sensors.

Actuators.

Operational aspects.

Programming aspects.

These are discussed in the following sections.

15.3 Prefiltering and Computational Delay

The interactions between the computer and its environment are important when implementing a control system. The sensor interface is discussed in this section.

The consequences of disturbances are discussed in Chapter 6. The importance of using an analog prefilter to avoid aliasing is treated in Chapter 2. It is also clear from Sec. 15.2 that there is always a time delay associated with the computations.

The prefilter and the computational delay give rise to additional dynamics, which may be important when implementing a digital controller. These effects are now discussed.

Analog Prefiltering

To avoid aliasing (see Sec. 2.5), it is necessary to use an analog prefilter for elimination of disturbances with frequencies higher than the Nyquist frequency associated with the sampling rate. Different prefilters are discussed from the signal-processing point of view in Sec. 2.5. The discussion is based on knowledge of the frequency content of the signal. In a control problem there is normally much more information available about the signals in terms of differential equations for the process models and possibly also for the disturbances. An analog Kalman filter would be a very good prefilter because it can be based on a detailed description of the signal. There are several advantages in implementing the Kalman filter in a computer. In such a case, it is useful to sample the analog signals at a comparatively high rate and to avoid aliasing by an ordinary analog prefilter designed from the signal processing point of view. The precise choice depends on the order of the filter and on the character of the measured signal. The dynamics of the prefilter should be taken into account when designing the system. Compare the discussion in Section 10.10.

If the sampling rate is changed, the prefilter must also be changed. With reasonable component values, it is possible to construct analog prefilters for sampling periods shorter than a few seconds. For slower sampling rates, it is often simpler to sample once per second or faster with an appropriate analog prefilter and apply digital filtering to the sampled signal. This approach also makes it possible to change the sampling period of the control calculations by software only.

Because the analog prefilter has dynamics, it is necessary to include the filter dynamics in the process model. If the prefilter or the sampling rate is changed, the coefficients of the control law must be recomputed. The following example illustrates a simple way of estimating when it is necessary to consider the prefilter dynamics in the control design.

Example 15.1—When prefilter dynamics can be neglected

Consider a servo with bandwidth ω_B (rad/s), which is implemented using a digital computer with the sampling rate ω_s (rad/s). Let the prefilter be a second-order Butterworth filter. The transfer function is thus (compare Table 2.1)

$$G(s) = \frac{\omega_f^2}{s^2 + 2\zeta\omega_f s + \omega_f^2}$$

with $\zeta = 1/\sqrt{2}$. At the frequency ω_B, this filter gives a phase lag of

$$\alpha = \arctan \frac{2\zeta\omega_f\omega_B}{\omega_f^2 - \omega_B^2} \approx \frac{2\zeta\omega_B}{\omega_f} \qquad \text{when } \omega_f \gg \omega_B$$

The attenuation at the Nyquist frequency is

$$n = \frac{1}{|G(i\omega_N)|} \approx \left(\frac{\omega_N}{\omega_f}\right)^2 = \left(\frac{\omega_s}{2\omega_f}\right)^2$$

Elimination of ω_f between these equations gives

$$\omega_s = \omega_B \frac{4\zeta\sqrt{n}}{\alpha} \tag{15.4}$$

With $\alpha = 0.1$ rad (5.7°) and $n = 10$, Formula (15.4) gives a sampling rate ω_s that is 90 times larger than the bandwidth. $\qquad\qquad\qquad\qquad\qquad\qquad\qquad$ □

The example indicates that with normal sampling rates—i.e., 15–45 times per period—it is indeed necessary to consider the prefilter dynamics (compare with Sections 2.5 and 10.10).

Computational Delay

Since A-D and D-A conversions and computations take time, there will always be a delay when a control law is implemented using a computer. The delay, which is called the *computational delay*, depends on how the control algorithm is implemented. There are basically two different ways to do this (see Fig. 15.2). In case A, the measured variables read at time t_k may be used to compute the control signal to be applied at time t_{k+1}. Another possibility, case B, is to read the measured variables at time t_k and to make the D-A conversion as soon as possible.

The disadvantage of case A is that the control actions are delayed unnecessarily; the disadvantage of case B is that the delay will be variable depending upon the programming. In both cases it is necessary to take the computational delay into account when computing the control law. This is easily done by including a time delay of h (case A) or τ (case B) in the process model. A good rule

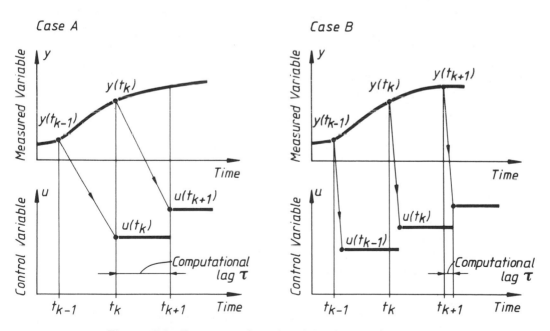

Figure 15.2 Two ways of synchronizing inputs and outputs. In case A the signals measured at time t_k are used to compute the control signal to be applied at time t_{k+1}. In case B the control signals are changed as soon as they are computed.

LISTING 15.2 Computer code skeleton that implements the control algorithm (15.2). This code has a smaller computational delay than the code in Listing 15.1.

```
        Procedure Regulate
        begin
1         Adin  y  uc
2         u := u1 + D*y + Dc*uc
3         Daout u
4         x := F*x + G*y + Gc*uc
5         u1 := C*x
        end
```

is to read the inputs before the outputs are set out. If this is not done, there is always the risk of electrical cross-coupling.

In case B it is desirable to make the computational delay as small as possible. This can be done by only making as few operations as possible between the A-D and D-A conversions.

Consider the program in Listing 15.1. Since the control signal u is available after executing the second line of code, the D-A conversion can be done before the state is updated. The delay may be reduced further by calculating the product $C*x$ after the D-A conversion. The algorithm in Listing 15.1 is then modified to Listing 15.2.

It is useful to have good estimates of computing times for different control algorithms. A good way to obtain these is to run test programs. For linear control laws, it is often possible to estimate times from results of a scalar-product computation. Since the key operations in linear control laws are matrix operations, it is often possible to get good estimates of computing times by timing the following scalar-product computation:

```
        Procedure Scapro;
        begin
          s := 0;
          for i := 1 to n do
          s := s + a[i]*b[i];
        end
```

The algorithm requires computation of the addresses of the array elements as well as additions and multiplications.

On simple microcomputers, which do not have floating-point arithmetic in hardware, there will be a substantial difference in computing time between fixed-point and floating-point operations. The difference is much less if there is hardware for floating-point operations.

To judge the consequences of computational delays, it is also useful to know the sensitivity of the closed-loop system with respect to a time delay. This may be evaluated from a root locus with respect to a time delay. A simpler way is to evaluate how much the closed-loop poles change when a time delay of one sampling period is introduced.

Detection of Outliers and Measurement Malfunctions

The linear filtering theory presented in Sec. 11.3 is very useful in reducing the influence of measurement noise. However, there may also be other types of errors, such as instrument malfunction and conversion errors. These are typically characterized by large deviations, which occur with low probabilities. It is very important to try to eliminate such errors so that they do not enter into the control-law calculations. There are many good ways to achieve this when using computer control.

The errors may be detected at the source. In systems with high reliability requirements, this is done by duplication of the sensors. Two sensors are then combined with a simple logic, which gives an alarm if the difference between the sensor signals is larger than a threshold. A pair of redundant sensors may be regarded as one sensor that gives either a reliable measurement or a signal that it does not work.

In more extreme cases three sensors may be used. A measurement is then accepted as long as two out of the three sensors agree (two-out-of-three logic). It is of course also possible to use even more elaborate combinations of sensors and filters.

It is also possible to use a Kalman filter for error detection. For example, consider the control algorithm of (15.1) with an explicit observer. Notice that the one-step prediction error

$$\epsilon(k) = y(k) - \hat{y}(k \mid k - 1) = y(k) - C\hat{x}(k \mid k - 1) \qquad (15.5)$$

appears explicitly in the algorithm. If estimates of the covariance matrix of the prediction error are available, it is easy to test if a particular measurement is reasonable.

One way to obtain the error covariance is to update the covariance equation of the Kalman filter on-line.

For a filter of the form of (15.1), the equations are given by

$$R(k) = CP(k \mid k - 1)C^T + R_2$$
$$K(k) = P(k \mid k - 1)C^T R^{-1}(k) \qquad (15.6)$$
$$P(k \mid k) = P(k \mid k - 1) - P(k \mid k - 1)C^T R^{-1}(k)CP(k \mid k - 1)$$
$$P(k + 1 \mid k) = \Phi^T P(k \mid k)\Phi + R_1$$

where $R(k)$ is the covariance of $\epsilon(k)$, $P(k \mid k - 1)$ is the covariance of $\tilde{x}(k \mid k - 1)$, and $P(k \mid k)$ is the covariance of $\tilde{x}(k \mid k)$ (see Sec. 11.3). If a measurement error is detected, the updating equations must be modified. If all measurements are rejected, the second term in the equation for updating $P(k \mid k)$ should not be subtracted, because it represents the reduction in variance due to the measurement $y(k)$. It is also possible to reject measurements of the different components of $y(k)$ selectively. Kalman filters and pairs of redundant sensors may also be combined. If measurement errors are checked in this way, it is possible to obtain a very flexible system. The scheme should be augmented with tests to ensure observability. It is then possible to obtain a system which can provide the different types of diagnosis.

In computer control there are also many other possibilities for detecting different types of hardware and software errors. A few extra channels in the A-D converter, which are connected to fixed voltages, may be used for testing and calibration. By connecting a D-A channel to an A-D channel, the D-A converter may also be tested and calibrated. The computer may be checked by performing calculations whose results are known and comparing the results with the known values.

15.4 Nonlinear Actuators

The design methods of Chapters 9 to 12 are all based on the assumption that the process can be described by a linear model. Although linear theory has a wide applicability, there are often some nonlinearities that must be taken into account. For example, it frequently happens that the actuators are nonlinear, as is shown in Fig. 15.3. Valves are commonly used as actuators in process-control systems. This corresponds to a nonlinearity of the saturation type where the limits correspond to a fully open or closed valve. The system shown in Fig. 15.3 can be described linearly when the valve does not saturate. The nonlinearity is thus important when large changes are made. There may be difficulties with the control system during start-up and shutdown, as well as during large changes, if the nonlinearities are not considered. A typical example is of integrator windup, which was illustrated in Listing 8.1.

The rational way to deal with the saturation is to develop a design theory that takes the nonlinearity into account. This can be done using optimal-control theory. However, such a design method is quite complicated. The corresponding control law is also complex. Therefore, it is practical to use simple heuristic methods.

Difficulties occur because the regulator is a dynamic system. When the control variable saturates, it is necessary to make sure that the state of the regulator behaves properly. Different ways of achieving this are discussed below.

State Space Regulators with an Explicit Observer

Consider first the case when the control law is described as an observer combined with a state feedback (15.1). The regulator is a dynamic system, whose state is

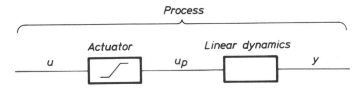

Figure 15.3 Block diagram of a process with a nonlinear actuator having saturation characteristics.

represented by the estimated state $\hat{x}$ in (15.1). In this case it is straightforward to see how the difficulties with the saturation may be avoided.

The estimator of (15.1) will give the correct estimate if the variable u in (15.1) is chosen as the actual control variable u_p in Fig. 15.3. If the variable u_p is measured, the estimate given by (15.1) and the state of the regulator are thus correct even if the control variable saturates. If the actuator output is not measured, it can be estimated—provided that the nonlinear characteristics are known. For the case of a simple saturation, the control law can be written as

$$\hat{x}(k \mid k) = \hat{x}(k \mid k - 1) + K[y(k) - C\hat{x}(k \mid k - 1)]$$

$$= [\Phi - KC]\hat{x}(k - 1 \mid k - 1) + Ky(k) + \Gamma \hat{u}_p(k - 1) \quad (15.7)$$

$$\hat{u}_p(k) = \text{sat}\{L[x_m(k) - \hat{x}(k \mid k)] + Du_c(k)\}$$

$$\hat{x}(k + 1 \mid k) = \Phi\hat{x}(k \mid k) + \Gamma \hat{u}_p(k)$$

where the function sat is defined as

$$\text{sat } u = \begin{cases} u_{\text{low}} & u \le u_{\text{low}} \\ u & u_{\text{low}} < u < u_{\text{high}} \\ u_{\text{high}} & u \ge u_{\text{high}} \end{cases} \quad (15.8)$$

for a scalar and

$$\text{sat } u = \begin{bmatrix} \text{sat } u_1 \\ \text{sat } u_2 \\ \vdots \\ \text{sat } u_n \end{bmatrix} \quad (15.9)$$

for a vector. The values u_{low} and u_{high} are chosen to correspond to the actuator limitations. A block diagram of a regulator with a model for the actuator nonlinearity is shown in Fig. 15.4. Observe that even if the transfer function from y to u for (15.1) is unstable, the state of the system in (15.7) will always be bounded if the matrix $\Phi - KC$ is stable. It is also clear that $\hat{x}$ will be a good estimate of the process state even if the valve saturates, provided that u_{low} and u_{high} are chosen properly.

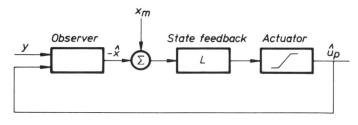

Figure 15.4 Regulator based on an observer and state-feedback with antiwindup compensation.

The General State-Space Model

The regulator may also be specified as a state-space model of the form in (15.2):

$$x(k + 1) = Fx(k) + Gy(k) \qquad (15.10)$$

$$u(k) = Cx(k) + Dy(k) \qquad (15.11)$$

which does not include an explicit observer. The command signals have been neglected for simplicity. If the matrix F has eigenvalues outside the unit disc and the control variable saturates, it is clear that windup may occur. Assume, for example, that the output is at its limit and there is a control error y. The state and the control signal will then continue to grow, although the influence on the process is restricted because of the saturation.

To avoid this difficulty, it is desirable to make sure that the state of (15.10) assumes the proper value when the control variable saturates. In conventional process controllers, this is accomplished by introducing a special *tracking mode*, which makes sure that the state of the system corresponds to the input-output sequence $\{u_p(k), y(k)\}$. The design of a tracking mode may be formulated as an observer problem. In the case of state-feedback with an explicit observer, the tracking is done automatically by providing the observer with the actuator output u_p or its estimate $\hat{u}_p$. In the regulator of (15.10) and (15.11), there is no explicit observer. To get a regulator that avoids the windup problem, the solution for the regulator with an explicit observer will be imitated. The control law is first re-written as indicated in Fig. 15.5. The systems in (a) and (b) have the same input-output relation. The system S_B is also stable. By introducing a saturation in the feedback loop in (b), the state of the system S_B is always bounded if y and u are bounded. This argument may formally be expressed as follows. Multiply (15.11) by K and add to (15.10). This gives

$$x(k + 1) = Fx(k) + Gy(k) + K[u(k) - Cx(k) - Dy(k)]$$
$$= [F - KC]x(k) + [G - KD]y(k) + Ku(k)$$
$$= F_0x(k) + G_0y(k) + Ku(k)$$

If the system of (15.10), and (15.11) is observable, the matrix K can always be chosen so that $F_0 = F - KC$ has prescribed eigenvalues inside the unit disc. Notice that this equation is analogous to (15.7). Applying the same arguments as

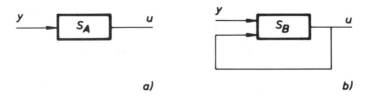

a)   b)

Figure 15.5 Different representations of the control law.

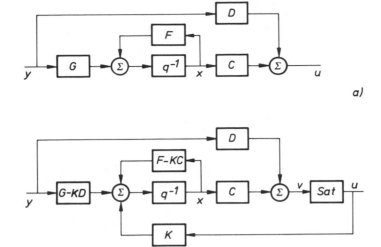

a)

Figure 15.6 Block diagram of the regulator (15.2) and the modification in (15.12) that avoids windup.

for the regulator with an explicit observer, the control law becomes

$$x(k + 1) = F_0x(k) + G_0y(k) + Ku(k) \qquad (15.12)$$
$$u(k) = \text{sat}[Cx(k) + Dy(k)]$$

The saturation function is chosen to correspond to the actual saturation in the actuator. A comparison with the case of an explicit observer shows that (15.12) corresponds to an observer with dynamics given by the matrix F_0. The system of (15.12) is also equivalent to (15.2) for small signals.

A block diagram of the regulator with antireset windup compensation is shown in Fig. 15.6.

Input-Output Form

The corresponding construction can also be carried out for regulators characterized by input-output models. Consider a regulator described by

$$R(q)u(k) = T(q)u_c(k) - S(q)y(k) \qquad (15.13)$$

where R, S, and T are polynomials in the shift operator. The problem is to rewrite the equation so that it looks like a dynamic system with the observer dynamics driven by three inputs, the command signal u_c, the process output y, and the control signal u. This is accomplished as follows.

Let $A_o(q)$ be the desired characteristic polynomial of the observer. Adding $A_o(q)u(k)$ to both sides of (15.13) gives

$$A_ou = Tu_c - Sy + (A_o - R)u$$

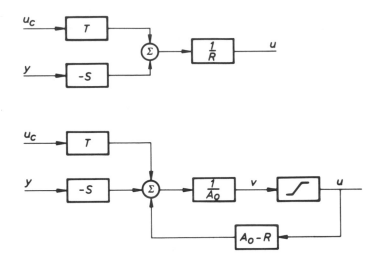

Figure 15.7 Block diagram of the regulator of (15.13) and the modification in (15.14) that avoids windup.

A regulator with antiwindup compensation is then given by

$$\begin{cases} A_o v = Tu_c - Sy + (A_o - R)u \\ \quad u = \text{sat } v \end{cases} \tag{15.14}$$

This regulator is equivalent to (15.13) when it does not saturate. When the control variable saturates, it can be interpreted as an observer with dynamics given by polynomial A_o.

A block diagram of the linear regulator of (15.13) and the nonlinear modification of (15.14) that avoids windup is shown in Fig. 15.7. A particularly simple case is that of a deadbeat observer, i.e., $A_o^* = 1$. The regulator can then be written as

$$u(k) = \text{sat}[T^*(q^{-1})u_c(k) - S^*(q^{-1})y(k) + (1 - R^*(q^{-1}))u(k)] \tag{15.15}$$

An example illustrates the implementation.

Example 15.2—PI-regulator with antireset windup

A discrete PI-regulator has the transfer function

$$H(z) = \frac{s_0 z + s_1}{z - 1} = \frac{s_0(z - 1) + s_0 + s_1}{z - 1} \tag{15.16}$$

where

$$s_0 = k$$

$$s_1 = k\left(-1 + \frac{h}{t_i}\right)$$

Let the control error be e and the control variable be u. Start with the state-space

Implementation of Digital Controllers   Chap. 15

representation

$$u(k) = s_0 e(k) + i(k)$$

$$i(k + 1) = i(k) + (s_0 + s_1)e(k)$$

of (15.16). Using (15.12) with a deadbeat observer—i.e., $K = 1$—gives

$$u(k) = \text{sat}[s_0 e(k) + i(k)]$$

$$i(k + 1) = s_1 e(k) + u(k)$$

(15.17)

The same result can be obtained by starting with (15.15). Compare Section 8.3. □

15.5 Operational Aspects

The interface between the regulator and the operator is discussed in this section. This includes an evaluation of the information displayed to the operator and the mechanisms for the operator to change the parameters of the regulator. In the case of conventional analog regulators for process control, it is customary to display the set point, the measured output, and the control signal. The regulator may also be switched from manual to automatic control. The operator may change the gain (or proportional band), the integration time, and the derivative time. This organization was motivated by properties of early analog hardware. When computers are used to implement the regulators, there are many other possibilities. So far the potentials of the computer have been used only to a very modest degree.

To discuss the operator interface, it is necessary to consider how the system will be used operationally. This is mentioned in Sec. 7.2 and a few additional comments are given here. First, it is important to realize the wide variety of applications of control systems. There is no way to give a comprehensive treatment, so a few examples are given.

Example 15.3—Autopilot for aircraft

Consider an autopilot for a high performance aircraft. In this case the operator interacts with the system by giving command signals via a joystick and pedals. There is a mode switch to select manual or automatic mode. In manual mode, the autopilot may perform stability argumentation to increase the natural damping of the aircraft. There may be several automatic modes, such as heading hold and altitude hold. The pilot has little interest in control errors and the parameters of the regulator. □

Example 15.4—Process control

Consider a process-control system with a few loops that are keeping process variables at desired levels. Traditionally the regulation error is displayed and stored. It is also common to handle large upsets and changes of operating conditions by manual control. The regulators are therefore provided with a mode switch for choosing automatic or manual operation.

To aid in the normal operation, there are also displays for set points, measured value, and the control variable. In the process-control field there are also several different actors: operators running the plant, instrument engineers maintaining and tuning the control system, and process engineers designing and modifying plants and instruments. □

Example 15.5—Pilot plant

In a pilot plant operation, there is need for frequent changes and modifications of the process and the control system. It is highly desirable to have a flexible system that allows data to be logged, as well as easy changes and modifications of the control system. □

Operating Modes

It is often desirable to have the possibility of running a system under manual control. A simple way to do this is to have the arrangement shown in Fig. 15.8, where the control variable may be adjusted manually. The manual control is often done with push buttons for increasing or decreasing the control variable. It is often practical to have some arrangement that allows the rate of change of the set point to be modified.

Because the regulator is a dynamic system, the state of the regulator must have the correct value when the mode is switched from manual to automatic. If this is not the case, there will be a switching transient. A smooth transition is called *bumpless transfer*, or *bumpless transition*.

In conventional analog regulators, it is customary to handle bumpless transition by introducing a *tracking mode*, which adjusts the regulator state so that it is compatible with the given inputs and outputs of the regulator. A tracking mode may be viewed as an implementation of an observer.

A tracking mode is obtained automatically in the regulators of (15.7), (15.12), and (15.15) because they have an observer built into them. To run them in a tracking mode, simply put

$$u_{\text{low}} = u_{\text{high}} = u_{\text{manual}}$$

This implies that the control signal is always equal to the manual input signal. The state of the regulator will be reset automatically because of the internal feedback in the regulator. The saturation introduced in the regulator to handle actuator saturation will automatically give bumpless transfer. There are also ways to have modes for semiautomatic control by keeping some feedback paths for stabilization.

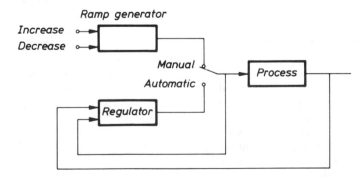

Figure 15.8 Control system with manual and automatic control modes.

With computer control, it is also possible to have many other operating modes. For example, it is possible to include parameter estimation and control-design algorithms in the regulator. An *estimation mode*, in which a model of the process is estimated, may be introduced. The estimated model may be used in the design algorithm to give an update of the parameters of the regulator in a *tuning mode*. Adaptive control modes, in which the parameters are updated continuously, may also be added. Compare this with the self-tuning regulator shown in Fig. 14.1.

Initialization

Since a regulator is a dynamic system, it is important to set the regulator state appropriately when the regulator is switched on. If this is not done, there may be large switching transients. In conventional process-control with PI-regulators, the regulator has one state only—namely, the integrator. It is customary to initialize such a regulator by operating it in manual control until the process output comes close to its desired value.

For an algorithm with an explicit observer, the regulator state may be initialized by keeping the control signal fixed for the time required for the observer to settle. A regulator with antiwindup may also be initialized by running it in manual mode during a period that corresponds to the settling time of the observer.

Information Display and Performance Monitoring

Computer control offers many interesting possibilities. As an example, the performance of the control loop may be displayed instead of set point and error.

It follows from Theorem 12.2 that the output of a control loop under minimum-variance control is a moving average. Thus it is possible to determine if a control loop is actually under minimum-variance control simply by looking at the covariance function of the regulation error. For control loops of this type, it is reasonable to display the covariance function of the error to the operator.

Similarly, it is shown in Sec. 12.5 that the output of an LQG-regulator is given by Equation (12.55). Let y be the process output. The quantity

$$\epsilon(t) = \frac{P(q)}{R(q)} y(k)$$

where P and R are known, should be white noise if the model is correct and the regulator is correctly tuned. Notice that the polynomials $P(q)$ and $R(q)$ are known from the design, so it would make sense to display the covariance function of the signal ϵ. It is, of course, also useful to provide simple statistics such as the mean value, the variance, and the largest and the smallest values. Similarly, it is useful to provide facilities for storing inputs and outputs.

Parameterization and Parameter Changes

In conventional process controllers, the operator can manipulate the set point and the basic parameters in the control law (gain, integration time, and derivative time). With computer control, there are many other interesting alternatives. Because of the simplicity of making computations, it is possible to use one parameterization in the control algorithm and another in the operator communication. The parameters displayed to the operator may then be related to the performance of the system rather than to the details of the control algorithm. The conversion between the parameters is made by an algorithm in the computer.

To illustrate the idea of performance-related parameters, consider design of a servo using the pole-placement method described in Chapter 10. The closed-loop properties may then be specified in terms of the relative damping ζ and the bandwidth ω_B. To perform the design, it is also necessary to have a model of the open-loop dynamics. One possibility is to have the process engineer enter the desired bandwidth and damping and a continuous-time model. The computer can then make the necessary conversions in order to obtain the control law. If the computer also has a recursive estimation algorithm, it is not necessary to introduce the model. Clearly, there are many interesting possibilities if estimation and design algorithms are included in the regulator.

There are two operational problems with on-line parameter changes. One problem is related to real-time programming. Data representing parameters are shared among different programs. It is then necessary to make sure that one program is not using data that is being changed by another program. This is discussed in Sec. 15.8.

The other problem is algorithmic. There may be switching transients when the parameters are changed in a control algorithm. To get some insight into what can happen, consider the simple PI-algorithm

```
e := uc − y
u := k*(e + i/ti)
i := i + e*h
```

It is clear that a change of the integration time ti will cause a step in the control signal unless the integral part, i, is zero. The problem can be avoided by changing the state from i to $i*ti/ti'$, where ti' is the new value of the integration time. Another simpler way is to write the algorithm as

```
e := uc − y
u := k*e + i
i := i + k*e*h/ti
```

The need for changing the state when parameters are changed is dictated by the fact that the state of the regulator depends on its parameters. One way to obtain bumpless parameter changes is to store a set of past input-output data and to run an observer when the regulator parameters are changed. However, it is often possible to use a simpler solution.

To see what should be done, consider the algorithm of (15.7) with an explicit observer and state feedback. First a realization should be chosen so that the matrices C and D do not depend on the adjustable parameters. If the state x represents an estimate of physical state variables, there are very few difficulties because the estimated state will not change drastically when model parameters are changed. Transients due to changes in the feedback gain cannot be avoided if there is a nonzero error $e = x - \hat{x}$. Similarly, there will not be any switching transients with the algorithm in (15.12), provided that there is a representation in which the matrices C and D do not contain any parameters that are modified.

It is more complicated to see what should be done with the algorithm of (15.14). In the representation of (15.14), the state is delayed inputs and outputs. This state is not minimal. Although the state does not depend on the coefficients of the polynomials R, S, and T, there is no guarantee that the given R, S, T, u, and y are compatible with Equation (15.14). With the representation of (15.14), there will be switching transients when the parameters are changed.

Security

It is very important to make sure that a computer-control system operates safely. Ideally, this means that the system should either give the correct result or an alarm if it is not functioning properly. Systems with extremely high requirements may be tripled (or quadrupled) and the output accepted if two subsystems give the same result. For simpler systems, it may be sufficient to rely on self-checking. There are many ways to do this using computer control. Arithmetic units may be checked by computing functions with known results. Memory and data transmission may be checked through checksums. A-D and D-A converters may be checked by using a few extra connected channels. A D-A conversion is commanded and the result of an A-D conversion of the same channel is checked. Timing may similarly be investigated by connecting a network with a known time constant between a D-A and an A-D converter.

15.6 Numerics

When implementing a computer-control system it is necessary to answer questions such as: How accurate should the converters be? What precision is required in the computations? Should computations be made in fixed-point or floating-point arithmetic? To answer these questions, it is necessary to understand the effects of the limitations and to estimate their consequences for the closed-loop system. This is not a trivial question, because the answer depends on a complex interaction of the feedback, the algorithm, and the sampling rate. Fortunately, only crude estimates have to be done. For instance, should the resolution be 10 or 12 bits and should the word length be 24 or 32 bits? Such questions may be answered using simplified analysis.

Error Sources

The major sources of error are the following:

> Quantization in A-D converters.
> Quantization of parameters.

Roundoff, overflow, and underflow in addition, subtraction, multiplication, division, function evaluation, and other operations.

Quantization in D-A converters.

Common types of A-D converters have accuracies of 8, 10, 12, and 14 bits, which correspond to a resolution of 0.4%, 0.1%, 0.025%, and 0.006%. The percentages are in relation to full scale. The D-A converters also have a limited precision. An accuracy of 10 bits is typical. The error due to the quantization of the parameters depends critically on the sampling period and on the chosen realization of the control law.

Word-Length

Digital-control algorithms are typically implemented on microcomputers and minicomputers, which have word lengths of 8, 16, or 32 bits. Digital signal processors (DSP) are now commonly used to implement computer-controlled systems when short sampling periods are required. The low-cost signal processors are all using fixed-point calculations. For the TMS family from Texas Instruments, the standard word length is 16 bits but the accumulator is 32 bits wide. A 16-bit DSP from AT&T has a 36-bit accumulator, and Motorola has a DSP with 24-bit word length and a 56-bit accumulator. The architecture with a long accumulator is ideal for computing scalar products, which is the key operation when implementing linear filters, because the products of the terms can be accumulated in double precision. The signal processors are very fast. The operation of multiply and accumulate (MAP) typically takes 100 nanoseconds. There are also more expensive signal processors with floating-point hardware.

There is also an increased use of computer-controlled systems implemented using special-purpose VLSI circuits. In these applications the word length is a design parameter which can be chosen freely. Such a choice naturally requires a more detailed calculation than a simple choice between single or double precision. There are applications of custom VLSI both in the aerospace industry and for mass-produced consumer goods like VCRs and CD players. For these applications it is of major concern to minimize chip area. A typical example is a CD player where both audio and servo functions are implemented on one chip. For a stationary CD player there are fewer demands on the servo than for a CD player for a car. The chip area for the control system can thus be smaller for the stationary player.

There are many number representations used in digital computers. Integers are typically 16, 32, or 48 bits. For a long time there were many representations of floating-point numbers. IEEE did, however, take the initiative to standardize them, and a standard ANSI-IEEE 754 was published in 1985. In this standard the numbers are represented as

$$\pm a2^b$$

where $0 \leq a < 2$ is the *significand*, also called mantissa, and b is the *exponent*.

In the standard there are three types of floating-point numbers:

short real (32 bits)	1 sign	8 exponent	23 significand
long real (64 bits)	1 sign	11 exponent	52 significand
short temporary real (80 bits)	1 sign	15 exponent	64 significand

The IEEE standard has gained widespread acceptance, and the floating-point chips from Intel and Motorola are based on it. The essential numerical difficulty is illustrated by the example.

Example 15.6—Scalar-product calculations

Consider the vectors

$$a = [100 \quad 1 \quad 100]$$

$$b = [100 \quad 1 \quad -100]$$

The scalar product is $\langle a, b \rangle = 1$. If the scalar product is computed in floating-point representation with a precision corresponding to three decimal places, the result will be zero because $100 \cdot 100 + 1 \cdot 1$ is rounded to 10,000. Notice that the result obtained depends on the order of the operations. Finite word-length operations are neither associative nor distributive. □

The difficulty may be avoided without using complete double-precision calculation by adding the terms in double precision and rounding to single precision afterwards. This method can be applied to fixed-point and floating-point calculations. Notice that the multiply instruction for many computers is implemented so that the product is available in double precision. Many high-level languages also have constructions that support this type of calculation. Generally speaking, roundoff and quantization will give rise to small errors, while the effects of overflow will be disastrous.

Overview of Effects of Roundoff and Quantization

The consequences of roundoff and quantization depend on the feedback system and on the details of the algorithm. The properties may be influenced considerably by changing the representation of the control law or the details of the algorithm. Thus it is important to understand the phenomena.

A detailed description of roundoff and quantization leads to a complicated nonlinear model, which is very difficult to analyze. Investigation of very simple cases shows, however, that quantization and roundoff may lead to limit-cycle oscillations. Such examples are presented later, together with approximative analysis. Limit-cycle oscillations have also been observed in more complex cases.

Some properties of quantization and roundoff in a feedback system may also be captured by linear analysis. Quantization and roundoff are then modeled as ideal operations with additive or multiplicative disturbances. The disturbances may be either deterministic or stochastic. This type of analysis is particularly useful for order-of-magnitude estimation. It allows investigation of complex systems and it is useful when comparing different algorithms.

Techniques from sensitivity analysis and numerical analysis are also useful in finding the sensitivity of algorithms to changes of parameters. Such methods may be used to compare and screen different algorithms. However, the methods are limited to comparison of the open-loop performances of the algorithms. It is also necessary to compare the effects of quantization and roundoff with the other disturbances in the system.

Nonlinear Analysis

The nonlinear aspects of roundoff and quantization errors are illustrated using a simple example. This shows that quantization and roundoff may give rise to multiple equilibria and limit cycles.

Example 15.7—Limit cycles due to roundoff

Consider the following model of a simple system with quantization:

$$y(k + 1) = u(k) + Q[ay(k)]$$

where the function Q represents roundoff to integers with the normal rules. Assuming that the input u is zero. The response of the system is governed by

$$y(k + 1) = Q[ay(k)] \qquad (15.18)$$

If there is no quantization—i.e., $Q(y) = y$—the equation has one equilibrium, $y = 0$. This equilibrium is stable if $|a| < 1$. A graph aids in finding the equilibria when there is quantization. Figure 15.9(a) shows the graph of ay for $a > 0$ and the inverse image of $Q[y]$. The equilibria correspond to the intersection of the graphs. It follows that $y = 0$ is an equilibrium. If there are integers $k > 0$ such that

$$\begin{cases} ak > k - 0.5 & k \text{ odd} \\ ak \geq k - 0.5 & k \text{ even} \end{cases}$$

there are also other equilibria.

A similar analysis for $-1 < a \leq 0$ is illustrated in Fig. 15.9(b). The figure shows that there may be limit cycles with a period of two sampling intervals if

$$\begin{cases} ak > -k + 0.5 & k \text{ odd} \\ ak \geq -k + 0.5 & k \text{ even} \end{cases} \qquad \square$$

The example illustrates some effects of roundoff in a system. In general, there may be multiple equilibria and periodic solutions with different amplitudes

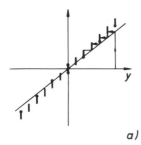

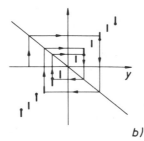

a)

b)

Figure 15.9 Graph for finding the equilibria of Equation (15.18). The parameter a has the value 0.8 in (a) and -0.8 in (b).

and periods. It is difficult to analyze more complicated systems exactly. There are, however, approximate methods that give some insight.

Describing Function Analysis

If there is only one nonlinearity in the loop, it is possible to use the method of describing functions to determine limit cycles approximately.

Consider the system in Fig. 15.10(a). The method of describing functions can be regarded as a generalization of the Nyquist criterion. The critical point -1 is replaced by $-1/Y_c(A)$, where $Y_c(A)$ is the describing function of the nonlinearity. The describing function characterizes the transmission of a sinusoidal signal with amplitude A through the nonlinearity. The method predicts a limit cycle if

$$H(e^{i\omega h}) = -1/Y_c(A)$$

[compare to Fig. 15.10(b)]. The frequency, ω_1, and the amplitude, A_1, at the intersection are the estimated frequency and the estimated amplitude of the limit cycle.

The describing function of a roundoff quantizer is

$$Y_c(A) = \begin{cases} 0 & 0 < A < \dfrac{\delta}{2} \\ \dfrac{4\delta}{A} \displaystyle\sum_{i=1}^{n} \sqrt{1 - \left[\dfrac{2i-1}{2A}\delta\right]^2} & \dfrac{2n-1}{2}\delta < A < \dfrac{2n+1}{2}\delta \end{cases}$$

The function Y_c only takes real values. Its smallest value is zero and its largest value is $4/\pi \approx 1.27$. The function is graphed in Fig. 15.11. This means that the critical part for quantization consists of the part of the negative real axis from $-\infty$ to -0.78. Describing function analysis thus predicts oscillations due to quantization if the Nyquist curve of the loop gain intersects this line segment. For stable systems this means that quantization will not give rise to oscillations if the amplitude margin is larger than 1.27. Describing function analysis predicts oscillations for all systems that are open-loop unstable.

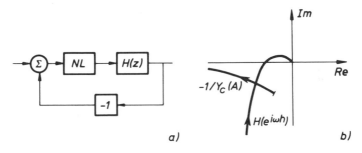

Figure 15.10 a. Discrete-time system with one nonlinearity NL.
b. Using the method of describing functions.

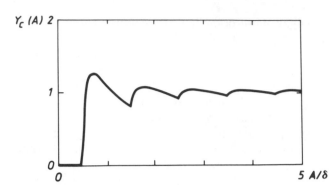

Figure 15.11 The describing function of roundoff.

Example 15.8—Quantization effects

Consider the system in Example 5.3 with the pulse-transfer function

$$H(z) = \frac{0.25K}{(z - 1)(z - 0.5)}$$

In Example 5.3 it is shown that the closed-loop system without quantization is asymp-

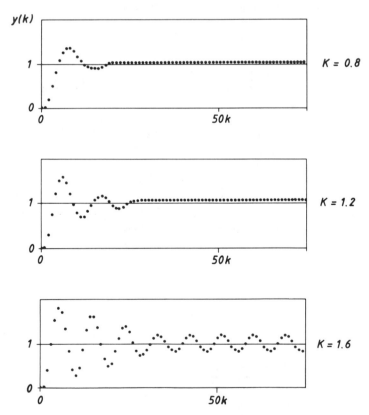

Figure 15.12 The output of the system in Example 15.8 when $\delta = 0.2$ and $K = 0.8$, 1.2, and 1.6.

Implementation of Digital Controllers   Chap. 15

totically stable if $K < 2$. With quantization of the error signal, the method of describing functions predicts that there will be a limit cycle if K is greater than about 1.3. Figure 15.12 shows the behavior of the system for a quantization level of $\delta = 0.2$. The limit cycle is clearly noticeable for $K = 1.6$. $\qquad\qquad\square$

Linear Analysis

The effects of quantization and roundoff may also be estimated by linear analysis. The idea is to represent the operations by their ideal models and an additive disturbance e_a. The D-A and A-D converters are then simply represented as linear gains with a disturbance that models the quantization. With fixed-point calculations, the additions are exact. There will, however, be errors in multiplications. These are represented as exact multiplications, with an additive error, which represent the roundoff. These models are illustrated in Fig. 15.13.

The errors may be modeled as deterministic or stochastic signals. In a deterministic model, the error is modeled as constants having the sizes of quantization errors and with the resolution in the arithmetic calculations. In the stochastic model, the error introduced by rounding or quantization is then described as additive white noise with a rectangular distribution. The errors at different sampling times are thus assumed to be uncorrelated. If the quantization is done as rounding, then the error is equally distributed over the interval $(-\delta/2, \delta/2)$, where δ is the quantization step. If the quantization is done as truncation, the error is equally distributed over $(0, \delta)$. A rectangular noise distributed over an interval of length δ has a variance of $\delta^2/12$.

Using the linear models of roundoff and quantization, it is possible to reduce the problem of estimating the effect of the quantization and roundoff to the problem of calculating responses of a linear system to deterministic or stochastic inputs. Methods for this are discussed in Sec. 6.5. Using the linear model, it is possible to assess the effects of quantization qualitatively without going into detailed calculation. It is also easy to compare roundoff with other disturbances in the system. In the linear model, the effect of roundoff in the A-D converter is the

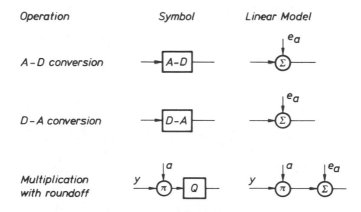

Figure 15.13 Linear models for quantization and roundoff.

same as the effect of measurement noise. The effect on the control signal may be substantial for those frequencies where the regulator has high gain. The effect of roundoff in the D-A converter is the same as a disturbance in the process input. Because the process normally attenuates high frequencies, the effect on the process output is normally small. Remember, however, that the linear model does not capture all aspects of roundoff.

Having obtained some tools to analyze the effects of quantization, we will now illustrate how they can be applied.

Example 15.9—Effects of A-D quantization for the double integrator

Figure 15.14 shows a simulation of digital control of a double integrator where the A-D converter is quantized with the level 0.02. The controller is the same as the one used in Section 10.9. It is given by

$$R(q)u(k) = T(q)u_c(k) - S(q)y(k)$$

with

$$R(q) = (q - 1)(q - 0.441)$$

$$S(q) = 1.140q^2 - 2.022q + 0.908$$

$$T(q) = (3.472q^2 - 0.940q + 0.064) \cdot 10^{-2}$$

and the sampling period 1 s. The simulation clearly shows that there is a limit-cycle oscillation where the output changes one quantization level. The period is 15 s. The describing function analysis predicts a limit cycle with period 36 s. See Figure 15.15, which shows the Nyquist curve of the sampled loop gain. Describing function anal-

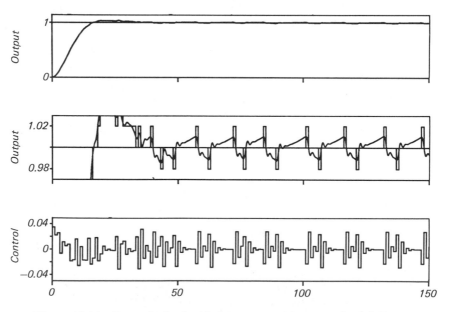

Figure 15.14 Control of a double integrator with a quantized A-D converter. The sampling period is 1 s and the quantization level is 0.02.

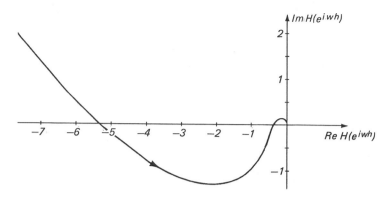

Figure 15.15 Nyquist curve for the sampled loop gain for the double integrator.

ysis gives correct qualitative results in this case, but the prediction of the period is poor because the signals deviate significantly from sinusoids.

Since the describing function analysis gives such poor results, we will use another method to estimate the amplitudes of the fluctuations caused by the quantization. To do this, first observe that the output signal oscillates with one quantization level up or down at widely spaced intervals. This means that the regulator output is given by the pulse response of the controller, i.e.,

$$H_c(z) = \frac{S(z)}{R(z)} = 1.14 - 1.38z^{-1} + 0.64z^{-2} \cdots \qquad (15.19)$$

multiplied by the quantization level. This gives an excellent prediction of the fluctuations in the control signal. Compare with Figure 15.14. Notice that the first coefficient in the expansion of $H_c(z)$ is equal to s_0. □

This example shows that the ripple in the control signal due to quantization in the A-D converter can be estimated from a simple calculation. In the next example we will analyze the effect of quantization in the D-A converter.

Example 15.10—Effects of D-A quantization for the double integrator

Figure 15.16 shows a simulation of the double integrator with a D-A converter with quantization level 0.01. The quantization causes a limit-cycle oscillation. The process output is, however, much more sinusoidal than with A-D quantization. The reason for this is that the nonlinearity is just before the process which attenuates high frequencies. This means that the describing function can also be expected to give better result. As before, the describing function predicts an oscillation with the period 36 s while the actual period is 26 s. The amplitude of the oscillation in the process output can be estimated by evaluating the magnitude of the pulse-transfer function of the process at the period of oscillation. The process gain is approximately 15. With $\delta = 0.01$, the amplitude can be estimated at 0.15, which is about 50% too large. The reason the amplitude is overestimated is that the first harmonic of the input is smaller than δ because of the waveform. See Fig. 15.16.

Notice that the oscillations due to the quantization in the D-A converter can

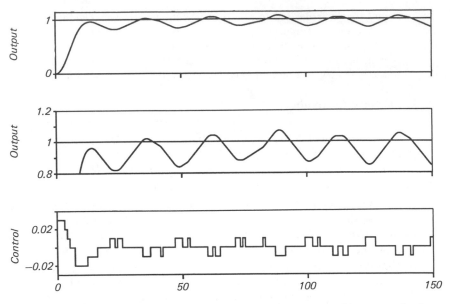

Figure 15.16 Digital control of the double integrator with a quantized D-A converter.

be avoided if the output of the D-A converter is fed back into the control law in the same way as was done to avoid windup. □

Selection of Resolution of A-D and D-A Converters

Using the insight obtained from the examples, some recommendation on the selection of resolution of the converters can now be given.

The resolution of the A-D converter must obviously be chosen so that it at least gives the desired precision in the process output. One should investigate whether quantization can give rise to limit-cycle oscillations. Further the magnitude of the ripple in the control signal caused by the A-D quantization should be investigated. This can be estimated simply from Equation (15.19). If the ripple in the control signal is too large, better resolution of the A-D converter is required.

To determine the required resolution of the D-A converter, the frequency of a possible limit cycle is first determined. If there is a limit-cycle oscillation, the amplitude can be estimated crudely from the process gain at the oscillation frequency, or more accurately using the theory of relay oscillations quoted in the references. The estimates obtained in this way will typically give the order of magnitude. It is recommended to use simulation to get more accurate results. The procedure is illustrated by an example.

Example 15.11—Choosing resolution in D-A and A-D converters

Consider the double integrator that we have investigated. Assume that the process output is in the range $[-1, 1]$, that the range of the control signal is $[-0.04, 0.04]$, and that it is desired to control the output with a precision of 1%. If we let each converter contribute 0.5%, the A-D converter must have a resolution of at least 0.005, which is equivalent to 9 bits. Since the gain of the process at the limit cycle is about 15, the resolution of the D-A converter must be better than 0.00033. With the given signal range, this corresponds to 1 part in 240, or 8 bits. □

Sensitivity Analysis

Both the nonlinear and the linear analysis consider the effects of roundoff and quantization in a feedback loop. There are other, simpler methods for estimating the errors in an open-loop setting. Although these approaches disregard the feedback, they allow comparison of different algorithms.

Crude estimates of errors in computations may be derived using the idea of conditioning numbers in numerical analysis. This notion may be explained as follows. Consider the problem of evaluating a function f. The conditioning number, or the sensitivity of the problem, is defined as the number of C_p in the inequality

$$\frac{\| f(x + \delta x) - f(x) \|}{\| f(x) \|} \le C_p \frac{\| \delta x \|}{\| x \|}$$

The conditioning number makes it possible to relate the relative error in the result to the relative error in the data.

An algorithm for solving a problem is a finite sequence of specified elementary operations (additions, multiplications, and so on) that will generate a solution to a problem. An algorithm a for evaluating the function f may be viewed as a finite sequence of mappings

$$a = a_{(r)} \cdot a_{(r-1)} \cdots a_{(1)} \cdot a_{(0)}$$

where each $a_{(i)}$ represents an elementary operation. The conditioning number for the algorithm a is defined as follows:

$$C_a = \sup_x \frac{\| \delta \xi \|}{\| x \|}$$

where

$$a(x) = f(x + \delta \xi)$$

The conditioning number of the algorithm thus gives an error estimate measured in units of relative error in the input data. The conditioning numbers C_p and C_a may be used to estimate the relative error when evaluating the function f using the algorithm a with the formula

$$\frac{\| a(x + \delta x) - f(x) \|}{\| f(x) \|} \le C_p \left\{ \frac{\| \delta x \|}{\| x \|} + C_a \right\}$$

The conditioning numbers are useful in estimating how sensitive problems and algorithms are to small changes in the data. The error in a computation thus depends on both the problem and the algorithm.

When implementing a control law using a digital computer, the problem is evaluation of the output of a dynamic system from the input. The data correspond to the input signal and the parameters. The conditioning number of the problem may be influenced significantly by the chosen realization and the parameterization. Thus it is important to consider realizations, parameterizations, and algorithms.

15.7 Realization of Digital Controllers

The previous section illustrated how quantization in A-D and D-A converters influence the behavior of the system. Roundoff errors in the computations of the control law also cause quantization, which can be modeled and analyzed in the same way as converter quantization. The quantization arising from the computations depends critically on how the computations are organized, e.g., on how the sampled data controller is realized. This section discusses different realizations. Some advantages and disadvantages of different methods are given.

A control law is a dynamic system. Different realizations can be obtained by different choice of the state variables. The different representations are equivalent from an input-output point of view if we assume that the calculations are done with infinite precision. With finite precision in the calculations, the choice of the state-space representation is very important. A bad choice of the representation may give a controller that is very sensitive to errors in the computations. Assume that we want to realize the controller

$$y(k) = H(q^{-1})u(k) = \frac{b_0 + b_1 q^{-1} + \cdots + b_m q^{-m}}{1 + a_1 q^{-1} + a_2 q^{-2} + \cdots + a_n q^{-n}} u(k) \qquad (15.20)$$

where $n \geq m$. Some different realizations are:

- Companion form
- Direct form
- Series or diagonal form
- Parallel form
- Lattice or ladder form
- δ-operator form

Coefficient-Pole Sensitivity

Finite precision in the representation of the coefficients of the controller gives a distortion of the poles and zeros of the controller. The following analysis gives quantitative results for the sensitivity of the roots of a polynomial with respect to changes in the coefficients. Consider a linear filter with distinct poles in p_i and the characteristic polynomial

$$A(z) = (z - p_1) \cdots (z - p_n) = z^n + a_1 z^{n-1} + \cdots + a_n$$

The characteristic polynomial A can be regarded as a function of z and a_i. When the parameter a_i is changed to $a_i + \delta a_i$, the poles are changed from p_k to $p_k + \delta p_k$. Hence

$$0 = A(p_k + \delta p_k, a_i + \delta a_i) \approx A(p_k, a_i) + \left.\frac{\delta A}{\delta z}\right|_{p_k} \delta p_k + \left.\frac{\delta A}{\delta a_i}\right|_{p_k} \delta a_i + \cdots$$

The first term on the right-hand side is zero. If terms of second-order and higher are neglected, it follows that

$$\delta p_k \approx -\left.\frac{\delta A/\delta a_i}{\delta A/\delta z}\right|_{z=p_k} \cdot \delta a_i$$

Because

$$\left.\frac{\delta A}{\delta a_i}\right|_{z=p_k} = p_k^{n-i}$$

and

$$\left.\frac{\delta A}{\delta z}\right|_{z=p_k} = \prod_{j\neq k} (p_k - p_j)$$

the following estimate is obtained:

$$\delta p_k \approx -\frac{p_k^{n-i}}{\prod_{j\neq k} (p_k - p_j)} \delta a_i \qquad (15.21)$$

If the polynomial has a root p_k with multiplicity m, Equation (15.21) becomes

$$\delta p_k \approx -\frac{p_k^{n-i}}{\prod_{j\neq k} (p_k - p_j)} (\delta a_i)^{1/m} \qquad (15.22)$$

If the filter is stable, then $|p_k| < 1$ and the numerator of (15.21) will be largest in magnitude for $i = n$. The coefficient a_n is thus the most sensitive parameter. Furthermore, the denominator will be small if the poles are close, which then makes the system sensitive to changes in the coefficients. Equation (15.22) shows that the sensitivity is even higher if the polynomial has multiple roots. Equations (15.21) and (15.22) may be used to determine the conditioning numbers for the transformation from the diagonal form to the companion form. It follows from the equations that the computation of companion forms may be poorly conditioned.

Direct- and Companion-Form Realizations

The most straightforward way to realize (15.20) is to write it in the *direct form*

$$y(k) = \sum_{i=0}^{m} b_i u(k - i) - \sum_{i=1}^{n} a_i y(k - i)$$

It is then necessary to store $y(k - 1)$, $y(k - 2)$, . . . , $y(k - n)$, and $u(k - 1)$, . . . , $u(k - m)$, i.e., $n + m$ variables. The direct realization has $n + m$ states and thus is not a minimal realization. The controllable or observable canonical forms, see Section 3.3, have n states. The direct form has the advantage that the state variables are simply delayed versions of the input and output signals. This means that the state does not have to be recomputed when the parameters are changed. Both the direct form and the companion forms have the disadvantage that the coefficients in the characteristic polynomial are the coefficients in the realizations. This makes the realizations extremely sensitive to computational errors if the system is of high order and if the poles or zeros are close to one as discussed above.

Well-Conditioned Realizations

The difficulty associated with the companion form can be avoided simply by representing the system as a combination of first- and second-order systems (see Fig. 15.17).

If the dynamic system representing the regulator has nr distinct real poles and nc complex-pole pairs, the control algorithm may be transformed to the model form

$$
\begin{cases}
z_i(k + 1) = \lambda_i z_i(k) + \beta_i y(k) & i = 1, \ldots, nr \\[2mm]
v_i(k + 1) = \begin{bmatrix} \sigma_i & \omega_i \\ -\omega_i & \sigma_i \end{bmatrix} v_i(k) + \begin{bmatrix} \gamma_{i1} \\ \gamma_{i2} \end{bmatrix} y(k) & i = 1, \ldots, nc \quad (15.23) \\[4mm]
u(k) = Dy(k) + \displaystyle\sum_{i=1}^{nr} \gamma_i z_i(k) + \sum_{i=1}^{nc} \delta_i^T v_i(k)
\end{cases}
$$

where the complex poles are represented using real variables. Notice that z_i are scalars and v_i are vectors with two elements. To avoid numerical difficulties, the control law should be transformed to the form of (15.23), which is then implemented in the control computer. The transformation may easily be done in a package for computer-aided design. It is easy to use fixed-point calculations and scaling for equations in the form of (15.23).

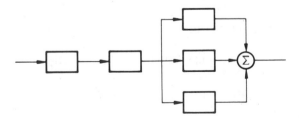

Figure 15.17 Realization of a control law as series or parallel connections of first- and second-order blocks, i.e., each block in the figure is a system of first or second order.

Implementation of Digital Controllers   Chap. 15

If the control law has multiple eigenvalues, a Jordan canonical form replaces (15.23). An eigenvalue λ of multiplicity 3 thus corresponds to a block

$$z(k+1) = \begin{bmatrix} \lambda & 1 & 0 \\ 0 & \lambda & 1 \\ 0 & 0 & \lambda \end{bmatrix} z(k) + \begin{bmatrix} \beta_1 \\ \beta_2 \\ \beta_3 \end{bmatrix} y(k)$$

Ladder Realizations

Ladder realizations are other representations that avoid the coefficient sensitivity in the implementations. One representation of the ladder network is obtained by making a continued-fraction expansion in the pulse-transfer operator in the following way

$$H(z) = \alpha_0 + \cfrac{1}{\beta_1 z + \cfrac{1}{\alpha_1 + \cfrac{1}{\beta_2 z + \cfrac{1}{\vdots \atop \alpha_{n-1} + \cfrac{1}{\beta_n z + \cfrac{1}{\alpha_n}}}}}} \qquad (15.24)$$

Another realization is obtained by making the continued-fraction expansion in z^{-1}. The ladder forms have low sensitivity against coefficient errors and roundoff errors. If

$$H(z) = \frac{B(z)}{A(z)}$$

where $\deg A(z) = \deg B(z) = n$, the coefficients α_i and β_i can be computed using the following iteration:

$$\alpha_0 = B(z) \text{ div } A(z)$$

$$A_1(z) = A(z)$$

$$B_1(z) = B(z) \bmod A(z)$$

repeat for $i = 1$ to n

$$\beta_i = A_i \text{ div } z B_i$$

$$A_{i+1} = A_i \bmod zB_i$$

$$\alpha_i = B_i \text{ div } A_{i+1}$$

$$B_{i+1} = B_i \bmod A_{i+1}$$

end

The ladder network representation can be expressed by the following state equations:

$$\beta_1 x_1(k + 1) = \frac{1}{\alpha_1} (x_2(k) - x_1(k)) + u(k)$$

$$\beta_2 x_2(k + 1) = \frac{1}{\alpha_1} (x_1(k) - x_2(k)) + \frac{1}{\alpha_2} (x_3(k) - x_2(k))$$

$$\vdots$$

$$\beta_i x_i(k + 1) = \frac{1}{\alpha_{i-1}} (x_{i-1}(k) - x_i(k)) + \frac{1}{\alpha_i} (x_{i+1}(k) - x_i(k)) \quad (15.25)$$

$$\vdots$$

$$\beta_n x_n(k + 1) = \frac{1}{\alpha_{n-1}} (x_{n-1}(k) - x_n(k)) - \frac{1}{\alpha_n} x_n(k)$$

$$y(k) = x_1(k) + \alpha_0 u(k)$$

A block diagram of the representation is shown in Figure 15.18. The name ladder network derives from the shape of the graph.

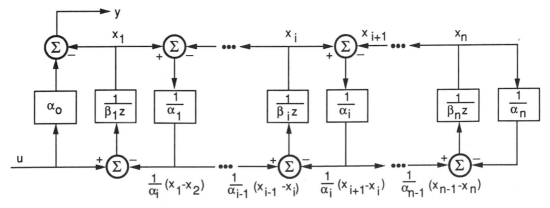

Figure 15.18 Block diagram of a ladder network representation of the transfer function (15.24).

Implementation of Digital Controllers   Chap. 15

Lattice Filters

In some special cases it is convenient to express the controller as a finite-impulse response. Such a system can be represented as a lattice filter which has good numerical properties. A block diagram of the filter is shown in Figure 15.19. The name derives from the shape of the block diagram. The filter can be described by the equations

$$f_{i+1}(k) = f_i(k) + r_i b_i(k - 1) \quad i = 0, \ldots, n - 1$$
$$b_{i+1}(k) = r_i f_i(k) + b_i(k - 1) \quad i = 0, \ldots, n - 1 \tag{15.26}$$
$$f_0(k) = b_0(k) = u(k)$$
$$y(k) = f_n(k)$$

where the parameters r_i are called the reflection coefficients. The input-output relation of the filter can be described by

$$y(k) = F_n(q^{-1})u(k)$$
$$= (1 + f_1 q^{-1} + \cdots + f_n q^{-n})u(k)$$

where

$$f_i(k) = F_i(q^{-1})u(k) \quad i = 0, \ldots, n - 1$$
$$b_i(k) = B_i(q^{-1})u(k) \quad i = 0, \ldots, n - 1$$

It can be shown that B_i and F_i are reciprocial polynomials, i.e.,

$$B_i(q^{-1}) = F_i^*(q^{-1}) \tag{15.27}$$

To compute the reflection coefficients r_i, we observe that equation (15.26) implies that

$$F_{i+1}(z^{-1}) = F_i(z^{-1}) + r_i z^{-1} B_i(z^{-1})$$
$$B_{i+1}(z^{-1}) = r_i F_i(z^{-1}) + z^{-1} B_i(z^{-1}) \tag{15.28}$$

The reflection coefficient r_i is thus the coefficient of the power z^{-i} in the poly-

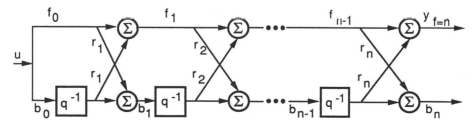

Figure 15.19 Block diagram of a lattice filter.

nomial $F_i(z^{-1})$. It follows from (15.27) and (15.28) that

$$F_{i-1}(z^{-1}) = F_i(z^{-1}) - r_{i-1}z^{-1}F_i^*(z^{-1})$$

$$r_n = f_n$$

The lattice form is often used in signal processing as whitening filters, since it has the property that it expands the original signal in a sequence of orthogonal signals that are spanning the same space.

Short Sampling Interval Modification

We have shown that it is useful to transform the system to a well-conditioned form before it is implemented on a digital computer. An additional modification that is useful when the sampling period is short will now be discussed. Consider the compensator described by the general-state model (15.10) and (15.11). For short sampling periods the matrix F is close to the unit matrix and the matrix G is proportional to the sampling period. With a short sampling period, the matrices F and G may therefore differ by several orders of magnitude. It is then convenient to rewrite equation (15.10) as

$$x(k + 1) = x(k) + (F - I)x(k) + Gy(k) \tag{15.29}$$

where the matrix $F - I$ is also proportional to the sampling period h. The numerical representation of $F - I$ requires fewer decimals than the representation of F itself. The term $(F - I)x(k) + Gy(k)$ represents a correction to the state, which will be small if the sampling period is short. This representation is particularly useful in fixed-word-length computations. The state is stored in double precision. The change in the control is calculated using single-precision multiplication, and the product is then added to the state. Compare with the discussion of the scalar product computation in Section 15.3.

The δ-Operator

The system (15.29) can be written as

$$x(k + 1) = x(k) + h(\overline{F}x(k) + \overline{G}y(k)) \tag{15.30}$$

where

$$h\overline{F} = F - I$$

$$h\overline{G} = G$$

Instead of the shift operator we can now introduce the δ-operator defined by

$$\delta = \frac{q - 1}{h} \tag{15.31}$$

Equation (15.30) can now be written as

$$\delta x(k) = \overline{F}x(k) + \overline{G}y(k)$$

A general pulse-transfer operator can be transformed from shift form to δ form as

$$H(q) = \frac{B(q)}{A(q)} = \frac{B(\delta h + 1)}{A(\delta h + 1)} = \frac{\overline{B}(\delta)}{\overline{A}(\delta)} = \overline{H}(\delta)$$

The δ operator is thus equivalent to the shift operator. All the analysis done for the shift operator can be translated into δ form. The δ operator has the property

$$\delta f(kh) = \frac{f(kh + h) - f(kh)}{h}$$

i.e., it can be interpreted as the forward-difference approximation of the differential operator $p = \dfrac{d}{dt}$. In this respect the δ operator is 'closer' to the continuous-time domain than the shift operator. For instance, the stability region in the δ form is a circle with radius $1/h$ and with the origin in $-1/h$. When $h \to 0$ the stability region becomes the left half-plane.

The δ operator representation has the property that it translates into the corresponding continuous-time system when the sampling interval approaches zero. Hence

$$\lim_{h \to 0} \overline{H}(\delta) = G(\delta)$$

where G is the continuous-time transfer function. This implies, for instance, that the zeros of the transfer function in the δ form approach the zeros (finite as well as infinite) of the continuous-time transfer function.

Example 15.12—Double integrator in δ form

Consider the double integrator $G(s) = 1/s^2$. Then

$$H(q) = \frac{h^2(q + 1)}{2(q - 1)^2} = \frac{h^2(\delta h + 2)}{2h^2 \delta^2} = \frac{1 + \delta h/2}{\delta^2} = \overline{H}(\delta)$$

When h goes to zero we get

$$\lim_{h \to 0} \overline{H}(\delta) = \frac{1}{\delta^2} = G(\delta)$$

Notice that the δ form also has 'sampling zeros,' $\delta = -2/h$. This zero will approach $-\infty$ when $h \to 0$. $\square$

Heuristically we can interpret the δ operator as a shift of origin and scaling. This is a common trick in numerical analysis and has the consequence that the δ form can obtain better numerical properties than the shift operator. A controller in the δ form can be described by the state equations

$$\delta x(kh) = \overline{F}x(kh) + \overline{G}y(kh) = d(kh) \tag{15.32}$$
$$u(kh) = Cx(kh) + Dy(kh)$$

The shift operator and its inverse are implemented exactly using an assignment statement. To make a realization in δ form we must implement the operator δ^{-1}. Solving (15.32) for $x(kh)$ gives

$$x(kh) = \delta^{-1}d(kh) = x(kh - h) + hd(kh)$$

The extra amount of computations compared to the shift form is marginal. One extra vector addition is necessary. Notice that since $hd(kh)$ is normally much smaller than $x(kh - h)$, it is necessary to represent $x(kh - h)$ with a sufficient word length.

An Example

An example will now be used to illustrate the properties of the different realizations. Assume that a system with the pulse-transfer function

$$H(z) = \frac{b^4}{(z + a)^4} \tag{15.33}$$

where $b = 1 + a$ is to be implemented.

To obtain the computer program, a state-space realization of the pulse-transfer function is first determined. The computer code is then obtained as a direct implementation of the difference equations. There are many possible choices of the coordinate system in the state-space realization. A controllable canonical form in shift and δ operator, a Jordan canonical form, and a ladder form are chosen to demonstrate that the numerical properties may differ considerably. The codes for the representations are given in Listings 15.3–15.7. For the shift-operator, controllable form we implement

$$\frac{b^4}{(z + a)^4} = \frac{b^4}{z^4 + 4az^3 + 6a^2z^2 + 4a^3z + a^4}$$

The code is given in Listing 15.3. The numerical values of the parameters for the

Listing 15.3 Computer code for implementation of (15.33) based on the shift-controllable canonical form.

```
begin
  y: = b*b*b*b*x4
  s: = − a1*x1 − a2*x2 − a3*x3 − a4*x4 + u
  x4: = x3
  x3: = x2
  x2: = x1
  x1: = s
end
```

controllable canonical form when $a = -0.99$ are given by

$$a_1 = 4a = -3.96 \qquad a_2 = 6a^2 = 5.8806$$

$$a_3 = 4a^2 = -3.881196 \qquad a_4 = a^4 = 0.96059601$$

Listing 15.4 gives an implementation based on the Jordan canonical form.

Listing 15.4 Computer code for implementing (15.33) based on the Jordan canonical form.

```
begin
    x4: = - a*x4 + b*u
    x3: = - a*x3 + b*x4
    x2: = - a*x2 + b*x3
    x1: = - a*x1 + b*x2
    y: = x1
end
```

Listing 15.5 is obtained by rewriting the Jordan form as (15.30). Notice that this slight modification is a significant improvement over the form in Listing 15.4, because the state is obtained by adding a small correction to the previous state.

Listing 15.5 Rearrangement of the code in Listing 15.4.

```
begin
    x4: = x4 + b*(u - x4)
    x3: = x3 + b*(x4 - x3)
    x2: = x2 + b*(x3 - x2)
    x1: = x1 + b*(x2 - x1)
    y: = x1
end
```

For the δ form we implement (in controllable canonical form)

$$\frac{b^4}{(\delta + b)^4} = \frac{b^4}{\delta^4 + 4b\delta^3 + 6b^2\delta^2 + 4b^3\delta + b^4} = \frac{b_4}{\delta^4 + b_1\delta^3 + b_2\delta^2 + b_3\delta + b_4}$$

where $b = 1 + a$. Notice that b is a small number when a is close to -1. The system is implemented using (15.30) where $\overline{F}$ and $\overline{G}$ have the same form as for the shift-operator companion form in Listing 15.3. The implementation in δ companion form is given in Listing 15.6, where b_i are the coefficients in the characteristic polynomial.

Listing 15.6 Computer code for implementing (15.33) based on δ-operator, controllable canonical form.

```
begin
    y: = b*b*b*b*x4
    s: = -b1*x1 - b2*x2 - b3*x3 - b4*x4 + u
    x4: = x4 + x3
    x3: = x3 + x2
    x2: = x2 + x1
    x1: = x1 + s
end
```

Finally the ladder form can be implemented using Listing 15.7. Notice that the ladder form implements a filter with direct term. The output has to be shifted to give the signal with delay. The realization is thus not minimal.

The implementation of the discrete-time system also includes a monitor system that runs the program each sampling period. Notice that the code contains only addition, multiplication, and assignment statements; thus it can easily be implemented using many computer languages. Because assignment statements only transfer data, they will not introduce any numerical errors. This means that 0 and 1 in a standard matrix representation are represented exactly.

Listing 15.7 Computer code for implementing (15.33) based on ladder form.

```
begin
    y: = y1
    y1: = y2
    y2: = b*b*b*b*x1
    nx1: = ((x2 - x1)/alpha1 + u)/beta1
    nx2: = ((x1 - x2)/alpha1 + (x3 - x2)/alpha2)/beta2
    nx3: = ((x2 - x3)/alpha2 + (x4 - x3)/alpha3)/beta3
    nx4: = ((x3 - x4)/alpha3 - x4/alpha4)/beta4
    x1: = nx1
    x2: = nx2
    x3: = nx3
    x4: = nx4
end
```

Figure 15.20 shows simulations using Simnon on a Vax 11/780 computer. The simulation is simply an iteration of the state equations in Listings 15.3–4 and 15.6. The results are obtained using single-precision arithmetic in FORTRAN. The figure shows the results when different values of a are used. For $a = -0.9$ all the implementations give compatible results, as shown in Figure 15.20. When a is increased, the shift-operator, controllable, canonical form is very sensitive

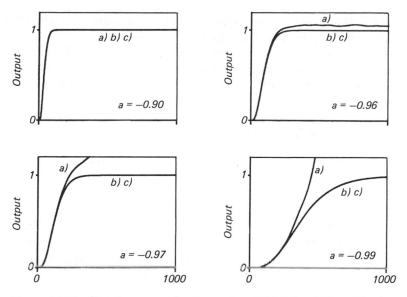

Figure 15.20 Step responses for the system (15.33) for different implementations and different values of a.
a. Shift-operator, controllable, canonical form.
b. Jordan form.
c. δ-operator, controllable, canonical form.

and the solution is inaccurate. The other two implementations give approximately the same results. They will, however, differ when short word length is used. The Jordan form is better than the δ controllable canonical form when a is decreased further.

The sensitivity of the shift-operator controllable form with respect to changes in the parameters is given by (15.22). Perturbing the characteristic equation with a constant term ϵ,

$$(z - 0.99)^4 + \epsilon = 0$$

gives the roots

$$z = 0.99 + (-\epsilon)^{1/4}$$

The roots are moved from 0.99 to a circle with origin at 0.99 and the radius $r = |\epsilon|^{1/4}$. If $\epsilon = 10^{-8}$ then $r - 10^{-2}$; i.e., the system can be unstable even if the perturbation is very small.

Making a similar calculation for the δ companion form we get

$$(\delta + 0.01)^4 + \epsilon = 0$$

which gives the roots

$$\delta = -0.01 + (-\epsilon)^{1/4}$$

The roots are moved from 0.01 to a circle with origin at 0.01 and the radius $r =$

$|\epsilon|^{1/4}$. If $\epsilon = 10^{-8}$ then $r = 10^{-2}$ which is the same as for the shift-operator case. Notice, however, that the relative variation in parameters required to make the system unstable is two orders of magnitude larger with the δ operator. The δ companion form is thus less sensitive to parameter perturbations than the shift companion form. Notice, however, that the Jordan realizations in shift or δ forms are superior.

Effects of the Sampling Period

The sampling period also has a considerable influence on the conditioning, as shown by the following examples.

Example 15.13—Effect of sampling period on coefficient precision

Consider a first-order system with time constant T. The discrete-time equivalent of such a system is

$$x(kh + h) = ax(kh) + bu(kh)$$

where

$$a = e^{-h/T}$$

Simple calculations show that

$$\frac{da}{a} = \frac{h}{T}\frac{dT}{T}$$

For a given relative precision, the equivalent time constant is thus inversely proportional to the sampling period. ☐

Example 15.14—Numerical precision required for PI-control

Consider the formula for updating the integral in a PI-regulator:

$$i(kh + h) = i(kh) + e(kh) * h/ti$$

If the sampling period is 0.03 s and the integration time is 15 min = 900 s, the ratio h/ti becomes $3 \cdot 10^{-5}$, which corresponds to about 15 bits. To avoid that, the quantity eh/ti is rounded. It is thus necessary to make computations with a longer word length. This is the reason why the integral is often implemented in 24-bit representation in dedicated PI-regulators. ☐

The examples show that a rapid sampling requires a high precision in the coefficients.

How to Choose Representations?

The selection of representations is crucial when implementing a control law using a digital signal processor or with custom VLSI. It is less crucial for implementations using microcomputers with floating-point hardware. The companion forms should be avoided, since so much is gained by using series or parallel forms as in Figure 15.17. Each block should be implemented on Jordan form. This is particularly important for high-order compensators and short sampling periods. For low-order controllers implemented with floating-point hardware and with poles well inside the stability area, the choice of realization is less crucial. It is, however, good practice to hedge against possible numerical problems.

15.8 Programming

Programming is an important aspect of the implementation of a control system, both with respect to the efficiency of the system and the time required for the implementation.

The effort required and the approaches used depend on the available software and the nature of the control problem. When computer control was first introduced, there were no high-level languages available for control computers. The computing power available was often so poor that assembly coding was necessary. The situation has now changed drastically, and there are many high-level languages that may be used, such as Forth, BASIC, FORTRAN, C, Jovial, Pascal, Concurrent Pascal, Modula, and Ada. Ada, which was developed by the U.S. Department of Defense for computer-control applications, is the first language designed and developed for real-time applications. There are many advantages in using a high-level language because it takes much less time to develop a system. The code is also much more readable, which makes it easier to modify. The character and the difficulty of the programming depend very much on the application. The requirements on operator communications are critical. The code required for operator communication is often much larger than the pure control code. A few examples illustrate this.

A Simple Dedicated Control System

Consider a simple control loop that has a few measured signals, a few outputs, and limited operator communication. Information may be displayed to the operator, and the operator may have a few buttons and a few dials.

The programming of such a system is very simple. If a real-time clock is available, the code is in essence given by Listing 15.8.

LISTING 15.8 Computer code skeleton for a simple control loop.

```
repeat
    Wait for clock interrupt
    Regulate
    Display
forever
```

The first line is simply a procedure that halts execution until a clock-interrupt occurs. The procedure *Regulate* is the code required to implement the desired control. A typical example is given in Listing 15.9.

LISTING 15.9 Computer code skeleton for the control algorithm with the command signals included.

```
        Procedure Regulate
        begin
1         Adin y uc umanual
2         Din manual
3         If manual then
4           u := ulow:= uhigh:= umanual
5         else
6           u := u1 + D * y + Dc * uc
7         u := sat(u, ulow, uhigh)
8         Daout u
9         x := Fo * x + Go * y + Gc * uc + K * u
10        u1 := C * x
        end
```

The statement *Din* on line 2 reads a digital signal to the Boolean variable manual. If this variable is true, the regulator operates in manual mode and the control variable is equal to the variable *umanual*, which is an analog input. Notice that the code is similar to Listing 15.2, except for the inputs from the operator and the saturation function, which has been added to avoid windup.

The procedure *Display* in Listing 15.8 computes some variables and displays them in analog or digital form. Notice that it is straightforward to introduce facilities for the operator to change parameters simply by introducing them as analog inputs.

The program in Listing 15.8 is fairly easy to debug. The procedures *Regulate* and *Display* are simple sequential procedures that can be tested off-line. It is also easy to check that the wait procedure gives an interrupt every sampling period. It will fail only if the time required to execute the procedures is longer than one sampling period. This may be tested by timing.

More Complicated Control Loops

The principles used in the program in Listing 15.8 may be extended to more complicated control systems with several loops having different sampling periods. A computer code, which may be represented by Fig. 15.21, is then obtained. The program P0 in Fig. 15.21 runs at the sampling rate given by the clock. The programs P1, P2, and P3 run each every third clock-pulse. In order to obtain the representation in Fig. 15.21, it is necessary that the time required to execute each path be shorter than the shortest sampling period in the system. This is easy to do for systems with long sampling periods. For systems with fast sampling, it may be necessary to split up the computations in a tedious, unnatural, and error-prone fashion.

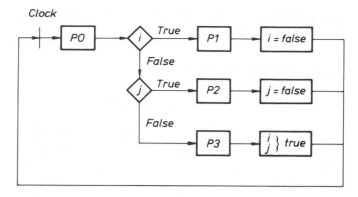

Figure 15.21 Flow chart for a multiloop control law with two sampling rates.

It is comparatively easy to debug the program shown in Fig. 15.21 if there are few paths and if the procedures are simple. The difficulty in debugging grows rapidly with increasing system complexity. New ideas and concepts are needed to handle such problems in a convenient way.

Concurrent, or Real-Time, Programming

It is natural to think about control loops as concurrent activities that are running in parallel. However, the digital computer operates sequentially in time. Ordinary programming languages can represent only sequential activities. Thus a key problem is to map a number of parallel, concurrent activities into a sequential program. This may be done manually, as shown in Fig. 15.21. There are also special-purpose software—*real-time operating systems*—that aid in this task. It would go far outside the scope of this book to discuss concurrent programs in detail. The basic ideas are given, together with a few examples.

The notions of *process* and *task* are fundamental concepts in real-time programming. They represent activities that may be thought of as running in parallel in time. Using these notions, it is possible to think of the computer running several activities in parallel. Hence, a real-time activity may be structured in the same way as a sequential activity is structured, using the notion of procedure or subroutine. The real-time operating system will organize the execution of the processes so that the desired result is obtained. To do this, a priority is associated with each process. Processes may also be scheduled to run periodically or in response to events such as interrupts or completion of other tasks.

The problem of *shared variables* and resources is one of the key problems in real-time programming. If different processes are using the same data, it is necessary to make sure that one process does not try to use data being modified by another process. If two processes may use the same resource, it is necessary to make sure that the system does not *deadlock* in a situation where both processes are waiting for each other.

Timing is a third problem. Computing power must be sufficient to allow all activities to be completed in the required time.

A Regulator with Operator Interaction

A single control loop with operator interaction is one of the simplest examples of real-time programming. The task could be to run a control loop like the one in Listing 15.4 with sampling rate of 20 ms and to provide an interface so that the operator may change parameters from a keyboard or a terminal. Assume that the operator changes parameters by typing in a character string on the terminal. Because the time required for this is considerably longer than one sampling period, it is necessary to break down the operation into many small pieces in order to use the solution shown in Fig. 15.21. This is both tedious and unnatural. It is much more natural to think of the problem in the form of two concurrent processes. One process, *regulation*, should be run once every sampling period. The other process, *operator communication*, may run whenever the process *regulation* is idle. To ensure that control actions are taken at regular sampling periods, it is necessary to impose the rule that the process *regulation* has priority over *operator communication* and that it may interrupt the operator communication at any time. For convenience, the rule that the process *regulation* runs to completion once it has started is also introduced. In a case like this, *regulation* is called the *foreground task*, or foreground process, and *operator communication* is called a *background task*, or background process.

Real-Time Operating Systems

For problems with only two processes, it is not difficult to write a monitor program that administrates the processes. Such a monitor may typically be written in less than 100 lines of assembly code.

The simple monitor may be extended to several processes. It is, however, a major task to make a monitor that can handle more complex situations. Such a monitor, which is also called a *real-time operating system*, may occupy anything from a few kilobytes to 20 kilobytes of code.

The real-time operating systems allow definition of tasks, or processes, in a high-level language such as FORTRAN or Pascal. It is also possible to run processes at regular intervals or in given relationships to other tasks. Processes may also be introduced, started, and removed on-line. Priorities between different tasks may be introduced and modified. The introduction of real-time operating systems was one of the major innovations when process-control computers were introduced in the mid-sixties. Examples of such operating systems are TSX from IBM and RSX from Digital Equipment Corporation. Processes, or tasks, have also been introduced into simple languages such as BASIC.

The real-time operating systems are large, general-purpose programs, which are often written in assembly code. They are difficult to maintain and modify.

There has been a need to have real-time operating systems that can be tailored to specific applications. Computer languages with facilities for real-time programming have therefore been developed. Concurrent Pascal and Modula are such languages. The language Ada is a standard tool for implementing computer-control systems in military systems.

DDC-Packages

There exist special techniques to program control systems consisting of a large number of identical control loops. The code is often structured as follows:

> Read all analog inputs and store in a table.
>
> Convert all signals to engineering units and store results in a table.
>
> Apply the control algorithm sequentially to all values in the table using regulator parameters stored in a parameter table.
>
> Perform D-A conversion to all variables stored in the output table.

Programs of this type are called DDC-packages. The choice of control algorithms is usually limited to different PID-algorithms. The packages are easy to use because all programming is reduced to enter the appropriate data in the tables. Programs of this type are called *table-driven*.

15.9 Conclusions

Implementation of control laws using a computer is discussed in this chapter. The key problem is to implement a discrete-time system. The principles for doing this have been covered in detail. It is straightforward to generate the code from the control algorithm. The importance of prefiltering to avoid aliasing has been mentioned. Sophisticated nonlinear digital filtering for removing outliers has also been discussed. The computational delay is influenced considerably by the organization of the computer code. Difficulties that arise from saturation in actuators and ways to avoid these difficulties are discussed. This also automatically gives a solution to mode switching and initialization.

Numerical problems and consequences of finite word length are also discussed. It is found to be very beneficial to transform the equations describing the control law to a form that is numerically well conditioned. Operational issues like mode switching and operator-machine interaction are discussed. There are many new possibilities in this area. Finally programming of control algorithms is discussed. Although the presentation is fairly short, the information given should be sufficient to implement control algorithms on minicomputers and microcomputers using high-level languages.

15.10 Problems

15.1. Consider control of a double integrator with a sampling period of 1 s. Calculate the deadbeat control for the system obtained using an antialiasing filter with the transfer function

$$G(s) = \frac{1}{s^2 + 1.4s + 1}$$

Compare the deadbeat strategy obtained with the deadbeat strategy for the pure double integrator using simulation.

15.2. Write a program for computing the scalar product of two arrays

```
begin
    s: = 0
    for i: = 1 to n do
    s: = s + a[i] * b[i]
end
```

in each case.

(a) s: integer; a, b: arrays of integers.
(b) s: double precision integer; a, b: arrays of integers.
(c) s: real; a, b: arrays of reals.
(d) s: double-precision real; a, b: arrays of real.

Compare the computing times and precision. Try to find computers that have floating-point calculations in software, as well as in hardware.

15.3. Consider Example 15.14. Discuss the possibilities of using two loops with different sampling periods in order to improve the precision in the calculations.

15.4. Write a code for a digital PI-regulator where the antiwindup is implemented as an observer with the time constant T_0.

15.5. Write a code for a digital PID-regulator where the antiwindup is implemented as a deadbeat observer.

15.6. Write a code in your favorite high-level language for a digital PID-algorithm where antiwindup is implemented as an observer with time constant T_0. Determine the number of operations required for one iteration. Compile the program. Determine how many memory calls it requires. Time the program. How do the measured computing times relate to the number of operations and the computing times given in the computer manual?

15.7. Consider the control algorithm of (15.1), where x_m is considered to be an input. Assume that the state, the control variable, and the process output have dimensions nx, nu, and ny and that the matrices are full. Determine the number of additions, multiplications, and divisions required for one iteration.

15.8. Consider the control algorithm of (15.2). Write a code that implements the control algorithm in your favorite high-level language. Compile the code, determine how much memory space the code occupies, and determine the execution time. Try to find a good, simple formula for determining the execution time.

15.9. Repeat Problem 15.8 but now use a subroutine to perform a scalar product. Discuss how computing time and storage requirements are influenced by the restructuring of the program.

15.10. Consider the control algorithm with rejection of outliers given by Equations (15.1) and (15.5). Make an estimate of the number of computations required for one iteration. (*Hint:* A matrix multiplication of an $n \times p$ matrix by a $p \times r$ matrix requires $N = npr$ operations, where an operation corresponds to one addition and one multiplication. Solution of the equation

$$Ax = B$$

where A is $n \times n$ and B is $n \times p$, requires approximately

$$N = \tfrac{1}{3}n^3 + \tfrac{1}{2}n^2 p$$

operations, where the major part of the calculations is the triangulation of the matrix A.)

15.11. Listed below is a program for a Kalman filter that can reject large measurement errors selectively. Study the algorithm and discuss how the filter is reconfigured after a faulty measurement. Notice that the code is written for the case in which the computations take approximately one sampling period.

```
begin
    e := y − C*x
    R := R2 + C*P*CT
    {Test what measurements are accepted}
    na := 0
    for i := 1 to ny do
        begin
            t = e[i] * e[i]/R[i,i]
            If t < test then
            begin
                na := na + 1
                ya[na] := y[i]
                ia[na] := i
            end
        end
    {The number of accepted measurements is na.
    The array ia[1 .. na] contains the indices of
    accepted measurements. The array ya[1 .. na]
    contains the accepted measurements.}
    for i := 1 to na do
    begin
        for j := 1 to na do
        Ra[i,j] := R[ia[i], ia[j]]
        for j := 1 to nx do
        Ca[i,j] := C[ia[i],j]
        ea[i] := e[ia[i]]
    end
        K := A * P * CaT * Inv(Ra)
        x := A * x + B * u + K * ea
        P := A * P * AT + R1 − K * Ca * P * AT
end
```

15.12. Consider a discrete-time system characterized by the pulse-transfer function

$$H(z) = \frac{1}{(z - a)^n}$$

Calculate the sensitivity of the poles with respect to the parameters using Equation (15.22) in each case.

(a) The filter is in companion form.

(b) The filter is in Jordan canonical form.

15.13. Make a flow chart similar to Fig. 15.2 for a system with loops having sampling periods of 1, 2, 5, and 60 s.

15.14. Consider a system with the transfer function

$$H(z) = \frac{1}{z^n + a_1 z^{n-1} + \cdots + a_n}$$

Assume that the system is realized with fixed-point arithmetic. Let the roundoff be described as normal rounding to integers. Show that the condition for a steady-state error k with no inputs is given by

$$k + \sum_{i=1}^{n} Q[a_i k] = 0$$

Furthermore, show that the condition for a limit-cycle oscillation with a period of two sampling periods is

$$k + \sum_{i=1}^{n} (-1)^i Q[a_i k] = 0$$

15.15. A PI-regulator with A-D and D-A converters is shown in Fig. 15.22(a). The linear model obtained with one realization of the PI-regulator is shown in Fig. 15.22(b). Show that the equivalent error reduced to the input of the D-A converter may be represented as

$$\epsilon = e_{ad} + \frac{1}{a + b}(e_a + e_b)$$

The linear model obtained with another realization of the PI-regulator is shown in Fig. 15.22(c). Give the corresponding error for this representation and discuss the differences between the implementations.

15.16. Consider the following algorithm for a PI-regulator:

```
Adin uc y
e := uc − y
v := k*e + i
u := max(min(512,v),0)
Daout u
i := u + k*h*e/ti
```

Assume that the A-D and D-A converters have a resolution of 8 bits and that all calculations are made using integers. What is the word length required to represent variable i if overflow should be avoided? Use $k = 50$ and (a) $h = 1$, $ti = 300$ or (b) $h = 0.01$, $ti = 1500$. Discuss how the result is influenced by the sampling period.

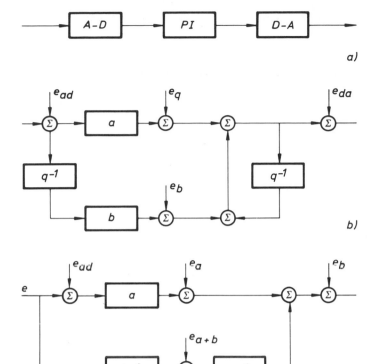

Figure 15.22

15.17. Three different algorithms for a PI-regulator are listed below. Use the linear model for roundoff to analyze the sensitivity of the algorithms to quantization in A-D and D-A converters and roundoff in the multiplications. Assume fixed-point calculations. Also, discuss the word lengths necessary for the algorithms.

Algorithm 1

$$e := uc - y$$
$$u := k*(e + h*i/ti)$$
$$i := i + e*h$$

Algorithm 2

$$e := uc - y$$
$$u := k*(e+i)$$
$$i := i + e*h/ti$$

Algorithm 3

$$e := uc - y$$
$$u := i + k*e$$
$$i := i + k*h*e/ti$$

15.18. Consider a dynamic system with the pulse-transfer function

$$H(z) = \frac{b_0 z^n + b_1 z^{n-1} + \cdots + b_n}{z^n + a_1 z^{n-1} + \cdots + a_n}$$

Show that the system has the following state descriptions:

(a)

$$x(k+1) = \begin{bmatrix} -a_1 & 1 & 0 & \cdots & 0 \\ -a_2 & 0 & 1 & \cdots & 0 \\ \vdots & & & & \\ -a_{n-1} & 0 & 0 & \cdots & 1 \\ -a_n & 0 & 0 & \cdots & 0 \end{bmatrix} x(k) + \begin{bmatrix} b_1 - b_0 a_1 \\ b_2 - b_0 a_2 \\ \vdots \\ b_{n-1} - b_0 a_{n-1} \\ b_n - b_0 a_n \end{bmatrix} u(k)$$

$$y(k) = \begin{bmatrix} 1 & 0 & 0 & \cdots & 0 \end{bmatrix} x(k) + b_0 u(k)$$

(b)

$$x(k+1) = \begin{bmatrix} -a_1 & 1 & 0 & \cdots & 0 \\ -a_2 & 0 & 1 & \cdots & 0 \\ \vdots & & & & \\ -a_n & 0 & 0 & \cdots & 1 \\ 0 & 0 & 0 & \cdots & 0 \end{bmatrix} x(k) + \begin{bmatrix} b_0 \\ b_1 \\ \vdots \\ b_{n-1} \\ b_n \end{bmatrix} u(k+1)$$

$$y(k) = \begin{bmatrix} 1 & 0 & 0 & \cdots & 0 \end{bmatrix} x(k)$$

(c)

$$x(k+1) = \begin{bmatrix} -a_1 & -a_2 & \cdots & -a_{n-1} & -a_n & b_1 & b_2 & \cdots & b_{n-1} & b_n \\ 1 & 0 & \cdots & 0 & 0 & 0 & 0 & \cdots & 0 & 0 \\ \vdots & & & & & & & & & \\ 0 & 0 & \cdots & 1 & 0 & 0 & 0 & \cdots & 0 & 0 \\ 0 & 0 & \cdots & 0 & 0 & 0 & 0 & \cdots & 0 & 0 \\ 0 & 0 & \cdots & 0 & 0 & 1 & 0 & \cdots & 0 & 0 \\ \vdots & & & & & & & & & \\ 0 & 0 & \cdots & 0 & 0 & 0 & 0 & \cdots & 1 & 0 \end{bmatrix} x(k)$$

$$+ \begin{bmatrix} b_0 \\ 0 \\ \vdots \\ 0 \\ 1 \\ 0 \\ \vdots \\ 0 \end{bmatrix} u(k+1)$$

$$y(k) = \begin{bmatrix} 1 & 0 & 0 & \cdots & 0 & 0 \end{bmatrix} x(k)$$

Assume that $H(z)$ represents a controller. Discuss the advantages and disadvantages with the different realizations of the controller.

15.19. A digital controller with the sampling period $h = 0.2$ has the pulse-transfer function

$$H(z) = \frac{6.25 z^2 - 11.5 z + 5.5}{z^2 - 1.5 z + 0.5}$$

The controller is used to control a process with the transfer function

$$G(s) = \frac{1}{s^2 + 1}$$

Discuss the effect of roundoff and quantization noise when different realizations are used to implement the controller on a computer having fixed word length.

15.20. Show that the continued-fraction representation (15.24) can be obtained recursively as

$$H_i(z) = \cfrac{1}{\beta_i z + \cfrac{1}{\alpha_i + H_{i+1}(z)}}$$

where $H_{n+1}(z) = 0$.

15.21. Determine the δ operator representations of the following continuous-time transfer functions:

(a) $1/(s^2 + 1)$
(b) $K/(1 + Ts)$
(c) $1/(s + a)^3$

Compute the poles and the zeros and investigate what happens when $h \rightarrow 0$.

15.22. Let $H(z)$ be the pulse transfer function obtained from stepinvariant sampling of the rational transfer function $G(s)$. Define

$$\overline{H}(\delta) = H(1 + \delta h)$$

Prove that

$$\lim_{h \rightarrow 0} \overline{H}(\delta) = G(\delta)$$

Show that this is true also for rampinvariant and impulseinvariant sampling.

15.11 References

Design of filters is covered in standard texts on networks. The books

Kuo, F. F. (1962): *Network Analysis and Synthesis.* New York: John Wiley.

Williams, A. B. (1981): *Electronic Filter Design Handbook.* New York: McGraw-Hill.

are good sources. Useful practical advice is also found in the handbooks published by manufacturers of operational amplifiers. Such handbooks are also useful for information about A-D and D-A converters. Make sure to get a new version of whatever handbook you use because the technology changes rapidly.

The problems associated with windup of PID-regulators is discussed in trade journals for the process industry. The general approach given in Sec. 15.4 is believed to be novel.

Possibilities for error detection and rejection of outliers are discussed in depth in

Willsky, A. (1976): "A Survey of Design Methods for Failure Detection in Dynamic Systems," *Automatica,* 12, 601–11.

which also contains many references.

A comprehensive text on the effects of quantization and roundoff in digital control systems is

MORONEY, P. (1983): "Issues in the Implementation of Digital Feedback Compensators," Cambridge, Mass.: MIT Press.

which contains many references. The following papers are classics in the area.

KNOWLES, J. B. and R. EDWARDS (1965): "Effect of a Finite-Word-Length Computer in a Sampled-Data Feedback System," *Proc. IEE,* 112, 1197–1207.

CURRY, E. E. (1967): "The Analysis of Round-Off and Truncation Errors in a Hybrid Control System," *IEEE Trans. Autom. Control,* AC-12, 601–4.

BERTRAM, J. E. (1958): "The Effect of Quantization in Sampled-Feedback Systems," *Trans. Amer. Inst. Elec. Engrs.,* 77, 177–82.

SLAUGHTER, J. B. (1964): "Quantization Errors Digital Control Systems," *IEEE Trans. Autom. Control,* AC-9, 70–74.

RINK, R. E. and H. Y. CHONG (1979): "Performance of State Regulator Systems with Floating-Point Computation," *IEEE Trans. Autom. Control,* AC-24, 411–21.

A review of digital signal processors is given in

LEE, E. A. (1988): "Programmable DSP Architectures: Part 1," *IEEE ASSP Magazine,* 5:4, October, 4–19.

Design of special-purpose signal processors in VLSI is described in

CATTHOOR, F., J. RABAEY, G. GOOSSENS, J. L. VAN MEERBERGEN, R. JAIN, H. J. DE MAN, and J. VANDEWALLE (1988): "Architectural Strategies for an Application-Specific Synchronous Multiprocessor Environment," *IEEE Trans. Acoustics, Speech and Signal Processing,* 36, 265–84.

and in a series of books entitled *VLSI Signal Processing I, II*, and *III*, published by IEEE. The books are based on presentations given at IEEE ASSIP workshops. The 1988 volume is

BRODERSEN, R. W. and H. S. MOSCOVITZ, eds., (1988): *VLSI Signal Processing III,* New York: IEEE Press.

A readable account of the IEEE standard and its impact on different high-level languages is found in

FATEMAN, R. J. (1982): "High-Level Language Implications of the Proposed IEEE Floating-Point Standard," *ACM Transactions on Programming Languages and Systems,* 4, no. 2, 239–57.

Problems associated with quantization, roundoff, and overflow are also discussed in the signal-processing literature. Overviews are found in

OPPENHEIM, A. V. and R. W. SCHAFER (1975): *Digital Signal Processing.* Englewood Cliffs, N.J.: Prentice Hall, Inc.

RABINER, L. R. and B. GOLD (1975): *Theory and Application of Digital Signal Processing.* Englewood Cliffs, N.J.: Prentice Hall, Inc.

Specialized issues are discussed in

JACKSON, L. B. (1970): "Roundoff Noise Analysis for Fixed-Point Digital Filters Realized in Cascade or Parallel Form," *IEEE Trans. Audio & Electroacoustics,* AU-18, 107–22.

Jackson, L. B. (1970): "On the Interaction of Roundoff Noise and Dynamic Range in Digital Filters," *Bell Syst. Tech. Journal,* 49, 159–84.

Jackson, L. B. (1979): "Limit Cycles in State-Space Structures for Digital Filters," *IEEE Trans. Circuits & Systems,* CAS-26, 67–68.

Parker, S. R. and S. F. Hess (1971): "Limit Cycle Oscillations in Digital Filters," *IEEE Trans. Circuit Theory,* CT-18, 687–97.

Willson, Jr., A. N. (1972): "Some Effects of Quantization and Adder Overflow on the Forced Response of Digital Filters," *Bell Syst. Tech. Journal,* 51, 863–87.

Buttner, M. (1977): "Elimination of Limit Cycles in Digital Filters with Very Low Increase in Quantization Noise," *IEEE Trans. Circuits & Systems,* CAS-24, 300–304.

Willson, Jr., A. N. (1972): "Limit Cycles Due to Adder Overflow in Digital Filters," *IEEE Trans. Circuit Theory,* CT-19, 342–46.

Limit cycles due to roundoff can be determined using the theory of relay oscillations. This is described in

Tsypkin, Ya. Z. (1984): *Relay Control Systems.* Cambridge, Mass.: Cambridge University Press

There are many standard texts on numerical analysis:

Björk, G., Å. Dahlqvist, and N. Andersson (1974): *Numerical Methods.* Englewood Cliffs, N.J.: Prentice Hall, Inc.

is a good source.

Concurrent programming is discussed in

Brinch-Hansen, P. (1973): *Operating System Principles.* Englewood Cliffs, N.J.: Prentice Hall, Inc.

Barnes, J. G. P. (1982): *Programming in Ada.* New York: Addison-Wesley.

Much useful information is also given in material from vendors of computer-control systems. The δ operator is an old idea. See

Tschauner, J. (1963): "A General Formulation of the Stability Constraints for Sampled-Data Control Systems," *Proc. IEEE,* 51, 619–20.

Tschauner, J. (1963): "Stability of Sampled Data Systems," *Proc. IEEE,* 51, 621–22.

The δ operator has recently been given much attention because of its numerical properties. See

Gawthrop, P. J. (1980): "Hybrid Self-Tuning Control," *Proc. IEE,* 127, pt. D, 229–36.

Peterka, V. (1984): "Predictor-Based Self-Tuning Control," *Automatica,* 20, 39–50.

Middleton, R. H. and G. C. Goodwin (1987): "Improved Finite Word Length Characteristics in Digital Control Using Delta Operators," *IEEE Trans. Automat. Control,* AC-31, 1015–21.

Middleton, R. H. and G. C. Goodwin (1989): *Digital Control and Estimation: A Unified Approach* (forthcoming). Englewood Cliffs, N.J.: Prentice Hall, Inc.

EXAMPLES

A

Examples used in the book as "standard processes" are presented in this appendix.

Example A.1—Double integrator

The double integrator is used throughout the book as a main example to illustrate the theories presented. The process is described by the differential equation

$$\frac{d^2y}{dt^2} = u \tag{A.1}$$

The transfer function is $G(s) = 1/s^2$. Introduce y and $\dot{y}$ as the states x_1 and x_2, respectively, of the system. The state-space representation is then

$$\dot{x} = \begin{bmatrix} 0 & 1 \\ 0 & 0 \end{bmatrix} x + \begin{bmatrix} 0 \\ 1 \end{bmatrix} u$$

$$y = [1 \quad 0]x \tag{A.2}$$

Sampling (A.2) using a zero-order hold with the sampling period h gives the discrete-time system (see Example 3.2)

$$x(kh + h) = \begin{bmatrix} 1 & h \\ 0 & 1 \end{bmatrix} x(kh) + \begin{bmatrix} h^2/2 \\ h \end{bmatrix} u(kh)$$

$$y(kh) = [1 \quad 0]x(kh) \tag{A.3}$$

The pulse-transfer operator of (A.3) is given by

$$H(q) = \frac{h^2(q + 1)}{2(q - 1)^2} \tag{A.4}$$

There are several physical processes that can be described as double integrators. One such is the attitude of a satellite, which can be described by the equation

$$J \frac{d^2\theta}{dt^2} = M_c + M_d$$

where θ is the attitude angle, M_c is the control torque, M_d is the disturbing torque, and J is the moment of inertia.

Another example that can be described by the double integrator is a rolling ball on a tilting beam (see Fig. A.1). The equation of the ball and beam can be described by

$$J \frac{d^2\theta}{dt^2} = mgr \sin \varphi \approx mgr \, \varphi$$

$$x = r\theta$$

or

$$\frac{d^2x}{dt^2} = mgr^2\varphi/J$$

where θ is the angle of the ball, g is the normal acceleration, x is the position of the ball, and φ is the tilting angle of the beam. ☐

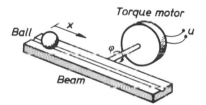

Figure A.1 Schematic illustration of the ball and beam.

Example A.2—Motor

A DC motor can be described by a second-order model with one integrator and one time constant (see Fig. A.2). The input is the voltage to the motor and the output is the shaft position. The time constant is due to the mechanical parts of the system, while the dynamics due to the electrical parts are neglected. A normalized model of the process is then given by

$$Y(s) = \frac{1}{s(s + 1)} U(s)$$

Introduce the velocity and the position of the motor shaft as states (see Fig. A.2).

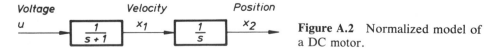

Figure A.2 Normalized model of a DC motor.

The state-space model of the motor is then given by

$$\dot{x} = \begin{bmatrix} -1 & 0 \\ 1 & 0 \end{bmatrix} x + \begin{bmatrix} 1 \\ 0 \end{bmatrix} u$$

$$y = [0 \quad 1]x(t)$$

(A.5)

Sampling (A.5) using a zero-order hold gives the discrete-time model

$$x(kh + h) = \begin{bmatrix} e^{-h} & 0 \\ 1 - e^{-h} & 1 \end{bmatrix} x(kh) + \begin{bmatrix} 1 - e^{-h} \\ h - 1 + e^{-h} \end{bmatrix} u(kh)$$

$$y(kh) = [0 \quad 1]x(kh)$$

(A.6)

(see Example 3.3). A current-controlled DC motor with the shaft velocity as output can also be described by the model of (A.5). Still another example that can be characterized by an integrator and a single pole is a ship. Let the input be the rudder angle and the output be the heading. The ship can then be described by the transfer function

$$G(s) = \frac{K}{s(1 + Ts)}$$

where the time constant may be positive or negative depending on the type of the ship. For instance, large tankers are unstable. □

Example A.3—Harmonic oscillator

Consider a pendulum (see Fig. A.3). The acceleration of the pivot point is the input and the angle y is the output. The system is then described by the normalized nonlinear equations

$$\dot{x}_1 = x_2$$

$$\dot{x}_2 = -\sin x_1 + u \cos x_1$$

$$y = x_1$$

where x_1 is the angle and x_2 is the angular velocity. Linearizing around $u = x_1 = 0$ gives

$$\dot{x} = \begin{bmatrix} 0 & 1 \\ -1 & 0 \end{bmatrix} x + \begin{bmatrix} 0 \\ 1 \end{bmatrix} u$$

$$y = [1 \quad 0]x$$

(A.7)

The transfer function of (A.7) is given by

$$G(s) = \frac{1}{s^2 + 1}$$

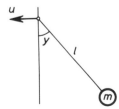

Figure A.3 Pendulum.

This transfer function can be generalized to

$$G(s) = \frac{\omega^2}{s^2 + \omega^2}$$

One state-space representation for this transfer function is

$$\dot{x} = \begin{bmatrix} 0 & \omega \\ -\omega & 0 \end{bmatrix} x + \begin{bmatrix} 0 \\ \omega \end{bmatrix} u$$

$$y = [1 \quad 0]x \tag{A.8}$$

Sampling (A.8) using a zero-order hold gives the discrete-time system

$$x(kh + h) = \begin{bmatrix} \cos \omega h & \sin \omega h \\ -\sin \omega h & \cos \omega h \end{bmatrix} x(kh) + \begin{bmatrix} 1 - \cos \omega h \\ \sin \omega h \end{bmatrix} u(kh)$$

$$y(kh) = [1 \quad 0]x(kh)$$

An overhead crane can also be modeled by (A.8). ▢

Example A.4—Time-delay process

Many industrial processes can be approximated by first-order dynamics and a time delay. One example is a paper machine (see Fig. A.4). The input is the thick stock flow, i.e., the amount of pulp. The output is the basic weight, i.e., the thickness of the paper. The equations describing the system can be normalized to the transfer function

$$G(s) = \frac{1}{s + 1} e^{-s\tau} \tag{A.9}$$

Another physical process that can be described by (A.9) is a mixing system with long pipes. Example 3.6 gives the zero-order-hold sampling of (A.9). ▢

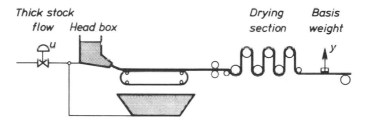

Figure A.4 Schematic picture of a paper machine.

Example A.5—An inventory model

An inventory is a typical example that can naturally be described as a discrete-time system. Orders and deliveries are obtained at regular intervals tied to the calendar—e.g., each day or week.

Let $y(k)$ be the inventory at time k before any transaction is started. The deliveries to the inventory that are ordered at time k are $u(k)$. It is assumed that there is a delay of one period from the order until the goods start coming into the inventory. Finally, the delivery from the inventory is $v(k)$. Introduce the state variables $x_1(k) = y(k)$ and $x_2(k) = u(k - 1)$. The inventory can be described by the

Examples

following discrete-time state equations:

$$x_1(k + 1) = x_1(k) + x_2(k) - v(k)$$

$$x_2(k + 1) = u(k)$$

or

$$x(k + 1) = \begin{bmatrix} 1 & 1 \\ 0 & 0 \end{bmatrix} x(k) + \begin{bmatrix} 0 \\ 1 \end{bmatrix} u(k) + \begin{bmatrix} -1 \\ 0 \end{bmatrix} v(k)$$

$$y(k) = [1 \quad 0]x(k)$$

(A.10)

The input-output relation is given by

$$y(k) - y(k - 1) = u(k - 2) - v(k)$$

(A.11)

□

MATRIX FUNCTIONS

In connection with sampled-data systems functions like exp A and ln A, where A is a matrix, are of interest. The matrix exponential and matrix logarithm are both *matrix functions*. This appendix gives some properties of matrix functions and discusses some ways to compute them.

A useful property of a square matrix is given by Theorem B.1.

Theorem B.1—The Cayley-Hamilton theorem.   Let

$$a(\lambda) = \lambda^n + a_1\lambda^{n-1} + \cdots + a_n = 0$$

be the characteristic equation of the square matrix A. Then A satisfies the following equation

$$a(A) - A^n + u_1A^{n-1} + \cdots + a_nI = 0$$

That is, the matrix satisfies its own characteristic equation.   □

Let A be an $n \times n$ square matrix and $f(\lambda)$ a scalar function of a scalar argument λ. We now want to extend the function $f(\lambda)$ to a function with a matrix argument, i.e., $f(A)$. If $f(\lambda)$ is a polynomial

$$f(\lambda) = \alpha_0\lambda^m + \alpha_1\lambda^{m-1} + \cdots + \alpha_m$$

then the matrix function $f(A)$ is defined as

$$f(A) = \alpha_0 A^m + \alpha_1 A^{m-1} + \cdots + \alpha_m I$$

The eigenvalues of $f(A)$ can be found using the following theorem.

Theorem B.2. If $f(A)$ is a polynomial in A and e_i the eigenvectors of A associated with the eigenvalues λ_i, then

$$f(A)e_i = f(\lambda_i)e_i$$

so $f(\lambda_i)$ is an eigenvalue of $f(A)$ and e_i is the corresponding eigenvector. □

Further, if $f(\lambda)$ can be defined by the power series

$$f(\lambda) = \sum_{i=0}^{\infty} c_i \lambda^i$$

which is assumed to be convergent for $|\lambda| < R$, then the matrix function

$$f(A) = \sum_{i=0}^{\infty} c_i A^i$$

is convergent if all the eigenvalues of A, λ_i, satisfy $|\lambda_i| < R$.

Using the Cayley-Hamilton theorem, it can be shown that for every function f there is a polynomial p of degree less than n such that

$$f(A) = p(A) = \alpha_0 A^{n-1} + \alpha_1 A^{n-2} + \cdots + \alpha_{n-1} I \tag{B.1}$$

From Theorem B.2 we get

$$f(\lambda_i) = p(\lambda_i), \qquad i = 1, \ldots, n \tag{B.2(a)}$$

If the eigenvalues are distinct, then these conditions are sufficient to determine α_i, $i = 0, \ldots, n - 1$. If there is a multiple eigenvalue with multiplicity m, then the additional conditions

$$f^{(1)}(\lambda_i) = p^{(1)}(\lambda_i)$$
$$\vdots \tag{B.2(b)}$$
$$f^{(m-1)}(\lambda_i) = p^{(m-1)}(\lambda_i)$$

hold, where $f^{(i)}$ is the ith derivative with respect to λ.

Using (B.1) and the conditions in (B.2), it is possible to compute matrix functions. For low-order systems, this is a very convenient method for hand calculations.

Example B.1

Let

$$A = \begin{bmatrix} 0 & 1 \\ -1 & 0 \end{bmatrix}$$

and determine

$$\exp Ah = \alpha_0 Ah + \alpha_1 I$$

Ah has the eigenvalues $\pm ih$; the system of equations

$$e^{ih} = \alpha_0 ih + \alpha_1$$

$$e^{-ih} = -\alpha_0 ih + \alpha_1$$

holds, giving

$$\alpha_0 = \frac{1}{2ih}(e^{ih} - e^{-ih}) = \frac{\sin h}{h}$$

$$\alpha_1 = \frac{1}{2}(e^{ih} + e^{-ih}) = \cos h$$

Finally,

$$\exp Ah = \sin h \begin{bmatrix} 0 & 1 \\ -1 & 0 \end{bmatrix} + \cos h \begin{bmatrix} 1 & 0 \\ 0 & 1 \end{bmatrix} = \begin{bmatrix} \cos h & \sin h \\ -\sin h & \cos h \end{bmatrix} \qquad \square$$

Example B.2

Let

$$\Phi = \begin{bmatrix} 1 & h \\ 0 & 1 \end{bmatrix}$$

and compute $\ln \Phi$. The eigenvalues are given by $(\lambda - 1)^2 = 0$—i.e., multiple eigenvalues exist. The matrix logarithm can now be written as

$$\ln \Phi = \alpha_0 \Phi + \alpha_1 I$$

where α_0 and α_1 are given by

$$\begin{cases} \ln 1 = \alpha_0 + \alpha_1 \\ \left. \frac{\partial}{\partial \lambda}(\ln \lambda) \right|_{\lambda = 1} = \alpha_0 \end{cases} \quad \text{or} \quad \begin{cases} 0 = \alpha_0 + \alpha_1 \\ 1 = \alpha_0 \end{cases}$$

Finally,

$$\ln \Phi = \begin{bmatrix} 1 & h \\ 0 & 1 \end{bmatrix} - \begin{bmatrix} 1 & 0 \\ 0 & 1 \end{bmatrix} = \begin{bmatrix} 0 & h \\ 0 & 0 \end{bmatrix} \qquad \square$$

Remark. Instead of starting with the characteristic polynomial, it is possible to use the minimal polynomial of the matrix. The degree of the series in (B.1) will then be the degree of the minimal polynomial minus one. In general, this will not reduce the computing time because the minimal polynomial—or, alternatively, the Jordan form—has to be computed.

Further properties of matrix functions can be found in

GANTMACHER, F. R. (1960): *The Theory of Matrices,* vols. I and II. New York: Chelsea.
BELLMAN, R. (1970): *Introduction to Matrix Analysis.* New York: McGraw-Hill.

SIMNON—An Interactive Simulation Program

C

A good simulation program is an indispensable tool for investigating a computer-controlled system. There are many programs available. This Appendix describes the program Simnon, which was used in this book. The MS™-DOS version of the language is described.

Simnon is a command-driven interactive program for simulation of systems described by nonlinear ordinary differential and difference equations. The program was developed at the Department of Automatic Control, Lund Institute of Technology. The work has been supported by the Swedish Institute of Applied Mathematics (ITM) and the Swedish Board of Technical Development (STU). A more detailed description of the program is given in Elmqvist (1975) and Åström (1983). This appendix is based on Elmqvist (1977) and Elmqvist et al. (1986).

Production

Simnon is an interactive simulation program. It has been used extensively since 1974 for research and education and to solve industrial control and simulation problems. The interactivity makes it possible to perform simulation studies in a short time. The simple model language and the extensive error-checking makes Simnon well suited for use in education.

The user interacts with the program using commands. For example, there are commands to change parameters of the model, to perform simulation, to plot results from simulations, and to modify the model. The user can construct personal high-level commands by means of a macro facility.

The model can be described either in a special model language or in FORTRAN. The model language is simple and easy to learn and permits extensive error-checking. The compiler is included in the program and works in parallel with an editor. This enables the user to correct erroneous lines immediately.

Simnon allows decomposition of the system into subsystems, which can be described separately. The subsystems are then connected using their inputs and outputs. This makes it possible to maintain a library of process models. Further, it is easy to change the structure of the process or the controller. When simulating computer-controlled processes, it is natural to describe the physical process by ordinary differential equations and the computer with its control algorithms by difference equations. For this reason Simnon allows description of both continuous-time subsystems and discrete-time subsystems.

Model Structure

There are three types of systems in Simnon: continuous, discrete, and connecting.

A continuous subsystem is defined mathematically as

$$\dot{x}(t) = f(x(t), t, u(t), p)$$

$$y(t) = g(x(t), t, u(t), p)$$

$$x(t_0) = x_0$$

The following notation is used:

- t: time (independent variable).
- x: vector of state variables (dependent variables).
- y: vector of output variables.
- p: parameter vector.
- t_0: start time for simulation.
- x_0: initial values for states.

A discrete subsystem is defined mathematically as

$$x(t_{k+1}) = f(x(t_k), t_k, u(t_k), p)$$

$$y(t_k) = g(x(t_k), t_k, u(t_k), p)$$

$$x(t_0) = x_0$$

The kth sampling instant is denoted by t_k. The sampling instants need not be equidistant and need not be equal in all discrete-time subsystems.

The equations of the discrete-time subsystems do not define the outputs between the sampling instants and before the first sampling instant. In order to allow for connection of continuous systems and discrete systems with different

SIMNON—An Interactive Simulation Program

sampling instants, the definition of the outputs must be extended. This is done by introducing zero-order-hold circuits at the outputs. The states also have zero-order-hold circuits because they also can be used as outputs.

The connections of the subsystems are done in a *connecting system.* The connecting system defines how the inputs and outputs of the different subsystems are interconnected. The connecting system can be represented as a static system. It may be time-varying, but there are no dynamics.

If the system does not contain any algebraic loop, then the equations in the system can be solved sequentially by sorting them appropriately. This is done automatically.

The sampling instants for each discrete subsystem are specified by a special variable in the system description. This variable is updated at each sampling instant to contain the time for the next sampling. Thus, after the activation of a discrete subsystem, it is known when the next sampling should be performed. The differential equations, which are parameterized by the discrete states and outputs, can be solved over the sampling interval by an ordinary integration routine. At present, there are a Hammings predictor-corrector algorithm, a Runge-Kutta algorithm, and a routine for stiff differential equations.

Model Language

Simnon includes a special language for describing subsystems and connections. The equations are entered using an assignment statement such as the one in Algol-60. The if-then-else construction has been shown to be very useful. The model language is very simple, which has made it possible to do extensive error-checking. If complicated models are needed, there is a possibility to use FORTRAN or Pascal subroutines for model descriptions.

In the following, the Simnon keywords are written in capital letters. Capital and lowercase letters can be used in the model descriptions. Descriptions of the model language statements are obtained through the command HELP SIMNON.

Continuous and discrete systems have the following structure:

```
CONTINUOUS SYSTEM ⟨name⟩      DISCRETE SYSTEM ⟨name⟩
declarations                  declarations
assignments                   assignments
END                           END
```

A system heading gives the type of the system and its name. There are declarations to specify variable types. The types are INPUT, OUTPUT, TIME, STATE, DER, NEW, and TSAMP. The type DER is used in continuous-time systems to associate a variable as derivative of a state variable. The declaration NEW is used in the same way in discrete-time systems for the update of a state variable. One variable of type TSAMP is used in each discrete system to specify the next sampling instant. The system description can also contain parameters and auxiliary variables, which are not declared.

Parameters are assigned by a statement of the following form:

⟨parameter⟩:⟨number⟩

Initial values of state variables can be assigned in the same way. Variables are assigned as

⟨variable⟩ = ⟨expression⟩

In the assignment section, the variables are given values. For instance, the derivatives, the updated states, and the outputs are given values. The assignments are automatically sorted by the compiler into appropriate computing order.

The connecting system has the following structure:

```
CONNECTING SYSTEM ⟨name⟩
declarations
connect section
END
```

The connect section contains assignment statements for the INPUT variables of the subsystems. The same identifier may be used for variables in different subsystems. Therefore, the following notation is used in a connecting system to reference variables in the subsystems:

⟨variable⟩[⟨subsystem⟩]

The right-hand part of the assignments may contain STATE and OUTPUT variables, which are referenced in the same way.

Standard Functions and Systems

Several standard functions are available in the package. These functions can be used in assignments in the subsystems or in the connecting system. The following standard functions are available:

ABS(x):	Absolute value of x.
ARCCOS(x):	Arc cosine of x.
ARCSIN(x):	Arc sine of x.
ATAN(x):	Arc tangent of x. The result is in radians in the interval $(-\pi/2, \pi/2]$.
ATAN2(x, y):	Arc tangent of x/y. The result is in radians in $(-\pi, \pi]$.
COS(x):	Cosine of x, x in radians.
COSH(x):	Hyperbolic cosine of x.
EXP(x):	Exponential function of x.
INT(x):	Integer part of x.

LN(x):	Natural logarithm of x, $x > 0$.
LOG(x):	Logarithm base 10 of x, $x > 0$.
MAX(x, y):	Maximal value of x and y.
MIN(x, y):	Minimum value of x and y.
MOD(x, y):	The remainder when dividing x by y.
NORM(x):	Normal distribution with mean value 0 and standard deviation 1.
RECT(x):	Rectangular distribution in the interval [0, 1].
SIGN(x):	Sign of x, $= -1$ for negative numbers, else 1.
SIN(x):	Sine of x, x in radians.
SQRT(x):	Square root of x, $x \geq 0$.
TAN(x):	Tangent of x, x in radians.
TANH(x):	Hyperbolic tangent of x.

Interactive Facilities

The model descriptions are stored in files on mass storage. The manipulation of the models is done using commands with arguments. The commands are normally entered from the terminal, but they can also be read from files. The results of the simulations are plotted on a graphical display.

A brief description of the commands is given below. For a complete description, see Elmqvist (1975). The structure of each command is flexible. In some cases, arguments can be omitted. Default values are then used.

One common situation when running the program is that the same sequence of commands is given several times. The user can then define a macro containing the commands. This macro is then used as a new command, possibly with different values of the arguments. There are provisions for jumps and input-output as there are in programming languages.

The user may invoke an editor or any operating-system command from within Simnon. The command should be preceded by $.

Help Function

The simulation package includes a help facility, which can be used during a simulation session. The command HELP gives a menu of available Simnon commands and concepts. Additional information about each command is obtained by typing HELP ⟨command⟩. It is also possible to get information about particular concepts, e.g., language constructs and examples.

Example

Some of the features of Simnon will be illustrated by an example. The system considered is the tank system shown in Fig. C.1.

The valve at the inlet is controlled by a regulator to keep the level constant. The valve at the outlet is externally manipulated. When studying this system the outlet valve area is considered as a disturbance.

SIMNON—An Interactive Simulation Program

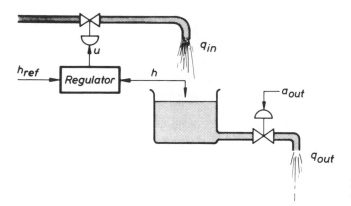

Figure C.1 A tank system with valves and a controller.

The simulation study is used to determine a suitable regulator and to find the regulator settings.

The model descriptions are shown in Listing C.1. The tank and the valves are described in the system called *tank*. The system has two inputs: the signal v to the inlet valve and the area of the outlet valve a_{out}. The level h is the state and is also considered as output. The tank is controlled by a process computer, which contains a PI-regulator. The subsystem describing the regulator is called *dpi*.

The connection of tank and dpi is made in *regtank*. The disturbance at the outlet is also given in regtank. When time = 100, the outlet area is increased from 0.01 to 0.05. The tank is initially empty.

The interaction with Simnon is shown below. Comments are written after quotation marks.

```
)SYST tank dpi regtank   " Compile model.
)STORE h[tank] href qin   " Prepare for storage of results.
)"
)" Try a proportional regulator with k = 1.
)"
)PAR h:5   " Set sampling interval.
)SIMU 0 200   " Simulate 200 s.
)AXES H 0 200 V 0 3   " Draw axis
)SHOW h href qin-MARK   " Plot results.
)HCOPY   " Make a hard copy (see Fig. C.2).
)"
)" Eliminate the static error with a PI-regulator.
)"
)PAR ti:25   " Change integration time
)SIMU   " Simulate.
)AXES
)SHOW h href qin-MARK
)HCOPY   " (See Fig. C.3).
)"
)" Too large overshoot, introduce antireset windup.
)"
)SAVE pipar   " Save parameters in file called pipar.
```

SIMNON—An Interactive Simulation Program

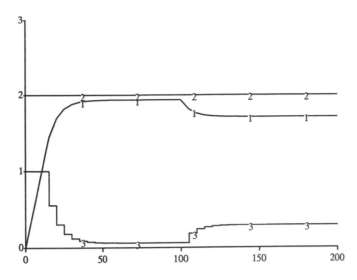

Figure C.2 Simulation with proportional regulator, $k = 1$.

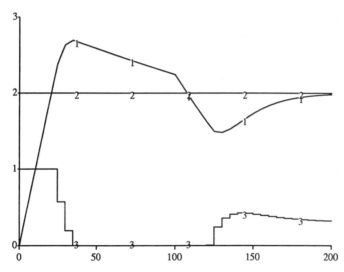

Figure C.3 Simulation with PI-controller, $k = 1$, $ti = 25$.

```
)EDIT dpi   "Edit the file dpi. This can also be replaced
               " by a call to system editor
        EDIT
     )L u =    " Locate string.
  u = k*e + i1
     )C /u/v/   " Change string.
  v = k*e + i1
     )I u = IF v<0 THEN 0 ELSE IF v<1 THEN v ELSE 1   " Insert line.
     )N   " Next line.
     )C /i1/i1 + u − v/
  ninte = i1 + u − v
     )E   " Leave editor.
```

SIMNON—An Interactive Simulation Program

```
)SYST tank dpi regtank
)GET pipar   '' Restore parameters.
)STORE h[tank] href qin
)SIMU
)AXES
)SHOW h href qin-MARK
)HCOPY   '' (See Fig. C.4).
```

LISTING C.1   The subsystems *tank, dpi,* and *regtank.*

```
CONTINUOUS SYSTEM tank
INPUT v aout
STATE h
DER dh
valve = IF v<0 THEN 0 ELSE IF v>1 THEN 1 ELSE v
qin = qmax*valve
qout = aout*SQRT(2*g*MAX(h,0))
dh = (qin - qout)/area
qmax:1
g:9.81
area:10
END

DISCRETE SYSTEM dpi
INPUT y yref
OUTPUT u
STATE inte
NEW ninte
TIME t
TSAMP ts
e = yref - y
i1 = inte + k*e*h/ti
u = k*e + i1
ninte = i1
ts = t + h
k:1
ti:1E10
h:1
END

CONNECTING SYSTEM regtank
TIME t
aout[tank] = IF t<100 THEN a1 ELSE a2
a1:0.01
a2:0.05
yref[dpi] = href
href:2
y[dpi] = h[tank]
v[tank] = u[dpi]
END
```

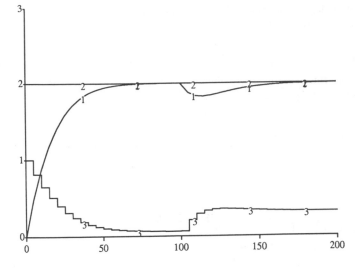

Figure C.4 Simulation using a PI-controller with antireset windup, $k = 1$, $ti = 25$.

Summary of Simnon Commands

A list of the commands, their syntax, and brief descriptions are given below. Some Intrac commands are also included in the list. More details about the commands are obtained using the command HELP. The list is valid for implementations on MS™-DOS systems. The following notation is used.

{opl | . . . | opn}	Defines different alternatives, of which one must be given.
[. . . .]	Parts within squared brackets are optional and can be omitted.
{. . . .}*	A star indicates that the previous part can be repeated.
⟨. . . .⟩	Defines arguments for the commands.

ALGOR{HAMPC\|RK\|RKFIX\|DAS\|EULER\|RKF23\|RKF45\|DOPRI45\|DOPRI45R\|?}.
    To select integration routine.

HAMPC	Hammings Predictor-Corrector method (default).
RK	Classical fourth-order Runge-Kutta method with variable step-size.
RKFIX	Classical fourth-order Runge-Kutta method with fixed step-size.
DAS	DASP3 stiff system solver with possibility for manual partitioning into FAST and SLOW subsystems. See the STATE command.
EULER	Forward Euler method with fixed step-size.
RKF23	2/3-order Runge-Kutta-Fehlberg method (3 function evaluations per step and 2 per failure) with variable step-size.

       SIMNON—An Interactive Simulation Program

RKF45	4/5-Order Runge-Kutta-Fehlberg method (6 function evaluations per step and 5 per failure) with variable step-size.
DOPRI45	4/5-order Runge-Kutta-Dormand-Prince method (similar to RKF45, but has a more realistic error estimation).
DOPRI45R	Essentially the same as DOPRI45, but features a step-size regulator with proportional and integral action.
?	Displays the name of the chosen integration routine.

AREA ⟨rows⟩ ⟨columns⟩.   To define the plotting area to be used next (see SPLIT).

ASHOW   To plot stored variables with automatic scaling of axes. For syntax, see SHOW.

AXES [⟨axis spec⟩ [⟨axis spec⟩]].   To draw axes.

⟨axis spec⟩: = {H|V} ⟨min value⟩ ⟨max value⟩
H, horizontal; V, vertical

DISP [({DIS|TP|LP}{FF|LF})] [⟨variable⟩]∗.   To display variables.

DIS	Display (default).
TP	Terminal.
LP	Line printer.
FF	Form feed (default when no variables are specified).
LF	Line feed (default when variables are specified).

If no variables are given, all varibles are displayed.

EDIT ⟨filename⟩.   To edit a file. The editor has two modes, INPUT and EDIT. The mode is changed by entering an empty line. In the edit mode it is possible to change the current file. The following commands are available:

A[PPEND] ⟨string⟩	Append string to line.
B[OTT]	Line pointer to bottom.
C[HANG] /x/y/	Replace string x by string y.
D[EL] ⟨integer⟩	Delete n lines.
DIS [ON|OFF]	Echo check enable/disable.
E[XIT]	Leave editor with update.
F[IND] ⟨string⟩	Find string beginning line.
I[NS] ⟨string⟩	Insert a line below current line.
L[OC] ⟨string⟩	Locate string anywhere in a line.
LEAVE	Leave the editor without update.
N[EXT] ⟨integer⟩	Move line pointer down n lines.

O[VERL] ⟨integer⟩	Overlay *n* lines by new text.
P[RINT] ⟨integer⟩	Print *n* lines.
R[ETYP] ⟨string⟩	Retype current line.
T[OP]	Move line pointer to top.

Any editor in the system can be invoked from within Simnon. The command should be preceeded by $.

ERROR ⟨error bound⟩. To choose error bound for integration routine. Default value is 0.001.

GET ⟨filename⟩. To get parameter values and initial values from a file that previously has been stored using SAVE.

HCOPY [⟨scale factor⟩] "⟨comment⟩. Make a hard copy of curves on display. The hard copy is scaled with ⟨scale factor⟩, which may be in the interval (0.5, 1.6). The comment is obtained on the hard copy, too.

HELP [{⟨command⟩|⟨concepts⟩}]. To get more information about the commands or concepts. A menu of commands is obtained by typing HELP.

INIT ⟨state variable⟩: {⟨number⟩|⟨variable⟩}. To change the initial value of a state variable.

LET ⟨variable⟩ = ⟨number⟩. Assign command variables.

LIST [({DIS|TP|LP}[FF|LF])]{⟨filename⟩}∗. To list files. The same as DISP but for files—e.g., subsystems.

LP Initiate listing on line printer of variables and files that have been ordered by DISP, LIST, or PRINT commands.

MARK To introduce text into a plot. For syntax, type HELP INTRAC MARK.

PAR ⟨parameter⟩ : {⟨number⟩|⟨variable⟩}. To change a parameter value.

PLOT [{⟨variable⟩}∗[(⟨variable⟩)]]. To select variables to be plotted when making the command SIMU. Examples: PLOT X1 X2 gives X1 and X2 as functions of time, while PLOT X1(X2) gives X1 as function of X2.

PRINT [({DIS|TP|LP} [FF|LF])] ⟨filename⟩ [⟨lines⟩] [/⟨start time⟩]. To list file generated with the commands STORE + SIMU.

⟨lines⟩ lines starting from ⟨start line⟩ will be printed. The other parameters are the same as for DISP.

SAVE ⟨filename⟩ [⟨systemid⟩] [−{PAR|INIT}]. To save parameter values and initial values on a file named ⟨filename⟩ to be used by the command GET. If only parameters or initial values should be saved, the options PAR and INIT, respectively, should be used.

SHOW {[⟨start⟩ ⟨stop⟩] {⟨variable⟩}∗ [(⟨variable⟩)] [−MARK] |−LIST} [/⟨filename⟩]. To plot stored variables from file ⟨filename⟩. To be used with the command STORE. The specified variables are plotted from ⟨start⟩ to ⟨stop⟩ time. If MARK is used, the different variables are numbered on the plot. The LIST option lists the names of all variables.

SIMU [⟨start time⟩ ⟨stop time⟩ [⟨increment⟩]] [−⟨option⟩]
   [/⟨filename⟩ [⟨increment⟩]]
   ⟨option⟩: = {MARK|CONT}

To simulate the system from ⟨start time⟩ to ⟨stop time⟩ using the maximum step size ⟨increment⟩ [default (stop time − start time)/100]. Using MARK, the variables defined by PLOT are numbered. With CONT, the simulation is continued with the previously obtained state variables as initial values. When specifying ⟨filename⟩, the plotted variables are stored in ⟨filename⟩ with ⟨increment⟩ as sampling interval.

SPLIT ⟨rows⟩ ⟨columns⟩.   To split the screen into maximum six plotting areas.
   ⟨rows⟩: = {1|2|3} default 1
   ⟨columns⟩: = {1|2} default 1

STATE   Use with the integration routine DAS. For syntax, write HELP STATE.

STOP   To leave Simnon.

STORE [⟨variable⟩]∗ [−ADD].   Use to select varibles to be stored at each simulation. With ADD, new variables can be added to a previously defined list of variables. The variables can be displayed using ASHOW or SHOW and printed using PRINT.

SWITCH ⟨option⟩ {ON|OFF}
   ⟨option⟩: = {CLOCK|DATE|ECHO|EXEC|LOG|TRACE}

To control the execution of Intrac.

   CLOCK   Adds time to hardcopy output. Default OFF.
    DATE   Adds date to hardcopy output. Default ON.
    ECHO   Macro commands are echoed. Default OFF.
    EXEC   Commands executed while typed during macro generatoin. Default OFF.
     LOG   Executed commands in macro logged on line printer. Default OFF.
   TRACE   Affects ECHO and LOG.

SYST {⟨subsystem⟩}∗ [−⟨option⟩] [/⟨filename⟩]
   ⟨option⟩: = {EDIT|EXIT}

To define the system. The subsystems are compiled. If there are several subsystems, the last one has to be a connecting system: EDIT means that the compiler goes into the editor for each file. If filename is specified, the sorted equations are written into a text file.

TEXT ⟨any string not containing single quote⟩'.   To include text on paper plot.

TURN ⟨option⟩ {ON|OFF}
   ⟨option⟩: = {DARK}

To turn on and off switches.

DARK   ON means that plotted curves will not be connected between the sampling instants. Default OFF.

References

ELMQVIST, H. (1975): *Simnon-User's Guide*. Department of Automatic Control, Lund Inst. of Techn., CODEN: LUTFD2/(TFRE-3091)/(1975).

ELMQVIST, H. (1977): *Simnon:* An Interactive Simulation Program for Nonlinear Systems. Proceedings Simulation: '77, Montreux 1977.

ÅSTRÖM, K. J. (1983): "Computer-Aided Modeling, Analysis, and Design of Control Systems: A perspective," IEEE Control Systems Magazine, 3, No. 2, 4–16.

ELMQVIST, H., K. J. ÅSTRÖM, and T. SCHÖNTHAL (1986): *Simnon-User's Guide for MS™-DOS Computers*. Department of Automatic Control, Lund Inst. of Techn.

PROOF OF THEOREM 10.1

D

Theorem 10.1. Let A, B, and C be polynomials with real coefficients. Then the equation

$$AX + BY = C \qquad (D.1)$$

has a solution if and only if the greatest common factor of A and B divides C.

Proof. Without loss of generality, it may be assumed that $\deg A \geq \deg B$. Introduce $A_0 = A$ and $A_1 = B$. Let Q_1 be the quotient and A_2 the remainder when A_0 is divided by A_1, i.e.,

$$A_0 = A_1 Q_1 + A_2$$

where $\deg A_2 < \deg A_1$. Divide A_1 by A_2:

$$A_1 = A_2 Q_2 + A_3$$

Continuing in the same way gives

$$\begin{aligned}
A_0 &= A_1 Q_1 + A_2 \\
A_1 &= A_2 Q_2 + A_3 \\
&\;\;\vdots \\
A_{n-2} &= A_{n-1} Q_{n-1} + A_n \\
A_{n-1} &= A_n Q_n + A_{n+1} \\
A_{n+1} &= 0
\end{aligned} \qquad (D.2)$$

where A_{n+1} is the first polynomial that is zero. Since

$$\deg A_0 > \deg A_1 > \deg A_2 > \cdots > \deg A_{n+1}$$

the algorithm must terminate in a finite number of steps. The algorithm is called Euclid's algorithm.

Because $A_{n+1} = 0$, it follows that A_{n-1} is divisible by A_n. The polynomial A_n is thus the largest common factor of A_n and A_{n-1}. It follows from (D.2) that

$$A_{n-2} = A_{n-1}Q_{n-1} + A_n = A_nQ_nQ_{n-1} + A_n$$

Hence, the polynomial A_n is a factor of A_{n-1} and A_{n-2}. It is also the greatest common factor because if it were a larger common factor Q, then it follows that

$$A_{n-2} = A_{n-1}Q$$

This implies, however, that $A_n = 0$, which contradicts the assumption that the algorithm terminates with $A_{n+1} = 0$. Proceeding recursively, A_n is found to be the largest common factor of $A_0 = A$ and $B_0 = B$. Equation (D.1) can then be written as

$$AX + BY = (A'X + B'Y)A_n = C \tag{D.3}$$

This shows that the equation does not have a solution unless A_n is also a factor of C.

If A_n is a factor of C, both sides of (D.3) can be divided by C; this gives Equation (D.1), where A and B do not have any common factors. In this case Euclid's algorithm gives $A_n = 1$ as the greatest common factor of A and B. The algorithm of (D.2) gives

$$1 = A_n = A_{n-2} - Q_{n-1}A_{n-1}$$

It follows from (D.2) that

$$A_{n-1} = A_{n-3} - A_{n-2}Q_{n-2}$$

Hence

$$1 = A_{n-2} - Q_{n-1}(A_{n-3} - Q_{n-2}A_{n-2})$$

$$= -Q_{n-1}A_{n-3} + (1 + Q_{n-1}Q_{n-2})A_{n-2}$$

$$= U_{n-3}A_{n-3} + V_{n-2}A_{n-2}$$

Proceeding recursively, it is found that

$$1 = U_0A_0 + V_1A_1 = U_0A + V_1B$$

A solution to Equation (D.1) is then

$$X = U_0C$$

$$Y = V_1C$$

INDEX

SIMNON

For further information mail this card to:

In USA: Engineering Software Concepts
436 Palo Alto Avenue
Palo Alto, CA 94301

Outside USA: SSPA Maritime Consulting
Box 24001
S-400 22 Göteborg
SWEDEN

SIMNON

Please send information and prices for the simulation package SIMNON for

____ Vax computers

____ MS-DOS computers

____ Sun workstations

Name _____

Company/University _____

Title/Position _____

City _____ State _____ Zip _____

SIMNON

For further information mail this card to:

In USA: Engineering Software Concepts
436 Palo Alto Avenue
Palo Alto, CA 94301

Outside USA: SSPA Maritime Consulting
Box 24001
S-400 22 Göteborg
SWEDEN

SIMNON

Please send information and prices for the simulation package SIMNON for

____ Vax computers

____ MS-DOS computers

____ Sun workstations

Name _____

Company/University _____

Title/Position _____

City _____ State _____ Zip _____